CALCULUS FOR BUSINESS

The IEP Series in Mathematics
under the consulting editorship of

RICHARD D. ANDERSON
Louisiana State University

ALEX ROSENBERG
Cornell University

CALCULUS FOR BUSINESS

RICHARD D. ANDERSON
Louisiana State University

CECIL L. SMITH
Louisiana State University

IEP
A DUN-DONNELLEY PUBLISHER
New York

Published simultaneously in Canada by Fitzhenry & Whiteside, Ltd., Toronto.

Library of Congress Cataloging in Publication Data

Anderson, Richard D
Calculus for business.

(The IEP series in mathematics)
Includes index.
1. Calculus. I. Smith, Cecil L., joint author.
II. Title.
QA303.A477 1976 515 75-35861
ISBN 0-7002-2487-4

IEP A Dun-Donnelley Publisher
666 Fifth Avenue
New York, New York 10019

Design by William Frost, Dimensions.

TURED IN THE UNITED STATES OF AMERICA

TO JEANETTE AND CHARLOTTE

CONTENTS

* Asterisks identify optional sections

CHAPTER 3 THE DERIVATIVE

CHAPTER 4 APPLICATIONS OF THE DERIVATIVE

CHAPTER 5 EXPONENTIAL AND LOGARITHMIC FUNCTIONS

CHAPTER 6 THE INTEGRAL

CHAPTER 7 MORE ON INTEGRATION

CHAPTER 8 PARTIAL DERIVATIVES

CHAPTER 9 DIFFERENTIAL EQUATIONS

PREFACE

We wrote this book to introduce the concepts and notation of calculus to students of business and economics and to others who require a reasonable introduction to the subject, with a minimum of technical complication. In the belief that nonscience students are more hindered than helped by involvement with physical science and engineering principles and applications, we avoid them as much as possible. Our illustrations and examples derive almost exclusively from business, economics, and everyday life.

Computing Concepts. We have given a modest computer orientation to our book, using *symbolic* flowcharts to give a basis for sound step-by-step analysis of many of the processes of calculus, and *computational* flowcharts to motivate and explain numerical methods. Flowcharting is difficult for some students, and we stress reading and analyzing flowcharts more than writing them. We intentionally omit those aspects of flowcharting that become significant only at the program-writing level.

An earlier mimeographed version of the text was used successfully by several teachers, two of whom had little or no prior experience with flowcharting. No "hands-on" computing is expected in using this book.

Numerical Awareness. One of the major thrusts of our approach is to emphasize awareness of numbers. We have observed that many college freshmen have little or no knack for estimating numerical quantities or feel for the role of numbers in the business world. Many have no awareness of the time it takes money to double at 5 percent simple interest, not to mention the time it takes to double at 5 percent compound interest. We believe that a calculus course can help students increase their awareness of numerical relationships; and we stress estimations throughout the book, in the text material and in problems.

Level and Type of Presentation. In general we tried to provide informal but honest treatments of the various concepts arising in introductory calculus. We avoid formal proofs and statements of theorems as such, referring to properties rather than theorems. However, the Intermediate Value Theorem and the Fundamental Theorem of the Calculus are explicitly treated.

We feel that ε, δ rigor is totally out of place for our readers. At the same time, students should be given valid explanations, and for those who will take further work in calculus, there should be no unlearning to do. We hope that students who study this text will come to understand something of the power and scope of calculus and especially of its applicability to real world problems.

Students are best served by a book in which the text is readable and necessary, not simply an adjunct to the problem sets. Although we include many examples, we tried to avoid a format that encourages students to work the exercises simply by following by rote the illustrative examples in the text.

Subject Matter Selection and Notation. Generally, we concentrate on "Main Line" material using standard notation and apparatus. However, because our treatment is concerned with numerical awareness and involves computer concepts, there is extra emphasis on such topics as exponential growth and the estimation of integrals as well as on sequences and sequential procedures.

In developing an informal approach to calculus, Σ notation could be left out; certainly Σ notation is a stumbling block for some students. However, since business and economics students almost inevitably are going to have to deal with statistics and since Σ notation naturally occurs there, we felt required to include an introduction to such notation as a contribution to students' future needs.

We deliberately omit trigonometric functions and avoid complicated algebraic expressions. But we do introduce topics naturally related to the ideas of calculus, such as linear programming and other max-min type problems.

Pace and the Length of the Course. The various sections are designed for use on a one-a-day basis. Sections marked by stars in the table of contents are optional and generally cover topics not used in later sections. It should be possible to cover the thirty-six unstarred sections of the first six chapters in a three-hour one-semester course. This course would include introductory material, sequences, derivatives and their applications, exponential and log functions, and integration. The optional sections in the early chapters together with the three final chapters provide ample material for a two-quarter course. For greater teaching flexibility, we include chapter review problem sets.

Level of Sophistication of Students. We assume that most students who will be studying from this book are not mathematically oriented or trained. Therefore, we emphasize technically simple material. One of the difficulties in teaching calculus to an academically mixed group of students is the need to introduce concepts rather rapidly. Where possible we re-use material introduced earlier at intervals throughout the text to give students a continuing review of important concepts and techniques. Individual sections are bite-sized, with a substantial number of exercises that students can do successfully.

Finally, we should say that we hope in our treatment calculus appears not as a disconnected formal system involving just derivatives and integrals and their evaluations but as part of the framework of business, of computing, of other mathematics, and of ordinary life.

Acknowledgments. We are very appreciative to Professors Marguerite Summers and Leonard Richardson who, with the authors, taught from a preliminary version of the text and who made particularly helpful suggestions for revising that version to make it more teachable. We also used frequent helpful comments from several anonymous reviewers provided by IEP. And we wish to sincerely thank Ms Miranda Medica, Ms Jean Johanson, Ms Mary Johnston, and numerous others for much help and hard work in typing various versions of the manuscript.

CHAPTER ONE

INTRODUCTION

1-1. COORDINATES AND LINES

We shall frequently resort to drawings in our presentation of the concepts of calculus. In this section and the next, we shall review a few of the graphical concepts that form the basis of concepts to come.

The association of the real numbers with points on a line leads to the concept of the number line. Beginning with a line, we choose any two points on that line and let one of these points be the *origin* and the distance between these points be a *unit*. We can then represent any point on the line in terms of its distance from the origin as measured in terms of the unit.

The usual convention for horizontal lines is to assign positive numbers to the points to the right of the origin and negative numbers to the points to the left. For a vertical line, positive numbers are assigned to points above the origin. The *number scale* or *number line* is usually presented by locating points at distances from the origin using integral multiples of the unit distance, as illustrated in Figure 1-1. The number $2\frac{1}{2}$ is assigned to the point midway between 2 and 3; the number -1.1 is assigned to the point a tenth unit to the left of -1.

FIGURE 1-1

In this way, we establish a correspondence between the real numbers and the points on a line. That is, to each real number, there corresponds a unique point on the number line, and vice versa.

An *open interval* is a subset of the set of real numbers consisting of all real numbers greater than some real number a and less than some other real number b (where $b > a$). The open interval is represented by the notation (a, b), where a and b are the end points of the interval. Observe that the open interval does *not* contain its end points. A *closed interval* is a set consisting of an open interval together with its end points. The notation $[a, b]$ designates a closed interval. We use the notation $[a, b)$ or $(a, b]$ to denote a *half-open* interval including the left- or right-hand end point as indicated. We can also use double inequalities to designate an interval. Thus $a < x \leq b$ denotes the half-open interval $(a, b]$, which includes b but does not include a. Similarly, $a \leq x \leq b$ denotes the closed interval $[a, b]$; $a < x < b$ denotes the open interval (a, b); and $a \leq x < b$ denotes the half-open interval that includes a but not b.

The correspondence of the real numbers to the points on a line can be extended to associate ordered pairs of real numbers with the points in a plane. We take two mutually perpendicular lines in the plane, calling one horizontal and the other vertical. The point of intersection is designated as the origin of each line. We call the horizontal line the x axis and the vertical line the y axis, and regard each as being a number line.

Once our axes are established, we can easily correspond points in the plane to ordered pairs of real numbers, and vice versa. For any point P in the plane we consider the lines through P perpendicular to the x axis and the y axis, as illustrated in Figure 1-2. The x coordinate of P is the x coordinate of the foot of the perpendicular from P to the x axis. The y coordinate of P is the y coordinate of the foot of the perpendicular from P to the y axis.

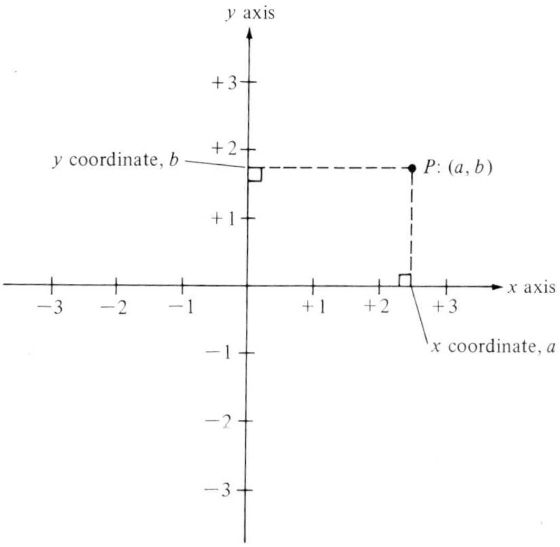

FIGURE 1-2

Therefore, each point of the plane corresponds to a unique ordered pair of real numbers, the x coordinate of the point and the y coordinate of the point.

Given an ordered pair of real numbers, we can locate the corresponding point in the plane. For the first real number we locate the corresponding point on the number line used as the x axis. For the second real number we locate the corresponding point on the y axis. We then construct the perpendiculars, and their point of intersection is the point in the plane corresponding to the ordered pair of real numbers.

The two coordinates of a point in a plane are generally denoted by enclosing in parentheses and separating by a comma. By convention, the x coordinate is given first, followed by the y coordinate. Thus, point P in Figure 1-2 could be denoted as the point (a, b). The order is significant. That is, point (a, b) will be different from point (b, a) unless $a = b$. Figure 1-3 on the next page shows several points in the plane. Note that point (1, 2) is different from point (2, 1) and that point (0, 4) is different from point (4, 0).

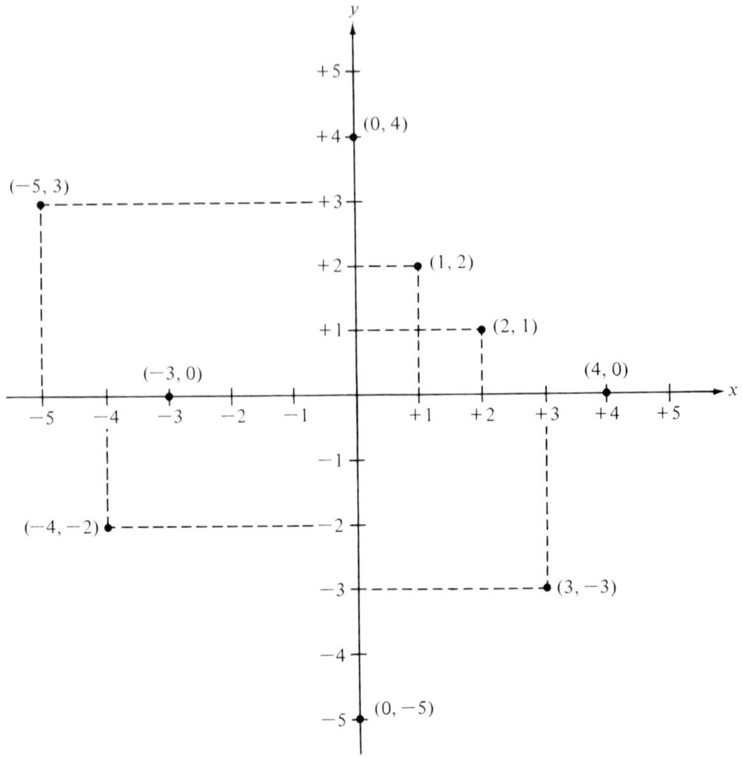

FIGURE 1-3

Given any two points, say P_1 and P_2, in a plane, there is a unique straight line containing them, as illustrated in Figure 1-4. A useful concept concerning straight lines is that of *slope*. To define the slope of a line, we choose any two points of the line. If the line is neither vertical nor horizontal, we construct a right triangle using lines parallel to the axes. For purposes of illustration, such a triangle is constructed in Figure 1-4, from points P_1 and P_2 whose coordinates are (x_1, y_1) and (x_2, y_2), respectively. The slope m is defined as follows:

$$m = \frac{y_2 - y_1}{x_2 - x_1}, \qquad x_2 - x_1 \neq 0$$

When $x_2 - x_1 = 0$, the line is vertical and its slope is undefined. When $y_2 - y_1 = 0$, the line is horizontal and its slope is zero.

In determining the slope of a line, the order in which the coordinates of the points are inserted into the formula is immaterial, since, for $x_1 \neq x_2$,

$$m = \frac{y_2 - y_1}{x_2 - x_1} = \frac{y_1 - y_2}{x_1 - x_2}$$

But the same order must be used for the y's as for the x's.

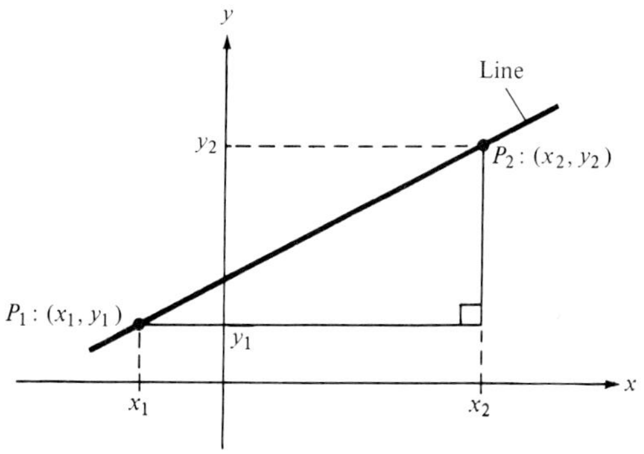

FIGURE 1-4

From elementary geometry for parallel lines and similar triangles, it follows that the slope of a line is independent of the choice of points whose coordinates are used in the formula. Using Figure 1-5 as an example, we can define the slope using points (x_1, y_1) and (x_2, y_2) or points (x_3, y_3) and (x_4, y_4):

$$m = \frac{y_2 - y_1}{x_2 - x_1} = \frac{y_4 - y_3}{x_4 - x_3}$$

Although any two points can be chosen, it is often convenient to choose two points so that $x_2 - x_1 = 1$. In this case, the slope is given by

$$m = \frac{y_2 - y_1}{1} = y_2 - y_1$$

Thus we see that the slope represents the algebraic change in the y coordinate when the x coordinate is increased by one unit.

Lines rising toward the right, such as the one in Figure 1-5, have a positive slope, i.e., $m > 0$. Lines rising toward the left have a negative slope.

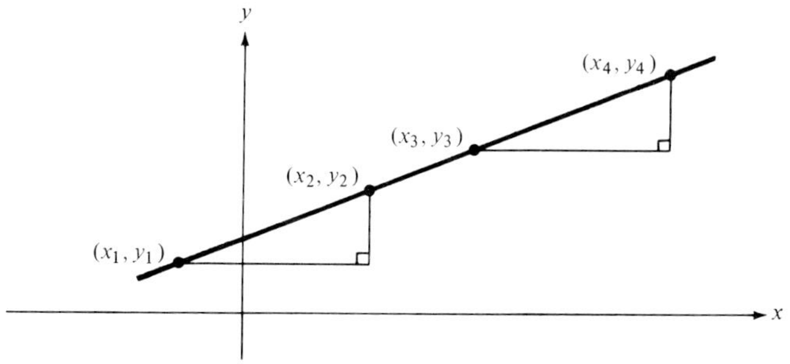

FIGURE 1-5

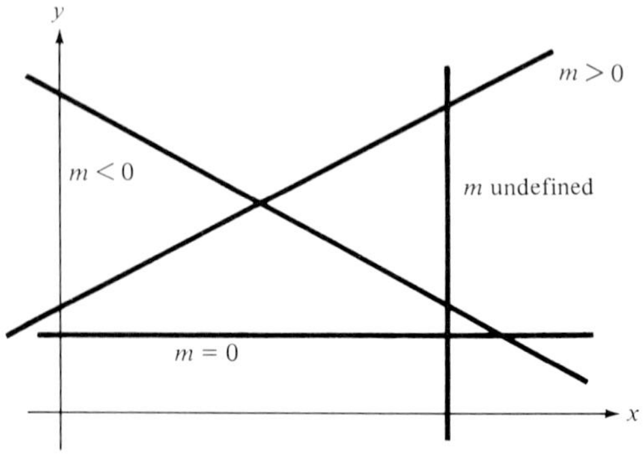

FIGURE 1-6

Lines parallel to the x axis have a zero slope. The slope is undefined for lines parallel to the y axis. These are illustrated in Figure 1-6.

As a specific example, Figure 1-7 illustrates the straight line drawn through points (1, 2) and (3, 5). Arbitrarily choosing point (1, 2) as (x_1, y_1) and point (3, 5) as (x_2, y_2), the slope is

$$m = \frac{y_2 - y_1}{x_2 - x_1} = \frac{5 - 2}{3 - 1} = \frac{3}{2}$$

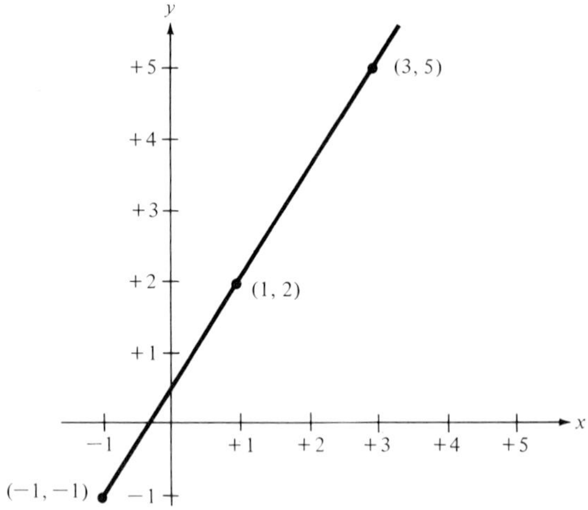

FIGURE 1-7

Reversing the roles of (x_1, y_1) and (x_2, y_2), we get

$$m = \frac{y_1 - y_2}{x_1 - x_2} = \frac{2 - 5}{1 - 3} = \frac{-3}{-2} = \frac{3}{2}$$

as we should. The point $(-1, -1)$ is also on the line. Using $(-1, -1)$ as (x_1, y_1) and $(3, 5)$ as (x_2, y_2), we get

$$m = \frac{5 - (-1)}{3 - (-1)} = \frac{6}{4} = \frac{3}{2}$$

which is the slope of the line.

We now consider how to find an equation of a line. An equation relating x and y will be an equation of a line if

1. Each ordered pair of numbers that satisfy the equation are the coordinates of a point on the line.
2. The coordinates of each point on the line satisfy the equation.

Suppose we have a line drawn through point (x_0, y_0), as shown in Figure 1-8. Furthermore, suppose we either know the slope of this line or

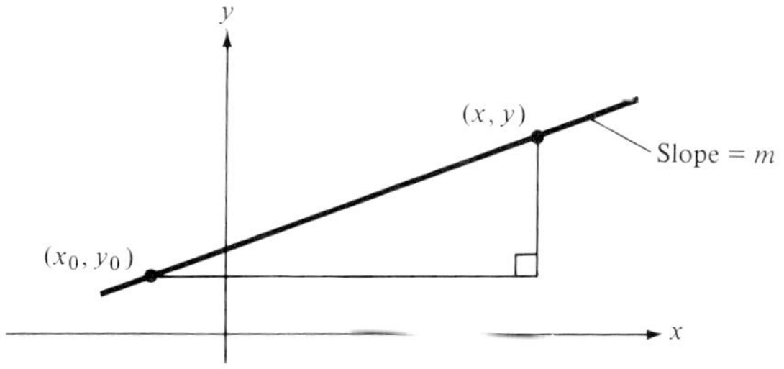

FIGURE 1-8

know the coordinates of another point on the line so that we can compute the slope. Then for any point (x, y) on the line other than (x_0, y_0), we have the formula for the slope:

$$m = \frac{y - y_0}{x - x_0}, \qquad x \neq x_0$$

This equation can be rearranged to

$$y - y_0 = m(x - x_0)$$

This equation is called the *point-slope form* of the equation of the line passing through point (x_0, y_0) with slope m.

As a specific example, we shall consider the line in Figure 1-7 whose slope was previously determined to be $\frac{3}{2}$. Choosing point (1, 2) as (x_0, y_0), an equation for the line is

$$y - y_0 = m(x - x_0)$$

$$y - 2 = \tfrac{3}{2}(x - 1)$$

Left in this form, we can immediately see that the line has slope $\frac{3}{2}$ and passes through the point (1, 2). If we wished, we could simplify the equation to

$$3x - 2y = -1$$

Alternatively, we could have chosen point (3, 5) as (x_0, y_0). The point-slope form of the equation for the line yields

$$y - 5 = \tfrac{3}{2}(x - 3)$$

This equation also can be simplified to

$$3x - 2y = -1$$

Indeed, any equation of the form $Ax + By + C = 0$, with A and B not both zero, is an equation of a straight line, and from the above, we see that any straight line has an equation that can be put in such a form.

If $B = 0$, then the equation has the form $x =$ constant and the straight line is vertical. If $B \neq 0$, we may solve for y and get $y = mx + b$, where m is the slope and b the y intercept, i.e., the y coordinate of the point $(0, b)$, where the line crosses the y axis. We call $y = mx + b$ the *slope-intercept* form for the line. Observe that $(y - b) = m(x - 0)$ easily converts into $y = mx + b$.

NOTE: In the exercises at the end of this section and elsewhere in the text, you are asked to sketch some graphs. The objective of sketching a graph is to obtain a visual interpretation of the problem at hand. Sketching a graph is not intended to be a tedious or time-consuming process. One should draw the coordinate axes (a straight edge is not absolutely necessary), indicate the scale for each, approximately locate points as necessary, and sketch the graph. The importance of the graph is that it must convey a message, namely, the visual interpretation. For our purposes it is not expected to be a refined, precise, or artistic product.

EXERCISES 1-1

One approach to representing intervals on the number line is illustrated in Figure A for the interval $[-\frac{3}{2}, \frac{4}{3})$.
The left end of this interval is closed (i.e., the point $-\frac{3}{2}$ is in the interval),

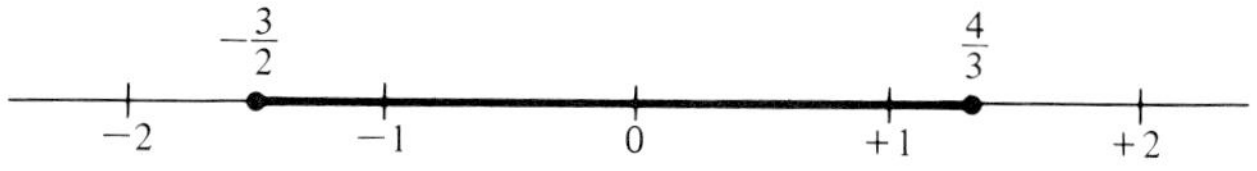

FIGURE A

which is designated by the shaded or closed dot. The right end is open (i.e., the point $\frac{4}{3}$ is not in the interval), which is designated by the unshaded or open dot.

Represent the following intervals on the number line by this scheme:

1. $(-1, 1)$ **2.** $(1, 2]$

3. $[-2, 0)$ **4.** $[-2, 2]$

5. $(-\frac{1}{2}, 3]$ **6.** $[\frac{3}{4}, 2)$

Intervals can also be interpreted via inequalities. For example, the real number x is in the interval $[-\frac{3}{2}, \frac{4}{3})$ if and only if both of the following inequalities are satisfied

$$-\tfrac{3}{2} \leq x$$

$$x < \tfrac{4}{3}$$

Customary practice is to combine these in a double inequality, as follows:

$$-\tfrac{3}{2} \leq x < \tfrac{4}{3}$$

Observe that the effect of having $-\frac{3}{2}$ in the interval is the inclusion of the "or equal" in the inequality.

Write the double inequality corresponding to the following intervals:

7. $(-1, 1)$ **8.** $(1, 2]$

9. $[-2, 0)$ **10.** $[-2, 2]$

11. $(-\frac{1}{2}, 3]$ **12.** $[\frac{3}{4}, 2)$

In each of Exercises 13 through 22, the coordinates of two points are given. (**a**) Determine the slope of the straight line drawn through the two points or state that it has no slope. (**b**) Using the first point, write the point-slope equation for the line if it has slope. Do not simplify. If it does not have slope, write an equation for the line in the form $x = a$. (**c**) Repeat (**b**) using the second point. (**d**) By solving (**b**) for y, convert to the slope-intercept equation for the line. (**e**) Determine the y coordinate of the point on the line whose x coordinate is 4, or if there is no such point, so state. (**f**) Determine the x coordinate of the point on the line whose y coordinate is -4, or if there is no such point, so state. (**g**) Graph the line.

13. $(1, 2); (3, 4)$ **14.** $(1, 5); (3, 3)$

15. (2, 5); (1, 4)

16. (−3, 1); (1, 5)

17. (−4, 2); (−2, −1)

18. (−3, −1); (−1, −3)

19. (1, 5); (1, 4)

20. (2, −2); (0, −3)

21. (2, −1); (−1, −1)

22. (0, 3); (2, −1)

In Exercises 23 through 28, a point on a line and the slope of the line are given. (**a**) Write an equation of the line in point-slope form. (**b**) Solving for y, convert to the slope-intercept form of an equation for the line. (**c**) Determine the y coordinate of the point on the line whose x coordinate is 8, or if there is no such point, so state. (**d**) Determine the x coordinate of the point on the line whose y coordinate is -8, or if there is no such point, so state. (**e**) Graph the line.

23. (−1, 2); slope = 2

24. (1, −1); slope = −1

25. (−2, −1); slope = $-\frac{1}{2}$

26. (1, 4); slope = $\frac{1}{4}$

27. (1, 2); slope = 0

28. (2, −1); slope = 1

In Figure B, six lines are drawn. For the line designated in Exercises 29 through 34, (**a**) determine the slope, if any; (**b**) using the point designated

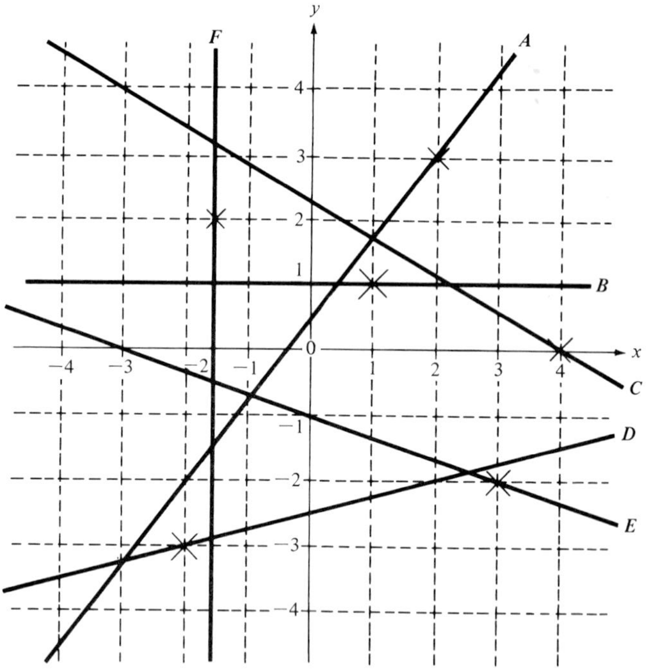

FIGURE B

by a cross, write the point-slope form of an equation for the line; and **(c)** write the slope-intercept form of an equation for the line.

29. Line A **30.** Line B

31. Line C **32.** Line D

33. Line E **34.** Line F

1-2. FUNCTIONS AND THEIR GRAPHS

By the statement "y is a function of x," we mean that given a value of x, there exists a unique value of y that corresponds to the value of x. Functions need not be mathematical in nature, nor must they involve numbers in any way. For example, for any book that is in the library and is authored by a single person, the functional relationship may designate the author. We can denote such relationships in the following manner:

$$\text{author} = f(\text{book})$$

In order to have a function, we must have three things:

1. A set called the domain.
2. A set called the range.
3. A rule or set of instructions which to each element of the domain assigns a particular element of the range.

In the example given, the domain is the set of books in the library having unique authorship. The range may be considered to be the set of authors of books and the rule assigns to any single-authored book in the library, the particular author of that book.

Most or all of the functions with which we shall be concerned in calculus will have domains and ranges that are sets of real numbers. Usually, the rule that assigns a unique element of the range to an element of the domain will be a formula. For example, we could use the formula $x^2 + 1$ to describe a function whose domain is the set of all real numbers and whose range is the set of all real numbers. For each number x, the assigned value is $x^2 + 1$. We call the value assigned to x the *functional value* at x.

For simplicity the formula for the function is sometimes called *the function.* Thus we may say "the function $x^2 + 1$." In such a case we either specify the domain explicitly or understand that the domain is the set of all real numbers for which the formula makes sense, i.e., produces a functional value which is a real number. For the function $x^2 + 1$, the understood domain is the set of all real numbers. For this function, the set of all functional values is the set of all real numbers greater than or equal to 1. As another example, the domain of the function $\sqrt{x + 2}$ is the set of all real numbers for which $x + 2$ is positive or zero. Since $x + 2 \geq 0$ is equivalent to $x \geq -2$, we can explicitly identify the domain as the set of all real numbers greater than or equal to -2. The set of functional values is the set of all nonnegative real numbers.

Functions are customarily denoted by letters such as f, g, h, or y or by $f(x)$, $g(x)$, $h(x)$, or $y(x)$. The latter form emphasizes that the symbol x stands for an element of the domain. The symbol $f(x)$ also denotes the functional value at x, i.e., the element in the range assigned to the element x in the domain by the function f.

In business applications we frequently label functions by letters suggested by the business terms. For example, the quantity of a product that will sell depends upon its selling price. In a simple case, the quantity q might be related to the price p in dollars by the formula $1000/p^2$, in other words $q = 1000/p^2$. We could alternatively write $q(p) = 1000/p^2$. By the nature of this problem, the domain of the function is the set of all positive values of p, i.e., the selling price is a positive real number. Either of these expressions shows that we can understand a function to be defined by an equation. Thus $p = 10d + \frac{1}{10}d^2$ gives p as a function of d, where we could interpret p as price in dollars and d as demand. When the demand is big enough, the price may be sky-high.

Functions whose domains are restricted occur frequently in business applications. Functions giving quantity-price relationships such as the one discussed earlier are often the result of an analysis of past data. If, in the past, the price has always been between \$1 and \$4, we cannot be sure that the quantity-price relationship is the same for prices below \$1 or above \$4. Therefore, we might state

$$q(p) = \frac{1000}{p^2} \qquad \text{for} \quad 1 \leq p \leq 4$$

The domain of this function is any value of p between 1 and 4, inclusive.

Zeros and Roots In many problems, one needs to know for what value or values of x, the function $f(x)$ is equal to zero. This is equivalent to asking what values of x satisfy the equation

$$f(x) = 0$$

These values of x are called the *zeros* of the function $f(x)$ or the *roots* of the equation $f(x) = 0$.

For example, consider the function $f(x) = x^2 - 1$. The function is zero for $x = 1$ and $x = -1$; therefore, 1 and -1 are zeros of $f(x)$. Similarly, $x = 1$ and $x = -1$ are roots of the equation $x^2 - 1 = 0$.

Graph of a Function The *graph of the function* $f(x)$ is the set of all points $(x, f(x))$ in the plane for x in the domain of the function. We "graph" a function by sketching its graph in the plane. If the domain consists of only finitely many elements, then we could (presumably) graph the function completely. However, if the domain has more than finitely many elements, we either cannot graph the function explicitly or, even if we can, we do the graphing by analysis of the function.

In practice, two approaches can be followed in graphing a function:

1. Find and plot a large number of points of the graph.
2. Analyze the behavior of the function coupled with plotting a few points.

We shall illustrate both approaches.

To illustrate the former, suppose that the function $f(x)$ is as follows:

$$f(x) = x^3 - 5x^2 + 8$$

To begin the process, we might arbitrarily choose values of x and compute the corresponding values of $f(x)$ to construct Table 1-1.

TABLE 1-1

x	$f(x)$
-2	-20
-1	$+2$
0	$+8$
1	$+4$
2	-4
3	-10
4	-8
5	$+8$

The points indicated are a subset of the points comprising the graph of the function. In Figure 1-9, we have plotted these points and then connected them by a curve that "looks smooth." Of course, the accuracy of the curve between the points is only as good as the intuition of the person who drew the curve.

The graph can always be made more accurate by simply plotting more points. This makes the process more time consuming, and not very intellectually stimulating. Some of the manual work can be eliminated by using a computer to calculate the coordinates of the points to be plotted. In fact, most large computer systems include a *plotter* that can actually draw the graph, including the axes and associated lettering.

In the previous section, we stated that any equation of the form $y = mx + b$ had a graph that was a straight line having slope m and passing through the point $(0, b)$. We may regard $mx + b$ as a *linear function* and to

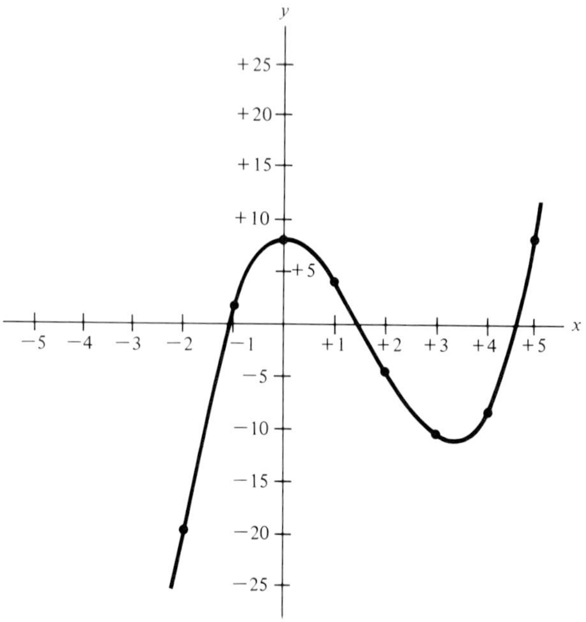

FIGURE 1-9

graph it we need merely find two points on the graph and draw the line through them, or alternatively, locate the point $(0, b)$ and draw a line with slope m through it. For example, the linear function $3x - 5$ has a graph that is a straight line with slope 3 and y intercept -5.

Certain functions can be recognized as related to linear functions. For example, the function $(x^2 - 1)/(x - 1)$ is defined for all $x \neq 1$. Since

$$\frac{x^2 - 1}{x - 1} = \frac{(x + 1)(x - 1)}{(x - 1)} = x + 1 \qquad \text{for all} \quad x \neq 1$$

we see that the graph of $(x^2 - 1)/(x - 1)$ is identical with the graph of $x + 1$ *except* for $x = 1$. The graph of $(x^2 - 1)/(x - 1)$ consists of all points on the line $x + 1$ except the point $(1, 2)$.

If the function is not of a simple form like $mx + b$, we may still be able to graph it quite accurately by analyzing the function and deciding what it must look like. We first ask the question

What is the domain of the function?

It is frequently easy to answer this question. Knowing the domain, we observe that the graph intersects a vertical line if and only if the x coordinate of a point on the vertical line is in the domain of the function. Furthermore, the graph of a function intersects any vertical line at most once (since for any value of x in the domain there is a *unique* functional value).

Having first determined the domain of the function, we then ask questions like the following.

1. For what values of x is the function positive (and therefore is the graph above the x axis)?
2. For what values of x is the function negative (and therefore is the graph below the x axis)?
3. For what values of x is $f(x)$ equal to zero (and therefore is the graph on the x axis)?
4. How does $f(x)$ behave for large positive values of x?
5. How does $f(x)$ behave for large negative values of x?

To illustrate, we shall first consider the function

$$y = f(x) = x^2$$

The domain is the set of all real values of x. Since $x^2 \geq 0$ for all x, the function is positive or zero everywhere. We note that $f(x) = 0$ if and only if $x = 0$. Thus the point (0, 0) is on the graph and all other points are above the x axis. For x large positively, or negatively, $f(x)$ is large positively. We might also observe that the points (1, 1), (−1, 1), (2, 4), and (−2, 4) are on the graph. Also for any $x, f(x) = f(-x)$; therefore, if $(x, f(x))$ is on the graph, so is $(-x, f(x))$. Hence the graph is symmetrical with respect to the y axis. With these observations, it should be apparent that the graph is similar to that in Figure 1-10.

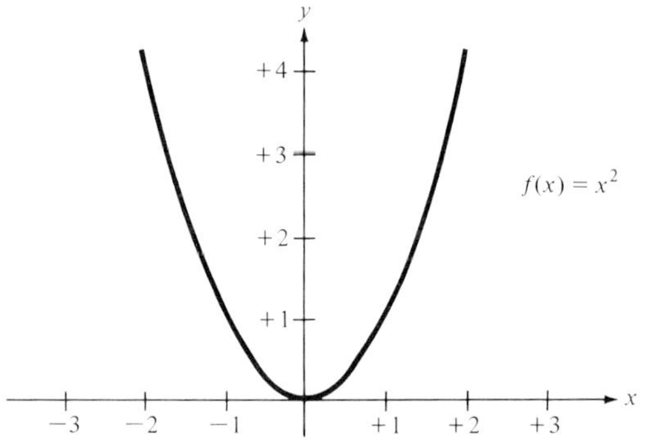

FIGURE 1-10

The appropriateness of the particular questions to be asked depends on the nature of the function. One of the "tricks of the trade" is to learn to ask (and answer) the "right" questions. Much of this book is concerned with posing and answering questions about functions and their graphs.

We next consider the function given by

$$y = f(x) = \frac{1}{x}$$

This function is defined for all $x \neq 0$. We might observe that for x positive, $f(x)$ is positive. For x negative, $f(x)$ is negative. For no value of x is $f(x)$ equal to zero. When x is a large positive number, $f(x)$ is a small positive number. When x is a large negative number, $f(x)$ is a small negative number. When $x = 1, f(x) = 1$. When $x = -1, f(x) = -1$. When x is a small positive number, $f(x)$ is a large positive number. When x is a small negative number, $f(x)$ is a large negative number. If $x > 0$ or if $x < 0$, as x increases y decreases. With these observations, it is plausible that the graph must be like that in Figure 1-11. You should check the graph against the above statements.

We previously introduced a function relating the quantity q of sales of a product to its selling price p, and we offered the example

$$q = \frac{1000}{p^2}$$

A logical question that a manager might ask is, "What must the price be in order to sell a given quantity of the product?" Instead of the function $q(p)$, we need the function $p(q)$. To obtain this function for our particular example, we need only solve the equation for p to obtain

$$p = \sqrt{\frac{1000}{q}}$$

and we have price as a function of quantity.

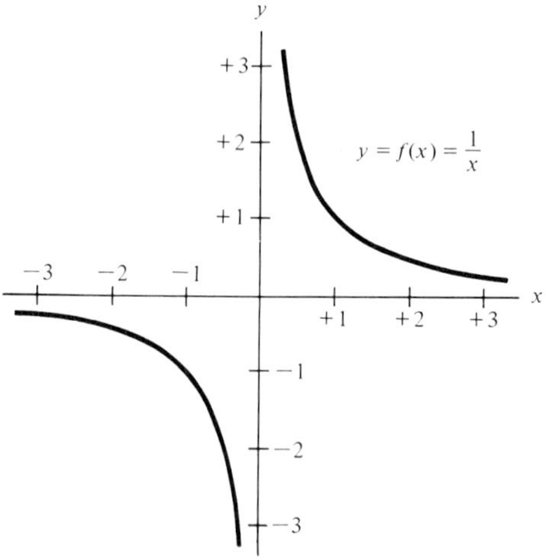

FIGURE 1-11

the solution process. Refer to the symbolic flowchart in Figure 1-12. In the card-shaped block (Block 1) we state the problem. The next block (Block 2) is a decision block. If $x_1 \neq x_0$, we obtain the answer to our problem via the operations in Blocks 3, 4, and 5. If $x_1 = x_0$, we obtain the answer via Block 6. Basically, there are two methods for solving this problem, one of which is applicable when $x_1 \neq x_0$ and the second of which is applicable when $x_1 = x_0$.

Although only one of these methods would be used for a specific problem, we do not generally know which one until we examine the numerical values of x_0 and x_1. Therefore, our flowchart must include both methods *along with* the decision process for choosing the proper method. The objective in preparing any flowchart is to include all possible conditions that might arise.

To illustrate, suppose we omit the decision block (Block 2) from the flowchart in Figure 1-12, retaining Blocks 1, 3, 4, and 5. The result is in Figure 1-13. If we are now presented with a problem for which $x_1 = x_0$, then in Block 2 we would attempt to divide by zero, an operation that we cannot do.

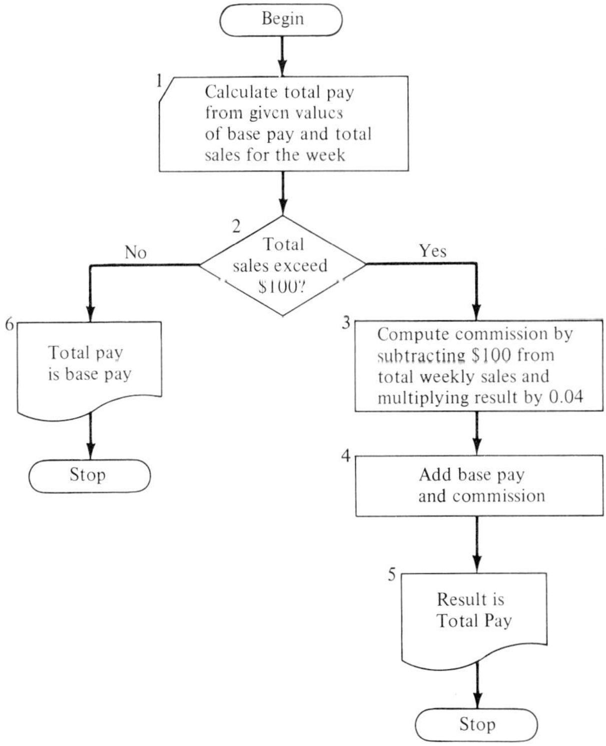

FIGURE 1-14 A Symbolic flowchart for the business problem discussed on page 22.

Example: A Business Problem As our other example of a symbolic flowchart, consider the problem of computing the total pay of a salesperson in our store. Our policy is that we pay a base salary plus a commission that is computed as follows:

1. Zero if the salesperson's total sales for the week amount to $100 or less.
2. If the salesperson's total sales for the week exceed $100, the commission is 4% of total weekly sales *in excess* of $100.

The symbolic flowchart for solving this problem is presented in Figure 1-14. This flowchart embodies two approaches for determining total pay, one applicable if weekly sales exceed $100 and the other applicable otherwise. The test in Block 2 determines which should be used.

EXERCISES 1-3

1. Suppose that you are given a point x_0 on the number line and are asked to determine if point x_1 is to the left of x_0, to the right of x_0, or the same point as x_0. In the symbolic flowchart in Figure H suppose that x_0 is 2.

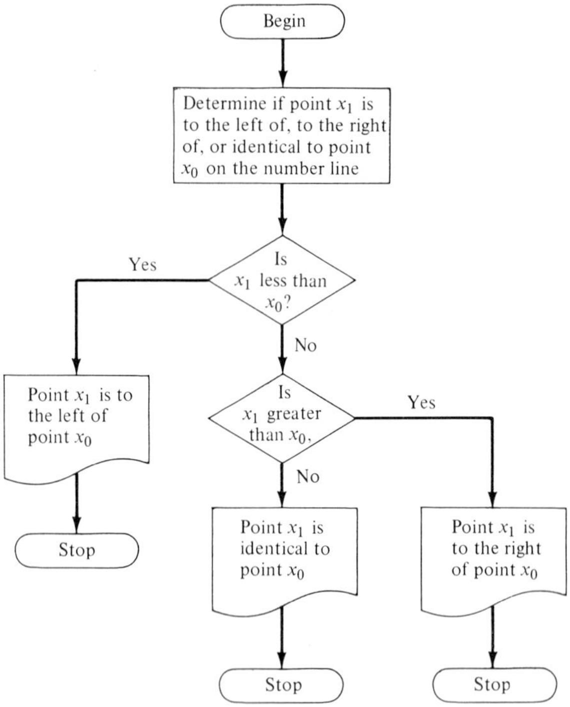

FIGURE H

(a) What result is indicated by the flowchart if x_1 is 1?
(b) If x_1 is 5?
(c) If x_1 is 2?
(d) Will this flowchart produce the correct answers for *any* given numerical values of x_0 and x_1?

2. Refer to the previous exercise. The flowchart in Figure I has been proposed. Suppose that x_0 is 2.

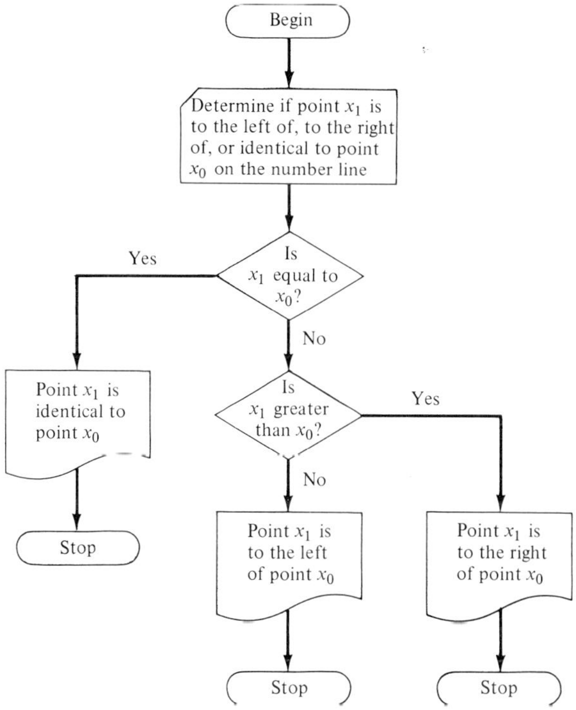

FIGURE I

(a) What result is indicated by the flowchart if x_1 is 1?
(b) If x_1 is 5?
(c) If x_1 is 2?
(d) Will this flowchart produce the correct answer for *any* given numerical values of x_0 and x_1?
(e) If you present the same values of x_0 and x_1 to this flowchart *and* to the flowchart in the previous exercise, will they produce the same results?

3. Suppose that you are given the end points of the open interval (a, b) and would like to know if point c is in the interval. The flowchart in Figure J is proposed. Suppose that the interval is (2, 4).

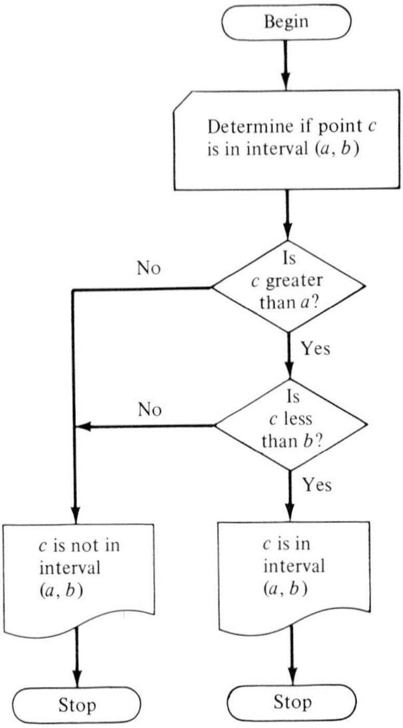

FIGURE J

(**a**) What result is indicated by the flowchart if c is 1?
(**b**) If c is 3?
(**c**) If c is 5?
(**d**) If c is 2?
(**e**) If c is 4?
(**f**) Is the flowchart correct?

4. Modify the flowchart in the previous exercise so that it applies to the closed interval $[a, b]$.

5. A professional society assesses annual dues of $25 for members under the age of 35, and $40 for members whose age is 35 or more. Prepare a symbolic flowchart to determine the dues from the age of the member.

6. A hotel charges $28 for a room for one person, plus $5 for each additional person occupying the room. Prepare a symbolic flowchart to determine the charge for the room from the number of occupants.

7. A bank charges a monthly service charge of $1 plus 10 cents per check for all checks over six. Prepare a symbolic flowchart to determine the monthly service charge from the number of checks cashed in a given month.

8. A bank computes service charges as in the above exercise. However, if the minimum monthly balance exceeds $400, no service charge is assessed. Prepare a symbolic flowchart to determine the monthly service charge from the number of checks cashed and the minimum monthly balance.

9. A straight line passes through point (x_0, y_0) and has slope m, where m is a real number (i.e., the line is not vertical). Prepare a symbolic flowchart to determine the y coordinate of the point *on the line* whose x coordinate is x_A.

10. Modify the previous exercise to determine the x coordinate of the point *on the line* whose y coordinate is y_A, assuming that the slope of the line is not zero.

11. Suppose that you are given a point in the plane and an equation for a straight line. We shall designate the point as (x_1, y_1) and use the form $y = mx + b$ for the equation of the line. Draw a symbolic flowchart to determine if point (x_1, y_1) is on the line.

12. Extend the previous exercise to determine if the point is on the line, above the line, or below the line.
HINT: Begin by determining the point on the line corresponding to x_1, i.e., determine the y coordinate, say y_p, using $y_p = m \cdot x_1 + b$. If $y_1 > y_p$, then the point is above the line.

1-4. COMPUTATIONAL FLOWCHARTS

In the previous section, we distinguished between symbolic flowcharts and computational flowcharts. The former type of flowchart is used to represent pictorially the approach to solving problems, and may involve any operation that you or I are able to perform. The computational flowchart is used to represent the computational process by which a numerical answer is obtained for a numerical problem. *Only those operations that can be performed by a computer are permissible in computational flowcharts.*

Although we shall consistently discuss computational flowcharts in terms of their use with computers, these flowcharts are almost equally useful with the small electronic hand calculators now widely available.

At the most elementary level, computers can perform the following numerical operations:

1. Arithmetic computations.
2. Decision making of the "Yes" or "No" type.
3. Input and output of numerical data.

We shall consider each of these.

Arithmetic Computations Computers can add, subtract, multiply, divide, raise numbers to powers (exponentiate), take square roots, take logarithms, and so on. In fact, computers can perform virtually all operations normally encountered in algebraic equations.

In performing these computations, computers operate on numbers. In a computational flowchart we use *variables* as symbolic representations of numbers. For example, we might represent the addition of two numbers as

$$x + y$$

which is known in computer terminology as an *expression*. The variables x and y represent the two numbers that are to be added.

What number is associated with a variable such as x or y? In a computer it is the last number that was *assigned* to the particular variable. For example, suppose we wish to assign the number 4.86 to x. We could do this by the *arithmetic assignment statement*

$$x \leftarrow 4.86$$

where the backward arrow $\leftarrow$ is called the assignment operator. Since the arithmetic assignment statement assigns values to variables, *only a variable may appear to the left of the assignment operator.*

Consider the following three arithmetic assignment statements:

$$x \leftarrow 4.86$$

$$y \leftarrow 3.12$$

$$h \leftarrow x + y$$

The order of arithmetic assignment statements is usually significant, so we consider them in the order as given. The statement $x \leftarrow 4.86$ assigns the numerical value 4.86 to variable x. The statement $y \leftarrow 3.12$ assigns the numerical value 3.12 to y. The statement $h \leftarrow x + y$ will assign a value to variable h. What value? It is the number obtained by adding the numerical values assigned to x and y. In this case, these values are 4.86 and 3.12, respectively; therefore, the value assigned to h is 4.86 + 3.12 or 7.98.

We often use the phrase "the value of h" to mean the numerical value that was last assigned to variable h. Whenever a new value is assigned to h, its value changes.

For the business problem at the end of Section 1-3, we wish to compute a commission as 4% of sales in excess of \$100 and add the resulting commission to the base pay to obtain the total pay. We might choose variable s for sales, variable c for commission, variable b for base pay, and variable p for total pay. Assuming s exceeds \$100, the appropriate arithmetic assignment statements would be

$$c \leftarrow 0.04 \cdot (s - 100)$$

$$p \leftarrow b + c$$

Although you and I might attach some significance to the name of a variable, the computer attaches absolutely none. The statement $c \leftarrow 0.04 \cdot (s - 100)$ instructs the computer to take the number currently assigned to s, subtract 100 from this number, multiply the result by 0.04, and assign the resulting number to c.

Decision Making In making decisions or "tests," the computer recognizes the conditions "greater than," "less than," "equal to," and combinations thereof. The six possibilities are

Greater than	$>$
Greater than or equal to	$\geq$
Less than	$<$
Less than or equal to	$\leq$
Equal to	$=$
Not equal to	$\neq$

These are called *relational operators.*

The relational operators are used to connect numbers, variables, or arithmetic expressions to form *relational expressions.* Examples of relational expressions are

$$s > 100$$

$$x \leq 0$$

$$a + b < c + 1$$

$$2n + 1 = m$$

In evaluating relational expressions, the computer first evaluates any arithmetic expressions using the values most recently assigned to the variables. Thus the computer will have two numbers that must be compared to determine if the relational expression is "true" or "false."

If we want to determine whether the value of variable s is greater than 100, we would use the following construction in a flowchart (Figure K).

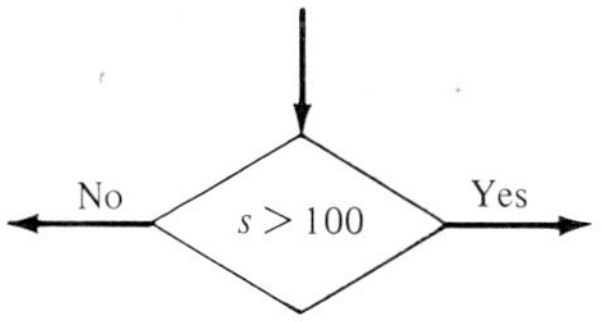

FIGURE K

If the value most recently assigned to s is greater than 100, the "Yes" exit is taken; otherwise, the "No" exit is taken.

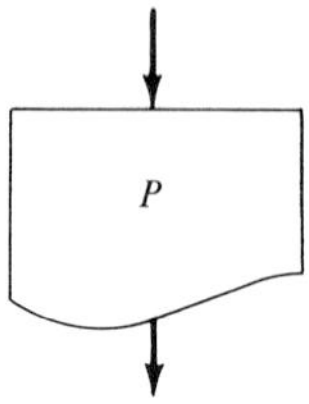

FIGURE L

Input/Output Whenever we work a numerical problem, the answer normally consists of one or more numbers. In order to obtain answers from a computer, it is necessary to specifically instruct it to print the desired numbers.

In flowchart notation, the block shown in Figure L specifies that the value of p is to be printed, i.e., the number most recently assigned to p is printed.

In the example in the previous paragraph, *only the number* is printed. In some cases it is desirable to print some messages or other explanatory information along with the number. In our flowchart we shall include such messages in the output block and customarily enclose them in quotes. For example, the block shown in Figure M prints the message "The total pay is," followed by the numerical value of p.

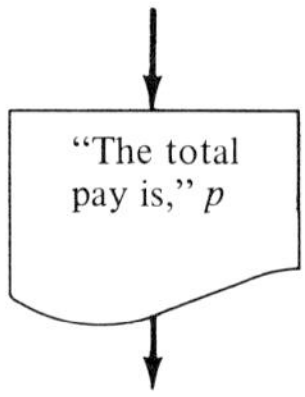

FIGURE M

For input, we shall always use the notation shown in Figure N. This block calls for the input of a number that is to be assigned to variable x.

More than one variable may be included in an input or output block. For example, the block in Figure O calls for the input of two numbers, the first of which will be assigned to variable s and the second to variable b.

Following our first example we shall discuss the use of the input block in more detail.

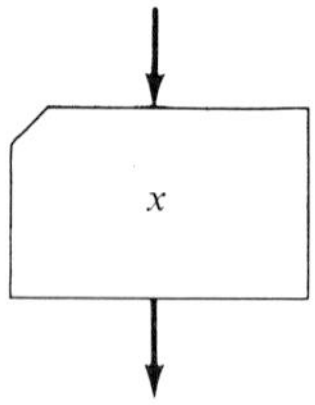

FIGURE N

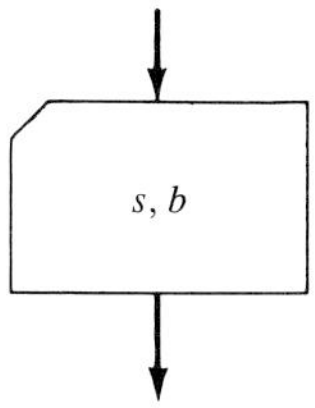

FIGURE O

Example 1. A company pays its salespersons on a weekly basis. Each salesperson is paid a base pay, to which is added his or her commission to obtain the total pay. The commission is calculated as 4% of his or her total weekly sales *in excess of* \$100; if the salesperson sold \$100 or less, no commission is paid. The symbolic flowchart for this problem was presented in Figure 1-14.

The computational flowchart in Figure 1-15 computes the total pay for a salesperson who sold \$471.28 last week and whose base pay is \$120. These values are assigned to variables s and b, respectively, in Block 1. In Block 2, we test s to see if the sales exceed \$100. If so, we compute the commission c as in Block 3. If not, we assign zero to c in Block 4. In either

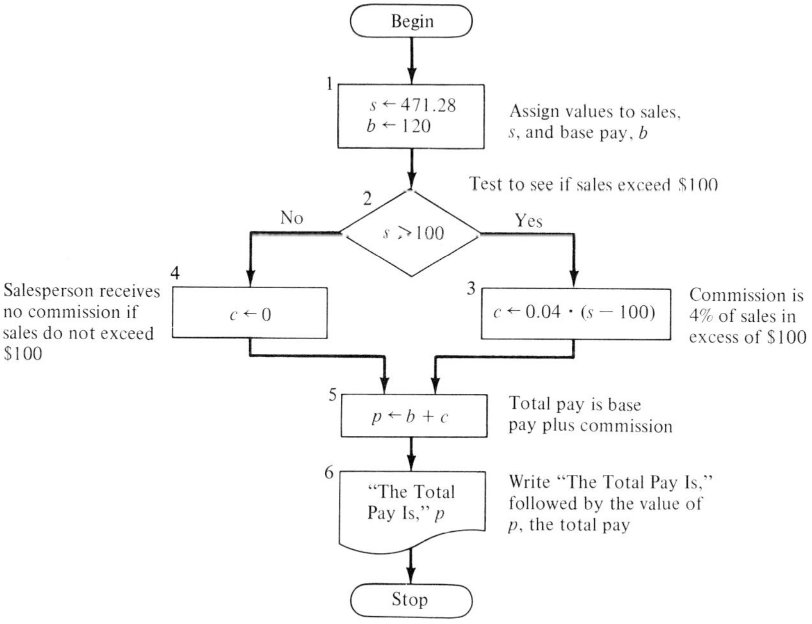

FIGURE 1-15

case, we add the base pay *b* and commission *c* in Block 5 and assign the result to variable *p*. In Block 6 we write the message "The total pay is," followed by the value of *p*.

What if we want to modify the program for Example 1 to compute the total pay of a salesperson who sold \$87.25 last week and whose base pay is \$128 per week? We need only change Block 1 in Figure 1-15 to that of Figure P.

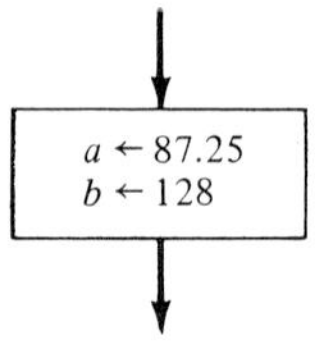

FIGURE P

In fact, we could apply the flowchart in Figure 1-15 to *any* salesperson, provided that we insert the proper arithmetic assignment statements into Block 1.

In changing these arithmetic assignment statements, we are essentially only providing the computer with the proper values of *s* and *b*. This is

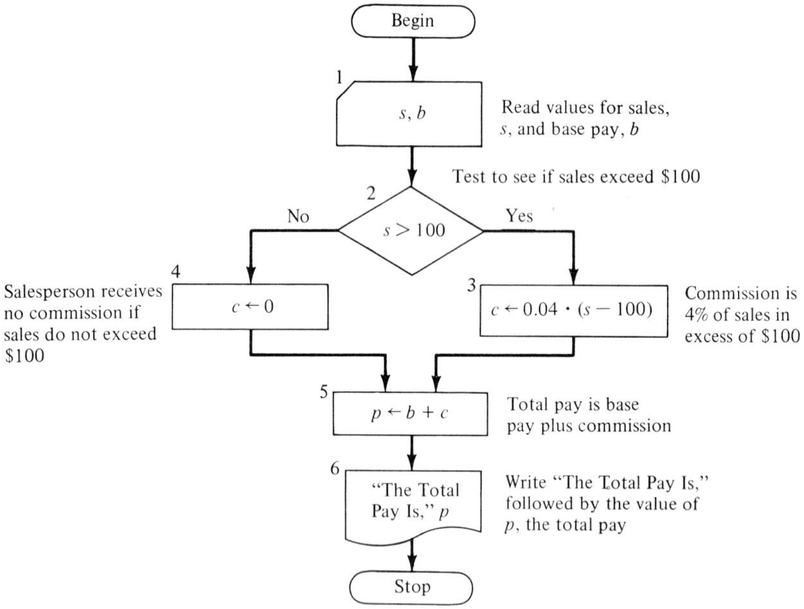

FIGURE 1-16

the type of situation in which the input block is more appropriate. The flowchart in Figure 1-16 is identical to the one in Figure 1-15 except that the input block is used for Block 1. The flowchart is now a *general* flowchart that works for any values supplied for s and b. That is, we can use the flowchart to compute the total pay for any salesperson by supplying the appropriate values for s and b.

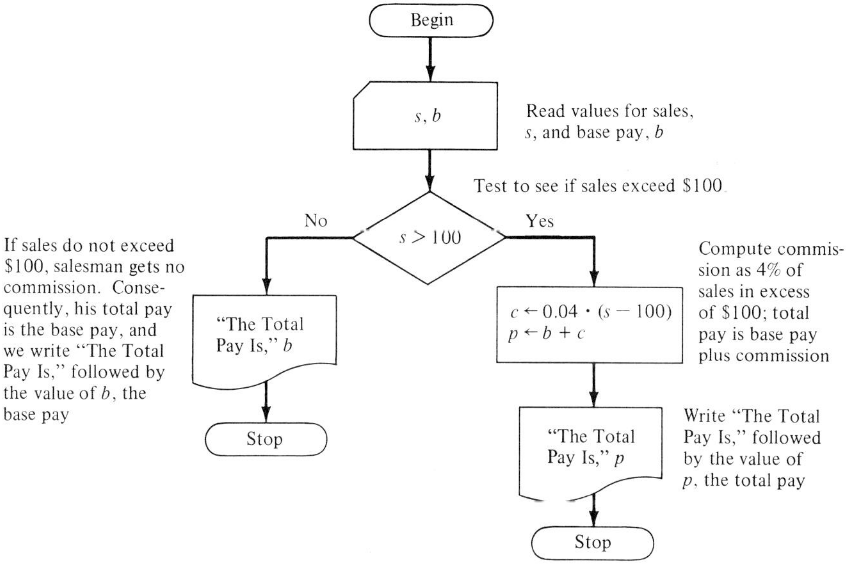

FIGURE 1-17

Except for the simplest of problems, several different flowcharts may be valid solutions. The only requirement is that when the operations indicated by the flowchart are followed in every detail, the answers produced are correct. To illustrate, the flowchart in Figure 1-17 is not identical to the one in Figure 1-16. However, for any entered values of s and b, these two flowcharts would produce exactly the same value for the total pay. Therefore, both are valid.

EXERCISES 1-4

In Exercises 1 through 6, what numerical value is assigned to z?

1. $x \leftarrow 5$
$y \leftarrow 8$
$z \leftarrow x + y$

2. $g \leftarrow 10$
$a \leftarrow 4$
$z \leftarrow g \cdot (a + 2)/5$

3. $y \leftarrow 5$
$b \leftarrow 2$
$c \leftarrow 10$
$z \leftarrow y^b/c$

4. $x \leftarrow 2$
$b \leftarrow x^2$
$z \leftarrow (b + 1)/(b - 1)$

5. $a \leftarrow 5$
$y \leftarrow 10$
$b \leftarrow y/a$

6. $a \leftarrow 120$
$b \leftarrow 4$
$c \leftarrow b \cdot (a - 100)/100$
$d \leftarrow 88$
$z \leftarrow c + d$

In Exercises 7 through 10, determine which exit is taken from the decision blocks.

7.

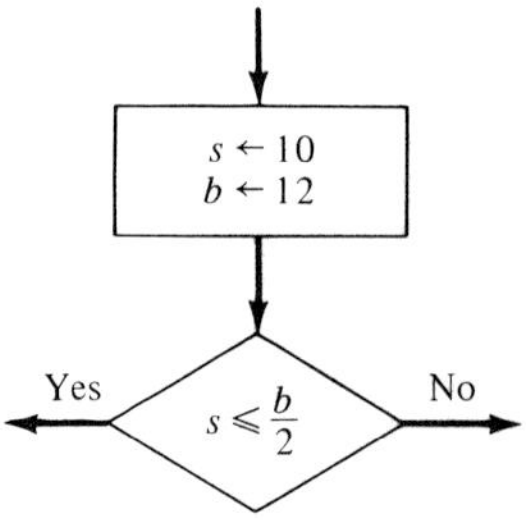

FIGURE Q

8.

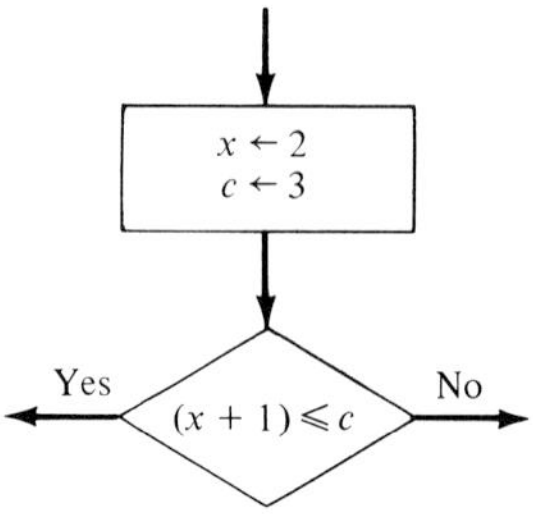

FIGURE R

9.

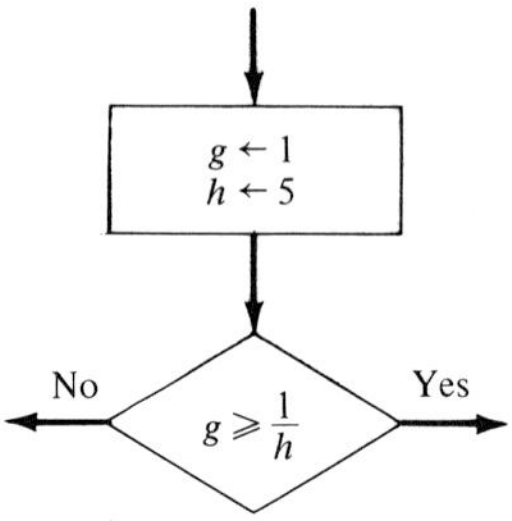

FIGURE S

10.

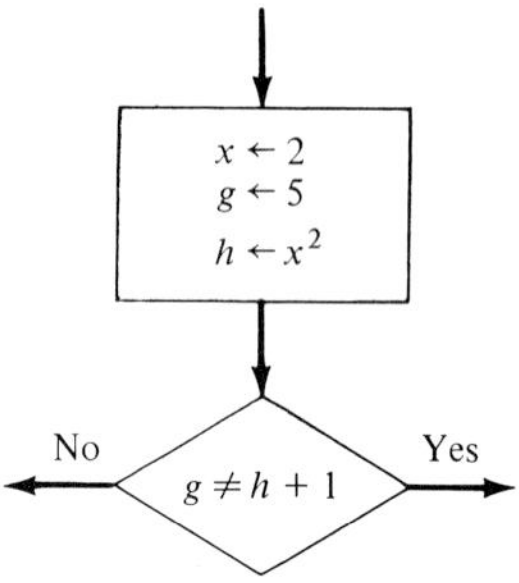

FIGURE T

11. **(a)** What is written by the flowchart in Figure U if a value of 2 is entered for s?
(b) If a value of 5 is entered for s?
(c) If a value of 11 is entered for s?

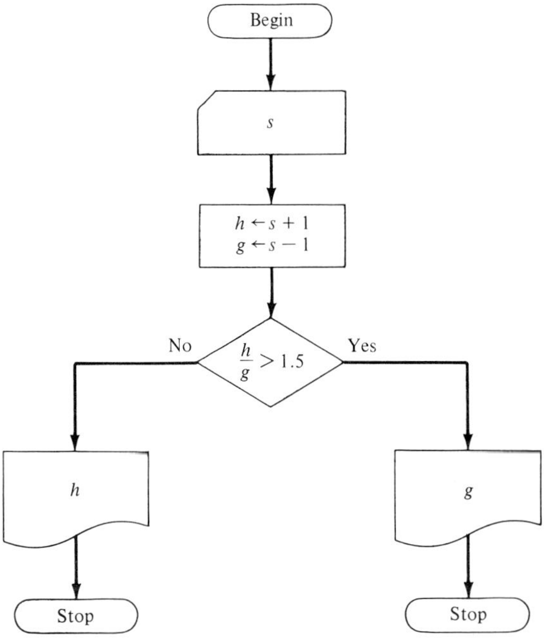

FIGURE U

12. **(a)** What is written by the flowchart in Figure V if values 2 and 5 are entered for p and q, respectively?
(b) If 3 and 1 are entered?
(c) If 2 and 2 are entered?

13. Consider the flowchart in Figure W, which is supposed to determine if the number entered for x is positive, negative, or equal to zero.
(a) What is written by the flowchart if 2.8 is entered for x?
(b) If 0 is entered for x?
(c) If -18.1 is entered for x?
(d) Is the flowchart correct, i.e., does it accomplish what it is supposed to accomplish?

14. The flowchart in Figure X is proposed to accomplish the same objective as the one in the previous exercise.
(a) What is written by the flowchart if 2.8 is entered for x?
(b) If 0 is entered for x?
(c) If -18.1 is entered for x?
(d) Is this flowchart correct?
(e) Is it equivalent (i.e., does it give the same answers) as the flowchart in the previous exercise?

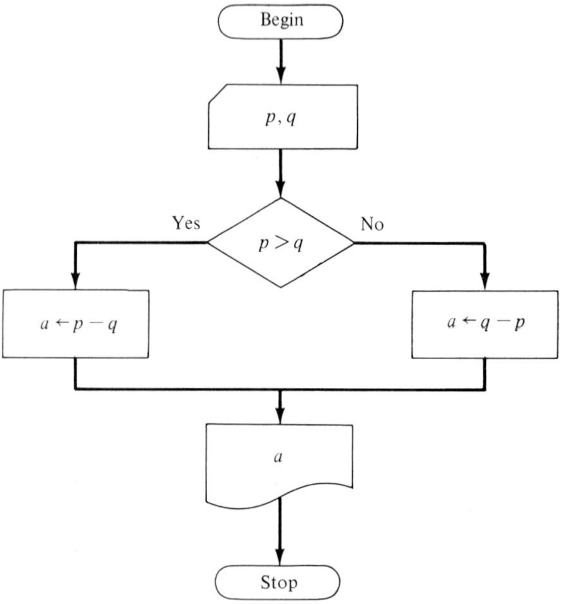

FIGURE V

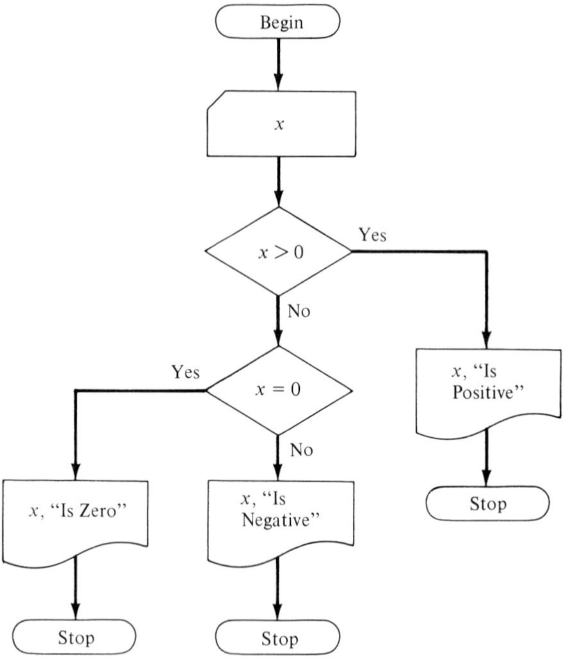

FIGURE W

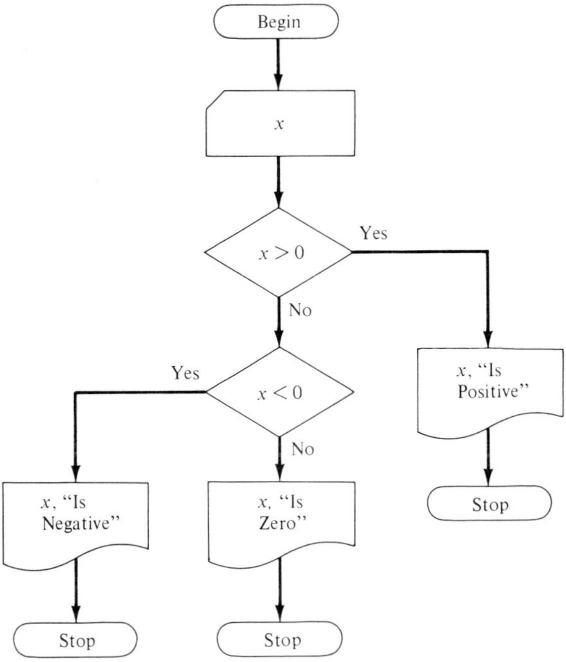

FIGURE X

15. Prepare a computational flowchart to determine if point x_1 is to the left of, to the right of, or identical to point x_0 on the number line. An input block should be provided for entry of the numerical values of x_0 and x_1. Refer to Exercise 2 of Section 1-3.

16. Prepare a computational flowchart to determine if point c is in the open interval (a, b). An input block should be provided for entry of the numerical values for a, b, and c. Refer to Exercise 3 of Section 1-3.

17. Modify the computational flowchart for the previous exercise to determine if point c is in the closed interval $[a, b]$.

18. Prepare a computational flowchart for Exercise 5 of Section 1-3. An input block should be provided for entry of the numerical value for the member's age.

19. Prepare a computational flowchart for Exercise 6 of Section 1-3. An input block should be provided for entry of the numerical value for the number of occupants in the room.

20. Prepare a computational flowchart for Exercise 7 of Section 1-3. An input block should be provided for entry of the numerical value for the number of checks.

21. Prepare a computational flowchart for Exercise 8 of Section 1-3. An input block should be provided for entry of the numerical values of the number of checks and the minimum monthly balance.

22. Prepare a computational flowchart for Exercise 9 of Section 1-3. An input block should be provided for entry of the numerical values of x_0, y_0, m, and x_A.

23. Prepare a computational flowchart for Exercise 10 of Section 1-3. An input block should be provided for entry of the numerical values of x_0, y_0, m, and y_A.

24. Prepare a computational flowchart for Exercise 11 of Section 1-3. An input block should be provided for entry of the numerical values of m, b, x_1, and y_1.

25. Prepare a computational flowchart for Exercise 12 of Section 1-3. An input block should be provided for entry of the numerical values of m, b, x_1, and y_1.

REVIEW EXERCISES: CHAPTER 1

In Exercises 1 through 5, determine an equation for the line passing through the two given points.

1. $(1, 5); (2, 7)$

2. $(1, 3); (3, 1)$

3. $(1, 5); (3, 5)$

4. $(-2, 3); (1, 2)$

5. $(1, -3); (1, 2)$

In Exercises 6 through 10, determine an equation for the line passing through the given point with the indicated slope.

6. $(1, 3)$; slope $= 2$

7. $(-1, 4)$; slope $= -5$

8. $(0, 5)$; slope $= 0$

9. $(-3, -2)$; slope $= -1$

10. $(-1, 3)$; slope is undefined

In Exercises 11 through 15, state the domain of $f(x)$.

11. $f(x) = 1 - x^2$

12. $f(x) = \dfrac{1}{(x+1)^2}$

13. $f(x) = \dfrac{1}{x^2 + 1}$

14. $f(x) = \sqrt{x^2 - 1}$

15. $f(x) = \sqrt{x^2 + 1}$

In Exercises 16 through 20, find the real numbers that are the zeros of $f(x)$, or if there are none, so state.

16. $f(x) = (x - 3)(x - 2)$

17. $f(x) = (1 - x)(1 - 2x)$

18. $f(x) = (x - 1)(x^2 + 1)$

19. $f(x) = \dfrac{1}{x} - 1$

20. $f(x) = x(x - 3)(x - 1)$

In Exercises 21 through 25, graph the function $f(x)$.

21. $f(x) = x - 2$

22. $f(x) = 3 - 2x$

23. $f(x) = \dfrac{1}{x} - 1$

24. $f(x) = 1 - x^2$

25. $f(x) = \dfrac{x^2 + 3x + 2}{x + 1}$ (Hint: Factor $x^2 + 3x + 2$.)

26. Suppose that our company produces custom uniforms and charges as follows:
(a) For orders under 25, the charge is \$150 plus \$30 per uniform.
(b) For orders 25 or over, the charge is \$300 plus \$24 per uniform.
Develop a symbolic flowchart to determine the total charge to be made for any given order.

27. Let a quadratic equation be represented as $ax^2 + bx + c = 0$. The roots of this equation are given by

$$r_1, r_2 = \frac{-b \pm \sqrt{b^2 - 4ac}}{2a}$$

In this book we shall intentionally avoid cases in which the roots have an imaginary part (which occurs whenever $b^2 - 4ac < 0$). Prepare a symbolic flowchart to calculate the numerical values of r_1 and r_2 for any given values of the coefficients a, b, and c. If the roots have an imaginary part, you need only so indicate; you do not need to calculate the roots.

28. In many problems we would like to know if a function $f(x)$ is close to zero at some specified value of x. Suppose we arbitrarily define "close to zero" to mean within 0.01 of zero. Then, $f(a_0)$ is close to zero if the double inequality $-0.01 < f(a_0) < 0.01$ is true. Observe that $-0.01 < f(a_0) < 0.01$ is equivalent to $|f(a_0)| < 0.01$. Prepare a symbolic flowchart to determine if $f(a_0)$ is close to zero in this sense.

29. Develop a computational flowchart for the problem in Exercise 26 above. An input block should be provided for entry of the numerical value for the number of uniforms ordered.

30. Prepare a computational flowchart for the problem in Exercise 27 above. Enter values for a, b, and c via an input block.

31. Prepare a computational flowchart for Exercise 28 above. Let $f(x) = x^5 - x + 1$. Enter the value for a_0 via an input block.

CHAPTER TWO

LIMITS OF SEQUENCES AND FUNCTIONS

2-1. SEQUENCES

The solution to many problems involves a sequence of numbers. To illustrate, suppose that a homeowner offers to pay an enterprising young lad \$10 each time he mows the lawn or to pay him 1 cent the first time he mows it and to double the pay on each subsequent mowing. If the latter option is selected, the remuneration is as shown in Table 2-1.

TABLE 2-1

Mowing	Pay
1	\$0.01
2	0.02
3	0.04
4	0.08
5	0.16
6	0.32
7	0.64
8	1.28
9	2.56
10	5.12
11	10.24

From Table 2-1, we see that on the eleventh mowing, the pay has finally reached \$10. However, if the lawn is mowed 20 times, the pay on the final mowing is \$5242.88.

The pay for the various mowings of the lawn constitutes a sequence. We let n be the times the lawn is mowed and p_n the pay in number of dollars received for the nth mowing. Then we can represent the sequence by the formula

$$p_n = 0.01 \cdot 2^{n-1}$$

Alternatively, we could note that

$$p_n = 2p_{n-1}$$

with the condition that $p_1 = 0.01$. The equation for relating p_n to p_{n-1} coupled with a specified value of p_1 constitutes a *recursion relationship*.

For the purposes of this book, a sequence will be defined as a function whose domain is either of the following:

1. The set of positive integers: 1, 2, 3, ...
2. The set of nonnegative integers: 0, 1, 2, 3, ...

Customarily, the domain will be the set of positive integers. We shall always assume such to be the case unless we specify the contrary. However, in some applications, it is convenient to include zero in the domain, and in those cases we shall specifically indicate that we are doing so.

The range of a sequence is understood to be a subset of the set of real numbers. In using the customary notation for functions, the real number corresponding to some integer n would be denoted by $p(n)$. However, for sequences the customary notation is p_n, which denotes the real number corresponding to the integer n. The real number p_n is said to be a *term* in the sequence, and the integer n is said to be the *index*. When the domain of the sequence is the set of positive integers, the first term in the sequence is p_1, the second term is p_2, and a general term or the nth term is represented by p_n.

When the domain of the sequence is the set of positive integers, we shall represent the sequence as $\langle p_1, p_2, p_3, \ldots \rangle$ or as $\langle p_n \rangle_{n=1, 2, 3, \ldots}$, or more briefly as $\langle p_n \rangle$. When the domain of the sequence is the set of nonnegative integers, we shall usually use the representation $\langle p_0, p_1, p_2, \ldots \rangle$ or $\langle p_n \rangle_{n=0, 1, 2, 3, \ldots}$.

In this chapter, we shall discuss three ways to represent sequences:

1. By enumeration of the terms, i.e., by listing the first several terms, or by stating a rule that tells us how the sequence is defined.
2. By a formula giving p_n as a function of n.
3. By a recursion relationship. As used in this book, we must have
   a. An equation relating p_n to p_{n-1}.
   b. A specified value either for p_0 or for p_1.

To illustrate a sequence that is best represented by a stated rule, we define term n of a sequence to be the decimal approximation to the number π represented by the first n digits in the usual decimal expansion of π. The first five terms in this sequence are 3, 3.1, 3.14, 3.141, and 3.1415. As another example, we could let term n be the nth digit to the right of the decimal point in the expansion of π. The first five terms of this sequence are 1, 4, 1, 5, and 9.

Although any sequence can be represented by enumerating the terms, such a representation is not especially convenient for most problems and is avoided whenever possible. For many sequences, a formula can be devised that expresses any term p_n in the sequence as a function of n. For example, the first five terms in the sequence whose terms are given by

$$p_n = \frac{1}{n}$$

are 1, $\frac{1}{2}$, $\frac{1}{3}$, $\frac{1}{4}$, $\frac{1}{5}$. In fact, we can readily construct Table 2-2.

As another example, the terms in the sequence consisting of the odd integers could be represented by the formula

$$p_n = 2n - 1$$

TABLE 2-2

Index of term	n	1	2	3	4	5	$\cdots$	10	$\cdots$	20	$\cdots$	100	$\cdots$	1000
Value of term	p_n	1	$\frac{1}{2}$	$\frac{1}{3}$	$\frac{1}{4}$	$\frac{1}{5}$	$\cdots$	$\frac{1}{10}$	$\cdots$	$\frac{1}{20}$	$\cdots$	$\frac{1}{100}$	$\cdots$	$\frac{1}{1000}$

when the index n for the first term is one. When the index of the first term is zero, the formula would be

$$p_n = 2n + 1$$

In either case, the first seven terms of this sequence are 1, 3, 5, 7, 9, 11, and 13.

It is also permissible for some or all of the terms in the sequence to be negative. For example, consider the sequence whose terms are given by the formula

$$p_n = \frac{(-1)^n}{n^2}$$

The first five terms of this sequence are $-1, \frac{1}{4}, -\frac{1}{9}, \frac{1}{16}, -\frac{1}{25}$. Observe the role of $(-1)^n$ in the formula for p_n. For n odd, it has the value -1 and for n even,

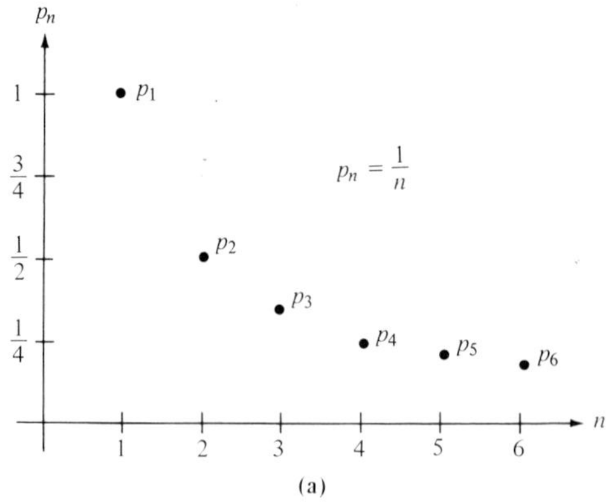

FIGURE 2-1a Graphing in rectangular coordinates

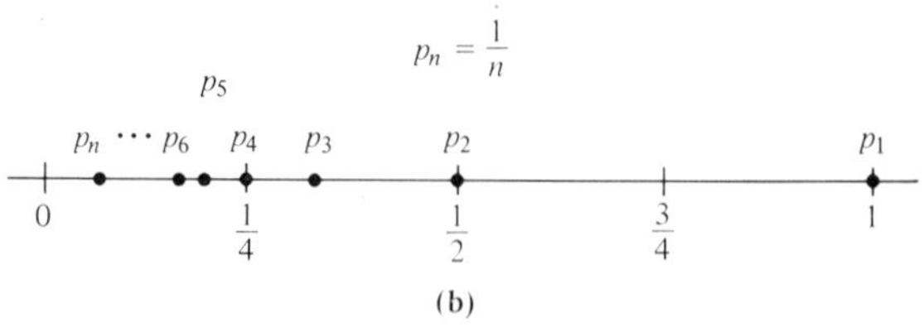

FIGURE 2-1b Plotting on the number line

it has the value $+1$. Thus, if we include $(-1)^n$ as a factor in a formula for p_n, we have a way of making the terms of the sequence alternate in sign starting with a negative sign for $n = 1$. The factor $(-1)^{n+1}$ can be used to make the first term positive and have the terms alternate in sign.

It is also possible to have a sequence represented by one formula for some terms and another formula for the others, as illustrated by

$$p_n = \begin{cases} \dfrac{n-1}{n}, & \text{for } n \text{ odd and positive} \\[2ex] \dfrac{1}{n}, & \text{for } n \text{ even and positive} \end{cases}$$

The first six terms of this sequence are $0, \frac{1}{2}, \frac{2}{3}, \frac{1}{4}, \frac{4}{5}$, and $\frac{1}{6}$.

As for other functions, a graphical representation of a sequence is often enlightening. For the sequence $\langle 1/n \rangle$, the terms could be graphed as a function, as illustrated in Figure 2-1(a). An alternate useful pictorial representation is obtained by plotting the terms as points on the number line, as illustrated in Figure 2-1(b). Using the latter approach, it is desirable to number the terms to avoid confusion. The sequence

$$\left\langle \frac{(-1)^n(n+4)}{2n+1} \right\rangle_{n=0,\,1,\,2,\,\ldots}$$

is illustrated in Figure 2-2(a) below and 2-2(b) on the next page.

If you know a formula for p_n, it is usually easy to compute the first several terms of $\langle p_n \rangle$. But if you know only the first several terms of $\langle p_n \rangle$, it is usually more difficult to find a formula for p_n that fits these known terms. There is no automatic procedure that will produce a satisfactory result. The

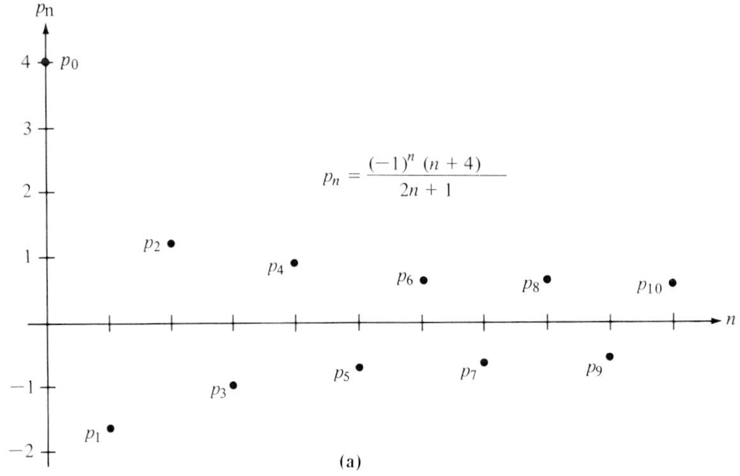

FIGURE 2-2a Graphing in rectangular coordinates

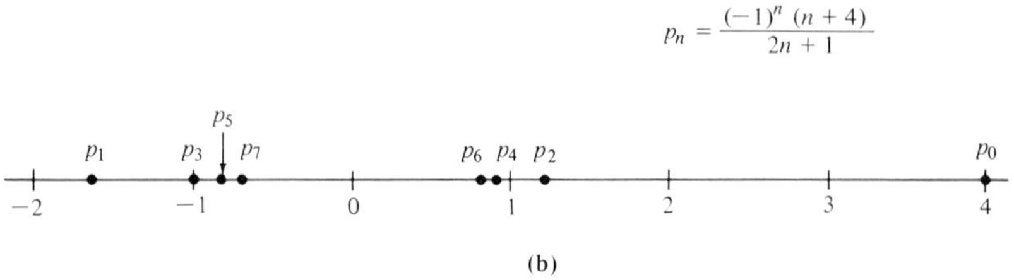

FIGURE 2-2b Plotting on the number line

student must observe relationships and use arithmetic ingenuity. For the sequence $\frac{1}{2}, \frac{1}{4}, \frac{1}{6}, \frac{1}{8}, \frac{1}{10}, \ldots$ we observe that the sequence consisting of the denominators only is 2, 4, 6, 8, 10, ... which is represented by $2n$ for $n = 1, 2, 3, 4, 5$. Therefore, the sequence $\frac{1}{2}, \frac{1}{4}, \frac{1}{6}, \frac{1}{8}, \frac{1}{10}, \ldots$ may be represented by $p_n = 1/(2n)$, since for $p_n = 1/(2n)$ we get the sequence $\frac{1}{2}, \frac{1}{4}, \frac{1}{6}, \frac{1}{8}, \frac{1}{10}, \ldots$.

By getting experience on easier problems, many students can learn to find formulas that "fit the facts." In general, there will be more than one possible correct formula that will fit the first several known terms. An answer is correct if it works, that is, if on substituting $n = 1, 2, 3$, etc., one actually gets the known terms.

EXERCISES 2-1

For Exercises 1 through 10, n is positive and a formula for p_n is given. (**a**) Write the first six terms. (**b**) Graph the first six terms of the sequence in rectangular coordinates. (**c**) Plot on the number line.

1. $\dfrac{1}{2n}$ **2.** $-n$

3. n^2 **4.** 2^n

5. $(2)^{-n}$ **6.** $\dfrac{n^2}{n+1}$

7. $\dfrac{(-1)^n}{n+1}$ **8.** $\dfrac{n+3}{n+2}$

9. $\dfrac{n-1}{n^2}$ **10.** $(n-1)(n-2)(n-3)$

For Exercises 11 through 20, n is nonnegative. (**a**) Write the first six terms. (**b**) Graph the first six terms in rectangular coordinates. (**c**) Plot on the number line.

11. $\dfrac{1}{2n+1}$ **12.** $n+2$

13. n^3 **14.** $(-1)^n(n+1)$

15. $(-2)^n$ **16.** $(-1)^n \cdot 2n$

17. $\dfrac{(-1)^n}{n+1}$ **18.** $\dfrac{2n+3}{n+2}$

19. $\dfrac{n-1}{n^2+1}$ **20.** $(n)(n-1)(n-2)$

21. In the text, $p_n = 2n - 1$ is given as a formula for generating the odd integers starting with $p_1 = 1$. Write a formula for p_n to generate the even integers starting with $p_1 = 2$.

22. In the text $p_n = 2n + 1$ is given as a formula generating the odd integers starting with $p_0 = 1$. Write a formula for p_n to generate the even integers starting with $p_0 = 2$.

For Exercises 23–34, the first six terms of a sequence are given. Devise a formula for p_n for n positive.

NOTE: For each of these exercises, several formulas can be developed to give the indicated first six terms, but they would presumably give different values for the remaining terms in the sequence. You need only find one formula in each case. For sequences of fractions, it is generally useful to consider the numerators and denominators separately and then combine them. For sequences with alternating signs, use the device suggested in the text.

23. $3, 5, 7, 9, 11, 13, \ldots$ **24.** $4, 6, 8, 10, 12, 14, \ldots$

25. $-1, -3, -5, -7, -9, -11, \ldots$

26. $-2, -4, -6, -8, -10, -12, \ldots$

27. $-1, 2, -3, 4, \quad 5, 6, \ldots$ **28.** $1, -2, 3, -4, 5, -6, \ldots$

29. $2, -4, 8, -16, 32, -64, \ldots$ **30.** $-\frac{1}{2}, \frac{1}{4}, -\frac{1}{8}, \frac{1}{16}, -\frac{1}{32}, \frac{1}{64}, \ldots$

31. $\frac{6}{4}, \frac{7}{5}, \frac{8}{6}, \frac{9}{7}, \frac{10}{8}, \frac{11}{9}, \ldots$ **32.** $2, \frac{3}{4}, \frac{4}{9}, \frac{5}{16}, \frac{6}{25}, \frac{7}{36}, \ldots$

33. $1, -8, +27, -64, 125, -216, \ldots$

34. $1 \cdot 2, 2 \cdot 3, 3 \cdot 4, 4 \cdot 5, 5 \cdot 6, 6 \cdot 7, \ldots$

35. The decimal representation of $\sqrt{2}$ is 1.41421356237 .... (**a**) Defining term p_n as the decimal representation to n decimal places, list the first seven terms in the sequence. (**b**) Defining term p_n as the nth digit following the decimal point in the decimal representation, list the first seven terms in the sequence.

36. Repeat Exercise 9 for $\sqrt[3]{5}$, whose decimal representation is 1.70997594667 ....

37. Do Exercises 23, 27, and 29 for n nonnegative.

38. Do Exercises 24, 28, and 30 for n nonnegative.

2-2. RECURSION RELATIONSHIPS

As mentioned earlier, an approach to representing sequences is to relate the nth term to the previous term in the sequence by what is known as a *recursion relationship*. For example, the sequence of odd integers results from the recursion relationship

$$p_n = p_{n-1} + 2$$

with the additional condition that the first term in the sequence is 1, or in other words,

$$p_1 = 1$$

The computation of the remaining terms in the sequence proceeds as shown in Table 2-3.

TABLE 2-3

Term	Equation
2	$p_2 = p_1 + 2 = 1 + 2 = 3$
3	$p_3 = p_2 + 2 = 3 + 2 = 5$
4	$p_4 = p_3 + 2 = 5 + 2 = 7$
5	$p_5 = p_4 + 2 = 7 + 2 = 9$
6	$p_6 = p_5 + 2 = 9 + 2 = 11$
⋮	⋮

The process can be carried out indefinitely, always generating the value for another term in the sequence by application of the recursion relationship.

The recursion relationship $p_n = p_{n-1} + 2$ with $p_1 = 1$ can be understood to mean that we can get any term from the preceding term by adding 2 to it. Starting with 1 and successively adding 2, we can directly write the first six terms as 1, 3, 5, 7, 9, and 11. In more complicated cases we may need to compute terms by the process indicated in Table 2-3.

Another example of a recursion relationship is

$$p_n = \frac{p_{n-1}}{2}$$

where the first term is given by

$$p_1 = 1$$

The recursion relationship implies that each term is computed by dividing the previous term by 2. By starting with 1 and successively dividing by 2, the terms in the sequence are easily seen to be 1, $\frac{1}{2}$, $\frac{1}{4}$, $\frac{1}{8}$, $\frac{1}{16}$, $\frac{1}{32}$.

A recursion relationship representation of a sequence embodies the following:

1. An equation relating a term in the sequence to the previous term, for example,

$$p_n = 2p_{n-1}$$

2. Initial specification of one term in the sequence, usually either p_0 or p_1. For example,

$$p_1 = 1$$

The above equation and the initial specification constitute a recursion relationship for the sequence $\langle 1, 2, 4, 8, 16, \ldots \rangle$ obtained by starting with 1 and successively multiplying by 2.

Many sequences can be conveniently represented by either a formula or a recursion relationship. For example, a sequence with 1, 1.5, 2.25, 3.375, 5.0625 as the first five terms could be represented by either the formula

$$p_n = (1.5)^{n-1}$$

or by the recursion relationship

$$p_n = (1.5)p_{n-1}$$

with the condition that

$$p_1 = 1$$

Obtaining Recursion Relationships from Formulas We wish to show how a recursion relationship can be found directly from a formula for the nth term, p_n. First, we note how the expression for p_n enables us to find expressions for p_{n-1} and p_{n+1}. Essentially, p_{n-1} is what we get by replacing n by $n - 1$ in the formula for p_n. Let $p_n = n(n + 2)$. Then

$$p_{n-1} = (n - 1)[(n - 1) + 2] = (n - 1)(n + 1)$$

Similarly,

$$p_{n+1} = (n + 1)(n + 3).$$

As an example of finding a recursion relationship, let the terms of a sequence be represented by the formula

$$p_n = \frac{1}{n}$$

The first term is 1, i.e., $p_1 = 1$. To determine the ratio of the nth term to the $(n-1)$st term, we first note that

$$p_{n-1} = \frac{1}{n-1}$$

Therefore, the ratio of p_n to p_{n-1} is

$$\frac{p_n}{p_{n-1}} = \frac{1/n}{1/(n-1)} = \frac{n-1}{n}$$

Solving for p_n yields the recursion relationship

$$p_n = \frac{n-1}{n} \cdot p_{n-1} \qquad \text{with} \quad p_1 = 1$$

The reader might want to calculate the first few terms to verify that this recursion relationship defines a sequence whose terms are given by the formula $1/n$.

The method of the example given is quite general. We can always use it, although sometimes the algebra in simplifying the ratio of the formulas may be messy.

So far we have used p_n to represent the nth term of a sequence. Of course, other symbols such as a_n, b_n, or q_n can also be used.

EXERCISES 2-2

For Exercises 1 through 6, write the first six terms of the following sequences.

1. $p_n = p_{n-1} + 1$
$p_1 = 1$

2. $p_n = 2p_{n-1} + 1$
$p_1 = 0$

3. $p_n = 3 \cdot p_{n-1}$
$p_1 = 2$

4. $p_n = p_{n-1} + n$
$p_0 = 0$

5. $p_n = n^2 - p_{n-1}$
$p_0 = 0$

6. $p_n = p_{n-1} + \frac{1}{n}$
$p_0 = 0$

For Exercises 7 through 10, write formulas for p_{n-1} and p_{n+1}.

7. $p_n = n^2 - 1$

8. $p_n = \frac{1}{n+1}$

9. $p_n = (-1)^n \cdot (3n)$

10. $p_n = (-1)^n \cdot (4n - 1)$

For Exercises 11 through 16, use the method of the text to write recursion relationships for the sequences:

11. $q_n = \dfrac{1}{2n}$, n positive

12. $q_n = (-2)^n$, n positive

13. $q_n = (n + 1)(n + 2)$, n nonnegative

14. $q_n = n^2$, n positive

15. $q_n = \dfrac{(-1)^n}{n + 1}$, n nonnegative

16. $q_n = (-1)^n \cdot (n + 2)$, n nonnegative

Writing a formula for a sequence specified by a recursion relationship can be approached by enumerating the first several terms and then trying to devise a formula by observation or even by trial and error. In Exercises 17 through 20, obtain a formula for p_n.

17. $p_n = p_{n-1} + 1$

$p_1 = 1$

18. $p_n = \dfrac{1}{3} p_{n-1}$

$p_1 = 2$

19. $p_n = (-3)p_{n-1}$

$p_0 = 1$

20. $p_n = 2 + p_{n-1}$

$p_0 = 3$

Instead of deriving a recursion relationship by determining the *ratio* of two successive terms, the *difference* between two successive terms can be used. Illustrating for the sequence whose terms are given by the formula $1/n$,

$$p_n - p_{n-1} = \frac{1}{n} - \frac{1}{n-1} = \frac{-1}{n(n-1)}$$

The recursion relationship becomes

$$p_n = p_{n-1} - \frac{1}{n(n-1)}$$

with the condition that

$$p_1 = 1$$

Use this approach to derive recursion relationships for the following:

21. $p_n = 2n + 1$, n positive

22. $p_n = -n$, n nonnegative

23. $p_n = 3 - n$, n positive

24. $p_n = 1 + n^2$, n nonnegative

2-3. APPLICATIONS OF SEQUENCES

In the previous two sections, the terms in the sequence were determined for values of the index beginning at either zero or one and continuing indefinitely. Such sequences might be thought of as infinite sequences. If the process is not carried out indefinitely but terminated at some value of n, the result might be thought of as a finite sequence. Many applications of sequences in business are of the latter type.

To illustrate an application of such sequences, suppose that a man invests an amount V_0 at an annual interest rate i compounded annually. At the end of the first year the man has the original investment plus the interest drawn upon it.

Original investment: V_0
Interest drawn: iV_0
Total at year end: $V_0 + iV_0 = (1 + i)V_0$

Suppose that we let V_1 be the value of the investment at the end of year 1, which is also the value at the beginning of year 2.

For year 2, we can follow a similar procedure to determine that V_2, the value at the end of year 2, is given by

$$V_2 = (1 + i)V_1 = (1 + i)[(1 + i)V_0] = (1 + i)^2 V_0$$

For the following years, we obtain

$$V_3 = (1 + i)V_2 = (1 + i) \cdot [(1 + i)^2 V_0] = (1 + i)^3 V_0$$

$$V_4 = (1 + i)V_3 = (1 + i) \cdot [(1 + i)^3 V_0] = (1 + i)^4 V_0$$

$$V_5 = (1 + i)V_4 = (1 + i) \cdot [(1 + i)^4 V_0] = (1 + i)^5 V_0$$

Generalizing to year n, we have

$$V_n = (1 + i)V_{n-1} = (1 + i)^n V_0$$

Thus we can specify the sequence in terms of the formula

$$V_n = (1 + i)^n V_0$$

or in terms of the recursion relationship

$$V_n = (1 + i)V_{n-1}$$

where V_0 must be given.

In many business applications, the terms in the sequence are calculated on a digital computer. For example, when a potential investor of a sum of money at some interest rate is contacted by a bank or other financial institution, he is usually presented with a statement of the worth of his investment for some years in the future. An example of such a statement is shown in Figure 2-3.

Initial deposit: \$1000.00

Interest rate: 7.50 percent per annum

Year	Value
0	1000.00
1	1075.00
2	1155.63
3	1242.30
4	1335.47
5	1435.63
6	1543.30
7	1659.05
8	1783.48
9	1917.24
10	2061.03

FIGURE 2-3

For any specified values of i and V_0, we may get a computer printout of the terms in this sequence using the formula $V_n = (1 + i)^n V_0$ or using the recursion relationship $V_n = (1 + i)V_{n-1}$. We shall first illustrate the use of the formula.

To compute the first 10 terms of the sequence, we must compute $(1 + i)^n \cdot V_0$ for $n = 1, 2, 3, \ldots, 10$. Let us assume that an input block will be provided in our flowchart for obtaining numerical values for i and V_0. Then, for any value of n, we can calculate and print the value of $(1 + i)^n \cdot V_0$ by the blocks in Figure Y.

We must first calculate for n equals 1, then for n equals 2, then for n equals 3, and so on through n equals 10. We could do this by writing these blocks 10 times, once for each value of n. A preferable alternative is to use

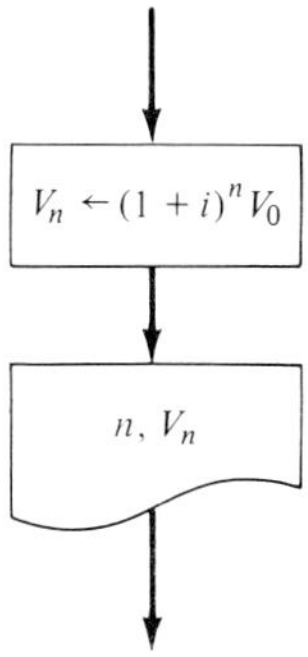

FIGURE Y

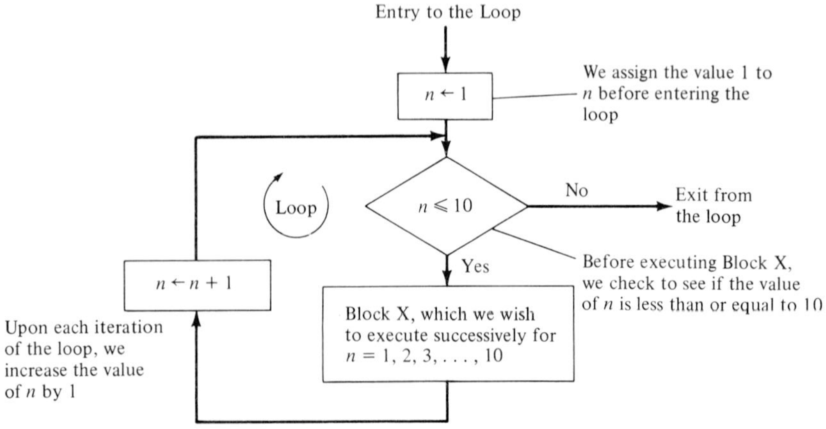

FIGURE 2-4

the same blocks but change the value of n between successive executions of these blocks.

To do this, we use a flowchart construction known as a *loop*. Suppose that we have a block, say Block X, which we wish to execute for $n = 1, 2, 3, \ldots, 10$. The arrangement of blocks illustrated in Figure 2-4 accomplishes this objective.

To construct the loop requires three blocks in addition to the one to be executed for the prescribed values of n. We wish to explain the purpose of each of these three blocks (refer to Figure 2-4). Since we wish to execute the block for $n = 1, 2, 3, \ldots, 10$, we must begin with a value of 1 for n. Therefore, entry to the loop in Figure 2-4 is via a block with the statement $n \leftarrow 1$ to assign a value of 1 to n.

Inside the loop structure in Figure 2-4, the following operations proceed on a repetitive cycle:

1. We test to see if the value of n is less than or equal to 10. If not, we exit the loop.
2. Block X is executed once.
3. The value of n is increased by 1.

Each repetition of these three operations is referred to as an *iteration* of the loop. In general, we could, of course, insert any sequence of blocks where Block X appears in the loop in Figure 2-4.

We want to explain the statement $n \leftarrow n + 1$ in more detail. In executing this statement, we take the numerical value currently assigned to n, add 1 to it, and assign the result to n. The net effect is to increase the value of n by 1.

The variable n used in the manner of Figure 2-4 is often called the *index* of the loop. Alternatively, some authors call it a *counter*, since it assumes the values 1, 2, 3, ..., 10.

For computing $(1 + i)^n V_0$ for $n = 1, 2, 3, \ldots, 10$, the complete flowchart is given in Figure 2-5. In Block 1, numerical values are read for the

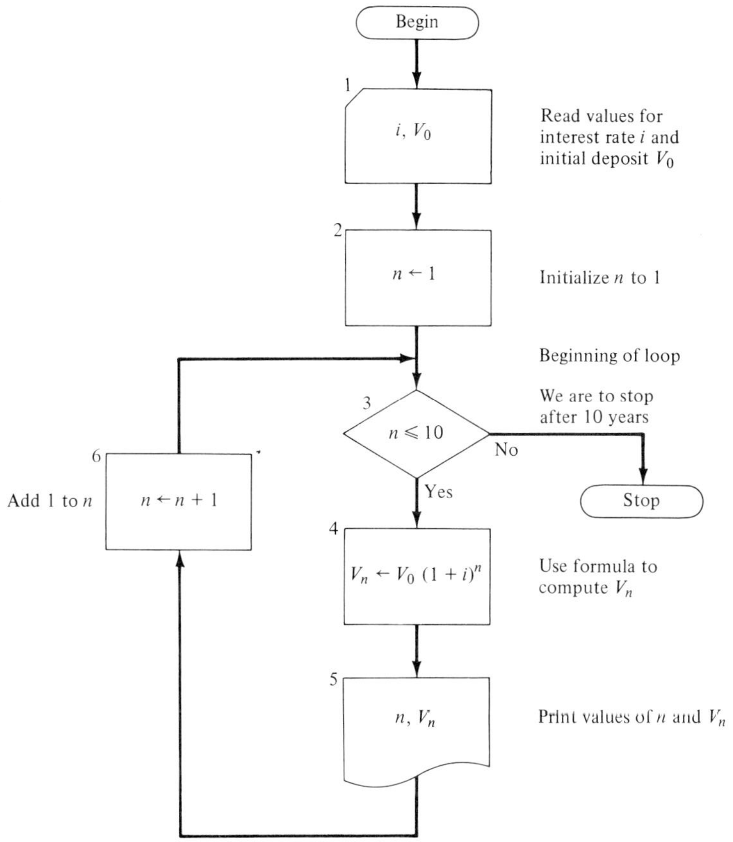

FIGURE 2-5

interest rate i (as a fraction or decimal) and the initial investment V_0. We then enter the loop through Block 2, which assigns the value 1 to n. In Block 3, we test to see if n is less than or equal to 10. If not, we exit the loop. In Block 4, we compute a value for V_n and then print the result (along with the value of n) in Block 5. In Block 6, we add 1 to n and return to the beginning of the loop.

In analyzing the flowchart in Figure 2-5, it is often beneficial to perform a *trace*. In this process we proceed to "execute" each block as the computer would do. We begin with Block 1.

Block 1: We read numerical values for i and V_0. Suppose these values are 0.075 (= 7.5%) and 1000, respectively.
Block 2: We assign the value 1 to n.
Block 3: The current value of n is 1. Since $1 \leq 10$, we proceed to Block 4.
Block 4: Since $n = 1$, we compute $V_1 = (1.075)^1 \cdot (1000) = 1075$.
Block 5: We print the values of n and V_n, namely, 1 (= n) and 1075 (= V_1).
Block 6: The current value of n is 1. After adding 1, the value of n is 2.
Block 3: The current value of n is 2. Since $2 \leq 10$, we proceed to Block 4.
Block 4: Since $n = 2$, we compute $V_2 = (1.075)^2 \cdot (1000) = 1153.63$.
Block 5: We print the values of n and V_n, namely, 2 (= n) and 1153.63 (= V_2).
Block 6: The current value of n is 2. After adding 1, the value of n is 3.

In executing the loop, Blocks 3, 4, 5, and 6 are executed repetitively until the exit in Block 3 is taken. To see how this happens, we begin to execute Block 3 with the value of 10 for n. Since $10 \leq 10$, we proceed to Block 4 and compute V_{10}. We print n (= 10) and V_{10} in Block 5, and in Block 6, we add 1 to n, getting 11 for n. When we test n in Block 3, we find that the value of n, namely 11, is *not* less than or equal to 10, so the exit is taken. Although the value of n is 11 when the exit is made, the value of n is 10 on the final execution of Blocks 4 and 5.

In order to use the recursion relationship instead of the formula, we need only to change Block 4 in Figure 2-5. The result is shown in Figure 2-6. Using this flowchart, we execute the statement $V_n \leftarrow (1 + i)V_{n-1}$ for $n = 1, 2, 3, \ldots, 10$. Therefore, we use the same loop structure.

To see how the recursion relationship is used, observe that the value of n is 1 when the statement $V_n \leftarrow (1 + i)V_{n-1}$ is executed the first time. Consequently, we calculate V_1 as $V_1 \leftarrow (1 + i)V_0$. We can do this, since numerical values for i and V_0 were obtained in Block 1. The next time the statement $V_n \leftarrow (1 + i)V_{n-1}$ is executed, the value of n is 2. Consequently, we calculate $V_2 \leftarrow (1 + i)V_1$, which we can do, since i was entered and V_1 was calculated on the previous execution of Block 4. On the next iteration of the loop, the value of n is 3 so that the statement $V_n \leftarrow (1 + i)V_{n-1}$ becomes $V_3 \leftarrow (1 + i)V_2$. This calculation is made using the entered value of i and the value of V_2 calculated on the previous execution of Block 4.

In creating flowcharts for problems like this, either the formula or recursion relationship may be used. Although we are not stressing efficiency, the calculations in the statement $(1 + i)V_{n-1}$ involve only one multiplication,

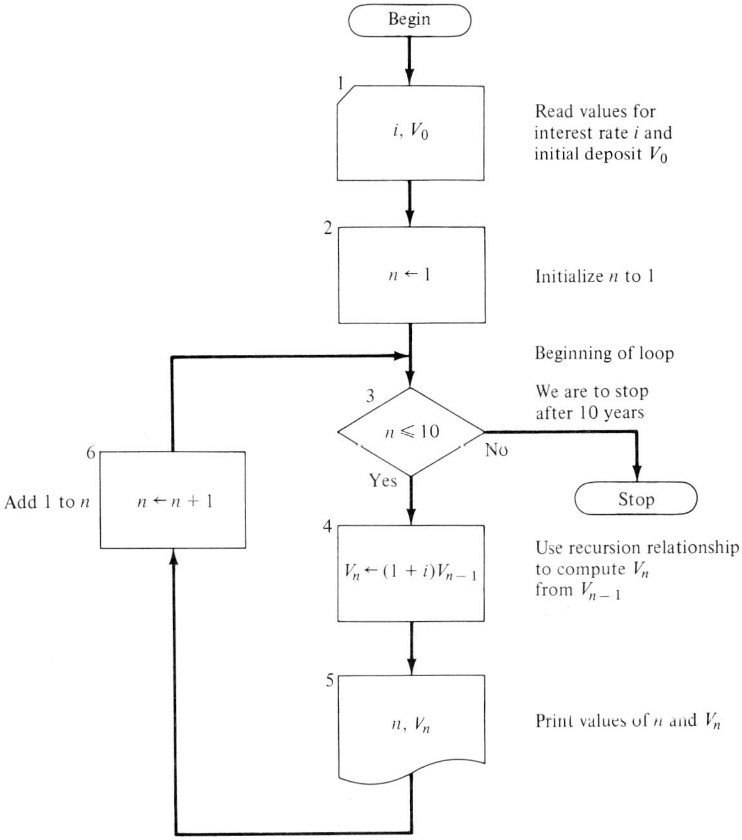

FIGURE 2-6

whereas the calculations in $(1 + i)^n V_0$ require raising a number to a power *and* a multiplication. Consequently, the recursion relationship form is preferred.

EXERCISES 2-3

1. Consider the computational flowchart in Figure Z. Suppose we read a value of 2 for x.

(**a**) For what values of n are Blocks 4 and 5 executed?

(**b**) What values are written by Block 5?

(**c**) Perform a "trace" of this flowchart, continuing until the "Stop" block is encountered.

(**d**) How should this flowchart be modified to compute Blocks 4 and 5 for $n = 1, 2, 3, 4, 5$?

(**e**) For $n = 3, 4, 5, 6$?

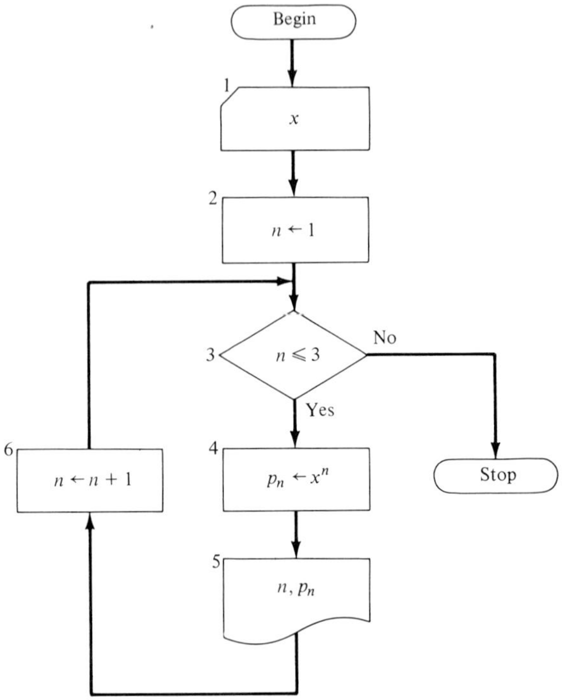

FIGURE Z

2. Consider the computational flowchart in Figure AA.
(a) For what values of n are Blocks 3 and 4 executed?
(b) What values are written by Block 4?
(c) Perform a "trace" of this flowchart, continuing until the "Stop" block is encountered.
(d) How should this flowchart be modified to compute Blocks 3 and 4 for $n = 0, 1, 2, 3, 4$?
(e) For $n = 4, 5, 6$?

3. Consider the computational flowchart in Figure BB.
(a) For what values of n are Blocks 4 and 5 executed?
(b) Suppose a value of 2 is entered for a in Block 1. What values are written in Block 5?
(c) How should this flowchart be modified to compute Blocks 4 and 5 for $n = 1, 2, 3, 4, 5, 6$?
(d) For $n = 20, 21, 22, \ldots, 30$?

4. Consider the computational flowchart in Figure CC. Suppose that we enter a value of 2 for x in Block 1.
(a) For what values of n are Blocks 4 and 5 executed?
(b) What values are written by Block 5?

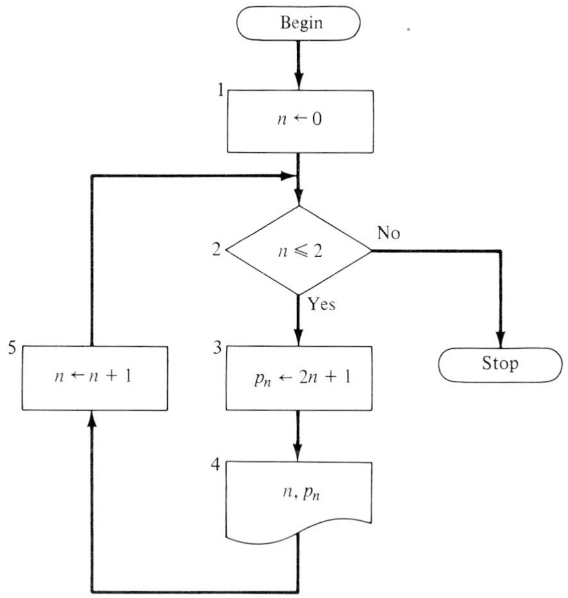

FIGURE AA

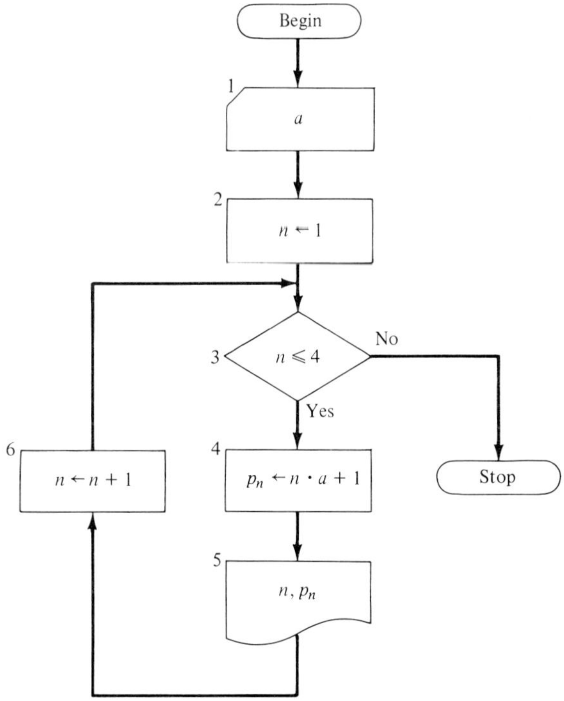

FIGURE BB

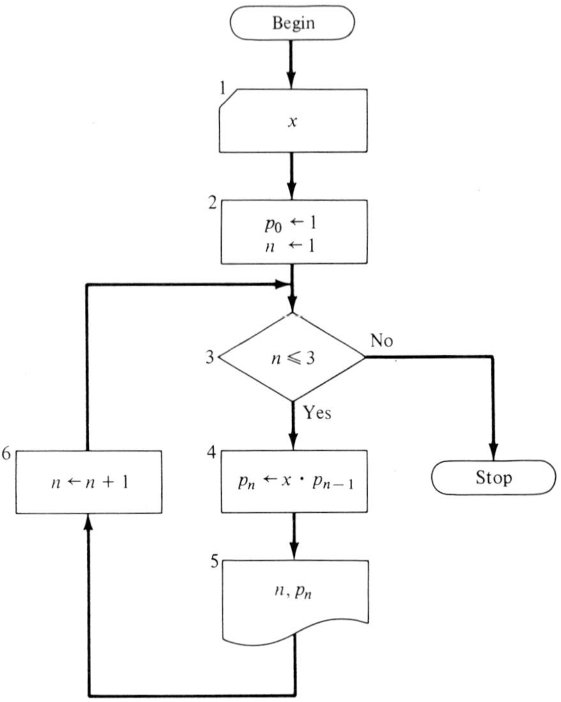

FIGURE CC

(**c**) Is this flowchart equivalent to the one in Exercise 1? That is, does it give the same results?

(**d**) How should it be modified to execute Blocks 4 and 5 for $n = 1, 2, 3, \ldots, 100$?

5. Consider the computational flowchart in Figure DD.

(**a**) For what values of n are Blocks 5 and 6 executed?

(**b**) What values are written by Block 6?

(**c**) Is this flowchart equivalent to the one in Exercise 2? That is, does it write the same numerical values?

6. The flowchart in Exercise 3 computes the terms in the sequence $p_n = n \cdot a + 1$ for $n = 1, 2, 3, 4$. The equivalent recursion relationship is $p_n = p_{n-1} + a$ with $p_1 = 1 + a$. Modify the flowchart in Exercise 3 to use the recursion relationship for p_n instead of the formula.

7. The flowchart in Figure EE is a modification of the flowchart for Exercise 1. Again, we read a value of 2 for x.

(**a**) For what values of n are Blocks 3 and 4 executed?

(**b**) What numerical values are written by Block 4?

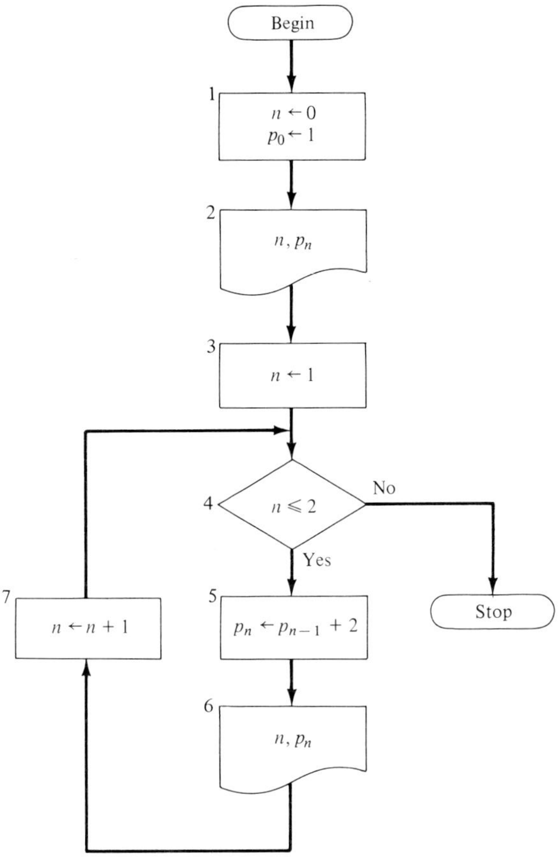

FIGURE DD

(**c**) How must this flowchart be modified to compute Blocks 3 and 4 for $n = 1, 2, 3, 4, 5$?
(**d**) For $n = 3, 4, 5, 6$?
(**e**) Is this flowchart equivalent to the one in Exercise 1?

In Exercises 8 through 11, prepare a computational flowchart for computing the indicated terms in the sequences. Your flowchart should be similar to those in Exercises 1 through 3.

8. $p_n = \dfrac{n-1}{n+1}, \; n = 1, 2, \ldots, 10$

9. $p_n = \dfrac{a^n - 1}{a^n + 1}, \; n = 1, 2, \ldots, 20$ NOTE: Use input block for a.

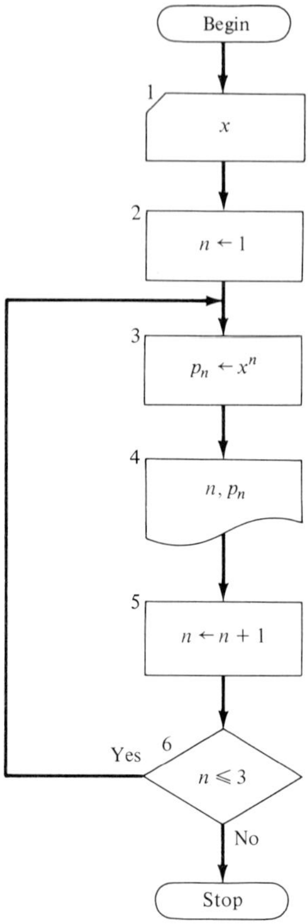

FIGURE EE

10. $p_n = (-1)^n \dfrac{(n+4)}{2n+1}$, $n = 0, 1, 2, \ldots, 50$

11. $p_n = (-1)^n(x^n - 1)$, $n = 1, 2, \ldots, 15$ NOTE: Use input block for x.

In Exercises 12 through 15, prepare a computational flowchart for computing the indicated terms in the sequences. Your flowchart should be similar to those in Exercises 5 and 6.

12. $p_n = p_{n-1} + n$; $p_0 = 0$; $n = 1, 2, \ldots, 10$

13. $p_n = -2p_{n-1}$; $p_0 = 1$; $n = 1, 2, \ldots, 5$

14. $p_n = \dfrac{x}{n} p_{n-1}$; $p_0 = 1$; $n = 1, 2, 3, \ldots, 50$ NOTE: Use input block for x.

15. $p_n = -\frac{x}{n} p_{n-1}; p_0 = 1; n = 1, 2, 3, \ldots, 25$ NOTE: Use input block for x.

2-4. LIMIT OF A SEQUENCE

Figure 2-7 presents the graph of the sequence whose nth term is given by

$$p_n = \frac{n+1}{2n}$$

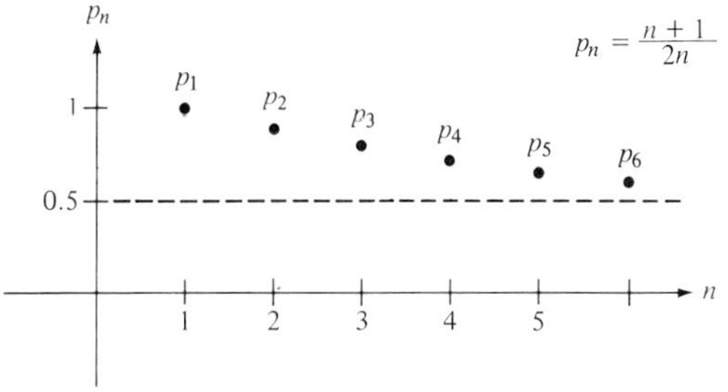

FIGURE 2-7

The first 10 terms are, 1, $\frac{3}{4}$, $\frac{4}{6}$, $\frac{5}{8}$, $\frac{6}{10}$, $\frac{7}{12}$, $\frac{8}{14}$, $\frac{9}{16}$, $\frac{10}{18}$, $\frac{11}{20}$. For this or any sequence, two important questions can be asked:

1. As the index n becomes large, do the terms get close to and stay close to any one number? In other words, do they approach any one number?
2. If the answer to the first question is "yes," then what is that number?

In this context, it is useful to introduce the phrase "high-indexed terms," which denotes all those terms in the sequence for which the index n is large.

From the graph in Figure 2-7, it appears that the high-indexed terms in the sequence $\langle (n+1)/2n \rangle$ are all close to the number $\frac{1}{2}$. For example, suppose that we would like to find an index N such that all terms of index greater than N are within 0.01 of the number $\frac{1}{2}$. We could take $N = 50$ or any number larger than 50. Thus

$$p_{51} = \frac{51+1}{102} = \frac{52}{102} = 0.5098$$

which is within 0.01 of $\frac{1}{2}$ and for $n > 51$, p_n is even closer to $\frac{1}{2}$. All sufficiently high-indexed terms are as close to the number $\frac{1}{2}$ as we wish; or said another way, the high-indexed terms get arbitrarily close to $\frac{1}{2}$.

When the high-indexed terms in a sequence $\langle p_n \rangle$ are all close to the one number C, we call C the *limit of the sequence* $\langle p_n \rangle$. In symbols, we may write

$$\lim_{n\to\infty} p_n = C \qquad \text{or} \qquad \lim_{n\to\infty} p_n = C \qquad \text{or} \qquad \operatorname*{limit}_{n\to\infty} p_n = C$$

We shall use the first two expressions, but some authors prefer the third. The symbol $n \to \infty$ means "as the index n becomes large," or we can say "as n approaches infinity." For the sequence $\langle (n+1)/2n \rangle$ graphed in Figure 2-7, we would write

$$\lim_{n\to\infty} \frac{n+1}{2n} = \frac{1}{2}$$

This expression states that as the index n becomes large (approaches infinity), the values of the terms get close to and stay close to the number $\frac{1}{2}$.

The concept that the high-indexed terms are as close to the limit as desired is extremely important. For example, each term p_{n+1} in the sequence $\langle (n+1)/2n \rangle$ graphed in Figure 2-7 is closer to the number 0 than is the term p_n. However, you cannot find an index so that all terms of index higher than that are within, say, 0.1 of the number 0. Therefore, 0 is not the limit of the sequence. Only for the number $\frac{1}{2}$ can you find all sufficiently high-indexed terms within any desired proximity.

As another example, consider the sequence

$$p_n = \frac{3n-1}{n+2}$$

The first five terms are $\frac{2}{3}, \frac{5}{4}, \frac{8}{5}, \frac{11}{6}, \frac{14}{7}$, and the graph is illustrated in Figure 2-8. Although it appears from the graph that the high-indexed terms are close to some value, it may not be obvious what this value is. Therefore we shall calculate a few high-indexed terms, as shown in Table 2-4.

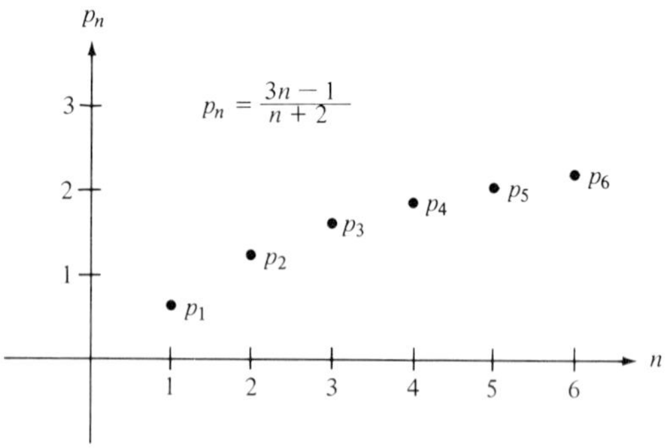

FIGURE 2-8

TABLE 2-4

n	p_n
18	$\frac{53}{20} = 2.65$
48	$\frac{143}{50} = 2.86$
98	$\frac{293}{100} = 2.93$
998	$\frac{2993}{1000} = 2.993$

You would probably now guess that the higher and higher-indexed terms are getting close to the number 3. In fact, all sufficiently high-indexed terms are as close to the number 3 as you like. Indeed,

$$\lim_{n \to \infty} \frac{3n - 1}{n + 2} = 3$$

Convergence and Divergence A sequence *converges* if it has a limit; otherwise, the sequence *diverges*. For sequences that converge, we say that the limit exists. For sequences that diverge, we say that the limit does not exist. Sequences may diverge for any of several reasons, and we shall illustrate only a few of the common types of behavior of sequences that diverge.

As our first illustration consider the sequence whose terms are given by

$$p_n = \frac{(-1)^n(n + 1)}{n}$$

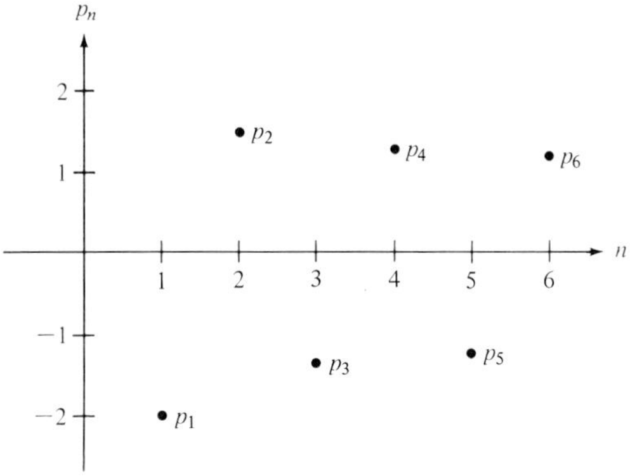

FIGURE 2-9

The first 10 terms are $-2, \frac{3}{2}, -\frac{4}{3}, \frac{5}{4}, -\frac{6}{5}, \frac{7}{6}, -\frac{8}{7}, \frac{9}{8}, -\frac{10}{9}$, and $\frac{11}{10}$. From the graph in Figure 2-9, it would seem apparent that the high-indexed terms with negative values are close to -1, whereas the high-indexed terms with positive values are close to $+1$. This sequence does not have a limit (i.e., there is no single number that all of its high-indexed terms are close to). Therefore, it diverges. This would be expressed as

$$\lim_{n\to\infty} \frac{(-1)^n(n+1)}{n} \qquad \text{does not exist}$$

In order for the limit to exist, *all* sufficiently high-indexed terms must be near the *same* value.

As the next example of a sequence that diverges, consider the sequence whose terms are given by

$$p_n = n^2$$

The first 10 terms are 1, 4, 9, 16, 25, 36, 49, 64, 81, and 100. As illustrated by the graph in Figure 2-10, the high-indexed terms become very large. Since the high-indexed terms are not all close to any one number, the limit does not exist and the sequence diverges. For situations in which the sufficiently high-indexed terms are arbitrarily large positive numbers, we write

$$\lim_{n\to\infty} n^2 = \infty$$

Although similar to the notation used to express the limit of a convergent sequence, the meaning is virtually the opposite. This notation can be interpreted as follows: "As the index n becomes large (approaches infinity), the terms in the sequence become large, positive numbers."

Some authors use the symbol "$+\infty$" to mean the same thing as our "∞." In our notation, saying $n \to \infty$ means that n becomes large positively.

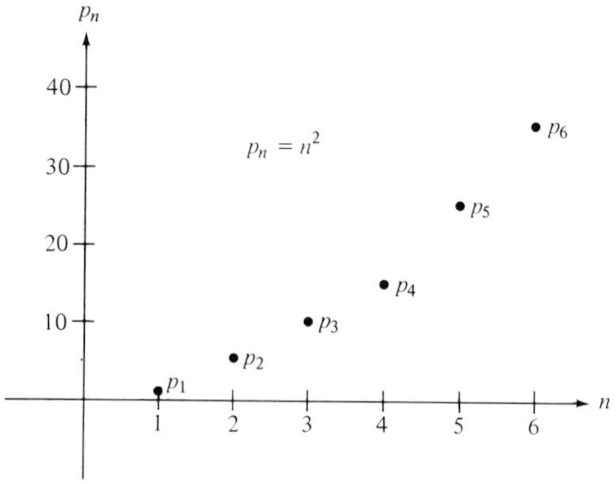

FIGURE 2-10

Writing $\lim_{n \to \infty} n^2 = \infty$ means that for large n, n^2 is large positively. We shall occasionally use "$+\infty$" for emphasis.

In some sequences, the high-indexed terms become large negatively. An example of such a sequence is

$$p_n = -(n^2)$$

The first 10 terms are -1, -4, -9, -16, -25, -36, -49, -64, -81, and -100, and the thousandth term is $-1{,}000{,}000$. Since the high-indexed terms are large negative numbers, we write

$$\lim_{n \to \infty} [-(n)^2] = -\infty$$

This sequence diverges, and the limit does not exist.

To summarize, questions involving the existence of the limit of a sequence $\langle p_n \rangle$ may have any one of the following answers.

1. *The sequence converges; in other words the limit exists.* All high-indexed terms are close to a *number* C, and we write

$$\lim_{n \to \infty} p_n = C$$

2. *The sequence diverges; in other words the limit does not exist.* We have the following possibilities:
   (a) All high-indexed terms are large, positive numbers, and we write

$$\lim_{n \to \infty} p_n = \infty \qquad \text{or} \qquad \lim_{n \to \infty} p_n = +\infty$$

   (b) All high-indexed terms are large, negative numbers, and we write

$$\lim_{n \to \infty} p_n = -\infty$$

   (c) The high-indexed terms are not all close to any one number, nor are they all large positive numbers or all large negative numbers.

The decision process associated with the existence or nonexistence of limits is illustrated by the symbolic flowchart in Figure 2-11.

NOTE: In deciding about convergence or divergence of a given sequence, it is sometimes useful to compute (approximately) or just to think about computing a high-indexed term like the 100th or the 1000th. However, in such cases we usually only need approximate values. Thus $(\frac{1}{1001})^2$ is approximately $(\frac{1}{1000})^2$ or $\frac{1}{1{,}000{,}000}$. There is no need to spend time or effort on precise evaluation.

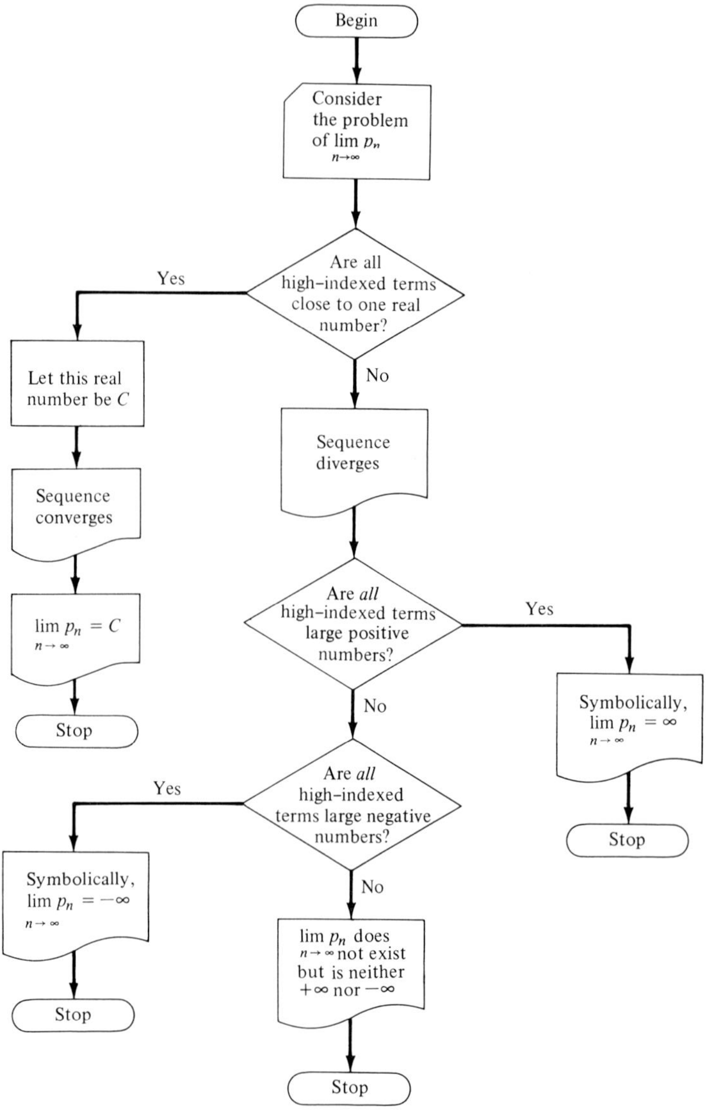
Begin
Consider the problem of $\lim_{n\to\infty} p_n$
Are all high-indexed terms close to one real number?
Yes
Let this real number be C
Sequence converges
$\lim_{n\to\infty} p_n = C$
Stop
No
Sequence diverges
Are *all* high-indexed terms large positive numbers?
Yes
Symbolically, $\lim_{n\to\infty} p_n = \infty$
Stop
No
Are *all* high-indexed terms large negative numbers?
Yes
Symbolically, $\lim_{n\to\infty} p_n = -\infty$
Stop
No
$\lim_{n\to\infty} p_n$ does not exist but is neither $+\infty$ nor $-\infty$
Stop

FIGURE 2-11

Using Common Sense about Numbers We can frequently use common sense about numbers to decide whether a given sequence converges or diverges and to what it converges or how it diverges.

If p_n is expressed as a ratio a_n/b_n and $\langle a_n \rangle$ converges, while $\langle b_n \rangle$ diverges to $+\infty$ or $-\infty$, then $\langle p_n \rangle$ converges to zero. For example,

$$\lim_{n\to\infty} \frac{5}{n^2+1} = 0$$

since the numerator is always 5, while the denominator gets large as n gets large; thus the ratio $5/(n^2+1)$ approaches zero.

If p_n is expressed as a ratio a_n/b_n and $\lim_{n\to\infty} a_n \neq 0$, whereas $\lim_{n\to\infty} b_n = 0$, then $\langle p_n \rangle$ diverges. For some large n, the denominator will be near zero, while the numerator is not near zero. Hence the ratio must be large positively or negatively; therefore, $\lim_{n\to\infty} p_n$ cannot exist. Other sequences can be similarly analyzed.

Here are some examples illustrating these ideas.

Let

$$p_n = \frac{2+(1/n)}{n^2}$$

For large n, the numerator $2 + (1/n)$ is close to 2, whereas the denominator is large. Therefore, the ratio is close to zero and we have

$$\lim_{n\to\infty} p_n = 0$$

Let

$$q_n = \frac{n+n^3}{1+(2/n)}.$$

Here, for large n, the numerator is large, while the denominator is close to 1. Therefore, the ratio is large and $\lim_{n\to\infty} q_n$ does not exist; in fact, $\lim_{n\to\infty} q_n = \infty$.

EXERCISES 2-4

Graph at least six terms of the following sequences and state whether they converge or diverge. If they converge, find their limits. If they diverge, state whether they diverge to ∞, to $-\infty$, or to neither. You may want to evaluate (or think about evaluating) a high-indexed term such as p_{100}.

1. $p_n = 4$

2. $p_n = \dfrac{1}{n^2}$

3. $p_n = \frac{1}{n^2} - 1$

4. $p_n = 1 - n$

5. $p_n = \frac{1-n}{n}$

6. $p_n = \frac{2n+6}{3n-1}$

7. $p_n = (-1)^n \frac{n}{n^2+2}$

8. $p_n = n^2 - n - 1$

9. $p_n = (\frac{1}{2})^n$

10. $p_n = (3)^n$

11. $p_n = (-1)^n \cdot \frac{n}{n+1}$

12. $p_n = \frac{n^2+1}{2n+4}$

For Exercises 13–24, use common sense about numbers to determine the convergence or divergence of the sequences indicated and to find the limit if it exists.

13. $p_n = \frac{n}{100}$

14. $p_n = \frac{100}{n}$

15. $p_n = \frac{2}{(n+1)^2}$

16. $p_n = \frac{(n-1)^3}{10}$

17. $p_n = \frac{n^2}{1 + (1/n)}$

18. $p_n = \frac{5 - (2/n)}{n \cdot (n+1)}$

19. $q_n = \frac{6}{1/(n+1)}$

20. $q_n = \frac{(n+2)^2}{2/n}$

21. $q_n = \frac{4}{n^2+n+6}$

22. $q_n = \frac{n^2+n+6}{2000}$

23. $q_n = \frac{-20}{n}$

24. $q_n = \frac{-n}{20}$

2-5 TECHNIQUES FOR DETERMINING CONVERGENCE OR DIVERGENCE

In this section, we consider two processes, either or both of which will enable us to answer many questions posed about the convergence or divergence of sequences. The processes are

1. Reduce to simpler or known cases.
2. Change the form of the function.

In the previous section, limits of a sequence were determined by graphing the sequence, by evaluating enough high-indexed terms so that the limit of the sequence became apparent, or by using common sense about numbers. After such considerations, the limits of many sequences should be obvious, two important examples being

1. $\lim_{n\to\infty} k = k$, where k is a constant.

2. $\lim_{n\to\infty} \frac{1}{n} = 0$.

However, for some sequences, it may not be apparent whether or not the limit exists, and if it does, what the limit is.

1. **Reduce to Simpler or Known Cases.** To illustrate, consider

$$\lim_{n\to\infty} \left(2 + \frac{1}{n}\right)$$

Many readers will recognize that the limit is the number 2. We get the result by reducing to simpler cases. The limit can be written as

$$\lim_{n\to\infty} \left(2 + \frac{1}{n}\right) = \lim_{n\to\infty} 2 + \lim_{n\to\infty} \frac{1}{n}$$

Since $\lim_{n\to\infty} 2 = 2$ and $\lim_{n\to\infty} 1/n = 0$, we conclude that

$$\lim_{n\to\infty} \left(2 + \frac{1}{n}\right) = 2 + 0 = 2$$

This example suggests that many sequences can be expressed in terms of more elementary sequences. To this end, we consider what we mean by the termwise sum, difference, product, or quotient of two sequences $\langle a_i \rangle$ and $\langle b_i \rangle$. The meaning is just what you think it should be. For example, the sequence $\langle a_i + b_i \rangle$ is the termwise sum of $\langle a_i \rangle$ and $\langle b_i \rangle$; the sequence $\langle a_i / b_i \rangle$ is the termwise quotient. In the case of such a termwise quotient, we must assume that the terms of $\langle b_i \rangle$ are not zero; otherwise we could not do the division.

Then, assuming that the sequences $\langle a_n \rangle$ and $\langle b_n \rangle$ converge, we have

Property 2-1 Any termwise sum, difference, product, or quotient of two convergent sequences $\langle a_n \rangle$ and $\langle b_n \rangle$ is a convergent sequence whose limit is the corresponding sum, difference, product or quotient of the limits of $\langle a_n \rangle$ and $\langle b_n \rangle$ (assuming that no division by zero occurs in the case of the quotient).

Sum: $$\lim_{n\to\infty} (a_n + b_n) = \lim_{n\to\infty} a_n + \lim_{n\to\infty} b_n$$

Difference: $$\lim_{n\to\infty} (a_n - b_n) = \lim_{n\to\infty} a_n - \lim_{n\to\infty} b_n$$

Product: $$\lim_{n\to\infty} (a_n \cdot b_n) = \lim_{n\to\infty} a_n \cdot \lim_{n\to\infty} b_n$$

Quotient: $$\lim_{n\to\infty} \left(\frac{a_n}{b_n}\right) = \frac{\lim_{n\to\infty} a_n}{\lim_{n\to\infty} b_n}$$

assuming the limits of $\langle a_n \rangle$ and $\langle b_n \rangle$ exist and no division by zero occurs.

A special case of a termwise product is the case where one of the sequences, say $\langle b_n \rangle$, is constant. Thus we have $b_n = k$ for each n. Then

$$\langle a_n \cdot b_n \rangle = \langle k \cdot a_n \rangle$$

and

$$\lim_{n\to\infty} \langle a_n \cdot b_n \rangle = \lim_{n\to\infty} \langle k \cdot a_n \rangle = k \lim_{n\to\infty} \langle a_n \rangle$$

In other words, the limit of a constant times a sequence is the constant times the limit of the sequence.

We use these properties to reduce complicated problems to simpler ones that we know how to solve. To illustrate the use of these relationships, consider the following examples.

$$\lim_{n\to\infty} \left(3 - \frac{1}{n}\right) = \lim_{n\to\infty} 3 - \lim_{n\to\infty} \frac{1}{n} = 3 - 0 = 3$$

Here we have used the basic facts on $\lim_{n\to\infty} k$ and $\lim_{n\to\infty} 1/n$ and the property on the limit of a difference. A more complicated example is

$$\lim_{n\to\infty} \frac{\left[\left(1 + \frac{1}{n}\right) \cdot \frac{1}{n}\right] + 3}{2 + \frac{1}{n}}$$

$$= \frac{\left[\lim_{n\to\infty} 1 + \lim_{n\to\infty} \frac{1}{n}\right] \cdot \lim_{n\to\infty} \left(\frac{1}{n}\right) + \lim_{n\to\infty} 3}{\lim_{n\to\infty} 2 + \lim_{n\to\infty} \frac{1}{n}}$$

$$= \frac{(1 + 0) \cdot 0 + 3}{2 + 0} = \frac{3}{2}$$

We are not going to prove the four relationships of Property 2 = 1, but let us informally explain why they are true. Let $\lim_{n\to\infty} a_n = A$ and $\lim_{n\to\infty} b_n = B$. Consider

$$\lim_{n\to\infty} (a_n + b_n) = \lim_{n\to\infty} a_n + \lim_{n\to\infty} b_n = A + B$$

for example. We can take a high enough index to make $a_n + b_n$ as close as we wish to $A + B$ by making a_n and b_n close enough to A and B, respectively. A similar statement may be made for the product relationship $\lim_{n\to\infty} (a_n \cdot b_n) = A \cdot B$. We can take a high enough index to make $a_n \cdot b_n$ as close as we wish to $A \cdot B$ by making a_n and b_n close enough to A and B, respectively.

We now consider the second process.

2. **Change the form of the function.** Consider the sequence

$$\left\langle \frac{2n}{n+1} \right\rangle$$

Does $\lim_{n\to\infty} [2n/(n+1)]$ exist? If so, what is $\lim_{n\to\infty} [2n/(n+1)]$? We cannot directly use the relationship just developed for the termwise quotient of two sequences, since $\lim_{n\to\infty} 2n = \infty$ and $\lim_{n\to\infty} (n+1) = \infty$, and ∞ is not a number. But observe that for each $n = 1, 2, 3, \ldots$, we have, on dividing the numerator and denominator by n,

$$\frac{2n}{n+1} = \frac{2}{1+(1/n)}$$

Therefore,

$$\lim_{n\to\infty} \left(\frac{2n}{n+1} \right) = \lim_{n\to\infty} \left[\frac{2}{1+(1/n)} \right] = \frac{2}{1+0} = 2$$

In effect, we have changed the form but not the values of the function p_n, which defines the sequence. We are always at liberty to do this if it helps [as it did for the function $p_n = 2n/(n+1)$].

Consider next the sequence defined by

$$p_n = \frac{2/(n+1)}{3/n}$$

Regarding $2/(n+1)$ as the numerator and $3/n$ as the denominator, we cannot apply the relationship for termwise quotients directly, since the limit of the denominator is zero. But for each $n = 1, 2, 3, \ldots$,

$$p_n = \frac{2/(n+1)}{3/n} = \frac{2n}{3n+3} = \frac{2}{3+(3/n)}$$

and hence

$$\lim_{n\to\infty} p_n = \lim_{n\to\infty} \frac{2}{3+(3/n)} = \frac{2}{3+0} = \frac{2}{3}$$

The three formulas just given for p_n illustrate that if one change of form does not help, a further change may help. The reader can use some algebraic ingenuity in finding a form that reveals the nature of the limit desired.

The two examples we have just considered illustrate two types of what are known as *indeterminate forms*, namely, ∞/∞ and $0/0$. For large n, each of the numerator and denominator of $2n/(n+1)$ was large and we might have been tempted to write $\lim_{n\to\infty} [2n/(n+1)] = \infty/\infty$. Similarly, for large n, each of the numerator and denominator of

$$\frac{2/(n+1)}{3/n}$$

was near zero and we might have been tempted to write

$$\lim_{n\to\infty} \frac{2/(n+1)}{3/n} = \frac{0}{0}$$

We have seen that in the two special cases considered, the desired limit actually existed and could be computed by changing the form of the function. However, there are other cases in which the limits do not exist when we have the ∞/∞ or $0/0$ forms.

Consider the sequence given by $p_n = n^2/(n+1)$. We seek $\lim_{n\to\infty} n^2/(n+1)$ and observe the case to be of the form ∞/∞. But,

$$\frac{n^2}{n+1} = \frac{1/n}{1/n} \cdot \frac{n^2}{n+1} = \frac{n}{1+(1/n)}$$

Since

$$\lim_{n\to\infty} \frac{n^2}{n+1} = \lim_{n\to\infty} \frac{n}{1+(1/n)} = \infty$$

the sequence diverges. Similarly, if we consider the sequence $(1/n)/(1/n^2)$, we have the indeterminate form $0/0$. Since $(1/n)/(1/n^2) = n$ for each $n = 1, 2, 3, \ldots$, we have

$$\lim_{n\to\infty} \frac{1/n}{1/n^2} = \lim_{n\to\infty} n = \infty$$

The simple knowledge that we have an indeterminate form $0/0$ or ∞/∞ does not tell us whether or not the limit exists, but we can always try to change the form of p_n in the hope that we can then recognize whether the limit exists and, if so, what the limit is.

EXERCISES 2-5

The rules for the termwise sum, product, etc., of two sequences can be applied directly to the convergent sequences in Exercises 1 through 10. Find the limit.

1. $p_n = \frac{1}{n} + 2$

2. $p_n = 27 \cdot \left(\frac{1}{n} - 1\right)$

3. $p_n = \left(2 + \frac{1}{n}\right)\left(3 - \frac{1}{n}\right)$

4. $p_n = \left(1 - \frac{1}{n}\right)^2$

5. $p_n = \dfrac{1 + (2/n^2)}{1 - (2/n^2)}$

6. $p_n = \dfrac{[4 + (1/n)]}{1 - (1/2n)}$

7. $p_n = \left(\frac{1}{n} + 1\right)\left(\frac{1}{n} - 2\right) + 4\left(1 - \frac{1}{n}\right)$

8. $p_n = \left(\frac{1}{n} + 2\right)\left(\frac{1}{n} - 3\right)\left(\frac{1}{n} - 1\right)$

9. $p_n = \dfrac{[(1/n) - 1][(1/n) - 2]}{[(1/n^2) + 1]}$

10. $p_n = \dfrac{[(1/n) + 1]}{[(1/n^2) - 3][(1/n^2) - 2]}$

For Exercises 11 through 18, an alternate form may be used to obtain the limit.

11. $p_n = \dfrac{2n - 1}{n + 1}$

12. $p_n = \dfrac{n^2 + n + 1}{n^2 - n - 1}$

13. $p_n = \dfrac{n(n - 1)}{3n^2 + 1}$

14. $p_n = \dfrac{1 - n + n^3}{4 - 2n^3}$

15. $p_n = (1 - n)^2\left(\frac{2}{n}\right)^2$

16. $p_n = (2n + 1)\left(\frac{n + 1}{n^2}\right)$

17. $p_n = \dfrac{n^2 - 4}{(n^2 + 2) - (4/n)}$

18. $p_n = \dfrac{n^2 - 1}{(n + 2)(n + 1)}$

For Exercises 19 through 30, determine if $\lim_{n\to\infty} q_n$ exists and, if so, evaluate it. If not, describe its divergence. Use whatever methods you have available.

19. $q_n = \dfrac{n^2 - 2n}{n^3 + 2n}$

20. $q_n = \dfrac{n^2 - n + 5}{6n - 7}$

21. $q_n = \left(\frac{1}{n}\right) \cdot \left(\frac{n + 1}{n}\right) \cdot (2n)$

22. $q_n = \dfrac{n}{\sqrt{n}}$

23. $q_n = \dfrac{1 - n - n^4}{n^3}$

24. $q_n = \dfrac{2 - (1/n)}{\sqrt{n}}$

25. $q_n = \dfrac{\sqrt{n}}{n^2 + 6}$

26. $q_n = (n)\left(\frac{4}{n^2 + 1}\right) \cdot (n - 1)$

27. $q_n = (2n + 1)\left(\frac{n + 1}{n^2}\right)$

28. $q_n = \left(\frac{n + 5}{n^2 - 5}\right)(n^4 + 1)$

29. $q_n = (1 - n)\left(\frac{1 + n}{n^2}\right)$

30. $q_n = \left(\frac{2}{n}\right) \cdot (n^2 - n)$

To determine limits for sequences whose terms alternate in sign, one approach is to determine the limit for the sequence of positive terms only and the limit for that of the negative terms only, and see if these two limits are the same. If so, the sequence converges to this value; if not, the sequence diverges and the limit does not exist. For example, consider

$$\lim_{n \to \infty} (-1)^n \left(1 - \frac{1}{n}\right)$$

For the positive terms, we have

$$\lim_{n \to \infty} \left(1 - \frac{1}{n}\right) = 1$$

For the negative terms, we have

$$\lim_{n \to \infty} \left[-\left(1 - \frac{1}{n}\right)\right] = -1$$

Since these limits are not equal, the sequence diverges and the limit does not exist.

NOTE: If both positive and negative terms approach zero, then the limit is zero. If either approaches a number different from zero, then the limit cannot exist, since a sequence of negative terms cannot approach a positive number or a sequence of positive terms a negative number.

Using this approach, determine whether the sequences in Exercises 31 through 36 diverge or converge, and obtain limits for them if they converge.

31. $p_n = (-1)^n \left(\frac{1 - n^2}{n}\right)$

32. $p_n = (-1)^n \left(\frac{n}{1 + n^2}\right)$

33. $p_n = (-1)^n \frac{(n - 1)^2}{(n + 1)^2}$

34. $p_n = (-1)^n (n^2 - n - 1)$

35. $p_n = 1 + \frac{(-1)^n}{n}$

36. $p_n = 2 - (-1)^n$

2-6. LIMIT OF A FUNCTION

One of the most fundamental concepts of calculus is that of the limit of a function. This concept will be used in developing the derivative in the next chapter, and is closely related to that of the limit of a sequence.

By the *limit of a function* $f(x)$, we mean the number, if any, that $f(x)$

gets close to as x approaches some value but is not equal to that value. For example, suppose that $f(x) = (x + 5)/(x + 1)$, and we would like to identify the limit of the function $f(x)$ as x approaches 1, which we shall denote as $\lim_{x \to 1} f(x)$.

When x is close to the number 1, the numerator $x + 5$ is close to the number 6 and the denominator $x + 1$ is close to the number 2. Therefore, the ratio is close to 3. We have

$$\lim_{x \to 1} \frac{x + 5}{x + 1} = 3$$

In this example, it turns out that

$$\lim_{x \to 1} f(x) = 3 \qquad \text{and} \qquad f(1) = 3$$

For this case, the limit is equal to the function evaluated at the point at which the limit is desired. However, such a statement is not true in general.

It is very important to distinguish between the functional value, such as $f(1)$, and $\lim_{x \to 1} f(x)$. Either may exist when the other does not. The concept of $\lim_{x \to c} f(x)$ is not concerned with $f(c)$; it is concerned with what number, if any, $f(x)$ is close to when x is in the domain of $f(x)$, is close to c, but is not equal to c.

Questions about limits of functions are very similar to questions about limits of sequences. We first ask: "Does the limit exist?" and then: "If the limit exists, what is it?" But observe the distinction: For the limit of a sequence we consider what happens when the index n is large or, in other words, as n approaches ∞; for the limit of a function $f(x)$ at a point c, we consider what happens when x is close to c but not equal to c, or, in other words, as x approaches c.

In determining if the limit of a function exists or, if it does, how to evaluate it, we follow essentially the same procedures as we followed for limits of sequences. We use common sense about numbers, and we can try either or both of the following:

1. Reduce to simpler or known cases.
2. Change the form of the function.

Reduce to Simpler or Known Cases For many elementary functions, the limit will be apparent. Two such limits that are important are the following:

I. $\lim_{x \to c} k = k$, where k is a constant.

II. $\lim_{x \to c} x = c$.

In the first example, the limit of the constant function $f(x) = k$ is the constant k, since $f(x)$ is close to k (in fact, is equal to k) for all values of x. In the second example, as x gets close to c, the value of the function $f(x) = x$ also, obviously, gets close to c, and thus the limit is c.

Although similar arguments can be made for many functions, they become more difficult as the complexity of the function increases. Therefore, we state some useful relationships like those for sequences.

Property 2-2 Assume that $\lim_{x \to c} f(x)$ exists and $\lim_{x \to c} g(x)$ exists. Then the limits of the sum, difference, product, and quotient of $f(x)$ and $g(x)$ exist and are equal to the corresponding sum, difference, product, and quotient of $\lim_{x \to c} f(x)$ and $\lim_{x \to c} g(x)$ (assuming that no division by zero occurs in the case of a quotient).

In symbols we have

Sum: $$\lim_{x \to c} [f(x) + g(x)] = \lim_{x \to c} f(x) + \lim_{x \to c} g(x)$$

Difference: $$\lim_{x \to c} [f(x) - g(x)] = \lim_{x \to c} f(x) - \lim_{x \to c} g(x)$$

Product: $$\lim_{x \to c} [f(x) \cdot g(x)] = \lim_{x \to c} f(x) \cdot \lim_{x \to c} g(x)$$

Quotient: $$\lim_{x \to c} \left[\frac{f(x)}{g(x)}\right] = \frac{\lim_{x \to c} f(x)}{\lim_{x \to c} g(x)}$$

assuming the limits of $f(x)$ and $g(x)$ exist and no division by zero occurs.

Here is an example:

$$\lim_{x \to 2} \frac{x^2 - x - 1}{x + 3} = \frac{\left(\lim_{x \to 2} x\right) \cdot \left(\lim_{x \to 2} x\right) - \lim_{x \to 2} x - \lim_{x \to 2} 1}{\lim_{x \to 2} x + \lim_{x \to 2} 3}$$

$$= \frac{2 \cdot 2 - 2 - 1}{2 + 3} = \frac{1}{5}$$

By the processes of addition, subtraction, multiplication, and division, we have reduced the problem to repeated use of the basic limits I and II given earlier in this section

A special case of the product formula implies that for any constant k, $\lim_{x \to c} [k \cdot f(x)] = k \cdot \lim_{x \to c} f(x)$, since $\lim_{x \to c} k = k$.

Change the Form of the Function We may change the form of the function to help us recognize whether the limit exists and, if it does, to evaluate it. Consider the function

$$f(x) = \frac{x(x^2 - 1)}{x - 1}$$

Since $\lim_{x \to 1} (x - 1) = 0$, we cannot directly use the form $x(x^2 - 1)/(x - 1)$ in considering $\lim_{x \to 1} f(x)$. So we change the form (but not the value). For all $x \neq 1$,

$$\frac{x(x^2 - 1)}{x - 1} = \frac{x(x + 1)(x - 1)}{(x - 1)} = x(x + 1)$$

Since for $\lim_{x \to 1} f(x)$, we are *never* concerned with $f(1)$ or even with whether it exists, we may write

$$\lim_{x \to 1} f(x) = \lim_{x \to 1} x(x + 1) = \left(\lim_{x \to 1} x\right) \cdot \left(\lim_{x \to 1} x + \lim_{x \to 1} 1\right)$$
$$= 1 \cdot (1 + 1) = 2$$

In Figure 2-12, we show the graph of $f(x)$, the break in the graph at the point (1, 2) symbolizing that $f(x)$ is not defined there.

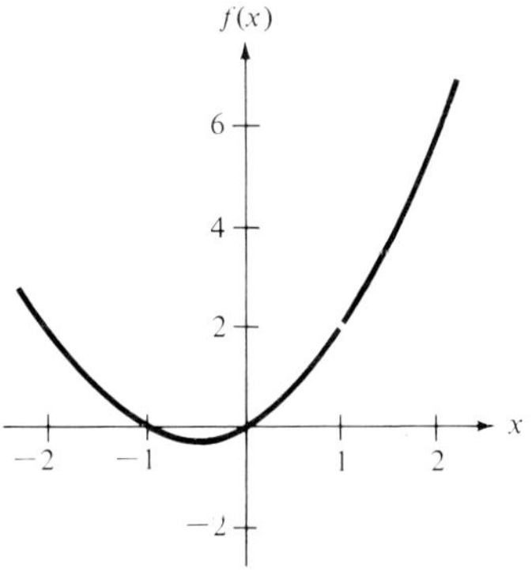

FIGURE 2-12

Notice that $\lim_{x \to 1} [x(x^2 - 1)]/(x - 1)$ produces an indeterminate form 0/0 in the sense of such forms for sequences. When we recognize such an indeterminate form, we try to find other expressions for the function concerned and evaluate the limit using one of these other expressions.

If $\lim_{x \to c} f(x)$ does not exist, then as with sequences, it may be that $\lim_{x \to c} f(x) = \infty$, $\lim_{x \to c} f(x) = -\infty$ or neither. For example,

$$\text{if } f(x) = 1/x^2 \text{ then } \lim_{x \to 0} f(x) = +\infty$$

$$\text{if } g(x) = -1/x^2\text{; then } \lim_{x \to 0} g(x) = -\infty$$

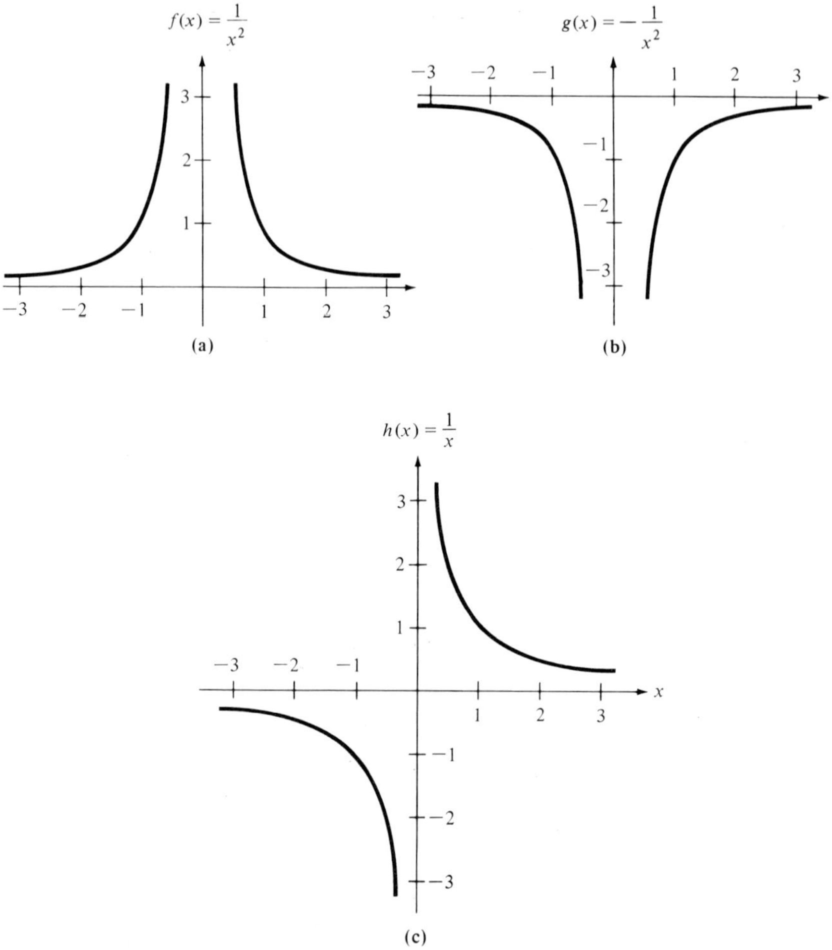

FIGURE 2-13

$$\text{if } h(x) = 1/x, \text{ then } \lim_{x \to 0} h(x) \neq \infty, \text{ and } \neq -\infty$$

Note that $h(x)$ is a large positive number for x, a small positive number, and $h(x)$ is a large negative number for x, a small negative number. We can use such facts to help us graph functions like $1/x^2$, $-1/x^2$, and $1/x$. See Figure 2-13.

For another interesting example, consider $\lim_{x \to 0} f(x)$ for the function

$$f(x) = \frac{x}{|x|}$$

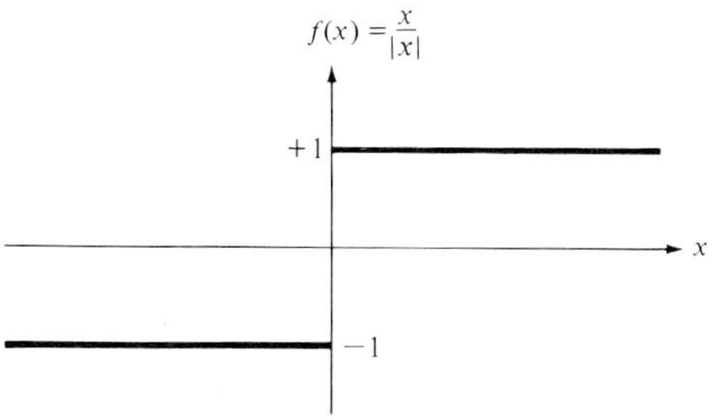

FIGURE 2-14

whose graph is given in Figure 2-14. When x is any positive number, $f(x) = x/|x|$ is equal to the number 1. But when x is any negative number, $f(x) = x/|x|$ is equal to the number -1. Therefore, $\lim_{x \to 0} f(x)$ does not exist.

We should use common sense about numbers in dealing with limits of functions, just as we did with limits of sequences. Thus we can immediately see that

$$\lim_{x \to 2} \frac{4 + [3/(x-2)]}{x-2}$$

does not exist, since the denominator is close to zero when x is close to 2, whereas the numerator is a large number (positively or negatively) when x is close to 2.

In graphing $f(x)$, we are frequently concerned with what happens when x is large. In observing that the graph of $f(x) = 1/x$ in Figure 2-13(c) is close to the x axis for x large, we are, in effect, saying that $\lim_{x \to \infty} 1/x = 0$. Therefore, it makes sense to consider $\lim_{x \to \infty} f(x)$ and $\lim_{x \to -\infty} f(x)$.

Both of these limits are evaluated following the procedures used for limits of sequences. This is illustrated by the following two examples:

$$\lim_{x \to \infty} \frac{3x + 2}{2 - x} = \lim_{x \to \infty} \frac{3 + (2/x)}{(2/x) - 1} = \frac{3 + 0}{0 - 1} = -3$$

and

$$\lim_{x \to \infty} \frac{x}{x^2 - x - 1} = \lim_{x \to \infty} \frac{1}{x - 1 - (1/x)} = 0$$

In the second form of the latter example, the denominator is large for x large; therefore, the ratio is near zero when x is large.

EXERCISES 2-6

In the following exercises, obtain the limits or state that they do not exist by observing the behavior of the functions in the domain and near the appropriate value of x. Graphing the functions may be useful in observing their behavior.

1. $\lim_{x \to 4} \sqrt{x}$

2. $\lim_{x \to 0} |x|$

3. $\lim_{x \to 0} \left(1 - \frac{x}{|x|}\right)$

4. $\lim_{x \to 2} \frac{x^2 - 3x + 2}{x - 2}$

5. $\lim_{x \to -1} \frac{x + 1}{x^2 - 1}$

6. $\lim_{x \to 0} \frac{x\sqrt{|x|}}{|x|}$

7. $\lim_{x \to 0} \frac{x + 2}{x^2}$

8. $\lim_{x \to 0} \frac{x + 2}{x}$

9. $\lim_{x \to 1} \frac{x}{(x - 1)^2}$

10. $\lim_{x \to 1} \frac{x}{(x - 1)(x - 2)}$

In Exercises 11–22 use the rules for the limits of $f + g, f - g, f \cdot g$, and f/g to evaluate the limits.

11. $\lim_{x \to 4} 3(x^2 - 1)$

12. $\lim_{x \to 2} \left[(x + 2) + \frac{1}{x - 1}\right]$

13. $\lim_{x \to 1} \sqrt{x} + \frac{1}{x}$

14. $\lim_{x \to 1} (3x^3 - 2x^2 + 4x + 1)$

15. $\lim_{x \to 1} (x^2 + x + 1)(x - 2)$

16. $\lim_{x \to 2} (x + 1)^2$

17. $\lim_{x \to 1} x(x + 2)(x - 3)$

18. $\lim_{x \to 1} \frac{1}{\sqrt{x + 3}}$

19. $\lim_{x \to 2} \frac{x + 2}{x - 1}$

20. $\lim_{x \to 0} \frac{x^2 - 2x - 2}{x + 1}$

21. $\lim_{x \to 0} \frac{(x + 2)(x - 1)}{x + 1}$

22. $\lim_{x \to -1} \left(\frac{x - 2}{x + 2}\right)^2$

In Exercises 23–32, either change the form or use common sense about numbers to obtain the limits or state they do not exist.

23. $\lim_{x \to 2} \frac{x^2 - 4}{x - 2}$

24. $\lim_{x \to \infty} \frac{2x + 3}{x - 1}$

25. $\lim_{x \to -\infty} \frac{2x + 3}{x - 1}$

26. $\lim_{x \to 0} \frac{x^2 + x}{x}$

27. $\lim_{x \to \infty} \frac{x^2 + 2x + 2}{3x^2 - 4x - 4}$

28. $\lim_{x \to \infty} \frac{2 + 2x}{2 - 3x}$

29. $\lim_{x \to \infty} (x^2 + x)$

30. $\lim_{x \to -\infty} (x^2 + x)$

31. $\lim_{x \to 0} \dfrac{x^2 + 1}{x}$ **32.** $\lim_{x \to 1} \dfrac{x^2}{x - 1}$

Access to an electronic calculator or to a computer can be helpful in understanding and evaluating limits. Consider

$$\lim_{x \to 2} \frac{1}{x + 3}$$

which we know by Property 2-2 is equal to $\frac{1}{5}$. To enhance our understanding, we might evaluate $1/(x + 3)$ for $x = 1.98$, 1.99, 2.01, and 2.02 [note that we evaluate for x close to 2 but not equal to 2. As all of the results are close to $\frac{1}{5}$, we should be more confident of our conclusion that $\lim_{x \to 2} 1/(x + 3) = \frac{1}{5}$. If any of these values were not close to $\frac{1}{5}$, we would presumably have made an error. It should be realized that it is not possible by this method to prove that the limit exists and is a certain value, but encountering values not close to the expected limit indicates a probable error somewhere in the process.

In the following exercises, state whether the limit exists or not, and if it exists, determine its value. Then evaluate (or approximately evaluate) the function at the indicated values and see if the results support your conclusions relevant to the limit. A calculator is useful in doing the arithmetic.

33. $\lim_{x \to 2} \dfrac{x^2 - 4}{x - 2}$ Evaluate at $x - 1.98$, 1.99, 2.01, and 2.02.

34. $\lim_{x \to 3} \left(x^2 + \dfrac{1}{x}\right)$ Evaluate at $x = 2.98$, 2.99, 3.01, and 3.02.

35. $\lim_{x \to -1} (x - 1)^2$ Evaluate at $x = -1.02$, -1.01, -0.99, and -0.98

36. $\lim_{x \to -2} \dfrac{1}{x + 2}$ Evaluate at $x = -2.02$, -2.01, -1.99, and -1.98

37. $\lim_{x \to 1} \dfrac{1}{(x - 1)^2}$ Evaluate at $x = 0.98$, 0.99, 1.01, and 1.02

2-7. CONTINUITY

An important concept of calculus using limits is the concept of continuity, which is defined as follows:

A function $f(x)$ is *continuous at point c* if

$$\lim_{x \to c} f(x) = f(c)$$

This definition implies that three conditions must be true:

1. The function is defined at c, i.e., $f(c)$ exists.
2. $\lim_{x\to c} f(x)$ exists.
3. $\lim_{x\to c} f(x) = f(c)$.

If $f(x)$ is not continuous at $x = c$, we say $f(x)$ is *discontinuous at point* c or there is a discontinuity at point c.

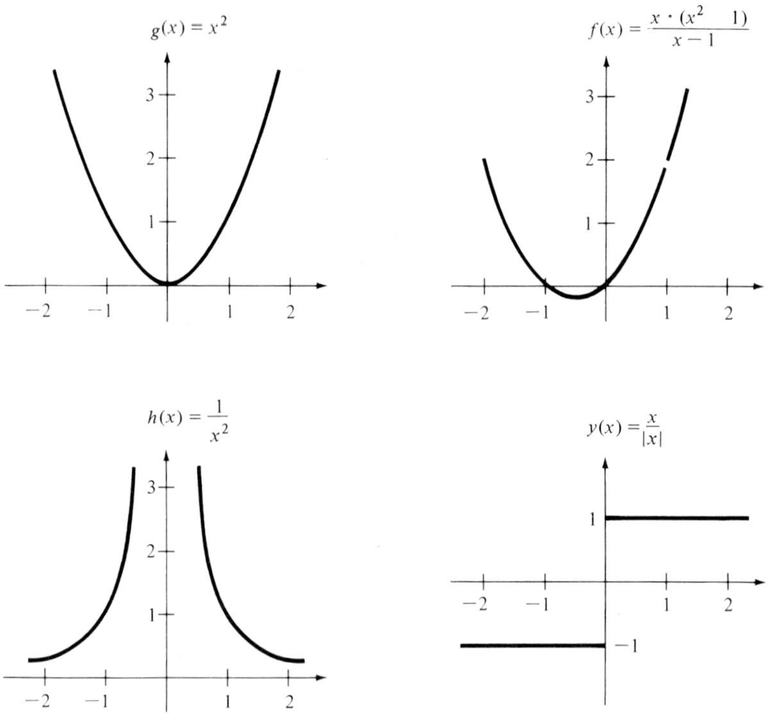

FIGURE 2-15

To illustrate the definition, consider the functions in Figure 2-15. The function $g(x) = x^2$ is continuous for all x. The function

$$f(x) = \frac{x(x^2 - 1)}{x - 1}$$

is not continuous at $x = 1$ because $f(1)$ does not exist. However, the function is continuous for all $x \neq 1$. The function $h(x) = 1/x^2$ is not continuous at $x = 0$; indeed, neither $h(0)$ nor $\lim_{x\to 0} h(x)$ exists. However, h is continuous for all $x \neq 0$. The function $y(x) = x/|x|$ is not continuous at $x = 0$; indeed neither $\lim_{x \to 0} y(x)$ nor $y(0)$ exists. However, this function is continuous for all $x \neq 0$.

A function $f(x)$ is *continuous over an interval* $[a, b]$ if it is continuous at each point of the interval. Intuitively, we can say that $f(x)$ is continuous over interval $[a, b]$ if it is possible to "draw the graph of $f(x)$ over this interval without picking up the pencil." The graph of $f(x) = x^2$ in Figure 2-15 can be completely drawn over $[-2, 2]$ without picking up the pencil. But the graph of $f(x) = [x(x^2 - 1)]/(x - 1)$ is missing a point at $x = 1$, which leads to a discontinuity at $x = 1$. For the functions $f(x) = 1/x^2$ or $f(x) = x/|x|$, we are able to draw either the positive part or the negative part without picking up the pencil, but we cannot draw both.

If it can be established that a function is continuous, a number of useful properties hold. One of these is that the limits of continuous functions can easily be determined. If $f(x)$ is continuous, then

$$\lim_{x \to c} f(x) = f(c)$$

Therefore, if $f(x)$ is known to be continuous at c, in order to determine $\lim_{x \to c} f(x)$, we need only evaluate $f(c)$.

We shall illustrate with an example. As will be shown shortly, the function

$$f(x) = x^3 + 2x - 1$$

belongs to a class of functions that are continuous for all x. Hence, to determine $\lim_{x \to 1} f(x)$, we need only evaluate $f(1) = 1^3 + 2 \cdot 1 - 1 = 2$. Therefore, we conclude that

$$\lim_{x \to 1} f(x) = f(1) = 2$$

Continuous functions possess many other nice properties in addition to permitting limits to be determined easily. This motivates us to address the question of what functions can be recognized as being continuous without having to apply the definition each time. By now, it should be apparent that functions such as $f(x) = x$ and $f(x) = k$, k a constant, are continuous. In the previous section, relationships were stated for the limits of sums, differences, products, quotients, etc. From these relationships and the definition of continuity, a parallel set of relationships apply, and we shall now state these.

If $f(x)$ and $g(x)$ are continuous functions at c, then the following properties follow directly from the corresponding properties of limits:

1. The function $k \cdot f(x)$, where k is a constant, is continuous at c.
2. The sum function $f(x) + g(x)$ is continuous at c. This property coupled with the previous property reveals that $f(x) - g(x)$ is also continuous at c.
3. The product function $f(x) \cdot g(x)$ is continuous at c.
4. The quotient function $f(x)/g(x)$ is continuous at c if $g(c) \neq 0$.

All these properties can also be stated for functions $f(x)$ and $g(x)$ continuous over $[a, b]$ with property (4) needing the condition that $g(x) \neq 0$ for x any point of $[a, b]$.

Beginning with a few functions known to be continuous, these relationships permit a variety of continuous functions to be constructed. For example, the functions $f(x) = x$ and $g(x) = x$ are continuous; therefore, their product $f(x) \cdot g(x) = x \cdot x = x^2$ is also continuous. This reasoning can be extended to show that the *power function*, $f(x) = x^n$, for any nonnegative integer n, is also continuous, as is kx^n, since k and x^n are both continuous. As sums of continuous functions are continuous, then the functions $f(x) = x + 3$, $f(x) = 2x^2 - 1$, $f(x) = 4x^2 + x$, $f(x) = x^{12} - 6x^5 + 8x - 17$, and so on, are also continuous.

Functions of this general class are referred to as *polynomials.* Any polynomial can be represented as a sum of terms, each of which is a constant times a power function. The functions $f(x) = 3$, $f(x) = x$, $f(x) = 2x^2$, etc., are special cases of polynomials.

Polynomials are frequently represented in the form of a product rather than a sum. Thus $(x - 1)(x - 3)$ is a polynomial in factored form. Since $(x - 1)(x - 3) = x^2 - 4x + 3$, we note that it can also be written in standard form as a sum.

As illustrated by examples in the exercises, we also note that the sum and the product of polynomials are also polynomials, i.e., the set of polynomials is closed under addition and multiplication.

Next, we wish to consider a function $f(x)$ that is the ratio of two polynomials:

$$f(x) = \frac{p(x)}{q(x)}, \qquad q(x) \neq 0$$

Such functions are called *rational functions.*

Among examples of rational functions are

$$\frac{x^6 - 1}{x + 5}, \quad \frac{x^{10} - 3x^5 + x^2}{1}, \quad \frac{1}{x^2 + 1} \quad \text{and} \quad \frac{x + 1}{x - 1}$$

From the properties of continuous functions we can conclude that rational functions are continuous for all values of x for which they are defined. Specifically, they are continuous except for values of x for which the denominator $q(x) = 0$, or, in other language, except for the zeros of $q(x)$.

Thus, knowing that $(x^3 - 2x + 1)/(x^4 - 3x + 6)$ is continuous at each point for which $x^4 - 3x + 6 \neq 0$, we have, for example,

$$\lim_{x \to 2} \frac{x^3 - 2x + 1}{x^4 - 3x + 6} = \frac{8 - 4 + 1}{16 - 6 + 6} = \frac{5}{16}$$

simply by evaluating the function at $x = 2$.

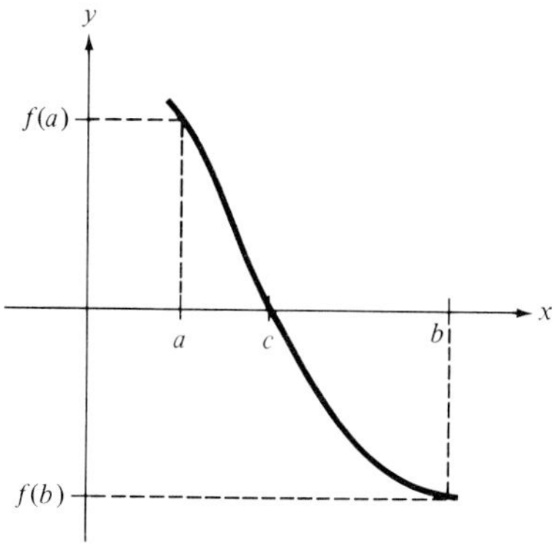

FIGURE 2-16

However, we should point out that we needed the concept of limit (or something almost equivalent to it) to define continuity. Thus we can justify the quick and easy evaluation of limits only after we first introduce limits the "hard way." Most of the limits that we shall need later in this course are limits for which the functions concerned are not continuous, indeed are not defined at the point in question.

A very important property of continuous functions is illustrated in Figure 2-16. Suppose $f(a)$ and $f(b)$ have opposite signs and we are looking for a number c between a and b for which $f(c) = 0$. If the function $f(x)$ is continuous over $[a, b]$, we may conclude that there must be a number c in the interval $[a, b]$ for which $f(c) = 0$. Observe that in Figure 2-17 there are three points c_1, c_2, and c_3 with $f(c_1) = 0$, $f(c_2) = 0$, and $f(c_3) = 0$.

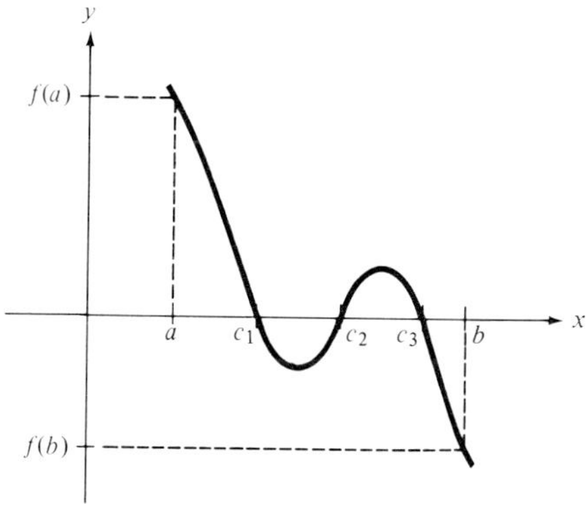

FIGURE 2-17

This very important property of continuous functions is generally referred to as the **Intermediate Value Theorem**:

> If $f(x)$ is continuous over $[a, b]$, and if $f(a)$ and $f(b)$ have opposite signs, then there exists at least one number c within $[a, b]$ for which $f(c) = 0$.

Although the proof of this theorem is beyond the scope of this book, the theorem is believable from our earlier pencil-drawing explanation of continuity. Recall that if a function is continuous over the interval $[a, b]$, we should be able to draw the graph from point $(a, f(a))$ to point $(b, f(b))$ "without picking up the pencil." This can be done only if the graph crosses the x axis between a and b.

A common alternate form of the intermediate value theorem is illustrated in Figure 2-18.

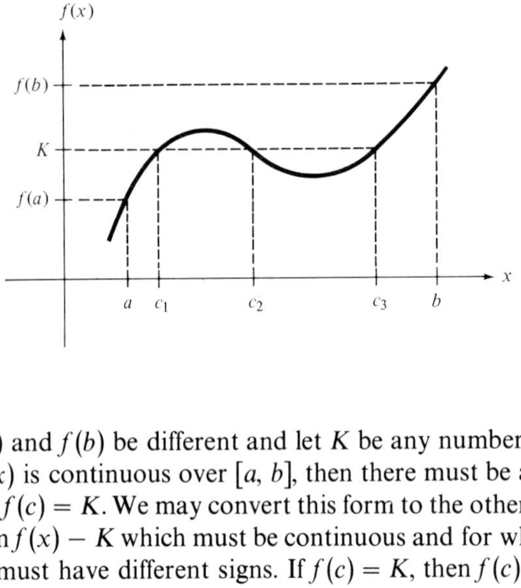

FIGURE 2-18

Let $f(a)$ and $f(b)$ be different and let K be any number between $f(a)$ and $f(b)$. If $f(x)$ is continuous over $[a, b]$, then there must be a number c in $[a, b]$ such that $f(c) = K$. We may convert this form to the other by considering the function $f(x) - K$ which must be continuous and for which $f(a) - K$ and $f(b) - K$ must have different signs. If $f(c) = K$, then $f(c) - K = 0$ and conversely. In Figure 2-18 any of the three numbers c_1, c_2, and c_3 would do for c.

EXERCISES 2-7

In Exercises 1–12, state the values, if any, at which the function is discontinuous.

1. $f(x) = 3x^3 - 2x - 5$

2. $f(x) = \dfrac{3}{x - 1}$

3. $f(x) = \frac{x-1}{x^2-9}$

4. $f(x) = \frac{x^2-4}{x+2}$

5. $f(x) = \frac{1}{x^2+1}$

6. $f(x) = \sqrt{x^2+1}$

7. $f(x) = \sqrt{x+1}$

8. $\frac{2x^2+3x+1}{x^2}$

In Exercises 9 through 11 and sometimes later in the text we describe functions by alternate formulas specified for various parts of the domain. Thus in Exercise 9, $f(x) = x$ whenever $x \geq 0$ and $f(x) = x^2$ whenever $x < 0$. The domain is determined by the description on the right, with the corresponding functional values by the formulas.

9. $f(x) = \begin{cases} x, & \text{whenever } x \geq 0 \\ x^2, & \text{whenever } x < 0 \end{cases}$

10. $f(x) = \begin{cases} x+1, & \text{whenever } x \geq 2 \\ 3, & \text{whenever } x < 2 \end{cases}$

11. $f(x) = \begin{cases} 2x-1, & \text{whenever } x > 1 \\ 1-2x, & \text{whenever } x \leq 1 \end{cases}$

12. $f(x) = x \cdot |x|$

For each of the following functions, state whether the intermediate value theorem indicates that a zero of the function exists on the indicated interval.

13. $f(x) = x^2 + 3x - 1$ $\quad [0, 5]$

14. $f(x) = x^3 - 6x^2 + 11x - 6$ $\quad [0, 5]$

15. $f(x) = \frac{x}{x^2-4}$ $\quad [1, 5]$

16. $f(x) = 1 - \frac{3}{x} - \frac{2}{x^2}$ $\quad [-2, 2]$

17. $f(x) = x^3 - 6x^2 - 7x - 6$ $\quad [0, 3]$

For the following polynomials $p(x)$ and $q(x)$, determine $p(x) + q(x)$ and $p(x) \cdot q(x)$. Express as polynomials in the standard form as a sum of terms each of which is a constant times a power of x.

18. $p(x) = x^3 + 2x + 1$
$q(x) = x^2 + 2$

19. $p(x) = x^4 - 1$
$q(x) = x^4 + 1$

20. $p(x) = x^2 + x + 1$
$q(x) = x^2 - x - 1$

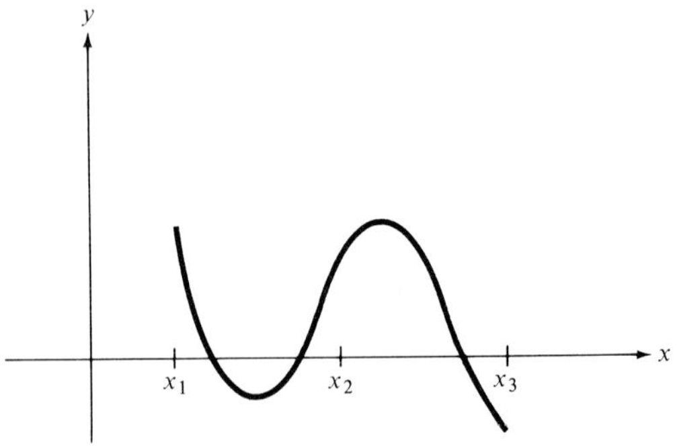

FIGURE FF

21. Consider the continuous function $f(x)$ to be graphed as indicated in Figure FF. On which interval(s) $[x_1\, x_2]$, $[x_2\, x_3]$, $[x_1, x_3]$ does the intermediate value theorem directly guarantee a zero of $f(x)$?

22. The function $(x^2 - 1)/(x - 1)$ is discontinuous at $x = 1$. Suppose we define a function as follows:

$$f(x) = \begin{cases} \dfrac{x^2 - 1}{x - 1}, & \text{whenever } x \neq 1 \\ 2, & \text{whenever } x = 1 \end{cases}$$

Is this function continuous at $x = 1$? Apply the definition, that is, evaluate $\lim_{x \to 1} f(x)$, evaluate $f(1)$, and see if they are equal.

23. The function $(x^2 - 3x + 2)/(x - 1)$ is discontinuous at $x = 1$. What value must be assigned at $x = 1$ so that a new function defined as in Exercise 22 will be continuous?

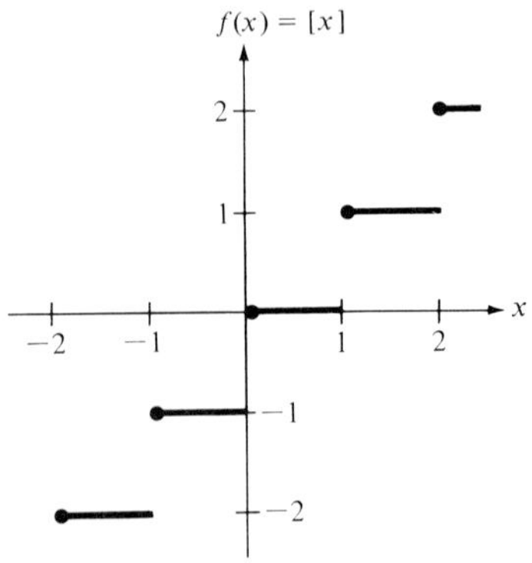

FIGURE GG

24. The greatest integer function is usually represented as $y = [x]$, and reads "y equals the greatest integer that does not exceed x." For example, $[5.2] = 5$, $[\frac{4}{3}] = 1$, $[2] = 2$, $[-\frac{1}{2}] = -1$, $[-5.2] = -6$. A part of the graph is illustrated in Figure GG. For this function,
(**a**) Does $[4]$ exist, and if so, what is it?
(**b**) Does $\lim_{x \to 4} [4]$ exist, and if so, what is it?
(**c**) Is the function continuous at $x = 4$?

*2-8. INTERVAL HALVING

So far we have used the intermediate value theorem only to determine the existence of a root of an equation within a prescribed interval. In this section we want to examine a method for approximating such a solution to any desired degree of accuracy.

As a specific example let us find $\sqrt[3]{5}$. The usual way to express such a problem is to ask for the real value of x for which

$$x^3 = 5$$

or equivalently for which $x^3 - 5 = 0$. Using the latter form, we can construct a graph for $y = f(x) = x^3 - 5$, as illustrated in Figure 2-19. The solution is the value of x for which $f(x) = 0$, which is also illustrated in Figure 2-19.

We observe that $f(x) = x^3 - 5$ is a polynomial and, therefore, continuous for all x. If we can find a number a for which $f(a) < 0$ and a number b for which $f(b) > 0$, the intermediate value theorem applies and we conclude that a solution c is between a and b.

We might first investigate using integer values for a and b. For $x = 1$, we observe that $f(1) = 1^3 - 5 = -4$; therefore, $f(1) < 0$. For $x = 2$, we observe that $f(2) = 2^3 - 5 = 3$; therefore, $f(2) > 0$. We conclude that a solution lies on the interval $[1, 2]$; i.e., we have bracketed the solution between successive integers.

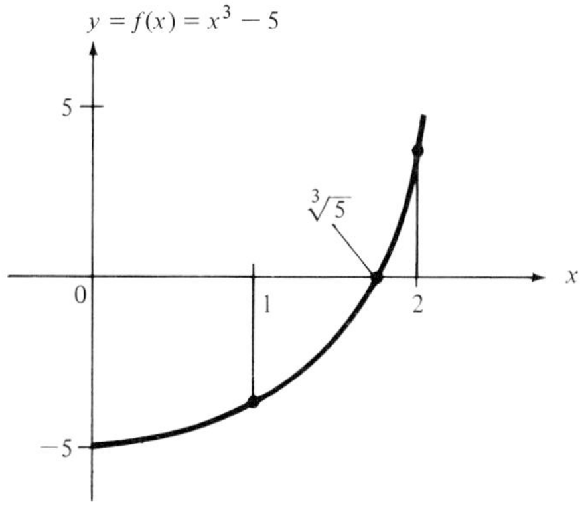

FIGURE 2-19

We could logically extend this approach by investigating $f(x)$ for values of x at 1.1, 1.2, 1.3, ..., 1.9. If we were to do so, we would find that $f(1.7) = -0.087 < 0$ and $f(1.8) = 0.832 > 0$. Therefore, we conclude that there is a solution between 1.7 and 1.8. We could extend this approach to hundredths, then to thousandths, and so on until a solution to any desired accuracy is obtained.

For machine computation, the above approach is not very convenient or efficient. Instead, an approach such as *interval halving* is a more logical choice. Using this method, we proceed as follows:

1. Begin with a given interval known to contain a solution.
2. Evaluate the function at the midpoint of the interval.
3. Determine which half of the interval must contain a solution.
4. Using this half-interval, repeat from step 2 until a solution to any desired accuracy is known.

This approach is represented by the symbolic flowchart in Figure 2-20.

You might have observed that the above procedure ignores the possibility that the midpoint of the interval could be a solution. For problems of practical interest, this result is extremely unlikely. The flowchart could, of course, be modified to identify this situation.

We shall now apply the interval halving method to our cube root problem. We shall begin the process with the interval [1, 2]. The graph of $f(x)$ over this interval is illustrated in Figure 2-21. For this interval, we proceed as follows ($a_0 = 1$, $b_0 = 2$):

1. Determine the midpoint z_0:

$$z_0 = \frac{a_0 + b_0}{2} = \frac{1+2}{2} = 1.5$$

2. Evaluate the function at the midpoint:

$$f(z_0) = z_0^3 - 5 = 1.5^3 - 5 = 3.375 - 5 = -1.625$$

3. We must now choose between the following two intervals:

$$[a_0, z_0] = [1, 1.5]; f(a_0) = f(1) = -4 < 0$$
$$f(z_0) = f(1.5) = -1.625 < 0$$
$$[z_0, b_0] = [1.5, 2]; f(z_0) = f(1.5) = -1.625 < 0$$
$$f(b_0) = f(2) = 3 > 0$$

Since $f(z_0) < 0$ and $f(b_0) > 0$, we choose the interval $[z_0, b_0]$ for the next iteration. To denote this interval as $[a_1, b_1]$, we let

$$a_1 = z_0 = 1.5$$
$$b_1 = b_0 = 2$$

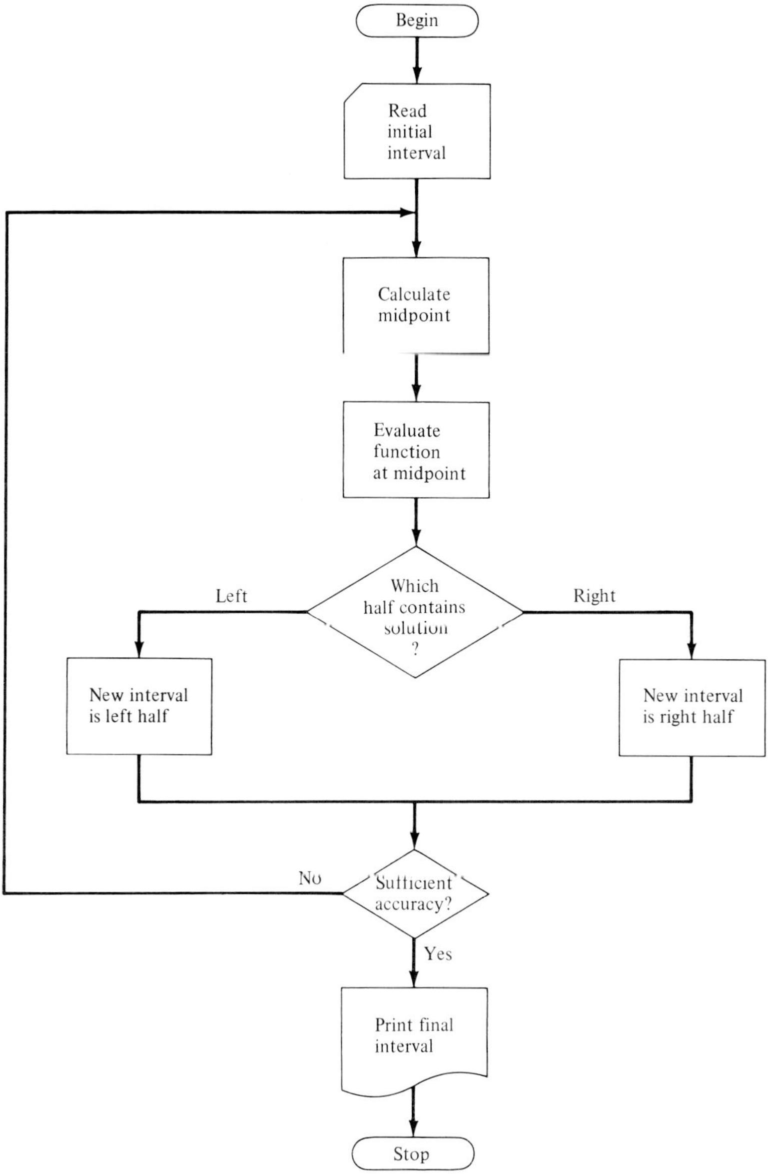

FIGURE 2-20 A symbolic flowchart.

This interval is illustrated in Figure 2-21.

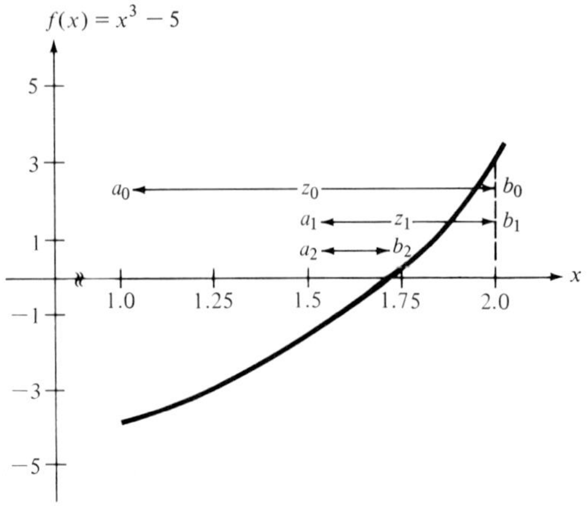

FIGURE 2-21

The same procedure is repeated starting with the interval [1.5, 2]. The steps are as follows:

1. Determine the midpoint:

$$z_1 = \frac{a_1 + b_1}{2} = \frac{1.5 + 2}{2} = 1.75$$

2. Determine $f(z_1)$

$$f(z_1) = z_1^3 = (1.75)^3 - 5 = 5.359375 - 5 = 0.359375$$

3. Since $f(z_1) > 0$, then there must be a solution within the interval $[a_1, z_1]$ or [1.5, 1.75]. We choose this interval for $[a_2, b_2]$.

This interval is also illustrated in Figure 2-21.

The first six iterations of the method produce the values listed in Table 2-5. To six decimal places, the value of c is 1.709976. In six iterations of

TABLE 2-5

n	a_n	b_n	z_n	$f(z_n) = z_n^3$
0	1	1.5	2	−1.625
1	1.5	1.75	2	+0.359
2	1.5	1.625	1.75	−0.719
3	1.625	1.6875	1.75	−0.195
4	1.6875	1.71875	1.75	+0.077
5	1.6875	1.703125	1.71875	−0.060

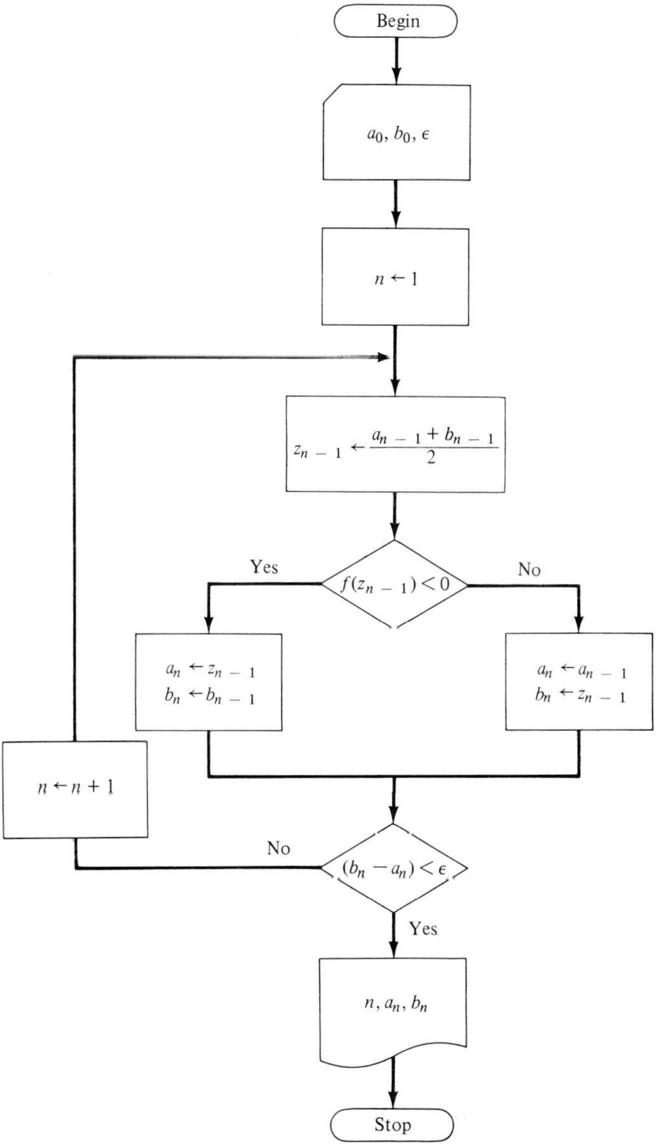

FIGURE 2-22 Note that this flowchart works only if $f(a_0) < 0$ and $f(b_0) > 0$.

interval halving, we have bracketed the solution between 1.703125 and 1.71875.

The flowchart for this example is given in Figure 2-22. The program terminates when the length of the interval containing c is less than some number ε, which is entered as data along with values for a_0 and b_0.

In the interval-halving method, the calculations produce a succession of intervals $[a_0, b_0]$, $[a_1, b_1]$, $[a_2, b_2]$, ... Each interval is contained in its predecessor, and thus we have a succession of *nested* intervals. The end points of these intervals form two sequences, namely, $\langle a_0, a_1, a_2, \ldots\rangle$ and $\langle b_0, b_1, b_2, \ldots\rangle$. The limit of each of these sequences is the solution c. A third sequence associated with this method is the sequence of the lengths of the intervals, namely, $\langle (b_0 - a_0), (b_1 - a_1), (b_2 - a_2), \ldots\rangle$. The limit of this sequence is zero.

EXERCISES 2-8

1. Let $f(x)$ be continuous over some interval containing numbers a and b. Suppose that

$$f(a) > 0$$

$$f(b) > 0$$

(a) Does the interval $[a, b]$ necessarily contain any number c for which $f(c) = 0$?

(b) Is it possible for the interval $[a, b]$ to contain a point c for which $f(c) = 0$? Draw some graphs to support your conclusion.

2. Let $f(x)$ be as in Figure HH. Which, if any, of the three values of c will the interval-halving method produce starting from the interval $[1, 3]$?

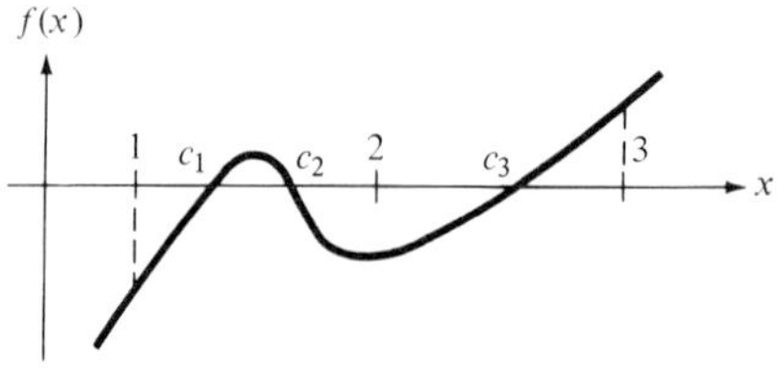

FIGURE HH

3. Find an interval of length 1 having integral end points that contains a value c such that $f(c) = 0$ for the function

$$f(x) = 2x^5 + 4x^3 - 9x^2 - 18x + 1$$

NOTE: There are several possible answers.

4. How must the flowchart in Figure 2-22 be modified for functions such as in Figure 2-16, i.e., where $f(a_0) > 0$ and $f(b_0) < 0$.

5. The flowchart in Figure 2-22 can be readily modified so that it will apply both to functions such as in Figure 2-16 and to functions such as in Figure 2-19. Observe that if

$$\left.\begin{matrix} f(a) < 0 \\ f(b) > 0 \end{matrix}\right\} \quad \text{or} \quad \left\{\begin{matrix} f(a) > 0 \\ f(b) < 0 \end{matrix}\right.$$

then $f(a) \cdot f(b) < 0$. Otherwise, the product is positive. On this basis, modify the decision block in Figure 2-22.

6. Suppose the length of the interval $[a_0, b_0]$ is 5.
(a) What is the length of the interval $[a_1, b_1]$?
(b) What is the length of the interval $[a_2, b_2]$?
(c) What is the length of the interval $[a_{10}, b_{10}]$?
(d) What is the length of the interval $[a_n, b_n]$?
(e) If the final interval is to be less than 0.000001 of the original interval, how many iterations are required?
[HINT: $(\frac{1}{2})^{10} = \frac{1}{1024} \approx 0.001$.]

For Exercises 7–12, a hand calculator is sufficient to apply interval halving to finding zeros in the indicated intervals. Show your work step by step. Stop when the length of the interval is less than 0.05.

7. $x^5 - 5 = 0$; $[a_0, b_0] = [0, 2]$

8. $x^3 + 2\sqrt{x} - 1 = 0$; $[a_0, b_0] = [0, 1]$

9. $x^2 + 2x - 2 = 0$; $[a_0, b_0] = [0, 1]$

10. $x^2 + 2x - 2 = 0$; $[a_0, b_0] = [-3, -2]$

[NOTE: A solution exists in each interval for Exercises 9 and 10.]

11. $x + (1/x) - 3 = 0$; $[a_0, b_0] = [\frac{1}{4}, 1]$

12. $x + (1/x) - 3 = 0$; $[a_0, b_0] = [2, 3]$

NOTE: A solution exists in each interval for Exercises 11 and 12.

REVIEW EXERCISES: CHAPTER 2

For Exercises 1 through 10, write the first six terms of the sequence.

1. $a_n = \frac{1}{n^2}$, n positive

2. $z_n = \frac{n-1}{n+1}$, n nonnegative

3. $q_n = \frac{(-1)^n}{2n}$, n positive

4. $p_n = (-\frac{1}{2})^n$, n nonnegative

5. $b_n = 1 + (-1)^n$, n positive

6. $c_n = c_{n-1} - n$
$c_0 = 0$

7. $a_n = (0.2)a_{n-1}$
$a_0 = 1$

8. $q_n = \frac{q_{n-1}}{n}$
$q_0 = 1$

9. $b_n = \frac{n}{b_{n-1}}$
$b_0 = 1$

10. $p_n = n - p_{n-1}$
$p_1 = 1$

For Exercises 11 through 15, write recursion relationships for the sequences. You may use either the ratio method presented in Section 2-2 or the difference method presented in the exercises for Section 2-2. For each exercise, one of these will yield a much simpler recursion relationship than will the other.

11. $q_n = 3n + 5$

12. $c_n = (-1)^n \frac{n-1}{n}$

13. $z_n = (-5)^n$

14. $a_n = 1 + (-1)^n$

15. $b_n = n^2$

For Exercises 16 through 18, write formulas for the sequences given by the recursion relationships. You will have to write the first several terms of the sequence and then look for a formula that would give the same terms.

16. $p_n = \frac{n-1}{n+1} p_{n-1}$
$p_1 = \frac{1}{2}$

17. $c_n = \frac{n-1}{n+2} c_{n-1}$
$c_1 = \frac{1}{6}$

18. $b_n = \frac{n^2}{n^2 - 1} b_{n-1}$
$b_1 = 1$

19. Consider the loop structure in Figure II.
(**a**) For what values of n is Block 3 executed?
(**b**) How should the flowchart be modified to execute Block 3 for $n = 2$, 4, 6, 8, 10?

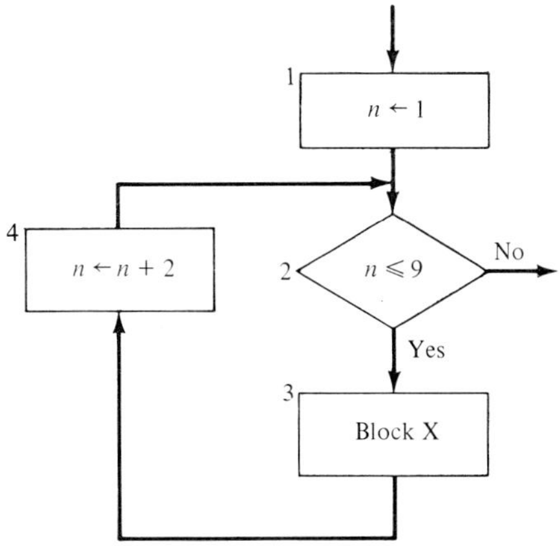

FIGURE II

(**c**) For $n = 0, 5, 10, 15, 20, 25$?
(**d**) For $n = 9, 7, 5, 3, 1$?

NOTE: Consider the decision block carefully for part (**d**).

20. Consider the flowchart in Figure JJ.
(**a**) This flowchart is supposed to compute and print the terms in the

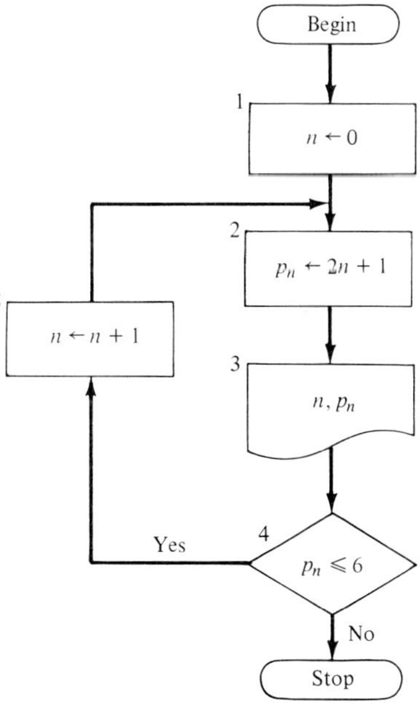

FIGURE JJ

sequence $p_n = 2n + 1$, stopping after encountering a term whose value exceeds 6. Does it do this?

(b) How must the flowchart be modified to stop after encountering a term whose value exceeds 100?

For Exercises 21 through 30, determine the limit of the sequences or state that they do not exist.

21. $p_n = \dfrac{6n - 2}{1 - 3n}$

22. $a_n = (-1)^n \cdot (n - 1)$

23. $c_n = \dfrac{n^2 - 1}{n + 2}$

24. $b_n = \sqrt{n}$

25. $q_n = \dfrac{n^4 - 1}{n^2 + 1}$

26. $c_n = \dfrac{1 - n^2}{n^2 + n^2}$

27. $p_n = \dfrac{1}{\sqrt{n}}$

28. $b_n = \dfrac{1 - n + n^2 - n^3}{(n + 1)^3}$

29. $q_n = (0.9)^n$

30. $p_n = \sqrt[10]{n}$

31. Suppose $\lim_{x \to a} p(x) = 4$ and $\lim_{x \to a} q(x) = 6$.

Determine:

(a) $\lim_{x \to a} [5p(x)]$

(b) $\lim_{x \to a} [p(x) + q(x)]$

(c) $\lim_{x \to a} [p(x) - q(x)]$

(d) $\lim_{x \to a} [p(x) \cdot q(x)]$

(e) $\lim_{x \to a} \left[\dfrac{p(x)}{q(x)}\right]$

(f) $\lim_{x \to a} [p(x)]^2$

(g) $\lim_{x \to a} \sqrt{p(x)}$

In Exercises 32 through 40, determine the limits of the functions or state that they do not exist.

32. $\lim_{a \to 1} \dfrac{a^2 - 1}{a - 1}$

33. $\lim_{x \to 2} \dfrac{x^2 - 3x + 2}{x}$

34. $\lim_{c \to \infty} \dfrac{c^2 - 2c - 5}{3 - c}$

35. $\lim_{b \to 1} \dfrac{b^2 - 1}{b^2 + 2b + 1}$

36. $\lim_{x \to 2} \sqrt{\dfrac{x^2 - x + 2}{x - 1}}$

37. $\lim_{x \to 3} \sqrt{(x - 3)(x + 1)}$

38. $\lim_{d \to \infty} \frac{1-d}{1+d}$

39. $\lim_{x \to 2} \left(\frac{x+2}{x+6}\right)^3$

40. $\lim_{x \to \infty} \frac{1 - 2x - x + x^3}{(1-x)^3}$

In Exercises 40 through 45, state the values of x for which $f(x)$ is not continuous.

41. $f(x) = \frac{x^2 + 2x + 1}{x}$

42. $f(x) = \frac{(x-1)^3}{x+2}$

43. $f(x) = \frac{x^2 - x - 1}{x^2 - 2x - 3}$

44. $f(x) = \sqrt{\frac{1}{x-1}}$

45. $f(x) = x^4 - x^3 + 9x^2 + 2x + \frac{1}{x} - \frac{1}{x+1}$

For Exercises 46 through 50, state if the intermediate value theorem indicates that a zero must exist within the indicated interval.

46. $f(x) = x^5 - x - 1$ $\quad [-1, 2]$

47. $f(x) = x^2 - 5$ $\quad [-3, 3]$

48. $f(x) = \frac{1}{x} \quad 1$ $\quad \left[\frac{1}{2}, 5\right]$

49. $f(x) = \frac{1}{x} - 1$ $\quad \left[-2, \frac{1}{2}\right]$

50. $f(x) = x^4 - 5x^3 - 2x^2 - x + 2$ $\quad [0, 1]$

51. The problem discussed in the body of Section 2-3 was for interest compounded annually. Suppose interest is compounded semiannually. Starting with an investment V_0, the value of the investment after 6 months is

$$V_{1/2} = V_0\left(1 + \frac{i}{2}\right)$$

Then after a year, the value V is

$$V_1 = V_{1/2}\left(1 + \frac{i}{2}\right) = V_0\left(1 + \frac{i}{2}\right)^2$$

(a) Write expressions for V_2, V_3, and V_4 (the values after 2, 3, and 4 years).

(b) Obtain a formula for V_n.

(c) Write a recursion relationship for V_n.

52. An investor decides to deposit an amount D each year into an account that draws interest at the annual rate i. Suppose that we let V_n be the amount at the beginning of each year (after the deposit D has been made). To illustrate, the amount at the beginning of year 1 is the deposit D; therefore,

$$V_0 = D$$

Since these funds draw interest over a year period, the value at the end of year 1 is $V_1 = V_0(1 + i)$. At the beginning of the next year, a deposit D is made, and then the amount $V_1 + D$ draws interest for an entire year. The value at the end of year 2 is

$$V_2 = (1 + i)(V_1 + D)$$

(**a**) Write expressions for V_3, V_4, and V_5.
(**b**) Write a recursion relationship for V_n.

53. Suppose that a man begins with an investment V_0 which draws interest at an annual rate i. He now wishes to withdraw an amount D at the end of each year. That is, at the end of year 1, the value of the investment is

$$V_0(1 + i) - D$$

(**a**) Write expressions for V_2, V_3, and V_4.
(**b**) Write a recursion relationship for V_n.

54. A fruit company ships a variety of perishable commodities (apples, oranges, pineapples, etc.) across the country. For a particular commodity, experience determines that 5% of the amount on hand at the beginning of a day must be discarded at the end of that day. For example, if we begin with Q_0 units, then after discarding the spoiled portion, the amount at the beginning of the next day is

$$Q_1 = Q_0(1 - 0.05)$$

(**a**) Write an expression for Q_2, Q_3, Q_4, and Q_5.
(**b**) Write a formula for Q_n.
(**c**) Write a recursion relationship for Q_n.

CHAPTER THREE

THE DERIVATIVE

3-1. TANGENT LINES

Suppose Johnny is driving straight down the road away from home, and that his position, measured as distance $s(t)$ from home, as a function of time, is illustrated in Figure 3-1. We can ask the questions:

1. How far is Johnny from home at time t_1?
2. What was his average speed between times t_1 and t_2?
3. What is his speed at time t_1?

The answer to the first question can be read directly from the graph for $s(t)$ in Figure 3-1. It is simply the height to the graph for the given value of t_1. In this section we wish to consider the remaining two questions in more detail.

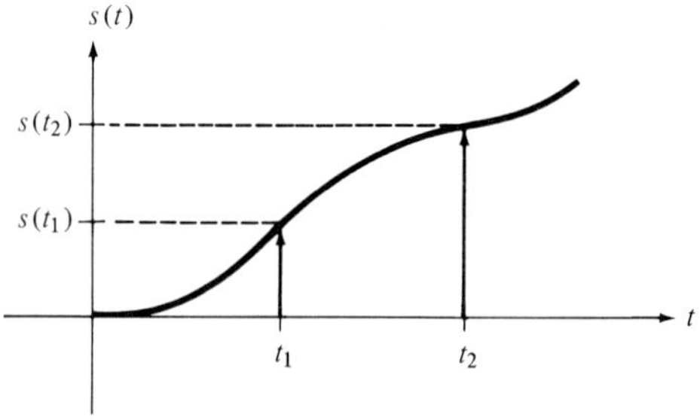

FIGURE 3-1

Average Speed To obtain the average speed over some interval of time, we divide the distance traveled during this interval of time by the time interval itself. At time t_1, Johnny is at a distance $s(t_1)$ from home. At time t_2, Johnny is at a distance $s(t_2)$ from home. Therefore, between times t_1 and t_2, the distance traveled is $s(t_2) - s(t_1)$. Since the interval of time is $t_2 - t_1$, the average speed M_{12} from time t_1 to time t_2 is given by

$$M_{12} = \frac{\text{distance traveled}}{\text{time elapsed}}$$

In Figure 3-2, we have drawn a straight line through the two points $(t_1, s(t_1))$ and $(t_2, s(t_2))$, both of which are on the graph of $s(t)$. The slope of this line is given by

$$\text{slope} = \frac{\text{change in } s(t)}{\text{change in } t} = \frac{s(t_2) - s(t_1)}{t_2 - t_1}$$

Thus we know that the average speed M_{12} is the slope of the line drawn through the points $(t_1, s(t_1))$ and $(t_2, s(t_2))$.

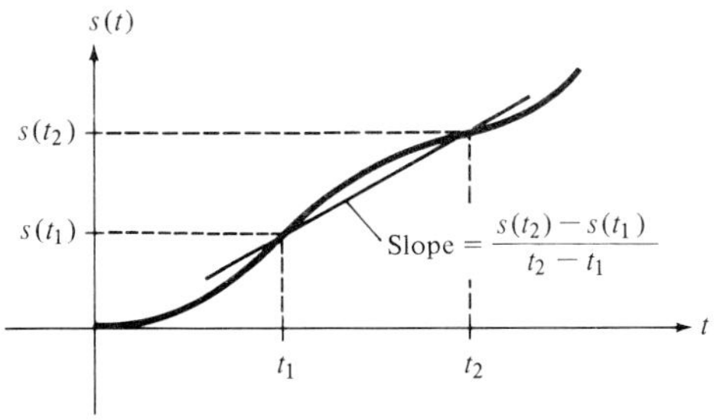

FIGURE 3-2

Difference Quotients The discussion regarding average speed can be extended to other functions. Consider the function $f(x)$ graphed in Figure 3-3. Over any interval $[x_1, x_2]$, the *difference quotient* is the net change in $f(x)$ over this interval divided by the length of the interval. Since the net change in $f(x)$ over the interval $[x_1, x_2]$ is $f(x_2) - f(x_1)$ and the length of the interval is $x_2 - x_1$, the difference quotient is

$$\text{difference quotient} = \frac{f(x_2) - f(x_1)}{x_2 - x_1}$$

The difference quotient is also the slope of the line drawn through points $(x_1, f(x_1))$ and $(x_2, f(x_2))$, as illustrated in Figure 3-3. Thus we typically think of the difference quotient as representing the ratio of the change in $f(x)$ to the change in x over some finite interval. If our variable x represents time and $f(x)$ represents distance, then the difference quotient is simply the average speed over $[x_1, x_2]$.

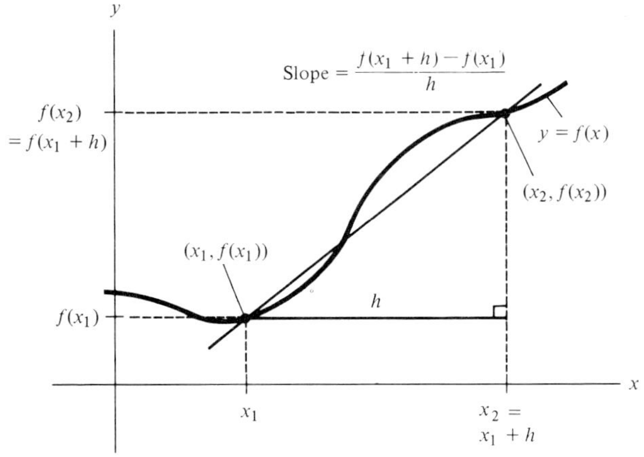

FIGURE 3-3

An alternate expression for the difference quotient is obtained by letting $h = x_2 - x_1$ or equivalently $x_2 = x_1 + h$. Thus h is the change in x or the length of the interval for which the difference quotient will be obtained. Then $f(x_2)$ is $f(x_1 + h)$, and the difference quotient becomes

$$\text{difference quotient} = \frac{f(x_1 + h) - f(x_1)}{h}$$

This is the form that we shall use most frequently.

Speed at Time t_1 In the discussion above, we obtained a formula for the average speed over some interval of time. As this time interval becomes smaller, the average speed should approach a limiting value, which we call the *instantaneous speed.* In an automobile the instantaneous speed is indicated by the speedometer.

Mathematically, we observe that when t_2 is very close to t_1, then the average speed M_{12} should be very close to the instantaneous speed $m(t_1)$ at t_1. In fact, by choosing t_2 sufficiently close to t_1, we should be able to make the average speed M_{12} as close to the instantaneous speed $m(t_1)$ as we like. Therefore, $m(t_1)$ should be the limiting value of M_{12} as t_2 approaches t_1 or

$$m(t_1) = \lim_{t_2 \to t_1} M_{12} = \lim_{t_2 \to t_1} \frac{s(t_2) - s(t_1)}{t_2 - t_1}$$

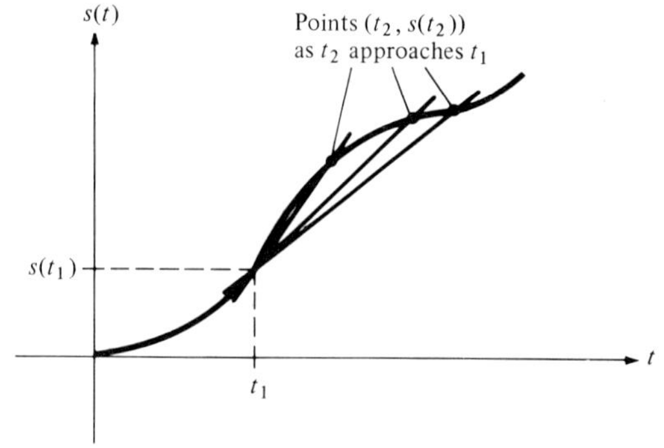

FIGURE 3-4

From a geometrical viewpoint, as t_2 approaches t_1, the point $(t_2, s(t_2))$ moves along the graph in Figure 3-4 toward the point $(t_1, s(t_1))$. Figure 3-4 illustrates lines drawn through point $(t_1, s(t_1))$ and through selected points $(t_2, s(t_2))$. As the second point approaches the first, the slopes of the lines appear to approach a limiting value, which is the slope of the line

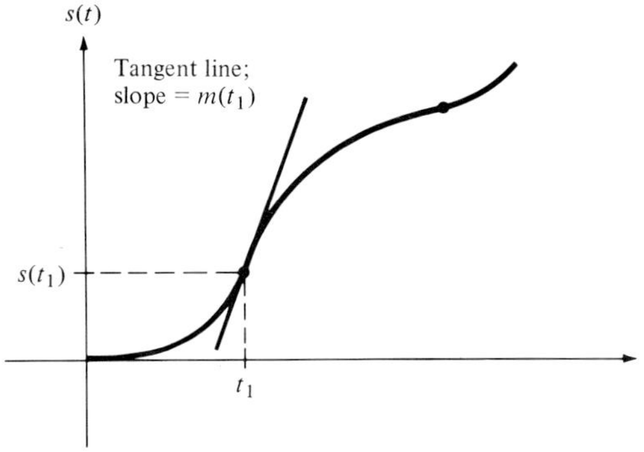

FIGURE 3-5

illustrated in Figure 3-5. This line is the tangent line to the graph at point $(t_1, s(t_1))$ and its slope is given by

$$m(t_1) = \lim_{t_2 \to t_1} \frac{s(t_2) - s(t_1)}{t_2 - t_1}$$

The equation for the tangent line can be written using the point-slope form of the equation of the line, the point being $(t_1, s(t_1))$ and the slope being $m(t_1)$:

$$y - s(t_1) = m(t_1) \cdot (t - t_1)$$

Tangent Lines The pattern we have observed for getting the tangent to the graph of $s(t)$ at the point $(t_1, s(t_1))$ is true in a general framework.

To obtain the slope of the tangent line to the graph of $f(x)$ at the point $(x_1, f(x_1))$, we must take the limit of the difference quotient $[f(x_1 + h) - f(x_1)]/h$ as h approaches zero. If we let $m(x_1)$ be the slope of the line tangent to the graph of $f(x)$ at $x = x_1$, then we have

$$m(x_1) = \lim_{h \to 0} \frac{f(x_1 + h) - f(x_1)}{h}$$

provided that the limit exists.

An alternate form of the limit is

$$m(x_1) = \lim_{x_2 \to x_1} \frac{f(x_2) - f(x_1)}{x_2 - x_1}$$

However, as remarked above, we shall normally use the expression involving h.

Once we know that the line passes through point $(x_1, f(x_1))$ and has

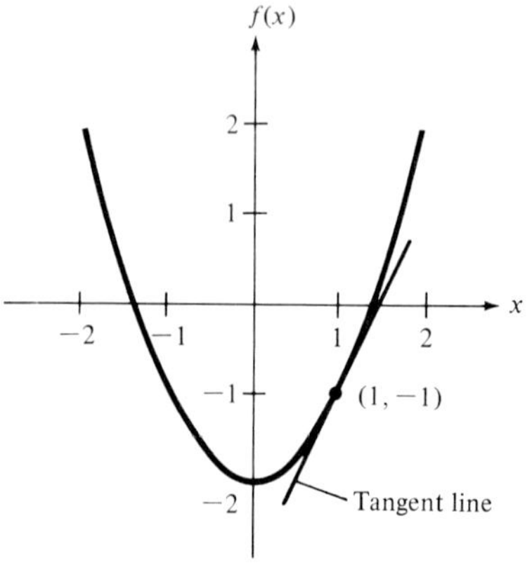

FIGURE 3-6

slope $m(x_1)$, we can write the equation of the tangent line using the point-slope form of the equation for a straight line:

$$y - f(x_1) = m(x_1) \cdot (x - x_1)$$

Let us now consider a specific example.

Suppose that we are given the graph of $f(x) = x^2 - 2$, as illustrated in Figure 3-6. We want to determine the equation of the line tangent to the graph at the point $(1, -1)$, which is on the graph. First, we must compute the slope. Since $f(x) = x^2 - 2$, then $f(1) = -1$ and $f(1 + h) = (1 + h)^2 - 2 = (1 + 2h + h^2) - 2 = -1 + 2h + h^2$. The slope m at the point $(1, -1)$ is

$$\begin{aligned} m(1) &= \lim_{h \to 0} \frac{f(1 + h) - f(1)}{h} \\ &= \lim_{h \to 0} \frac{(-1 + 2h + h^2) - (-1)}{h} \\ &= \lim_{h \to 0} \frac{2h + h^2}{h} \end{aligned}$$

We change the form of $(2h + h^2)/h$. For $h \neq 0$,

$$\frac{2h + h^2}{h} = 2 + h$$

Therefore,

$$m(1) = \lim_{h \to 0} (2 + h) = 2$$

Noting that the tangent line has slope 2 and passes through point $(1, -1)$, the point-slope form of the equation for a straight line gives:

$$y - (-1) = 2(x - 1)$$

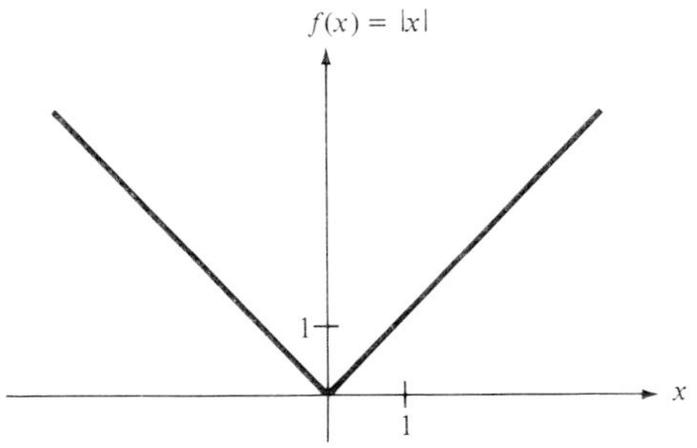

FIGURE 3-7

which reduces to

$$y = 2x - 3$$

Existence of the Limit Let us now examine the question of the existence of the limit in more detail. In effect, we have been studying the case where there is a tangent line *with slope*. The slope is the limit whose existence we have assumed. Here is an example in which the limit does not exist and there is no tangent line with slope.

Consider the graph of $f(x) = |x|$ illustrated in Figure 3-7. We wish to determine if there is a tangent line at $x = 0$. For any positive value of h, the difference quotient $[f(0+h) - f(0)]/h$ equals $+1$; for any negative value of h, the difference quotient is -1. Therefore,

$$\lim_{h \to 0} \frac{f(0+h) - f(0)}{h}$$

does not exist, and consequently there can be no tangent line at $x = 0$. Intuitively, one should not expect a unique tangent line to exist at a corner, and indeed it does not exist in a mathematical sense either.

Symbolic Flowchart In Figure 3-8 we have drawn a symbolic flowchart for the process of obtaining an equation of the line which has slope and is tangent to the graph of $f(x)$ at $x = x_1$. The steps in this process are summarized as follows:

1. Evaluate $f(x_1)$ and $f(x_1 + h)$ for $x_1 + h$ near x_1. This is represented by Block 2.
2. Evaluate
$$m(x_1) = \lim_{h \to 0} \frac{f(x_1 + h) - f(x_1)}{h}$$
if it exists. If this limit exists, the equation of the line is
$$y - f(x_1) = m(x_1) \cdot (x - x_1)$$
3. If $\lim_{h \to 0} [f(x_1 + h) - f(x_1)]/h$ does not exist, then the slope does not exist and there is no tangent line with slope.

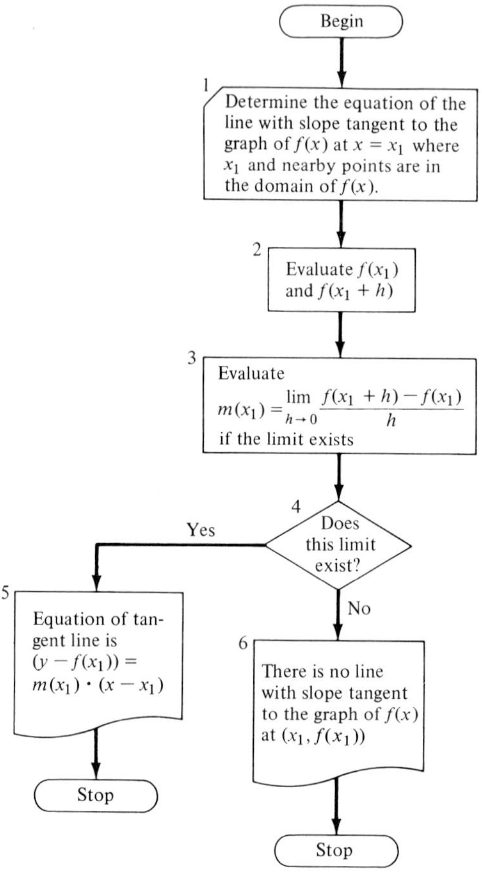

FIGURE 3-8

EXERCISES 3-1

For Exercises 1–10, consider the functions to be defined for all x for which the denominator, if any, is not zero. Determine **(a)** the difference quotient for the indicated interval; **(b)** the slope of the tangent line at the left end point of the indicated interval; and **(c)** the equation for this tangent line.

1. $f(x) = 3x^2 - 1$ $\quad [1, 2]$

2. $f(x) = 2 - x^2$ $\quad [-2, 0]$

3. $f(x) = 1 - x - x^2$ $\quad [-2, -2 + h]$

4. $f(x) = x^2 + 3x - 4$ $\quad [1, 1 + h]$

5. $f(x) = \frac{1}{x}$ $\quad [2, 4]$

6. $f(x) = \frac{1}{x^2}$ $\quad [1, 2]$

7. $f(x) = \frac{x + 2}{x - 1}$ $\quad \left[0, \frac{1}{2}\right]$

8. $f(x) = x|x|$ $\quad [-1, 0]$

NOTE: For $x \leq 0$, $x|x| = -x^2$.

9. $f(x) = 3x^3$ $\quad [1, 2]$

NOTE: $(x + h)^3 = x^3 + 3x^2h + 3xh^2 + h^3$.

10. $f(x) = 2x^3 - 1$ $\quad [0, 2]$

NOTE: Use the formula in Exercise 9.

One approach to approximating the slope of the tangent line is to observe that for h small, the limit is approximated by ($\approx$) the difference quotient.

$$\text{slope} = \lim_{h \to 0} \frac{f(x + h) - f(x)}{h}$$

$$\approx \frac{f(x + h) - f(x)}{h} \quad \text{(the difference quotient)}$$

To illustrate, let $f(x) = x^2 - 2$ and consider the case when $x = 1$. If we choose $h = 0.01$, we have

$$f(1 + h) = f(1.01) = (1.01)^2 - 2 = -0.9799$$

$$f(1) = 1^2 - 2 = -1$$

Then

$$\text{slope of tangent} \approx \frac{f(1 + h) - f(1)}{h} = \frac{(-0.9799) - (-1)}{0.01}$$

$$= \frac{0.0201}{0.01} = 2.01$$

The result is very close to the value 2 obtained by evaluation of the limit.

For Exercises 11 through 16, (**a**) approximate the slope at the indicated point, using the specified value of h, (**b**) evaluate the slope by taking the limit of the difference quotient as h approaches zero, and (**c**) determine the difference between the answers to parts (a) and (b).

11. $f(x) = \frac{1}{x}$ at $x = 2$; $h = 0.01$ NOTE: $\frac{1}{2.01} \approx 0.497512$

12. $f(x) = \frac{1}{x^2}$ at $x = 1$; $h = 0.1$ NOTE: $\frac{1}{(1.1)^2} \approx 0.826446$

13. $f(x) = \frac{1}{x^2}$ at $x = 1$; $h = -0.01$ NOTE: $\frac{1}{(0.99)^2} \approx 1.020304$

14. $f(x) = \sqrt{x}$ at $x = 4$; $h = 0.01$ NOTE: $\sqrt{4.01} \approx 2.002498$

15. $f(x) = 3x^3$ at $x = 1$; $h = 0.001$ NOTE: $(1.001)^3 \approx 1.003003$

16. $f(x) = 3x^3$ at $x = 1$; $h = -0.001$ NOTE: $(0.999)^3 \approx 0.997003$

3-2. THE DERIVATIVE

In the previous section we determined the slope $m(x_1)$ of the line tangent to the graph of $f(x)$ at $x = x_1$, using the limit

$$m(x_1) = \lim_{h \to 0} \frac{f(x_1 + h) - f(x_1)}{h}$$

provided that the limit exists. In our examples and in the exercises, we only considered specific numerical values for x_1. However, it is usually possible to take the limit for an arbitrary point x_1, as illustrated by the following example.

Example 1. Obtain the slope of the line tangent to $f(x) = x^2$ for any point $x = x_1$. To set up the difference quotient, we need $f(x_1)$ and $f(x_1 + h)$. We have

$$f(x_1) = x_1^2 \qquad \text{and} \qquad f(x_1 + h) = (x_1 + h)^2 = x_1^2 + 2x_1 h + h^2$$

We now form the difference quotient and then rewrite in an alternate form valid whenever $h \neq 0$:

$$\begin{aligned} \frac{f(x_1 + h) - f(x_1)}{h} &= \frac{(x_1^2 + 2x_1 h + h^2) - x_1^2}{h} \\ &= \frac{2x_1 h + h^2}{h} \\ &= 2x_1 + h \end{aligned}$$

We now take the limit:

$$m(x_1) = \lim_{h \to 0} \frac{f(x_1 + h) - f(x_1)}{h} = \lim_{h \to 0} (2x_1 + h) = 2x_1$$

For any point x_1, the slope of the line tangent to the graph of $f(x) = x^2$ is $2x_1$. For $x_1 = 2$, the slope is 4; for $x_1 = 5$, the slope is 10.

The concept of the tangent line leads directly to that of the derivative, which is simply the *slope of the tangent line*. The formal definition of the derivative is as follows.

Definition 3-1 The derivative of the function $f(x)$ at the point x_1 is

$$\lim_{h \to 0} \frac{f(x_1 + h) - f(x_1)}{h}$$

provided that the limit exists.

For notational purposes the derivative of a function $f(x)$ is generally denoted by

$$f'(x), \quad \frac{d}{dx}[f(x)], \quad \frac{df}{dx}, \quad \frac{df(x)}{dx}, \qquad \text{or } D_x f(x)$$

For the derivative at the point x_1, we write

$$f'(x_1) = \lim_{h \to 0} \frac{f(x_1 + h) - f(x_1)}{h}$$

Because we sometimes write $y = f(x)$, the notation dy/dx or y' is also frequently used to denote the derivative.

If a function $f(x)$ has a derivative at $x = x_1$, the function is said to be *differentiable* at x_1. We frequently do not bother with the use of x_1 and thus write

$$f'(x) = \lim_{h \to 0} \frac{f(x + h) - f(x)}{h}$$

where we regard x effectively as the constant x_1.

Using $f(x) = x^2$ as an example, we determined in Example 1 above that the slope of the tangent line at $x = x_1$ is $2x_1$. Thus the derivative $f'(x_1)$ equals $2x_1$, or as we shall customarily write, $f'(x) = 2x$.

If the derivative of a function $f(x)$ exists, it may customarily be obtained by substituting the formula for $f(x)$ directly into the definition of the derivative $f'(x)$ and evaluating the limit of the difference quotient $[f(x + h) - f(x)]/h$ as h approaches zero. However, putting $h = 0$ directly into the difference quotient gives

$$\frac{f(x) - f(x)}{0} = \frac{0}{0}$$

which is indeterminate. Therefore, we must seek an alternate form for $[f(x + h) - f(x)]/h$ in order to obtain the limit. The process for evaluating

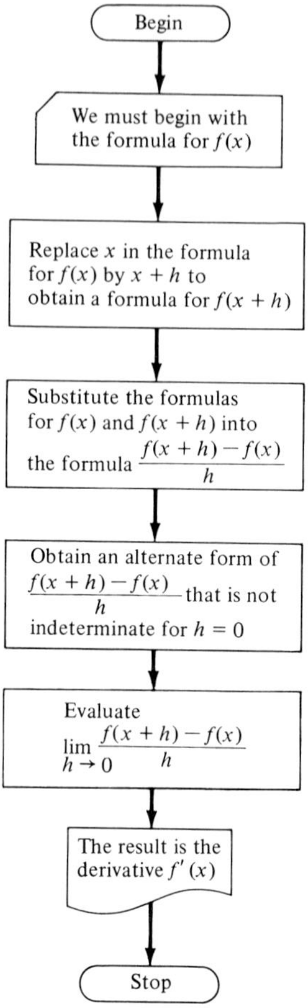

FIGURE 3-9

derivatives, if they exist, is represented by the symbolic flowchart in Figure 3-9 and is further illustrated by the following examples.

Example 2. $f(x) = x$.

We write $f(x+h)$:

$$f(x+h) = x + h$$

Then we form the difference quotient

$$\frac{f(x+h) - f(x)}{h} = \frac{(x+h) - x}{h}$$

Then we obtain the alternate form valid for $h \neq 0$

$$\frac{x + h - x}{h} = \frac{h}{h} = 1$$

and evaluate the limit

$$f'(x) = \lim_{h \to 0} (1) = 1$$

The derivative $f'(x)$ is 1 for each value of x.

Example 3. $f(x) = 1/x$.
We write $f(x + h)$:

$$f(x + h) = \frac{1}{x + h}$$

Then we form the difference quotient

$$\frac{f(x + h) - f(x)}{h} = \frac{\frac{1}{x + h} - \frac{1}{x}}{h}$$

and after simplifying obtain the desired alternate form valid for $h \neq 0$

$$\frac{\frac{1}{x + h} - \frac{1}{x}}{h} = \frac{\frac{x}{x(x + h)} - \frac{x + h}{x(x + h)}}{h} = \frac{x - (x + h)}{x(x + h)h} = \frac{-1}{x(x + h)}$$

Evaluating the limit, we find that

$$f'(x) = \lim_{h \to 0} \frac{-1}{x(x + h)} = -\frac{1}{x^2}$$

Since $f(x)$ and $f'(x)$ exist for all x except $x = 0$, we say that $f(x) = 1/x$ is differentiable for all $x \neq 0$.

Example 4. $f(x) = \sqrt{x}$, $x \geq 0$.
We write $f(x + h)$:

$$f(x + h) = \sqrt{x + h}$$

and form the difference quotient

$$\frac{f(x + h) - f(x)}{h} = \frac{\sqrt{x + h} - \sqrt{x}}{h}$$

Next we obtain the alternate form valid for $h \neq 0$ (using an algebraic trick):

$$\frac{\sqrt{x+h}-\sqrt{x}}{h}=\frac{\sqrt{x+h}-\sqrt{x}}{h}\cdot\frac{\sqrt{x+h}+\sqrt{x}}{\sqrt{x+h}+\sqrt{x}}$$

Since

$$(\sqrt{x+h}-\sqrt{x})\cdot(\sqrt{x+h}+\sqrt{x})=(x+h)-x,$$

we have

$$\frac{\sqrt{x+h}-\sqrt{x}}{h}=\frac{(x+h)-x}{h\cdot(\sqrt{x+h}+\sqrt{x})}=\frac{1}{\sqrt{x+h}+\sqrt{x}}$$

and then evaluate the limit:

$$f'(x)=\lim_{h\to 0}\frac{1}{(\sqrt{x+h}+\sqrt{x})}$$

$$=\frac{1}{\sqrt{x}+\sqrt{x}}=\frac{1}{2\sqrt{x}}$$

The derivative $f'(x)$ exists for all $x>0$. It is not defined for $x=0$.

In defining the derivative of $f(x)$ at $x=a$, we used $\lim_{h\to 0}[f(a+h)-f(a)]/h$ as the derivative. If the limit exists, then for small values of h, the difference quotient $[f(a+h)-f(a)]/h$ is approximately equal to

$$\lim_{h\to 0}\frac{f(a+h)-f(a)}{h}=f'(a)$$

or in other words,

$$\frac{f(a+h)-f(a)}{h}\approx f'(a)$$

Multiplying by h, we get

$$f(a+h)-f(a)\approx hf'(a)$$

Since $f'(a)$ is a fixed number, we note that if h is small, then $f(a+h)-f(a)$ is small; therefore, $f(a+h)\approx f(a)$. In effect, this is the condition that $f(x)$ be continuous at $x=a$. We state this fact explicitly.

Property 3-1 If $f(x)$ is differentiable at a, then $f(x)$ is continuous at a.

The converse of this property is not true. For example, the graph of $f(x) = |x|$ (refer to Figure 3-7 in the previous section) is continuous at $x = 0$, but the derivative does not exist there.

The derivative $f'(x)$ is itself a function, its domain being the set of all values of x for which $f(x)$ is differentiable. In Example 4 it was shown that if $f(x) = \sqrt{x}$, then $f'(x) = 1/(2\sqrt{x})$ for $x > 0$. From the graph of $f'(x)$, in Figure 3-10, it should be apparent that the derivative satisfies all requirements to be a function whose domain is the set of all positive x.

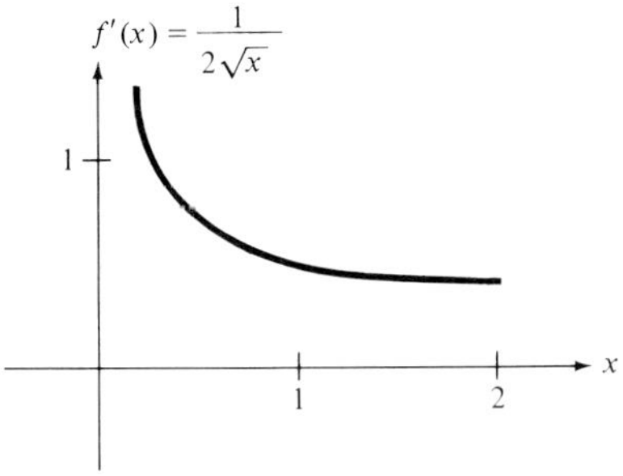

FIGURE 3-10

Although we shall virtually always apply the definition of the derivative with the limit expressed as

$$f'(x) = \lim_{h \to 0} \frac{f(x+h) - f(x)}{h}$$

there are other equivalent expressions of the limit that are in common use. Suppose that we let $x = a$ and $x + h = b$. Then $h = b - a$, and $h \to 0$ is equivalent to $b \to a$. Substituting into the previous formula for the limit gives

$$f'(a) = \lim_{b \to a} \frac{f(b) - f(a)}{b - a}$$

Several texts use this as the form for the definition of the derivative, but we shall use it only occasionally.

To solve certain problems, it is necessary to obtain a formula that gives the derivative $f'(x)$ at any point x at which $f(x)$ is differentiable. As determined earlier, the derivative of $f(x) = 1/x$ is $f'(x) = -1/x^2$ for all $x \neq 0$. In other problems it may only be necessary to obtain the derivative at a designated point, say $x = c$. We shall generally designate the derivative at a specific point c as $f'(c)$, although sometimes we also use notation such as

$$\left.\frac{df(x)}{dx}\right|_{x=c} \qquad \text{or} \qquad \left.\frac{df}{dx}\right|_{x=c}$$

to denote the derivative of f evaluated at c. As a specific example, the derivative of $f(x) = 1/x$ at $x = 2$ is

$$-\frac{1}{x^2}\bigg|_{x=2} = f'(2) = -\frac{1}{4}$$

Earlier in this section, we presented a symbolic flowchart of the process for obtaining derivatives. When it is necessary only to evaluate $f'(x)$ at a specific value of x, the computer can be used to calculate an approximation to the derivative at this point. To obtain the approximation, we recall that the definition of the derivative is

$$f'(x) = \lim_{h\to 0} \frac{f(x+h) - f(x)}{h}$$

The computer cannot be applied directly because the computer cannot evaluate limits. However, for values of h close to zero, the difference quotient $[f(x+h) - f(x)]/h$ is very close to the limit of the difference quotient

$$\lim_{h\to 0} \frac{f(x+h) - f(x)}{h}$$

Consequently

$$f'(x) \approx \frac{f(x+h) - f(x)}{h}$$

for h close to zero where $\approx$ is read "is approximately equal to."

To illustrate, in Example 3 earlier in this section, we concluded that for $f(x) = 1/x$, $f'(x) = -(1/x^2)$. Therefore, $f'(2) = -\frac{1}{4} = -0.25$. Using the approximation with $h = 0.1$, we have (to six decimal places)

$$f'(2) \approx \frac{\dfrac{1}{2.1} - \dfrac{1}{2}}{0.1} = -0.238095$$

With $h = 0.01$, we have (to six decimal places)

$$f'(2) \approx \frac{\dfrac{1}{2.01} - \dfrac{1}{2}}{0.01} = -0.248756$$

In general, as h becomes smaller, the approximation becomes more accurate.

For relatively simple functions, a hand calculator is entirely adequate. For more complicated functions, either a programmable calculator or a computer is desirable. The flowchart for the computer program is illustrated in Figure 3-11.

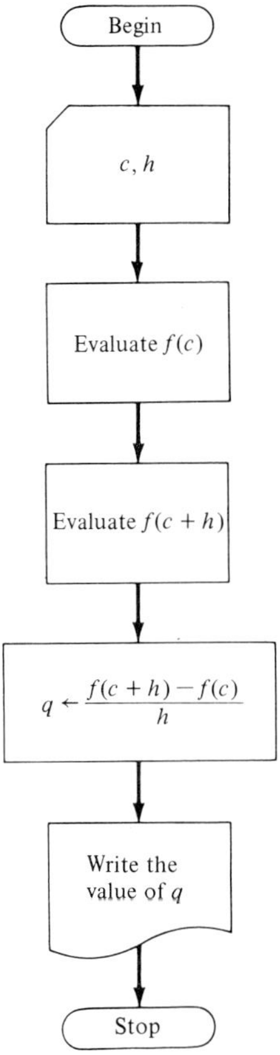

FIGURE 3-11

EXERCISES 3-2

For each of the following problems, **(a)** use the definition of the derivative to obtain $f'(x)$, and **(b)** state where the derivative is defined.

1. $f(x) = x + 1$

2. $f(x) = 1 - \dfrac{1}{x}$

3. $f(x) = x^3$

4. $f(x) = \dfrac{1}{x + 1}$

5. $f(x) = \dfrac{1}{x^2}$

6. $f(x) = \sqrt{x + 4}$ (Use a trick like that used in Example 4.)

7. $f(x) = \dfrac{1}{\sqrt{x}}$ (Use a trick like that used in Example 4.)

8. $f(x) = (x + 1)^2$

9. $f(x) = \dfrac{1}{(x - 1)^2}$

10. $f(x) = \dfrac{x}{x + 2}$

11. $f(x) = \dfrac{x + 2}{x}$

12. $f(x) = x|x|$

HINT: Note that for $x \geq 0$, $f(x) = x^2$; for $x < 0$, $f(x) = -x^2$.

For the following functions, determine (**a**) the derivative as a function of x, (**b**) the slope of the tangent line at the indicated value of x, and (**c**) an equation for that tangent line.

13. $x^2 + x$ at $x = 2$

14. $\dfrac{8}{x}$ at $x = 1$

15. $\dfrac{1}{x + 3}$ at $x = -2$

16. $\dfrac{2}{x - 1}$ at $x = 0$

17. $x + \dfrac{3}{x}$ at $x = 2$

For each of the following problems, (**a**) use the definition of the derivative to obtain $f'(x)$, (**b**) evaluate the derivative at the indicated value of x, and (**c**) for this value of x, determine the difference quotient as an approximation to the derivative for $h = 0.1$ and $h = 0.01$, and compare to the result of Step b.

For example, for parts (b) and (c), if $f(x) = x^2$, then $f'(x) = 2x$. At $x = 3$,

$$f'(x)\bigg|_{x=3} = 2 \cdot 3 = 6$$

For $x = 3$ and $h = 0.1$,

$$\frac{f(3 + 0.1) - f(3)}{0.1} = \frac{(3.1)^2 - 3^2}{0.1} = \frac{0.61}{0.1} = 6.1$$

18. $f(x) = x^2 - 1;\quad x = 2; f(2.1) = 3.41; f(2.01) = 3.0401$

19. $f(x) = \dfrac{1}{x - 1};\quad x = 0;\ f(0.1) \approx -1.1111;\ f(0.01) \approx -1.0101$

20. $f(x) = \dfrac{1}{(x + 1)^2};\ x = 1;\ f(1.1) \approx 0.2268;\ f(1.01) \approx 0.2475$

21. $f(x) = \dfrac{x+2}{x-2}$; $\quad x = -1;\ f(-0.9) \approx -0.3793;\ f(-0.99) \approx -0.3378$

22. $f(x) = \dfrac{(x+1)^2}{x}$; $x = 1;\ f(1.1) \approx 4.0091;\ f(1.01) \approx 4.0001$

3-3. DIFFERENTIATION FORMULAS

Although any derivative can be obtained by direct application of the definition, the process can be long and tedious, especially for functions such as $f(x) = (2x^4 - x)/(x^2 + 1)$. The process can be simplified by using differentiation formulas to express the derivative of a complicated function in terms of derivatives of simpler functions. To illustrate, let $f(x) = x$ and $g(x) = \sqrt{x}$. In the examples considered in the previous section, we used the definition to obtain the derivatives $f'(x) = 1$ and $g'(x) = 1/(2\sqrt{x})$. By using differentiation formulas, we can express the derivatives of functions such as $8\sqrt{x} = 8g(x)$ or $x + \sqrt{x} = f(x) + g(x)$ in terms of $f'(x)$ and $g'(x)$.

In the rules that follow, we assume that $f(x)$ and $g(x)$ are differentiable functions, i.e., that $f'(x)$ and $g'(x)$ exist, and we use the elementary properties of limits from chapter 2.

RULE 1 The derivative of a constant times a function is the constant times the derivative of the function:

$$\frac{d}{dx}[kf(x)] = kf'(x)$$

where k is a constant. This rule is obtained via the definition of the derivative as follows:

$$\begin{aligned}
\frac{d}{dx}[k\,f(x)] &= \lim_{h \to 0} \frac{k\,f(x+h) - k\,f(x)}{h} \\
&= \lim_{h \to 0} k\,\frac{f(x+h) - f(x)}{h} \\
&= k \lim_{h \to 0} \frac{f(x+h) - f(x)}{h} = kf'(x)
\end{aligned}$$

For example, this rule permits us to differentiate functions such as $7x$:

$$\frac{d}{dx}(7x) = 7\frac{d}{dx}(x) = 7 \cdot 1 = 7$$

RULE 2 (Sum Rule) The derivative of a sum of two functions is the sum of the derivatives:

$$\frac{d}{dx}[f(x) + g(x)] = f'(x) + g'(x)$$

This rule is obtained as follows:

$$\begin{aligned}
&\frac{d}{dx}[f(x) + g(x)] \\
&= \lim_{h\to 0} \frac{f(x+h) + g(x+h) - f(x) - g(x)}{h} \\
&= \lim_{h\to 0} \left\{\frac{f(x+h) - f(x)}{h} + \frac{g(x+h) - g(x)}{h}\right\} \\
&= \lim_{h\to 0} \frac{f(x+h) - f(x)}{h} + \lim_{h\to 0} \frac{g(x+h) - g(x)}{h} \\
&= f'(x) + g'(x)
\end{aligned}$$

This rule is illustrated by the following examples:

$$\frac{d}{dx}[x + 4] = \frac{d}{dx}(x) + \frac{d}{dx}(4) = 1 + 0 = 1$$

$$\frac{d}{dx}[3x - 2] = \frac{d}{dx}(3x) + \frac{d}{dx}(-2) = 3 + 0 = 3$$

The formula for the derivative of the difference of two functions follows from Rules 1 and 2:

$$\begin{aligned}
\frac{d}{dx}[f(x) - g(x)] &= \frac{d}{dx}[f(x)] + \frac{d}{dx}[(-1)\cdot g(x)] \\
&= f'(x) - g'(x)
\end{aligned}$$

RULE 3 (Product Rule) The derivative of the product of two functions is as follows:

$$\frac{d}{dx}[f(x)\,g(x)] = f(x)\,g'(x) + g(x)\,f'(x)$$

To obtain this rule, we begin by applying the definition of the derivative:

$$\frac{d}{dx}[f(x)\,g(x)] = \lim_{h\to 0} \frac{f(x+h)\,g(x+h) - f(x)\,g(x)}{h}$$

We next subtract and add $f(x+h)\,g(x)$ in the numerator:*

$$\frac{d}{dx}[f(x)g(x)]$$

$$= \lim_{h\to 0} \frac{f(x+h)g(x+h) - f(x+h)g(x) + f(x+h)g(x) - f(x)g(x)}{h}$$

$$= \lim_{h\to 0} \left\{ f(x+h)\frac{g(x+h)-g(x)}{h} + g(x)\frac{f(x+h)-f(x)}{h} \right\}$$

$$= \lim_{h\to 0} f(x+h) \lim_{h\to 0} \frac{g(x+h)-g(x)}{h}$$

$$+ g(x) \lim_{h\to 0} \frac{f(x+h)-f(x)}{h}$$

$$= f(x)g'(x) + g(x)f'(x)$$

This rule is illustrated by the following example:

$$\frac{d}{dx}[x(2x-3)] = x\frac{d}{dx}(2x-3) + (2x-3)\frac{d}{dx}(x)$$

$$= x \cdot 2 + (2x-3) \cdot 1 = 4x - 3$$

RULE 4 (Power Rule) The derivative of x^n is nx^{n-1} for any value of $n \neq 0$:

$$\frac{d}{dx}[x^n] = nx^{n-1}$$

We omit the proof but give the following examples:

$$\frac{d}{dx}(x^5) = 5x^4 \qquad (\textit{Note}: n = 5;\ n - 1 = 4)$$

$$\frac{d}{dx}(x^{-3}) = -3x^{-4} \qquad (\textit{Note}: n = -3;\ n - 1 = -4)$$

$$\frac{d}{dx}(x^{2.7}) = 2.7x^{1.7} \qquad (\textit{Note}: n = 2.7;\ n - 1 = 1.7)$$

$$\frac{d}{dx}(\sqrt{x}) = \frac{d}{dx}(x^{1/2}) = \frac{1}{2}x^{-1/2} = \frac{1}{2\sqrt{x}} \qquad (\textit{Note}: n = \tfrac{1}{2})$$

The combination of this rule with previous rules permits many complicated functions to be easily differentiated. In

* We do this because it gets us what we want. We are at liberty to add and subtract any expression we wish as long as we do not change the values of the function in the process.

the following example, we first apply Rule 2 on sums and then Rule 4 on powers of x.

$$\frac{d}{dx}[x^4 + x^2 + 1] = \frac{d}{dx}(x^4) + \frac{d}{dx}(x^2) + \frac{d}{dx}(1)$$
$$= 4x^3 + 2x + 0 = 4x^3 + 2x$$

In the following example, we first apply the product rule:

$$\frac{d}{dx}[(x^4 - 1)(x^2 + x)]$$
$$= \left[(x^4 - 1)\frac{d}{dx}(x^2 + x)\right] + \left[(x^2 + x)\frac{d}{dx}(x^4 - 1)\right]$$
$$= (x^4 - 1)(2x + 1) + (x^2 + x)(4x^3)$$
$$= 2x^5 - 2x + x^4 - 1 + 4x^5 + 4x^4$$
$$= 6x^5 + 5x^4 - 2x - 1$$

An alternate approach to this example is to obtain the product and then differentiate.

$$\frac{d}{dx}[(x^4 - 1)(x^2 + x)] = \frac{d}{dx}[x^6 + x^5 - x^2 - x]$$
$$= 6x^5 + 5x^4 - 2x - 1$$

RULE 5 (Reciprocal Rule) The derivative of $1/[f(x)]$ is $-f'(x)/[f(x)]^2$:

$$\frac{d}{dx}\left[\frac{1}{f(x)}\right] = -\frac{f'(x)}{[f(x)]^2}, \qquad \text{for} \quad f(x) \neq 0$$

This rule is illustrated by the following examples:

$$\frac{d}{dx}\left[\frac{1}{x}\right] = -\frac{\frac{d}{dx}(x)}{x \cdot x} = -\frac{1}{x^2}$$

$$\frac{d}{dx}\left[\frac{1}{x^2 + x + 1}\right] = -\frac{\frac{d}{dx}[x^2 + x + 1]}{(x^2 + x + 1)^2} = -\frac{2x + 1}{(x^2 + x + 1)^2}$$

In the first example observe that we could have written $1/x$ as x^{-1} and then applied the power rule.

$$\frac{d}{dx}\left[\frac{1}{x}\right] = \frac{d}{dx}(x^{-1}) = (-1)x^{-2} = -\frac{1}{x^2}$$

Using either approach, we obtain the same result.

Rules 3 and 5 can be readily combined to give a formula for the differentiation of a quotient.

$$\begin{aligned}\frac{d}{dx}\left[\frac{f(x)}{g(x)}\right] &= \frac{d}{dx}\left[f(x)\frac{1}{g(x)}\right]\\ &= \frac{1}{g(x)}\frac{d}{dx}[f(x)] + f(x)\frac{d}{dx}\left[\frac{1}{g(x)}\right]\\ &= \frac{f'(x)}{g(x)} + f(x)\left(-\frac{g'(x)}{[g(x)]^2}\right)\end{aligned}$$

On getting the common denominator and simplifying, we have

$$\frac{d}{dx}\left[\frac{f(x)}{g(x)}\right] = \frac{f'(x)g(x) - g'(x)f(x)}{[g(x)]^2}, \qquad \text{for} \quad g(x) \neq 0$$

This rule is illustrated by the following example:

$$\begin{aligned}\frac{d}{dx}\left(\frac{x+1}{x-1}\right) &= \frac{(x-1)\dfrac{d}{dx}(x+1) - (x+1)\dfrac{d}{dx}(x-1)}{(x-1)^2}\\ &= \frac{(x-1)\cdot(1) - (x+1)\cdot(1)}{(x-1)^2}\\ &= \frac{-2}{(x-1)^2}\end{aligned}$$

When differentiating quotients, one can use either the quotient formula or use Rule 3 (product formula) and Rule 5 (reciprocals). The disadvantage of the quotient formula is that it may be hard to remember (notably which of the numerator terms is to be subtracted), but the result from the quotient formula is in a more useful form. The product formula and reciprocal formula are easy to remember, but additional algebraic operations (e.g., taking the common denominator) are usually necessary to obtain the result in the most desirable form.

REMARK: Although we have used functions of x for our formulas and illustrations, we could just as well have used functions of t or some other variable. Thus, for example, $d(t^2)/dt = 2t$ and $d[f(t) + g(t)]/dt = f'(t) + g'(t)$.

EXERCISES 3-3

Using the differentiation formulas, obtain the derivative with respect to t of the following functions of t:

1. $3t^3$

2. $t^4 - t$

3. $8t^{-10}$

4. $8t^2 - \dfrac{1}{t}$

5. $t^5 + 4t^4 - 2t^2 + 5$

6. $\dfrac{1}{t^3} + \dfrac{1}{t^2} + 2$

For the following functions of x, obtain the derivative with respect to x by **(a)** applying the product rule, and **(b)** multiplying the factors to obtain a sum and then differentiating. Your answers should check.

7. $(x + 2)(x + 3)$

8. $(x - 4)(4 - x)$

9. $(x^2 + 2x + 2)(x + 8)$

10. $(x^4 + 1)(x^4 - 2)$

11. $(3x - 2)(-18x^4 + 1)$

12. $(8x^4 + 2x^2 + 2)(x + x^{-1})$

For the following functions, obtain the derivative with respect to x by **(a)** using the reciprocal rule and the product rule, and **(b)** using the quotient rule. Your answers should check.

13. $\dfrac{x + 4}{4 - x}$

14. $\dfrac{x^2 + 2x + 5}{x}$

15. $\dfrac{x}{x^2 + 2x + 5}$

16. $\dfrac{16x^2 - 9}{x^3 + 1}$

17. $\dfrac{8 - x^2}{8 + x^2}$

18. $\dfrac{3x^2}{x^2 + 2x + 9}$

Using any appropriate combination of the differentiation formulas, obtain derivatives of the following:

19. $(x + 4)(x + 6)(x + 8)$

HINT: You may multiply the factors together to write as a sum or let $g(x) = (x + 4) \cdot (x + 6)$, and determine $g'(x)$. The appropriate process is $\dfrac{d}{dx}[g(x)(x + 8)] = (x + 8)g'(x) + g(x)\dfrac{d}{dx}(x + 8)$

20. $\dfrac{(x + 4)(x - 1)}{x + 2}$

21. $\dfrac{x + 1}{(x - 1)(x + 5)}$

22. $\dfrac{x + 1}{x - 2} + x^2$

23. Let $f'(x)$ and $g'(x)$ exist, let k_1 and k_2 be constants, and let

$$h(x) = k_1 f(x) + k_2 g(x)$$

Express $h'(x)$ in terms of k_1, k_2, $f'(x)$, and $g'(x)$.

24. Let $f'(x)$, $g'(x)$, and $h'(x)$ exist, and let

$$u(x) = f(x) + g(x) + h(x)$$

Express $u'(x)$ in terms of $f'(x)$, $g'(x)$, and $h'(x)$.

25. Let $f'(x)$, $g'(x)$, and $h'(x)$ exist, and let

$$u(x) = f(x)g(x)h(x)$$

Express $u'(x)$ in terms of the functions $f(x)$, $g(x)$, $h(x)$, and their derivatives.

3-4. THE DERIVATIVE IN BUSINESS ANALYSIS

In the opening section of this chapter, the derivative was given two interpretations:

1. A physical interpretation, namely, as a rate of change.
2. A geometrical interpretation, namely, as the slope of the tangent line to a graph.

We shall now illustrate these concepts in terms of business examples.

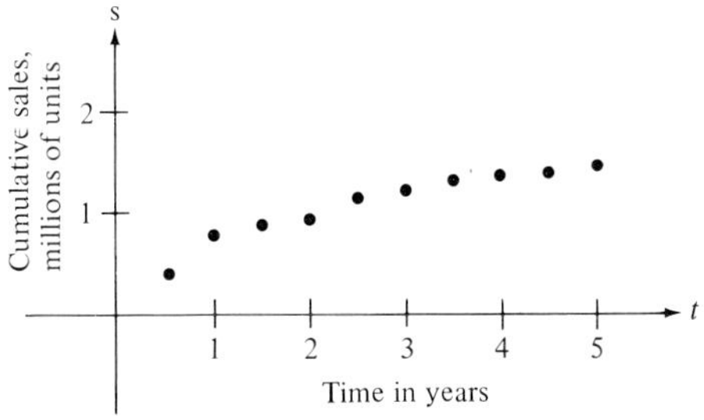

FIGURE 3-12

Rates from Cumulative Totals For product X, the cumulative or total sales since its introduction are given by the graph in Figure 3-12. Depending upon the situation, these sales might be reported daily, weekly, monthly, annually, or by some other appropriate interval of time. The graph of the function in

Figure 3-12 consists of finitely many points and is called a *discrete function.* Typical of many commercial products, sales are high when the product is first introduced, but they drop as competing products enter the marketplace.

In many applications, discrete functions are not especially convenient to use. In these cases the discrete function can be approximated by a continuous function of time as illustrated in Figure 3-13. Although beyond the

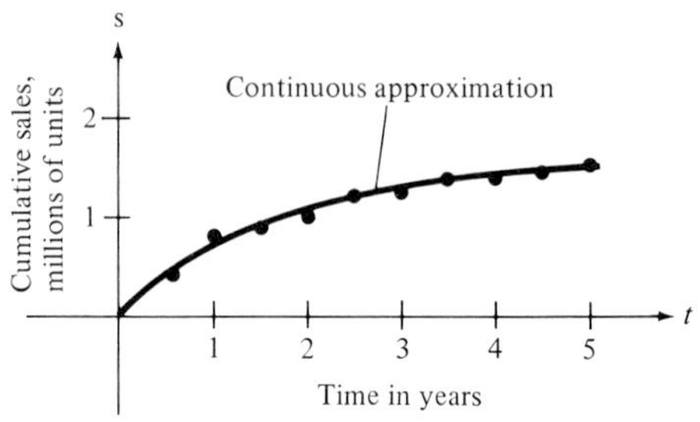

FIGURE 3-13

scope of this text, several methods are available to obtain an approximation in some " best " sense and to obtain a formula. For the graph in Figure 3-13, a continuous approximation might be given by the formula

$$s = f(t) = \frac{2t}{t+2}$$

with t given in years and s in units of a million.

Business decisions such as whether or not to continue to market the product may be based on the rate of sales. For example, if the cumulative sales for the first three years are approximated by the above formula, then the sales expected during the fourth year are given by

$$f(4) - f(3) = \frac{2 \cdot 4}{4+2} - \frac{2 \cdot 3}{3+2} = \frac{8}{6} - \frac{6}{5} = \frac{2}{15}$$

Therefore, we anticipate that the sales during the fourth year will be $\frac{2}{15}$ million.

From the shape of the graph in Figure 3-13, the rate of sales is higher at the beginning of the year than at the end. The instantaneous rate of sales, say r, is given by the derivative of the cumulative sales:

$$r = \frac{ds}{dt} = f'(t) = \frac{d}{dt}\left[\frac{2t}{t+2}\right] = \frac{4}{(t+2)^2}$$

Therefore, at the beginning of the fourth year, the rate of sales is given by

$$r = f'(3) = \frac{4}{(3+2)^2} = \frac{4}{25} \quad \text{million units per year}$$

At the end of year four, the rate of sales is given by

$$r = f'(4) = \frac{4}{(4+2)^2} = \frac{4}{36} = \frac{1}{9} \quad \text{million units per year}$$

Observe that $f'(3) = \frac{4}{25}$ is greater than the average rate of $\frac{2}{15} = \frac{4}{30}$ for the year, and that $f'(4) = \frac{1}{9} = \frac{4}{36}$ is less than the average rate for the year.

Suppose that the sale of each unit of the product returns \$2.90 above the manufacturing costs. However, advertising, warehousing, and other costs associated with offering the product for sale total \$350,000 per year. The optimum time to remove the product from the market is when the rate r of return from sales is at this value. Expressed mathematically, the break-even rate of return is given by

$$r \cdot (2.90) = 350{,}000$$

Substituting the formula for r and recalling that the formula gave the sales in units of a million, we have

$$\frac{4}{(t+2)^2} \cdot (1{,}000{,}000) \cdot (2.90) = 350{,}000$$

Rearranging yields

$$(t+2)^2 = \frac{(4) \cdot (1{,}000{,}000) \cdot (2.90)}{350{,}000} = 33.14$$

Taking the square root, we obtain

$$t + 2 = 5.76$$

or

$$t = 3.76$$

Therefore, the product should be discontinued after the third quarter of the fourth year.

Marginal Cost The marginal cost of a product is the cost of producing the $(n+1)$st unit after the nth unit has been produced. Let $C(x)$ be the cost of producing x units, as illustrated by the graph in Figure 3-14. The marginal cost is $C(n+1) - C(n)$.

In many business situations, the product is sold only in discrete units, and thus the function $C(x)$ is defined only for integer values of x. However, as discussed earlier, in practice this discrete function is often approximated by a continuous function, which would be defined when x is any positive real number. The use of continuous and, in fact, differentiable functions lets

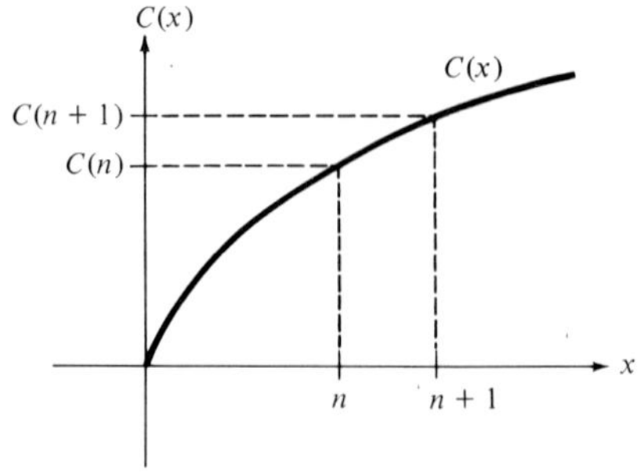

FIGURE 3-14

us utilize the derivative to describe real life phenomena. If we thought only in terms of discrete functions and integer values for x, the analysis would be much more complicated.

For most products, production costs decrease with volume, which means that the cost to produce unit $(n + 1)$ is less than the cost to produce unit n. This accounts for the shape of the graph in Figure 3-14.

When n is large, the marginal cost can be found using the derivative. From the definition of the derivative, we have

$$C'(x) = \lim_{h \to 0} \frac{C(x + h) - C(x)}{h}$$

When h is small compared to x, the limit can be approximated by the difference quotient.

$$\lim_{h \to 0} \frac{C(x + h) - C(x)}{h} \approx \frac{C(x + h) - C(x)}{h}$$

or

$$C'(x) \approx \frac{C(x + h) - C(x)}{h}$$

To relate this to the marginal cost $C(n + 1) - C(n)$, let $x = n$ and $h = 1$. Substituting into the approximation for $C'(x)$, we have

$$C'(n) \approx \frac{C(n + 1) - C(n)}{1} = C(n + 1) - C(n)$$

Therefore, for n large, the marginal cost $C(n + 1) - C(n)$ is approximately equal to the derivative $C'(n)$, and, in fact, economists now frequently define the marginal cost to be the derivative of the cost function.

To illustrate, assume that the cost to produce x units is given by

$$C(x) = 1 + \sqrt{x}$$

The marginal cost is

$$C(n+1) - C(n) = (1 + \sqrt{n+1}) - (1 + \sqrt{n})$$
$$= \sqrt{n+1} - \sqrt{n}$$

The derivative $C'(n)$ is

$$C'(n) = \frac{d}{dn}(1 + \sqrt{n}) = \frac{d}{dn}(\sqrt{n}) = \frac{1}{2\sqrt{n}}$$

When n is large, we expect these two results to be approximately equal. To see that this is true, we apply the same mathematical trick we used in Example 4 of Section 3-2:

$$C(n+1) - C(n) = (\sqrt{n+1} - \sqrt{n}) \cdot \frac{\sqrt{n+1} + \sqrt{n}}{\sqrt{n+1} + \sqrt{n}}$$
$$= \frac{n+1-n}{\sqrt{n+1} + \sqrt{n}} = \frac{1}{\sqrt{n+1} + \sqrt{n}}$$

For large values of n, this result is approximately $1/(2\sqrt{n})$, since $\sqrt{n+1}$ is approximately $\sqrt{n}$.

Marginal Revenue The marginal revenue is defined in the same manner as marginal cost. Specifically, the marginal revenue is the revenue obtained from the $(n + 1)$st unit after n units have been sold. That is, if $R(n)$ is the total revenue from selling n units, the marginal revenue is $R(n + 1) - R(n)$. As in the case for the marginal cost, the marginal revenue is approximated by the derivative $R'(n)$.

Marginal Profit The marginal profit is defined as the profit realized from the $(n + 1)$st unit after n units have been produced and sold. If $P(n)$ is the profit from the manufacture and sale of n units, the marginal profit is $P(n + 1) - P(n)$, which is approximated by $P'(n)$. Since profit is related to revenue and cost, marginal profit is related to marginal revenue and marginal cost. Specifically, profit is the revenue less the cost:

$$P(n) = R(n) - C(n)$$

The expression for marginal profit is obtained as follows:

$$P'(n) = \frac{d}{dn}[R(n) - C(n)] = R'(n) - C'(n)$$

Therefore, the marginal profit is the marginal revenue less the marginal cost.

EXERCISES 3-4

1. Since 1958, the enrollment at Podunk U. is approximated by the formula

$$e(t) = 7000 - 10(t^2 - 28t + 196)$$

where t represents the number of years since 1958. Determine the rate of growth of enrollment as a function of t.

2. Rework Exercise 1 using the following formula to approximate the enrollment:

$$e(t) = 10{,}000 + \frac{50{,}000}{t + 10}$$

3. Since 1940, the assets a of the Conservative National Bank are approximated by the function

$$a(t) = 2 + 0.1t - \frac{2}{t + 1}$$

where a is in millions of dollars and t is time in years since 1940. Determine the rate of growth in assets as an explicit function of t.

4. Rework Exercise 3 using the following approximation for the assets:

$$a(t) = 4 - \frac{t^2 - 40t + 400}{20(t + 20)}$$

5. For a product Z, the cost and revenue functions are as follows:

$$C(q) = \left(q + 1 - \frac{1}{q + 1}\right)$$

$$R(q) = 2q$$

where q is the quantity of Z that is sold. Determine the marginal cost, marginal revenue, and marginal profit.

6. Rework Exercise 5 for the following cost and revenue functions:

$$C(q) = 1 + \sqrt{q}$$

$$R(q) = \frac{5q}{q + 1}$$

7. Let the sales q for product Y and the price p of Y be related by the equation

$$p = \frac{1}{q + 1}$$

The revenue R from the sales of that product is given by

$$R = pq \qquad \text{or} \qquad R = \frac{1}{q+1} \cdot q$$

Obtain an expression for the marginal revenue, dR/dq.

8. Rework Exercise 7 when p is related to q by $p = q^2$.

9. Let the sales q of product Y and the price p of Y be related by the equation

$$q = \frac{1-p}{p}$$

The so-called *elasticity of demand* e is given by the equation

$$e = -\frac{p}{q}\frac{dq}{dp}$$

(**a**) Determine the value of e at a selling price of $\frac{1}{2}$.
(**b**) Determine e as an explicit function of p.

10. Rework Exercise 9 when q is related to p by

$$q = \sqrt{p}$$

11. A sensitivity function relates the change in one quantity to the change in another. For example, the number s of sales for a given product is a function of the selling price p. A manager often wants to know how "sensitive" the number of sales is to the selling price. Expressed another way, he wants to know how much the number of sales will change if he changes the selling price a small amount. The answer is approximated by the derivative ds/dp, and the result is often called a sensitivity function. Determine the sensitivity ds/dp if s is related to p by

$$s = \frac{4}{p+1}$$

12. The sales s of product Y is related to the daily advertising expenditure a in dollars by the formula

$$s = 4000\,\frac{a}{a+100}$$

Determine the sensitivity (see Exercise 11) of sales to advertising.

13. The demand s in pounds per week for steak is related to an individual's personal annual income i (in thousands of dollars per year) by the formula

$$s = \sqrt{\frac{i}{1000}}$$

Determine the sensitivity (see Exercise 11) of s to i.

3-5. THE CHAIN RULE

The concept of composite functions and a relationship known as the chain rule are useful in obtaining some derivatives. If y is a function of u and u is a function of x, then y is also a function of x. If $y = g(u)$ and $u = u(x)$, then $y = f(x)$ is considered to be a *composite* of the functions g and u, or we say that y is a *composition* of g and u.

To illustrate, let y be a function of x as represented by the formula

$$y = f(x) = (x^2 + 3x)^4$$

An alternate expression of this relationship is as a composition of the two functions

$$y = g(u) = u^4$$

$$u = u(x) = x^2 + 3x$$

The composition of these two functions gives y as a function of x:

$$y = f(x) = g(u(x)) = u^4 = (x^2 + 3x)^4$$

The problem of concern in this section is obtaining the derivative dy/dx. None of the differentiation formulas introduced in previous sections directly apply to the derivative in the form

$$\frac{dy}{dx} = f'(x) = \frac{d}{dx}(x^2 + 3x)^4$$

On the other hand, the derivatives dg/du and du/dx are readily obtained.

$$\frac{dy}{du} = g'(u) = \frac{d}{du}(u^4) = 4u^3$$

$$\frac{du}{dx} = u'(x) = \frac{d}{dx}(x^2 + 3x) = 2x + 3$$

Since $y(x)$ can be represented as a composition of the functions $g(u)$ and $u(x)$, it seems reasonable to suspect that dy/dx is somehow related to the derivatives dy/du and du/dx. This relationship, generally referred to as the chain rule, is as follows:

$$\frac{dy}{dx} = \frac{dy}{du} \cdot \frac{du}{dx}$$

The derivatives described above can be substituted directly:

$$\frac{dy}{dx} = \frac{dy}{du} \cdot \frac{du}{dx} = (4u^3)(2x + 3)$$

where $u = u(x) = x^2 + 3x$. To obtain dy/dx as an explicit function of x, we must substitute $x^2 + 3x$ for u:

$$\frac{dy}{dx} = 4(x^2 + 3x)^3(2x + 3)$$

Let us briefly explain why the chain rule is true. Observe that the derivatives dy/dx, dy/du, and du/dx are interpreted as follows:

$$\frac{dy}{dx} \approx \frac{\text{change in } y}{\text{change in } x}$$

$$\frac{dy}{du} \approx \frac{\text{change in } y}{\text{change in } u}$$

$$\frac{du}{dx} \approx \frac{\text{change in } u}{\text{change in } x}$$

With this interpretation, the product $(dy/du) \cdot (du/dx)$ is

$$\frac{dy}{du} \cdot \frac{du}{dx} \approx \frac{\text{change in } y}{\text{change in } u} \cdot \frac{\text{change in } u}{\text{change in } x} = \frac{\text{change in } y}{\text{change in } x}$$

since "change in u" appears in numerator and denominator. Since

$$\frac{\text{change in } y}{\text{change in } x} \approx \frac{dy}{dx}$$

it is plausible that

$$\frac{dy}{du} \cdot \frac{du}{dx} = \frac{dy}{dx}$$

which is the chain rule.

We shall now illustrate the use of the chain rule with three examples.

Example 1. Consider $d(\sqrt{x^2 + 1})/dx$. If we let $u = x^2 + 1$, we have

$$\frac{d}{dx}\sqrt{x^2 + 1} = \frac{d}{dx}(\sqrt{u})$$

We now apply the chain rule, observing that

$$\frac{d}{dx}(\sqrt{u}) = \frac{d}{du}(\sqrt{u}) \cdot \frac{du}{dx}$$

Therefore, we have

$$\frac{d}{dx}\sqrt{x^2+1} = \frac{d}{du}(\sqrt{u})\frac{du}{dx}$$

The derivative of $\sqrt{u}$ is obtained by the power rule:

$$\frac{d}{du}(\sqrt{u}) = \frac{d}{du}(u^{1/2}) = \frac{1}{2}u^{-1/2} = \frac{1}{2\sqrt{u}}$$

Since $u = x^2 + 1$, we have

$$\frac{du}{dx} = \frac{d}{dx}(x^2+1) = 2x$$

We now substitute $d(\sqrt{u})/du$ and du/dx into the expression for $d(\sqrt{x^2+1})/dx$:

$$\frac{d}{dx}\sqrt{x^2+1} = \frac{d}{du}(\sqrt{u}) \cdot \frac{du}{dx} = \frac{1}{2\sqrt{u}} \cdot 2x = \frac{x}{\sqrt{u}}$$

To obtain an explicit function of x, we substitute $x^2 + 1$ for u:

$$\frac{d}{dx}\sqrt{x^2+1} = \frac{x}{\sqrt{x^2+1}}$$

Example 2. Consider

$$\frac{d}{dx}\left[\frac{1}{\sqrt{x}+1}\right]$$

To attack this problem, we could either apply the quotient rule or use the chain rule. We shall use the chain rule, although the reader may wish to verify that the same result is obtained with the quotient rule.

To apply the chain rule, we write the derivative as follows:

$$\frac{d}{dx}\left[\frac{1}{\sqrt{x}+1}\right] = \frac{d}{dx}(\sqrt{x}+1)^{-1}$$

If we let $u = \sqrt{x} + 1$, we have, with the use of the chain rule,

$$\frac{d}{dx}\left[\frac{1}{\sqrt{x}+1}\right] = \frac{d}{dx}(u^{-1}) = \frac{d}{du}(u^{-1})\frac{du}{dx}$$

Since $u = \sqrt{x} + 1$,

$$\frac{du}{dx} = \frac{d}{dx}(\sqrt{x} + 1) = \frac{1}{2\sqrt{x}}$$

Furthermore,

$$\frac{d}{du}(u^{-1}) = -\frac{1}{u^2}$$

Substitution gives

$$\frac{d}{dx}\left[\frac{1}{\sqrt{x} + 1}\right] = \frac{d}{du}(u^{-1})\frac{du}{dx} = \left(-\frac{1}{u^2}\right)\left(\frac{1}{2\sqrt{x}}\right)$$

Then, substituting $\sqrt{x} + 1$ for u gives the derivative as an explicit function of x,

$$\frac{d}{dx}\left[\frac{1}{\sqrt{x} + 1}\right] = \left(-\frac{1}{(\sqrt{x} + 1)^2}\right)\cdot\left(\frac{1}{2\sqrt{x}}\right) = -\frac{1}{2\sqrt{x}(\sqrt{x} + 1)^2}$$

Example 3. Consider dy/dx where $y = (x^2 + x + 5)^{10}$. The steps are the same as in the prior examples, so we shall do them without explanation. We choose $u = x^2 + x + 5$ and note that $du/dx = 2x + 1$:

$$\begin{aligned}\frac{d}{dx}(x^2 + x + 5)^{10} &= \frac{d}{dx}(u^{10}) \\ &= \frac{d}{du}(u^{10})\frac{du}{dx} \\ &= (10u^9)\cdot(2x + 1) \\ &= 10(x^2 + x + 5)^9 \cdot (2x + 1)\end{aligned}$$

In using the chain rule, one needs only to write as many of these steps as is convenient. The objective is to obtain the correct answer, not to memorize a lengthy sequence of steps. After some practice, one should be able to solve Example 3 by writing as follows:

$$\begin{aligned}\frac{d}{dx}(x^2 + x + 5)^{10} &= 10(x^2 + x + 5)^9\frac{d}{dx}(x^2 + x + 5) \\ &= 10(x^2 + x + 5)^9 \cdot (2x + 1)\end{aligned}$$

We do not explicitly write an intermediate function u but simply understand that it is used. In fact, anyone well versed in the use of the chain rule should write the answer for this example directly without any intermediate steps.

Although we shall always use the chain rule in the form

$$\frac{dy}{dx} = \frac{dy}{du} \cdot \frac{du}{dx}$$

students should learn to recognize its other forms. As $y = g(u)$, then $dy/du = g'(u)$ is a function of u, which can be eliminated by substituting $u = u(x)$. Some authors indicate this explicitly by writing

$$\frac{dy}{dx} = \frac{dy}{du}\bigg|_{u=u(x)} \cdot \frac{du}{dx}$$

Other authors note that $dy/du = g'(u)$ and $du/dx = u'(x)$, and write

$$\frac{dy}{dx} = g'(u) \cdot u'(x)$$

Since $u = u(x)$, others write

$$\frac{dy}{dx} = g'(u(x))u'(x)$$

All of these forms of the chain rule are equivalent, and the use of one over the other is merely a matter of preference.

In the above examples, we applied the chain rule to obtain the derivative of a function raised to a power. This process can be applied to obtain a formula for the differentiation of the function $[f(x)]^n$. Applying the chain rule, we find that

$$\begin{aligned} \frac{d[f(x)]^n}{dx} &= n[f(x)]^{n-1} \cdot \frac{df(x)}{dx} \\ &= n[f(x)]^{n-1}f'(x) \end{aligned}$$

Some authors extend the differentiation formulas presented in Section 3-3 to include this relationship.

EXERCISES 3-5

In Exercises 1–8, find **(a)** a formula giving dy/dx as an explicit function of x by use of the chain rule and then substitution, and **(b)** by substitution first and then, if necessary, implicit use of the chain rule without the symbol u. For example, if $y = u^2$ and $u = (x^4 + 1)$ then by (a), $dy/dx = (2u) \cdot (4x^3) = 8 \cdot (x^4 + 1)x^3$ and by (b) $y = (x^4 + 1)^2$ and thus $y' = 2(x^4 + 1) \cdot 4x^3 = 8(x^4 + 1)x^3$.

1. $y = u^3, \quad u = x^4 - 2x$

2. $y = u^{1/2}, \quad u = 1 - x^2$

3. $y = \dfrac{1}{u}, \qquad u = \dfrac{x}{x - 1}$

4. $y = \dfrac{1}{u^2 + 1}, \qquad u = \sqrt{x}$

5. $y = u^2 - u + 1, \qquad u = x^{1/2} + 1$

6. $y = 1 - u^3, \qquad u = x^{1/3} + 1$

7. $y = u^2 - \dfrac{1}{u^2}, \qquad u = \sqrt{x}$

8. $y = u^4 - 3u^2 - 1, \qquad u = x + 1$

In Exercises 9–16, find the derivatives of the given functions of t by whatever mathematical means you wish.

9. $y = (t^2 + 1)^3 - \dfrac{1}{t^2 + 1}$

10. $y = \sqrt{t^3 + t + 1}$

11. $y = (t^3 + t^2 - 5)^{16}$

12. $y = \dfrac{1}{(t^4 - 2t + 1)^3}$

13. $y = \dfrac{(t - t^2)^3}{t + 1}$

14. $y = \dfrac{(t^3 - t)^2}{t - 2}$

15. $y = (2t^3 - t^2 + 5t)^{4/3}$

16. $y = (3t^2 - 5t + 1)^{5/6}$

17. Since 1950, assume the average personal income i (in \$1000 per year) is approximated by the formula

$$i(t) = 6 + \frac{t}{4} - \frac{10}{t + 10}$$

Assume the consumption of steak s (in pounds per week) is related to the annual income by

$$s(i) = \sqrt{\frac{i}{1000}}$$

(a) Obtain an explicit formula for s as a function of t.
(b) Differentiate this formula to obtain ds/dt.
(c) Use the chain rule to obtain ds/dt from ds/di and di/dt.

18. Given the following three functions, determine dy/dt:
(a) $y(u) = u^4$
(b) $u(x) = x^3 + 1$
(c) $x(t) = t - 1$

HINT: Observe that

$$\frac{dy}{dt} = \frac{dy}{du} \cdot \frac{du}{dt} \qquad \text{and} \qquad \frac{du}{dt} = \frac{du}{dx} \cdot \frac{dx}{dt}.$$

Therefore,

$$\frac{dy}{dt} = \frac{dy}{du} \cdot \frac{du}{dx} \cdot \frac{dx}{dt}.$$

3-6. HIGHER DERIVATIVES

As will be developed in the next chapter, there is a relationship between the derivative of a function and the geometry of the graph of the function. Furthermore, there is a relationship between the derivative of the derivative and the geometry of the graph. Thus we are led to consider the concept of the derivative of a derivative. We call the derivative itself the *first* derivative and the derivative of the derivative the *second* derivative.

For example, consider the function $f(x)$ given by

$$f(x) = x^2$$

The derivative $f'(x)$ is given by

$$f'(x) = 2x$$

The function $2x$ is also differentiable, the result obtained by differentiating $2x$ being the second derivative $f''(x)$:

$$f''(x) = 2$$

For notational purposes the second derivative is customarily designated by one of the following:

$$f''(x), \quad \frac{d^2}{dx^2}[f(x)], \quad \frac{d^2f}{dx^2}, \quad D_x^2 f(x)$$

In this book we shall generally use one of the first three.

For many functions, the second derivative is itself differentiable, giving the third derivative

$$f'''(x) \qquad \text{or} \qquad \frac{d^3}{dx^3}[f(x)]$$

Differentiating this gives the fourth derivative

$$f^{(4)}(x) \qquad \text{or} \qquad \frac{d^4}{dx^4}[f(x)]$$

This process can be extended further, the nth derivative being represented as

$$f^{(n)}(x) = \frac{d}{dx}[f^{(n-1)}(x)] = \frac{d^n}{dx^n}[f(x)]$$

For the function $f(x) = x^2$, we have $f''(x)$ as a constant and thus $f'''(x) = 0$. Each succeeding higher derivative of $f(x) = x^2$ is also zero.

Example 1. Determine the first four derivatives of $\sqrt{x}$. The first derivative is

$$f'(x) = \frac{d}{dx}(\sqrt{x}) = \frac{d}{dx}(x^{1/2}) = \frac{1}{2}x^{-1/2} = \frac{1}{2\sqrt{x}}$$

the second derivative,

$$f''(x) = \frac{d}{dx}\left(\frac{1}{2\sqrt{x}}\right) = \frac{1}{2}\frac{d}{dx}(x^{-1/2}) = -\frac{1}{4}x^{-3/2} = -\frac{1}{4x\sqrt{x}}$$

the third derivative,

$$f'''(x) = \frac{d}{dx}\left(-\frac{1}{4x\sqrt{x}}\right) = -\frac{1}{4}\frac{d}{dx}(x^{-3/2}) = \left(-\frac{1}{4}\right)\cdot\left(-\frac{3}{2}\right)x^{-5/2} = \frac{3}{8x^2\sqrt{x}}$$

and the fourth derivative,

$$f^{(4)}(x) = \frac{d}{dx}\left(\frac{3}{8x^2\sqrt{x}}\right) = \frac{3}{8}\frac{d}{dx}(x^{-5/2}) = \frac{3}{8}\cdot\left(-\frac{5}{2}\right)x^{-7/2} = -\frac{15}{16x^3\sqrt{x}}$$

Note that all of these derivatives are functions defined for all $x > 0$.

Example 2. Determine the first and second derivatives of the function

$$f(x) = \frac{1}{x^2 - 1}$$

The first derivative is obtained by applying the rule for reciprocals:

$$f'(x) = \frac{d}{dx}\left[\frac{1}{x^2 - 1}\right]$$

$$= \frac{-1}{(x^2 - 1)^2}\frac{d}{dx}[x^2 - 1] = \frac{-2x}{(x^2 - 1)^2}$$

The second derivative is obtained by applying the quotient rule:

$$f''(x) = \frac{d}{dx}\left[\frac{-2x}{(x^2-1)^2}\right]$$

$$= \frac{(x^2-1)^2 \dfrac{d}{dx}[-2x] - (-2x)\dfrac{d}{dx}[(x^2-1)^2]}{(x^2-1)^4}$$

$$= \frac{-2(x^2-1)^2 + 8x^2(x^2-1)}{(x^2-1)^4}$$

and on dividing out $x^2 - 1$

$$= \frac{-2(x^2-1) + 8x^2}{(x^2-1)^3} = \frac{6x^2+2}{(x^2-1)^3}$$

It should be noted that the first and second derivatives are functions, defined for all values of x except $+1$ and -1. Higher derivatives can also be determined for this function, but the algebra gets progressively more complicated.

An Example from Business At this point we wish to give an example illustrating the significance of the second derivative in a business problem. Suppose that the net assets a of two financial institutions are given by the following functions (a is in units of a billion dollars):

$$\text{Institution 1:} \quad a_1(t) = \frac{3t}{t+1}$$

$$\text{Institution 2:} \quad a_2(t) = \frac{t^2}{t+1} + 1$$

At time $t = 1$, the assets of the two institutions are equal:

$$a_1(1) = \frac{3 \cdot 1}{1+1} = \frac{3}{2}$$

$$a_2(1) = \frac{1^2}{1+1} + 1 = \frac{3}{2}$$

Next consider the rate of growth da/dt of the assets of these two institutions:

$$\frac{da_1(t)}{dt} = \frac{d}{dt}\left[\frac{3t}{t+1}\right] = \frac{3}{(t+1)^2}$$

$$\frac{da_2(t)}{dt} = \frac{d}{dt}\left[\frac{t^2}{t+1} + 1\right] = \frac{t^2+2t}{(t+1)^2}$$

At time $t = 1$, the rates of growth of the assets of the two institutions are also equal:

$$\left.\frac{da_1(t)}{dt}\right|_{t=1} = \frac{3}{(1+1)^2} = \frac{3}{4}$$

$$\left.\frac{da_2(t)}{dt}\right|_{t=1} = \frac{1^2 + 2 \cdot 1}{(1+1)^2} = \frac{3}{4}$$

Next we might consider how the growth rate is changing. The rate of change of the growth rate is

$$\frac{d}{dt}\left[\frac{da}{dt}\right] = \frac{d^2a}{dt^2}$$

or the second derivative. For the two financial institutions, we have

$$\frac{d^2a_1(t)}{dt^2} = \frac{d}{dt}\left[\frac{da_1(t)}{dt}\right] = \frac{d}{dt}\left[\frac{3}{(t+1)^2}\right] = -\frac{6}{(t+1)^3}$$

$$\frac{d^2a_2(t)}{dt^2} = \frac{d}{dt}\left[\frac{da_2(t)}{dt}\right] = \frac{d}{dt}\left[\frac{t^2+2t}{(t+1)^2}\right] = \frac{(t+1)(2t+2) - 2(t^2+2t)}{(t+1)^3}$$

At $t = 1$, the rate of change in the growth rate for each institution is

$$\left.\frac{d^2a_1(t)}{dt^2}\right|_{t=1} = -\frac{6}{(1+1)^3} = -\frac{6}{8} = -\frac{3}{4}$$

$$\left.\frac{d^2a_2(t)}{dt^2}\right|_{t=1} = \frac{2 \cdot 4 - 2 \cdot 3}{8} = \frac{1}{4}$$

For the first financial institution at $t = 1$, the growth rate da_1/dt is positive, but the rate of change of the growth rate d^2a_1/dt^2 is negative. We say that the growth rate of this institution is decelerating, which is analogous to the effect upon speed of taking one's foot off of the accelerator of an automobile. For the second financial institution at $t = 1$, the growth rate da_2/dt is positive and the rate of change of the growth rate d^2a_2/dt^2 is also positive. We say that the growth rate of this institution is accelerating, which is analogous to the effect upon speed of depressing the accelerator of an automobile.

If one had to choose between one of these institutions, we first observe that their assets a year from now will be the same and their growth rates are the same. However, the growth rate of the second institution is increasing with time, whereas the growth rate of the first is decreasing with time. This makes the second institution more attractive.

In practice, financial analysts often take into consideration a firm's current assets, its current growth rate, and how the growth rate is changing. Of course, practical decisions are never as easy as for our contrived example.

Does one prefer an organization with a currently high growth rate that is decreasing slowly, or one with a smaller growth rate that is increasing slowly?

EXERCISES 3-6

In Exercises 1–8, find **(a)** $f'(x)$, $f''(x)$, and $f'''(x)$ and **(b)** evaluate each at the indicated value of x.

1. $x^3 - 2x^2 + 5x - 1 \quad x = 2$

2. $(x - 1)^4 + x^2 \quad x = 1$

3. $2x^3 - \dfrac{1}{x} \quad x = -1$

4. $3x^2 - \dfrac{1}{x^2} \quad x = -1$

5. $x^{3/2} - x \quad x = 4$

6. $(3x - 1)^6 - (3x - 1)^3 \quad x = \frac{1}{3}$

7. $\dfrac{1}{x} - \dfrac{2}{x^2} + \dfrac{3}{x^3} \quad x = 1$

8. $x^{-1/2} + x^{-3/2} \quad x = 1$

In Exercises 9–12, find $f'(x)$ and $f''(x)$.

9. $(x^2 - x + 1)^4$

10. $\dfrac{1}{x^2 + 3x + 1}$

11. $\sqrt{x^3 - 2x + 4}$

12. $\dfrac{1}{\sqrt{3x^3 - 2x}}$

In Exercises 13–16, find $f'(x)$ and $f''(x)$ at the indicated points.

13. $\dfrac{x - 1}{x + 1} \quad x = 2$

14. $\dfrac{x^2 - 1}{x + 2} \quad x = 0$

15. $\dfrac{x}{x^2 + 1} \quad x = 0$

16. $\dfrac{2x}{x^2 + 4} \quad x = 0$

17. Let the current assets of a company be given by the formula

$$\frac{t}{4} + \frac{1}{t + 2}$$

What is the growth rate at $t = 1$? At $t = 4$? What is the rate of change in growth rate at $t = 1$? At $t = 4$?

18. The Gross National Product over a particular period was given by the formula

$$\sqrt{t} + \frac{1}{t + 1}$$

What are the growth rate and the rate of change in the growth rate at $t = 4$?

3-7. IMPLICIT DIFFERENTIATION

In the previous examples of this chapter we have considered only functions expressed in the form

$$y = f(x)$$

Such functions are called explicit functions; they give y explicitly as a function of x. For these functions, the derivative is

$$\frac{dy}{dx} = f'(x)$$

In many problems we have equations of the form

$$F(x, y) = 0$$

Such equations specify a relationship between x and y. One example of such an equation is

$$x^2 + y^2 - 1 = 0$$

The graph of the points (x, y) that satisfy this equation is a circle of radius 1 and center at the origin, as illustrated in Figure 3-15.

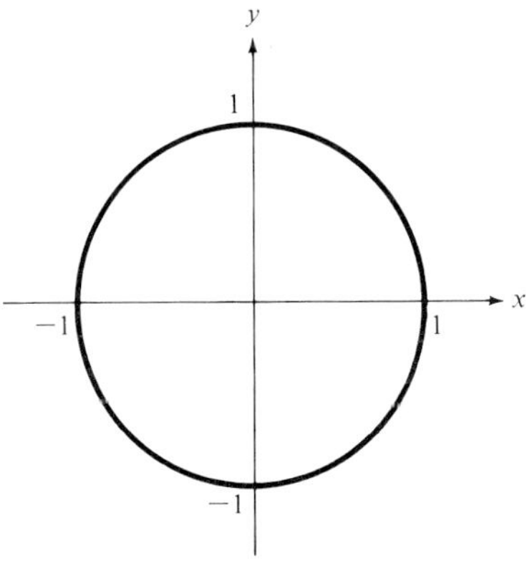

FIGURE 3-15

For this particular equation, it is possible to solve for y as an explicit function of x. The half of the circle corresponding to $y > 0$ is described by

$$y = \sqrt{1 - x^2}$$

The half of the circle corresponding to $y < 0$ is described by

$$y = -\sqrt{1 - x^2}$$

Either of these functions can be differentiated directly to give dy/dx for points on the appropriate half of the circle.

It is not always possible to solve equations of the type $F(x, y) = 0$ for y as an explicit function of x. However, the equation $F(x, y) = 0$ constitutes a relationship between x and y, and we may still consider y as a function of x. With y implicitly given as a function of x by $F(x, y) = 0$, we say that y is an *implicit function* of x. There is a complication because more than one value of y may correspond to a given value of x, as was the case with the equation $x^2 + y^2 - 1 = 0$. We shall consider this aspect of the problem shortly.

Although it may not be possible to solve for y as an explicit function of x, for many problems we can readily determine the derivative dy/dx by using a process known as implicit differentiation. We regard y to be a function of x. Using $x^2 + y^2 - 1 = 0$ as an example, $x^2 + y^2 - 1$ is also a function of x which is equal to 0 and, therefore, is a constant function of x. We differentiate both sides of the equation with respect to x:

$$\frac{d}{dx}(x^2 + y^2 - 1) = \frac{d}{dx}(0)$$

Since the derivative of a sum is the sum of the derivatives, we have

$$\frac{d}{dx}(x^2) + \frac{d}{dx}(y^2) - \frac{d}{dx}(1) = \frac{d}{dx}(0)$$

All derivatives in this equation are readily evaluated except for the derivative of y^2. However, we should recall that y is a function of x and apply the chain rule:

$$\begin{aligned}\frac{d}{dx}(y^2) &= \frac{d}{dy}(y^2)\frac{dy}{dx} \\ &= 2y\frac{dy}{dx} = 2yy'\end{aligned}$$

where y' is understood to be dy/dx.

Evaluating the remaining terms in the equation in the preceding paragraph, we have

$$\begin{aligned}\frac{d}{dx}(x^2) + \frac{d}{dx}(y^2) - \frac{d}{dx}(1) &= \frac{d}{dx}(0) \\ 2x + 2yy' &= 0\end{aligned}$$

Solving for y' gives

$$y' = -\frac{x}{y}$$

Although the formula for $y' = dy/dx$ contains both x and y, this is quite natural for such equations, as will be explained in the following paragraph.

As mentioned earlier, relationships of the type $F(x, y) = 0$ have the property that for a given value of x, there may be more than one value of y that will satisfy the relationship. Using $x^2 + y^2 - 1 = 0$ as an example, we note that to any x satisfying $-1 < x < 1$, there correspond two values of y, one positive and one negative. To each pair of values of x and y representing a point on the graph of $x^2 + y^2 - 1 = 0$, a line tangent to the graph at that point can be drawn, as illustrated in Figure 3-16. In this example, for $y > 0$

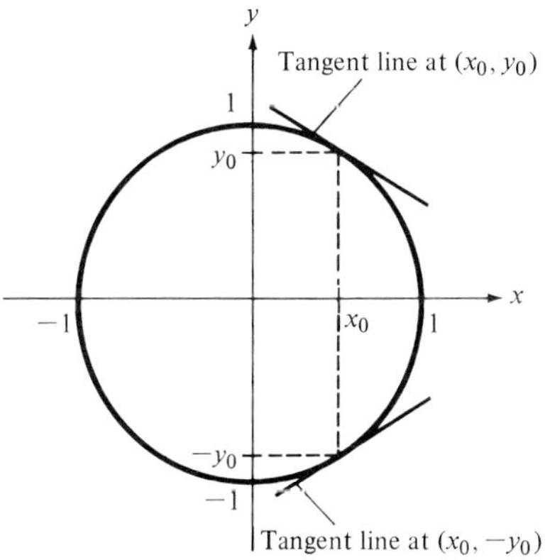

FIGURE 3-16

or for $y < 0$ the graph defines y as a function of x, which is also differentiable. The formula $y' = -x/y$ obtained by implicit differentiation works for any point on the circle for which y' exists. To specify a particular point on the circle, we note that both x and y appear in the formula for y', and we specify both coordinates of the point in question.

Thus for (x, y) on the circle $x^2 + y^2 - 1 = 0$ with x and y both positive, the tangent should have negative slope and indeed $y' = -(x/y)$ is negative for x and y positive. Similarly, for x and y both negative (in the third quadrant) the tangent also should have negative slope and indeed $-(x/y)$ is negative for x and y negative. If one of x and y is positive and the other is negative, then from the graph the slope should be positive and indeed $-(x/y)$ is positive in such cases.

In actually applying implicit differentiation to a given function $F(x, y) = 0$, in order to find y' we normally proceed in a straightforward way writing down only a few steps. This is illustrated by the following examples.

Example 1. Find dy/dx if $xy^2 - x^2y + y^4 + 1 = 0$. We begin by differentiating with respect to x using the sum, product, and power formulas:

$$\frac{d}{dx}(xy^2) - \frac{d}{dx}(x^2y) + \frac{d}{dx}(y^4) + \frac{d}{dx}(1) = \frac{d(0)}{dx}$$

$$(y^2 + 2xyy') - (2xy + x^2y') + 4y^3y' = 0$$

Solving for y', we get

$$y' = \frac{2xy - y^2}{2xy - x^2 + 4y^3}$$

which is the desired solution. For the point $(-1, 1)$, which is on the graph of $xy^2 - x^2y + y^4 + 1 = 0$, we have

$$y'\Big|_{\substack{x=-1\\ y=1}} = \frac{2(-1)\cdot 1 - 1^2}{(2(-1)\cdot 1 - (-1)^2 + 4\cdot 1^3} = \frac{-3}{1}$$

Since the tangent line passes through point $(-1, 1)$ and has slope -3, an equation for the tangent line may be obtained using the point-slope form of the equation for a line:

$$y - 1 = (-3)\cdot(x - (-1))$$

or, on simplifying,

$$y = -3x - 2$$

Example 2. $x^2y^3 + y^2x^2 + y + 2x - 2 = 0$
We first differentiate with respect to x:

$$\frac{d}{dx}(x^2y^3) + \frac{d}{dx}(y^2x^2) + \frac{d}{dx}(y) + \frac{d}{dx}(2x) - \frac{d}{dx}(2) = \frac{d(0)}{dx}$$

$$(2xy^3 + 3x^2y^2y') + (2x^2yy' + 2xy^2) + y' + 2 - 0 = 0$$

Solving for y', we get

$$y' = -\frac{2xy^3 + 2xy^2 + 2}{3x^2y^2 + 2x^2y + 1}$$

At the point $(1, 0)$ which is on the graph of $x^2y^3 + y^2x^2 + y + 2x - 2 = 0$, the value of y' is

$$y'\Big|_{\substack{x=1\\ y=0}} = -\frac{2}{1} = -2$$

Since the tangent line passes through point (1, 0) and has slope -2, an equation of the line is

$$y - 0 = (-2) \cdot (x - 1).$$

EXERCISES 3-7

---

1. For the example $x^2 + y^2 - 1 = 0$, we observed that for $y > 0$, the formula $y = \sqrt{1 - x^2}$ gives y as an explicit function of x. Differentiate this function and show that, for $y > 0$, the result is equivalent to $y' = -(x/y)$ obtained by implicit differentiation.

For each of Exercises 2 through 6, (**a**) determine y' by implicit differentiation; (**b**) with y assumed >0 solve for y as an explicit function of x and differentiate; and (**c**) show that the two results are equivalent.

2. $x^2 + y^2 - 4 = 0$ **3.** $x^2 - y^2 + 1 = 0$

4. $x^2y^2 - 1 = 0$ **5.** $\sqrt{x} + \sqrt{y} - 1 = 0$

6. $xy^4 - x^2 - 1 = 0$

In each Exercise 7 through 14, (**a**) find y' as a function of x and y, (**b**) verify that the indicated point is on the graph, (**c**) find y' at the particular point indicated, and (**d**) obtain an equation for the line tangent to the graph at the indicated point.

7. $x^2 + y^2 = 25$ — $x = -3,\ y = 4$

8. $x^2y^2 = 25$ — $x = 1,\ y = 5$

9. $(x + y)^2 - 1 = 0$ — $x = 1,\ y = 0$

10. $y^3 - x^3 - 7 = 0$ — $x = 1,\ y = 2$

11. $x^2y^3 - x = 0$ — $x = 1,\ y = 1$

12. $x^3y^2 - y = 10$ — $x = \sqrt[3]{3},\ y = 2$

13. $xy^3 + 2x^2y - 6y^2 = 4$ — $x = 2,\ y = 1$

14. $2y^4 - 3y^3x + 2yx^5 - 32 = 0$ — $x = 0,\ y = 2$

15. $\dfrac{x + y}{x - y} = 7$ — $x = 8,\ y = 6$

16. $\dfrac{xy}{x + y} - 8 = 0$ — $x = 16,\ y = 16$

17. $\dfrac{x}{x^2 + y^2} - 1 = 0 \qquad x = 1,\ y = 0$

18. $2\dfrac{y}{x} - 3\dfrac{y^2}{x^2} = -1 \qquad x = 1,\ y = 1$

*3-8. RELATED RATES

As customarily understood, a function relates one variable to another. For example, $y = f(x)$ relates x to y. Suppose, however, that we also have a rate of change of x with respect to t, namely, dx/dt.

By this notation, we imply that x is a function of t, and since $y = f(x)$, then y must also be a function of t. Furthermore, if x is changing with time t and $y = f(x)$, then y must also be changing with time t.

Can we determine dy/dt, and if so, what is it? Since $y = f(x)$, then

$$\frac{dy}{dt} = \frac{d}{dt}[f(x)]$$

By the chain rule,

$$\frac{d}{dt}[f(x)] = \frac{d}{dx}[f(x)]\frac{dx}{dt} = \frac{df}{dx}\frac{dx}{dt}$$

Consequently, we have

$$\frac{dy}{dt} = \frac{df}{dx} \cdot \frac{dx}{dt}$$

The rates dy/dt and dx/dt are called *related rates*. If we have a value of x and a value for dx/dt, then the above relationship can be used to determine dy/dt.

To illustrate, suppose that $f(x) = 1/x$ and that we specify that $x = 1$ and $dx/dt = 2$. Since $df/dx = -(1/x^2)$, then

$$\frac{dy}{dt} = \frac{df}{dx} \cdot \frac{dx}{dt} = -\frac{1}{x^2}\frac{dx}{dt}$$

If $x = 1$ and $dx/dt = 2$, then

$$\frac{dy}{dt} = -\frac{1}{(1)^2} \cdot 2 = -2$$

In business, we are frequently concerned about how a change in one variable affects another. For example, the demand function relates (1) the quantity of a product that will be bought and (2) the selling price of that product. An example of a simple demand function would be

$$q = \frac{10{,}000}{p + 1}$$

where q is the quantity and p is the price. If p changes, then q changes, and thus dp/dt and dq/dt are related rates.

For example, suppose the product is currently selling at a price of \$0.50. However, the manufacturer sees that his production costs are increasing at a rate of 10% annually, and he is considering increasing the selling price at the same rate. He wants to know how the demand will change.

If the price increases by 10% annually, this is equivalent to stating that $dp/dt = 0.1p$. By wanting to know how the demand will change as a function of time, the manufacturer wants to know dq/dt.

To obtain the relationship between dq/dt and dp/dt, we observe that both q and p are functions of time and proceed to differentiate the demand function with respect to time:

$$\frac{dq}{dt} = \frac{d}{dt}\left(\frac{10{,}000}{p + 1}\right)$$

By the chain rule, we have

$$\frac{d}{dt}\left(\frac{10{,}000}{p + 1}\right) = \frac{d}{dp}\left(\frac{10{,}000}{p + 1}\right)\frac{dp}{dt}$$

$$= -\frac{10{,}000}{(p + 1)^2}\frac{dp}{dt}$$

Therefore, the rate of change in q with respect to time is

$$\frac{dq}{dt} = -\frac{10{,}000}{(p + 1)^2}\frac{dp}{dt}$$

For our example, $dp/dt = 0.1p$ and $p = 0.5$. Substituting these values gives

$$\frac{dq}{dt} = -\frac{10{,}000}{(0.5 + 1)^2}(0.05) \approx -220$$

Therefore, the manufacturer should expect the demand for his product to decrease by 220 units annually, which is $(220/10{,}000) \cdot 100 = 2.2\%$ annually.

The concept of related rates also applies to problems in which one variable is related to two or more other variables. For example, suppose the demand q for a product is related to the price p and average personal income i (in \$1000 per year) by the formula.

$$q = \frac{\sqrt{i}}{p + 1}$$

where q is in units of a thousand. To relate dq/dt to di/dt and dp/dt, we differentiate with respect to t:

$$\frac{dq}{dt} = \frac{d}{dt}\left(\frac{\sqrt{i}}{p+1}\right)$$

Since i and p are functions of time, we apply the product rule:

$$\frac{dq}{dt} = \frac{d}{dt}\left(\frac{\sqrt{i}}{p+1}\right) = \sqrt{i}\frac{d}{dt}\left(\frac{1}{p+1}\right) + \frac{1}{p+1}\frac{d}{dt}(\sqrt{i})$$

Next we apply the chain rule:

$$\frac{dq}{dt} = \sqrt{i}\frac{d}{dp}\left(\frac{1}{p+1}\right)\frac{dp}{dt} + \frac{1}{p+1}\frac{d}{di}(\sqrt{i})\frac{di}{dt}$$

$$= -\frac{\sqrt{i}}{(p+1)^2}\frac{dp}{dt} + \frac{1}{2(p+1)\sqrt{i}}\frac{di}{dt}$$

If we were given or could figure out the values of dp/dt, di/dt, i, and p at a particular time t_0, we could then evaluate dq/dt at time t_0.

EXERCISES 3-8

In the following exercises, obtain a formula relating dy/dt to dx/dt (and du/dt for Exercises 4 and 5). Determine the value of dy/dt for the given values of x, dx/dt, u, and du/dt at time t_0.

1. $y = x^2 - 2x - 1$; $x = 6$ and $dx/dt = 1$ at time t_0

2. $y = \sqrt{25 - x^2}$; $x = 4$ and $dx/dt = -1$ at time t_0

3. $y = x/(1 + \sqrt{x})$; $x = 1$ and $dx/dt = 0.1$ at time t_0

4. $y = (x + 2)(u - 1)$; $x = 0$ and $dx/dt = 1$; $u = 3$ and $du/dt = -1$ at time t_0

5. $y = \sqrt{x^2 + u^2}$; $x = 6$ and $dx/dt = 1$; $u = 8$ and $du/dt = -1$ at time t_0

6. The consumption s of steak in pounds per year is related to the average annual personal income i by the formula

$$s = 52\sqrt{\frac{i}{10{,}000}}$$

If $i = \$4000$ and is increasing at a rate of 10% annually ($di/dt = 0.1i = 400$), determine the rate of increase in s, that is, determine ds/dt.

7. Suppose the amount of work W performed during a 24-hour period is related to the wage rate r by the formula

$$W = 1 - \frac{1}{r+1}$$

The wage rate is currently \$2.75 per hour, and is increasing by 5% per year, i.e., $dr/dt = 0.05r = 0.1375$. Determine dW/dt at this time.

8. The total cost C of producing product X is related to q, the quantity produced, by the equation

$$C = 0.04q^3 - q^2 + 12q + 6$$

Suppose that the current value of q is 20, and that currently $dq/dt = 0.2$. What is dC/dt at this time?

REVIEW EXERCISES: CHAPTER 3

In Exercises 1 through 10, **(a)** find $f'(x)$, and **(b)** determine the equation of the line tangent to the graph of $f(x)$ at the indicated point.

1. $f(x) = x^4$ at $x = 1$

2. $f(x) = \dfrac{1}{x-2}$ at $x = 3$

3. $f(x) = \sqrt{x}$ at $x = 1$

4. $f(x) = x^3 - 2x^2 + x - 1$ at $x = -1$

5. $f(x) = (x-1)^4$ at $x = 0$

6. $f(x) = \sqrt{x^2+1}$ at $x = 0$

7. $f(x) = \dfrac{x+2}{x^2-x-1}$ at $x = 2$

8. $f(x) = (x-2)^{20}$ at $x = 1$

9. $f(x) = x\sqrt{x+1}$ at $x - 3$

10. $f(x) = \dfrac{x^2(x+1)}{x-3}$ at $x = 2$

In Exercises 11 through 25, determine the derivative of the given function.

11. $f(t) = t^9 - 1$

12. $f(t) = t^{-1.2} + \dfrac{1}{t}$

13. $f(x) = 6x^{-20} - x^2$

14. $f(x) = \dfrac{x^5}{x^3 - 2x - 5}$

15. $f(z) = \dfrac{\sqrt{3}}{z+1}$

16. $f(t) = \dfrac{(t^2+2t+2)(t-1)}{t^3+1}$

17. $f(x) = \sqrt{t^3 - t^2 - t - 1}$

18. $f(x) = (x^2 + 2x + 5)^{1.9}$

19. $f(z) = (z^5 - 1)^{-5}$

20. $f(x) = (x+3)(x+2)\sqrt{x}$

21. $f(t) = t^2 + t + \dfrac{t+1}{t-1}$

22. $f(z) = \dfrac{z^2+1}{z^2-1} + z + \sqrt{z}$

23. $f(z) = (z+1)\sqrt{z^2-1}$

24. $f(x) = (x + \sqrt{x^2+1})^2$

25. $f(x) = \dfrac{2x}{x + \sqrt{x^2-1}}$

26. The population of a city is approximately represented by the function

$$f(t) = 200{,}000 - \frac{100{,}000}{t+1}$$

when t is time since 1950. Determine the population in 1959 and the rate of growth in population at the beginning of that year.

27. The cost function $C(q)$ and revenue function $R(q)$ for a product are given by

$$C(q) = \sqrt{q+1}$$

$$R(q) = \frac{10q}{(q+1)^2}$$

where q is the quantity sold. Determine the marginal cost, marginal revenue, and marginal profit.

For the functions in Exercises 28 through 33, determine the first, second, and third derivatives.

28. $f(x) = x^4 + 3x^3 - 9x^2 + 2x + 1$

29. $f(x) = \dfrac{1}{(x+3)^7}$

30. $f(x) = x^{2.8} - 5$

31. $f(x) = (x^2+2)^4$

32. $f(x) = (1 + \sqrt{x})^2$

33. $f(x) = \dfrac{x-2}{x+2}$

For Exercises 34 through 40, **(a)** determine y' as a function of x and y, **(b)** find y' at the indicated point, and **(c)** obtain an equation for the line tangent to the graph at the indicated point.

34. $x^2y^5 - y = 0 \qquad (1, 1)$

35. $\dfrac{x^2}{y^2} + xy = 0 \qquad (1, -1)$

36. $y^5 - y^2 + xy + \sqrt{x} - 6 = 0 \qquad (4, 1)$

37. $\dfrac{x+y}{x-y} - 3 = 0 \qquad (2, 1)$

38. $\dfrac{x^2}{y^2} + \dfrac{2y^2}{x^2} - 3 = 0 \qquad (1, 1)$

39. $2(x+y)^5 + xy = 0 \qquad (2, -1)$

40. $\sqrt{x^2+y^2} + y - 1 = 0 \qquad (3, -4)$

CHAPTER FOUR

APPLICATIONS OF THE DERIVATIVE

4-1. MAXIMA AND MINIMA

In this section we define the concept of a maximum (or minimum) value of a function and look at some examples. In the next section we shall examine the use of derivatives in identifying maxima and minima.

Definition 4-1 Let $f(x)$ be a function with specified domain D. If there is a number c in D such that $f(c) \geq f(x)$ for all x in D, then $f(c)$ is called the (*absolute*) *maximum* of $f(x)$ in D; and c is *a point at which* $f(x)$ *has a maximum.*

The word *absolute* is sometimes omitted. We similarly define the (*absolute*) *minimum* of a function. We may use the word *extremum* to refer to either a maximum or a minimum. Some people use the expressions "max" or "min" to refer to a maximum or a minimum. Problems that refer to maxima or minima are sometimes called max-min or min-max problems.

It is important to distinguish between the maximum of a function $f(x)$, which is a value of the function, and a number or point in the domain that yields the maximum, or, equivalently, at which the function has a maximum. There can be only one number that is the absolute maximum (or minimum) of a function, but there may be several or many points in the domain having the same maximum (or minimum) functional value. To illus-

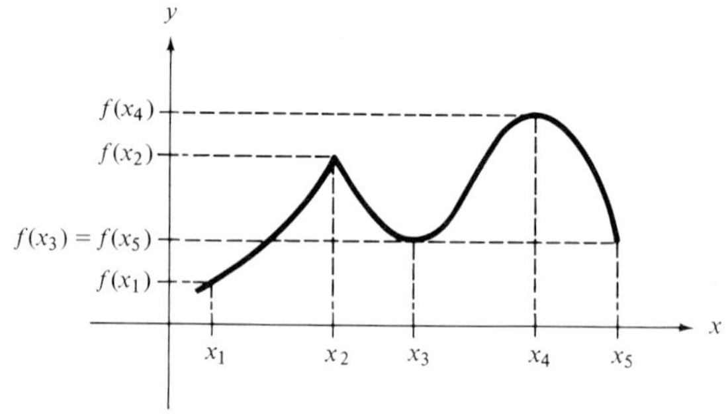

FIGURE 4-1

trate Definition 4-1, consider the function $f(x)$ graphed in Figure 4-1. Over the closed interval $[x_1, x_5]$, this function has an absolute maximum at $x = x_4$ and an absolute minimum at $x = x_1$. The maximum is the number $f(x_4)$ and the minimum is $f(x_1)$.

A concept related to that of absolute maximum is the following:

Definition 4-2 The number $f(c)$ is a *relative* (or *local*) *maximum* of the function $f(x)$ with domain D if there exists an open interval (a, b) containing c such that $f(c) \geq f(x)$ for every x which is both in (a, b) and in D.

A *relative* (or *local*) *minimum* is similarly defined.

In Figure 4-1 there are three relative minima and two relative maxima for the function $f(x)$ understood to have the domain $[x_1, x_5]$. The relative minima are $f(x_1)$, $f(x_3)$ and $f(x_5)$. The points at which the relative minima occur are x_1, x_3, and x_5. The relative maxima occur at x_2 and x_4 and are $f(x_2)$ and $f(x_4)$.

Clearly, if for any c, $f(c)$ is an absolute extremum, then it is also a relative extremum of the same type. Thus, in locating absolute extrema, we frequently locate all relative extrema and by comparing them decide which, if any, are the absolute ones. This point of view will be amplified in Section 4-4.

The specified domain of a function is important in questions of absolute extrema, and even of relative extrema. Thus, in Figure 4-1, if we specify the domain to be only the interval $[x_1, x_3]$, then the absolute maximum is $f(x_2)$ and the absolute minimum is $f(x_1)$. If we specify the domain to extend to the left of x_1, then $f(x_1)$ is neither an absolute nor a relative minimum.

If we specify the domain to be the closed interval $[x_2, x_5]$, then the absolute maximum is $f(x_4)$ and the absolute minimum is achieved twice, at x_3 and at x_5, and, therefore, is $f(x_3)$ or $f(x_5)$.

If we specify the domain to be the *open* interval (x_1, x_5), then there is no absolute minimum and the absolute maximum is $f(x_4)$. [To see that there is no absolute minimum, ask yourself what it would be if there were one. It could not be $f(x_1)$ since x_1 is not in the domain but could not be more than the number that would be $f(x_1)$ if x_1 were in the domain.] The relative extrema occur at x_2, x_3, and x_4. If we specify the domain to be the open interval (x_3, x_4), then there are no absolute or relative extrema on that interval.

For many functions it is possible to identify maxima and minima by inspection, by graphing, or by a little common sense.

Example 1. Let $f(x) = (x - 1)^2$ with the domain the set of all real x. See Figure 4-2 for the graph of $f(x)$. Since the square of any real number is positive or zero (and is zero only if the number is zero), we know that for each x, $f(x)$ is positive or zero. Also $f(x) = 0$ if and only if $x - 1 = 0$, or, in other words, if $x = 1$. Therefore, $x = 1$ is the point at which $f(x)$ has a minimum and $f(1) = 0$ is the minimum value of the function. There is no maximum value of $f(x)$, since $f(x)$ gets larger and larger as x becomes large.

Example 2. Let $f(x) = (x - 1)^2$ with the domain the closed interval $[-1, 3]$. As in Example 1, the minimum value of $f(x)$ is 0 with $x = 1$ the point at

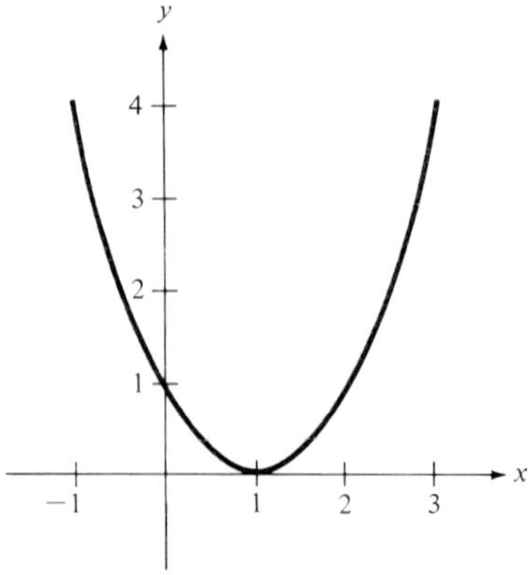

FIGURE 4-2

which $f(x)$ has a minimum. From the graph in Figure 4-2, it is clear that relative maxima occur at $x = -1$ and $x = 3$ and that since $f(-1) = f(3) = 4$, the absolute maximum of $f(x)$ over $[-1, 3]$ is 4.

Note that in this example there are two values of x that produce the absolute maximum. Note also that if the domain were specified as the *open* interval $(-1, 3)$, then there would be no absolute maximum. It would not be 4, since for no c in the open interval $(-1, 3)$ is $f(c) = 4$. It also could not be any number less than 4.

Example 3. Let $f(x) = 1/(x^2 + 1)$ with the domain the set of all real x. Then $f(x)$ is positive for all real x. Since the numerator is constant, the maximum value of $f(x)$ will occur for that value of x for which the denominator is a minimum. This value is $x = 0$, since at 0, $x^2 + 1 = 1$, and $x^2 + 1 \geq 1$ for all real x. The maximum value of $1/(x^2 + 1)$, therefore, is $f(0) = 1/(0 + 1)$. We note that as x gets large positively, $f(x)$ gets close to zero. There is no minimum value of $f(x)$. This knowledge, and the symmetry of the graph with respect to the y axis, help us to graph the function $1/(x^2 + 1)$ as in Figure 4-3.

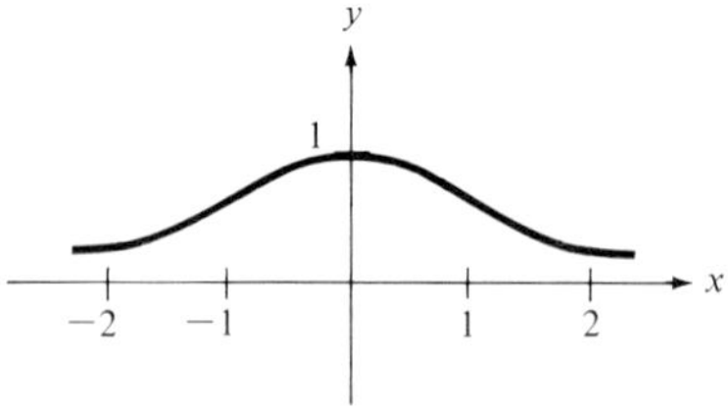

FIGURE 4-3

EXERCISES 4-1

1. Define the *absolute minimum* of a function $f(x)$ with domain D.

2. Define the *relative minimum* of a function $f(x)$ with domain D.

By observation, graphing, or common sense determine whether the following functions have absolute maxima or absolute minima; if they exist, find them and all points at which the extrema occur.

	$f(x)$	Domain
3.	$\frac{1}{x^2+4}$	$[-2, 2]$
4.	$x^2(x-2)^2-4$	all x
5.	x^4+1	all x
6.	$\frac{1}{\lvert x\rvert+1}$	$[-10, 10]$
7.	$\frac{1}{x^4+x^2+1}$	all x
8.	x^8-2	$[-1, 1]$
9.	$\lvert\sqrt[3]{x}\rvert$	$[-2, 1]$
10.	$\frac{1}{\sqrt{x}}$	$(0, 2]$

In Exercises 11–14, find an appropriate domain for which each function has -2 as a minimum and 7 as a maximum. There may not be a unique answer.

11. $3x-1$

12. x^2-2

13. x^3

14. $\frac{1}{x}-3$

In Exercises 15–20 let $f(x)$ be as graphed in Figure KK with the domain as specified in the individual exercises. For each exercise find **(a)** all relative maxima and minima and the points that yield them, and **(b)** all absolute maxima and minima and the points that yield them.

15. The domain is $[x_1, x_5]$

16. The domain is $[x_2, x_4]$

17. The domain is $x \leq x_5$

18. The domain is $x \geq x_1$

19. The domain is the open interval (x_2, x_5)

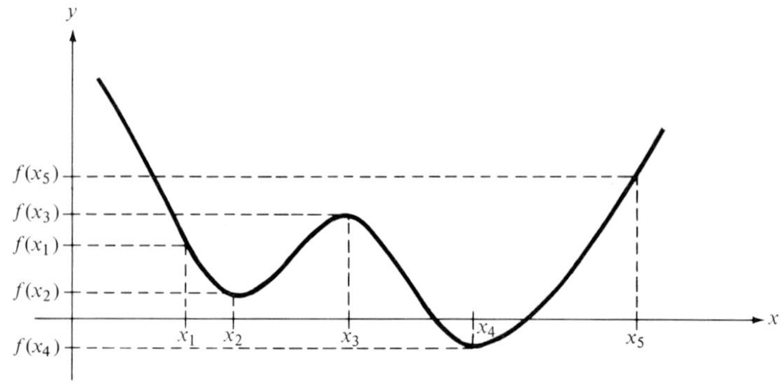

FIGURE KK

20. The domain is the open interval (x_1, x_4)

In Exercises 21 and 22, graph the (noncontinuous) function over the interval [0, 2] and determine its extrema if any.

21. $f(x) = \begin{cases} x^2 & \text{whenever } 0 \leq x < 1 \\ \frac{1}{2} & \text{whenever } 1 \leq x \leq 2 \end{cases}$

22. $g(x) = \begin{cases} 1 - x & \text{whenever } 0 \leq x \leq 1 \\ 3 - x & \text{whenever } 1 < x \leq 2 \end{cases}$

4-2. CRITICAL POINTS AND RELATIVE OR LOCAL MAXIMA AND MINIMA

Consider the function

$$f(x) = (x - 3)^2$$

whose graph is given in Figure 4-4. For $x < 3$, this function is decreasing at each point, since $f(x)$ becomes smaller as x increases. Similarly for $x > 3$, the function is increasing at each point, since $f(x)$ grows larger as x increases. As also indicated in Figure 4-4, the slope of the tangent line is negative where $f(x)$ is decreasing, and hence the derivative is negative there. Similarly, the slope of the tangent line is positive where $f(x)$ is increasing and hence the derivative is positive there.

Figure 4-5 illustrates a function increasing at c.

Definition 4-3 The function $f(x)$ is *increasing at the point c* if for some open interval (a, b) with $a < c < b$, $f(x) > f(c)$ for any x between c and b and $f(x) < f(c)$ for any x between a and c.

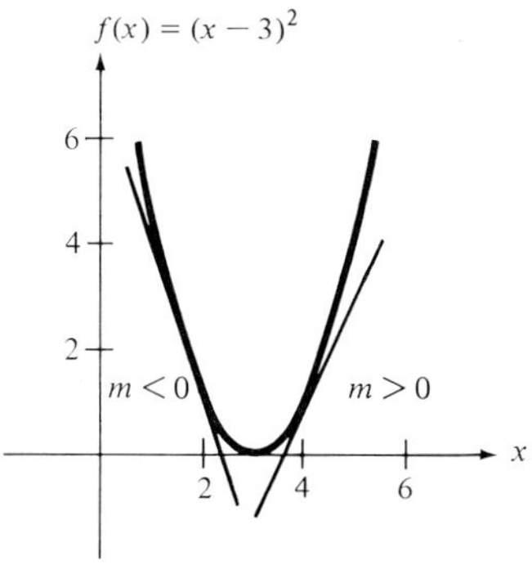

FIGURE 4-4

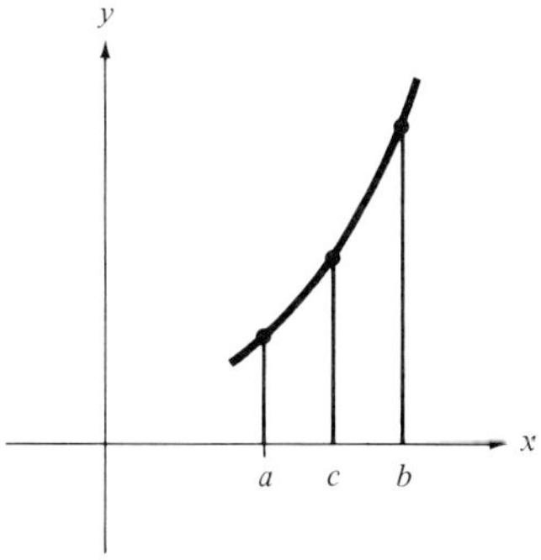

FIGURE 4-5

We may similarly define the function to be *decreasing at c*.

Our definition implicitly requires that if a function is considered increasing or decreasing at a point, then the function must be considered defined in an open interval containing the point. At an end point of the domain we do not consider the function to be either increasing or decreasing.

The function whose graph is given in Figure 4-6 is increasing at each $x < x_1$ and at each $x > x_2$. It is decreasing at each x between x_1 and x_2. At x_1 and x_2 it is neither increasing nor decreasing.

The property relating the derivative to increasing and decreasing functions is the following (we assume c is not an end point of the specified domain):

Property 4-1 If $f'(c) > 0$, then $f(x)$ is increasing at c. If $f'(c) < 0$, then $f(x)$ is decreasing at c.

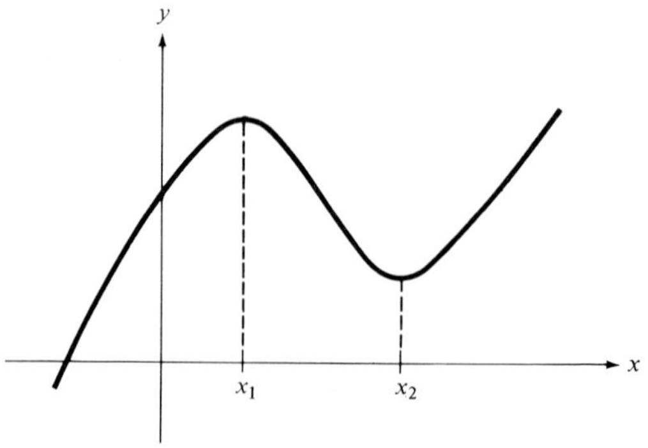

FIGURE 4-6

The explanation for this property is easy. We give it for the case $f'(c) > 0$. By the definition of the derivative

$$\lim_{x \to c} \frac{f(x) - f(c)}{x - c} > 0$$

Therefore, there must be an interval (a, b) containing c for which

$$\frac{f(x) - f(c)}{x - c} > 0 \qquad \text{for} \quad x \text{ in } (a, b) \text{ and } x \neq c$$

The numerator and denominator must have the same sign. Therefore,

$$f(x) > f(c) \qquad \text{whenever} \quad x > c \text{ and } x \text{ is in } (a, b)$$

$$f(c) > f(x) \qquad \text{whenever} \quad c > x \text{ and } x \text{ is in } (a, b)$$

which is what we mean by $f(x)$ increasing at c.

Clearly, if a function is increasing or decreasing at a point, the function cannot have a relative maximum or a relative minimum at that point. Thus, if we want to identify the values of x for which a function $f(x)$ can have a relative extremum, we need only consider the *critical points*, which are defined as follows:

Definition 4-4 The critical points of $f(x)$ are those values of x which are in the domain of the function and for which

1. x is an end point of the specified domain
2. $f'(x) = 0$, or
3. $f'(x)$ does not exist and $f(x)$ does exist.

Each relative extremum can occur only at a critical point of the function since at each other point $f'(x)$ exists and is positive or is negative.

However, critical points of a function do not necessarily yield local maxima or minima of the function. For example, consider the function $y = x^3$ whose domain is the whole x axis and whose graph is given in Figure 4-7. We have $y' = 3x^2$, which is defined for all values of x. Since $y'(x) = 0$ if and only if $x = 0$, we know that $x = 0$ is the only critical point of the function. However, the function is increasing at $x = 0$, since $f(0) = 0$, $f(x) > 0$ for x to the right of zero, and $f(x) < 0$ for x to the left of zero. Thus $f(x)$ has no extremum at $x = 0$.

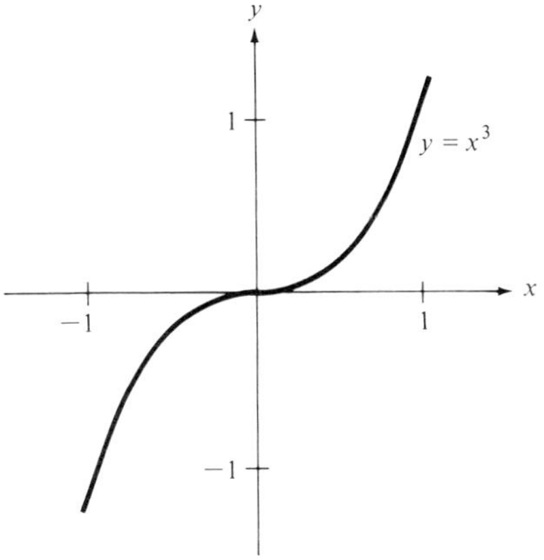

FIGURE 4-7

The important fact about critical points is that we can identify *all* possible relative extrema by finding all critical points, i.e., by finding all values of x in the domain of the function for which the derivative is zero or is not defined, or for which x is an end point of the specified domain of the function. Having identified such points, we then can try to determine whether they yield relative maxima or minima.

How do we identify whether a critical point yields a relative maximum, a relative minimum, or neither? For a point that is not an end point of the domain, there are really three standard tests. We give two here and the third in the next section.

Test 1 Compare $f(x)$ near the critical point c with $f(c)$. It may be apparent that for all x near c, $f(x) \geq f(c)$ or, alternatively, $f(x) \leq f(c)$. If so, we know that $f(c)$ is a local minimum or, alternatively, a local maximum.

Test 2 Examine $f'(x)$ for x near the critical point c. If $f'(x)$ is positive for x to the left of c and negative for x to the right of c, then $f(x)$

must have a relative maximum at c, since $f(x)$ is increasing to the left of c and decreasing to the right, as illustrated in Figure 4-8(a). Alternatively, if $f'(x)$ is negative to the left of c and positive to the right of c, then $f(x)$ must have a relative minimum at c, as illustrated in Figure 4-8(b).

If $f'(x)$ does not change sign at c, then $f(x)$ will have neither a maximum nor a minimum at c, since it will be either increasing on both sides of c or decreasing on both sides of c.

It is sometimes easy to use Test (2), since by inspection it may be easy to tell whether $f'(x)$ changes sign at c.

If c is a critical point of $f(x)$ because c is an end point of the specified domain D of $f(x)$, then it is frequently possible to tell whether c is a local maximum or local minimum by steps similar to (1) and (2) above. If $f(x) \geq f(c)$ for all x near c and in the domain D, then $f(c)$ is a local minimum; if $f(x) \leq f(c)$ for all such x, then $f(c)$ is a local maximum. If c is a left-hand end point of D, and $f'(x)$ is positive for x in D and near c, then $f(c)$ is a local minimum, since $f(c)$ is increasing to the right near c. By thinking about the function, the domain, and the derivative, one can give similar conditions for other cases of end points that are critical points.

Here are two examples illustrating these ideas.

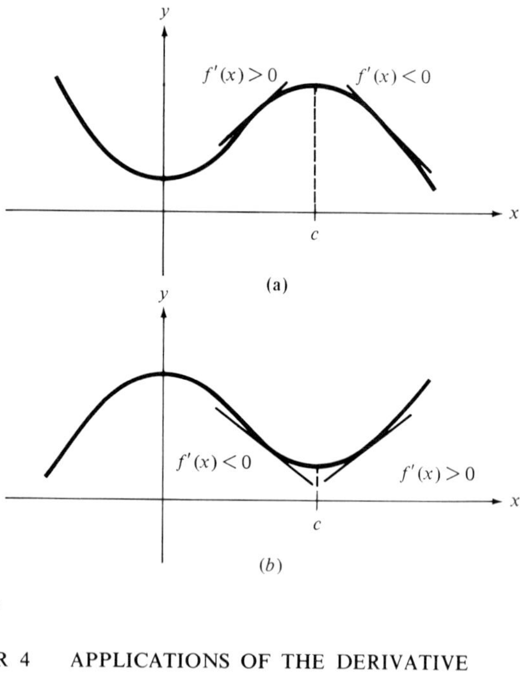

FIGURE 4-8a, b

Example 1. In Figure 4-4 we considered the function $f(x) = (x - 3)^2$ defined for all real values of x. The derivative is $f'(x) = 2(x - 3)$ for which

$$f'(x) > 0 \qquad \text{when } x > 3$$

$$f'(x) = 0 \qquad \text{when } x = 3$$

$$f'(x) < 0 \qquad \text{when } x < 3$$

Therefore, as we earlier concluded by inspection of the graph, $f(x)$ is decreasing for each $x < 3$ and increasing for each $x > 3$. The point $x = 3$ is a critical point, since $f'(3) = 0$. There are no other critical points. Clearly, $f(x)$ has a relative minimum at $x = 3$, since $f(3) = 0$ and $f(x) > 0$ for all $x \neq 3$. Indeed, $f(x)$ has an absolute minimum at $x = 3$, as is clear from this information.

Example 2. Let $f(x)$ be given by

$$f(x) = \frac{x^4}{4} - x^3 + x^2$$

with domain the set of all real x. We have

$$\begin{aligned} f'(x) &= x^3 - 3x^2 + 2x \\ &= x(x^2 - 3x + 2) \\ &= x(x - 1)(x - 2) \end{aligned}$$

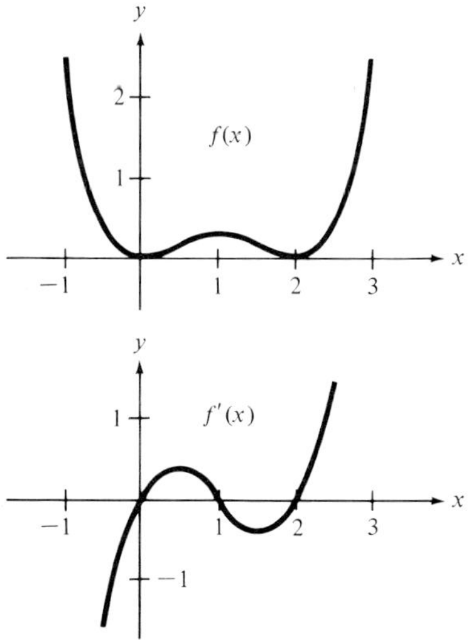

FIGURE 4-9

The graphs of $f(x)$ and $f'(x)$ are given in Figure 4-9. Since $f'(x)$ is defined for all x, the critical points of $f(x)$ are just the zeros of $f'(x)$, or in other words, the roots of $f'(x) = 0$. The critical points are $x = 0, 1, 2$. For each other value of x, $f'(x)$ exists and is either positive or negative. The only possible values of x that yield relative extrema are $x = 0, 1, 2$. From Figure 4-9 we observe that $x = 0$ and $x = 2$ yield relative minima each of which is zero, and that $x = 1$ yields a relative maximum that is $\frac{1}{4}$, since $f(1) = \frac{1}{4}$. By analysis of $f'(x) = x(x-1)(x-2)$ we could also have determined for example, that $x = 2$ yields a minimum. For $x > 2$, $f'(x) > 0$ (since each factor is positive) and for $1 < x < 2$, $f'(x) < 0$ (since exactly 2 of the 3 factors are positive). Hence $x = 2$ must yield a local minimum, since the function is decreasing on the left of $x = 2$ and increasing on the right of $x = 2$.

EXERCISES 4-2

For Exercises 1–10, (**a**) identify the domain, (**b**) give all critical points of the given function, and (**c**) give all relative maxima and minima, identifying each.

1. $f(x) = x^3$

2. $f(x) = (x+1)^2$

3. $f(x) = \dfrac{1}{2x-1}$

4. $f(x) = \sqrt{x}$

5. $f(x) = \sqrt[3]{x}$

6. $f(x) = \dfrac{1}{x^2+1}$

7. $f(x) = \sqrt{x^2-1}$

8. $f(x) = \sqrt{1-x^2}$

9. $f(x) = \dfrac{x+1}{x-1}$

10. $f(x) = \dfrac{x}{x^2-1}$

For Exercises 11–20, (**a**) determine the critical points for the specified domain; (**b**) state whether each critical point yields a relative maximum, relative minimum, or neither; and (**c**) find the absolute maximum and minimum.

11. $f(x) = x^2 + x$; $[-2, 8]$

12. $f(x) = 2x - 3$; $[0, 5]$

13. $f(x) = x^2 - 6x + 2$; $[0, 5]$

14. $f(x) = 4x^2 + 8x - 1$; $[-3, 3]$

15. $f(x) = x^3 + 3x^2 - 9x - 7$; $[-5, 5]$

16. $f(x) = x^4 - 2x^2$; $[-2, 2]$ HINT: Solve $f'(x) = 0$ by factoring.

17. $f(x) = \dfrac{1}{x^2}$; $[-1, 3]$, $x \neq 0$

18. $f(x) = \dfrac{x+1}{x-1}$; $[0, 3]$, $x \neq 1$

19. $f(x) = \dfrac{1}{x^2 - 1}$; $[-2, 3]$, $x \neq \pm 1$

20. $f(x) = \dfrac{x}{x^2 - 1}$; $[-2, 3]$, $x \neq \pm 1$

4-3. CONCAVITY

Concavity relates to the manner in which a graph is "bending." Consider the function $f(x) = x^3 - 3x^2 + 2x + 1$ whose graph is given in Figure 4-10. For $x > 1$, the graph is bending upward, or, as x increases, bending in a counterclockwise manner. Viewed another way, for $x > 1$, the graph appears to "hold water" and is concave upward. For $x < 1$, the graph is bending in a clockwise manner (as we draw it from the left), appears not to "hold water" and is concave downward.

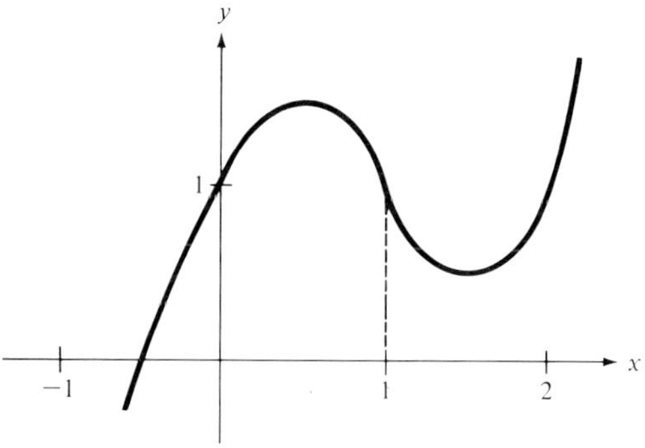

FIGURE 4-10

The manner in which a graph bends is closely related to changes in the slope of the tangent line. In Figure 4-11 we illustrate a graph that is bending upward for $x > x_1$. Several tangent lines have been drawn. For $x > x_1$, observe that near a point of tangency the graph always lies on or above the tangent line. Furthermore, for $x > x_1$, the slope of the tangent line increases as x increases, or in other words, the derivative $f'(x)$ is an increasing function. Such a graph is said to be *concave upward* for $x > x_1$.

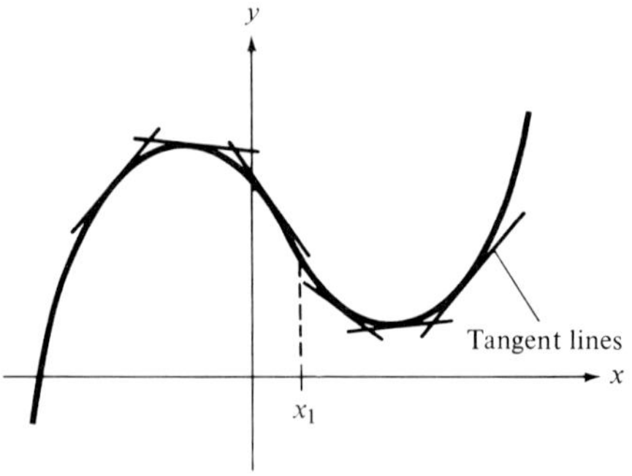

FIGURE 4-11

For $x < x_1$ the graph is bending downward. Near a point of tangency for $x < x_1$, the graph always lies on or below the tangent line. For $x < x_1$, the slope of the tangent line decreases as x increases, or in other words, the derivative $f'(x)$ is a decreasing function there. Such a graph is said to be *concave downward* for $x < x_1$.

Since the second derivative measures the rate of change of the first derivative, we note that the graph of $f(x)$ is concave upward at a point $x = x_0$ if $f''(x_0) > 0$, since this means that the first derivative is increasing at x_0. Similarly, the graph of $f(x)$ is concave downward at $x = x_0$ if $f''(x_0) < 0$, since this means that the first derivative is decreasing there.

For a function $f(x)$ for which $f''(x)$ exists we, therefore, have two easy criteria for concavity.

1. Look at the *first derivative*. If the first derivative is increasing at a point, the graph is concave upward there; if the first derivative is decreasing there, the graph is concave downward.

2. Look at the *second derivative*. If the second derivative is positive at a point, the graph is concave upward there; if the second derivative is negative at a point, the graph is concave downward there.

The graphs in Figures 4-10 and 4-11 illustrate extrema occurring at non-end points of the domain. We observe that at such a relative maximum the curve is concave downward, and that at such a relative minimum the curve is concave upward.

We can use these facts to develop our Test 3 or the so-called second derivative test for distinguishing between maxima and minima at non-end points of the domain.

Test 3 (The Second-Derivative Test): Let x_0 be any critical point for which $f'(x_0) = 0$. If $f''(x_0) < 0$, then $f(x_0)$ is a relative maximum; if $f''(x_0) > 0$, then $f(x_0)$ is a relative minimum.

This second derivative test is useful if it is not otherwise easy to tell whether a particular critical point yields a maximum or a minimum. It is also useful because it is a "mechanical" kind of test. We can apply it automatically when $f''(x_0)$ exists and is $\neq 0$. However, for many particular functions, we may have other information that lets us make decisions about extrema at least as easily as by use of the second derivative test. *Note:* If $f'(x_0)$ is zero and $f''(x_0)$ is also zero, the second derivative test fails. We might have a maximum, a minimum, or neither.

The use of the second derivative test is illustrated by the following examples.

Example 1. Let $f(x) = x^3 - 3x^2 + 2x - 1$. Then $f'(x) = 3x^2 - 6x + 2$ and $f''(x) = 6x - 6 = 6(x - 1)$. For all $x > 1, f''(x) > 0$ and the graph is concave upward. For all $x < 1, f''(x) < 0$ and the graph is concave downward. Any critical point to the right of 1 yields a relative minimum; any critical point to the left of 1 yields a relative maximum.

Example 2. Consider $f(x) = x^4 + 4x^3 + 6x^2 + 4x + 1$. Another form for $f(x)$ is $(x + 1)^4$. Then $f'(x) = 4x^3 + 12x^2 + 12x + 4 = 4(x + 1)^3$ and $f''(x) = 12x^2 + 24x + 12 = 12(x^2 + 2x + 1) = 12(x + 1)^2$. Observe that $f''(x) \geq 0$ for all x and the graph is concave upward everywhere. The function $f(x)$ can have no relative maxima. At $x = -1$, we have a relative (and absolute) minimum even though the formal second-derivative test fails.

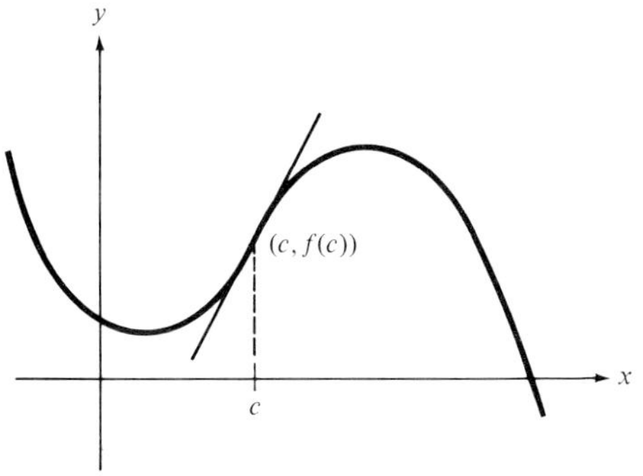

FIGURE 4-12

At some points on a graph, the tangent line may cross the graph, as illustrated in Figure 4-12 on p. 167. Such points are called *points of inflection,* and are defined as follows.

Definition 4-5 A point $(c, f(c))$ on the graph of $f(x)$ is called a point of inflection if $f'(c)$ exists and the sense of concavity of the graph changes at c.

For the graph in Figure 4-12, the graph is concave upward for $x < c$ and concave downward for $x > c$.

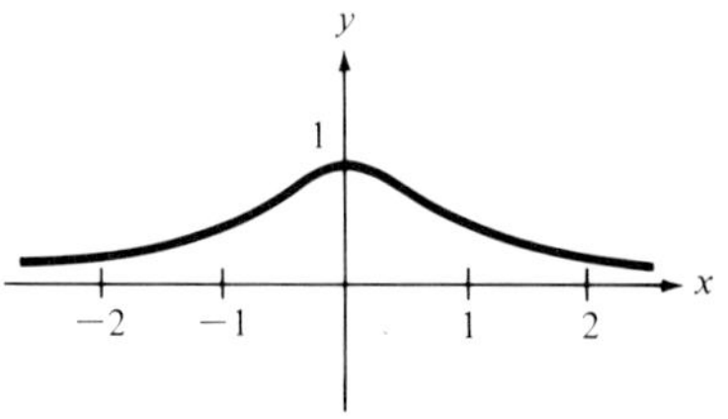

FIGURE 4-13

Example 3. Consider the function $f(x) = 1/(x^2 + 1)$ whose graph is shown in Figure 4-13 (this function was also considered in Section 4-1). We have $f'(x) = -2x/(x^2 + 1)^2$ and thus 0 is the only critical point of the function. We may recognize immediately that $f(0) = 1$ is both a relative and an absolute maximum of the function, and that the function has no relative minima. We compute $f''(x)$ to find the points of inflection and to check that $f''(0) \leq 0$ (which must be the case for $f(x)$ to have a maximum at $x = 0$).

$$f''(x) = \frac{-2(x^2 + 1)^2 - (-2x)(2)(x^2 + 1)(2x)}{(x^2 + 1)^4} = \frac{-2x^2 - 2 + 8x^2}{(x^2 + 1)^3}$$

$$= 2\frac{3x^2 - 1}{(x^2 + 1)^3}$$

For $x = 0$, we note that $f''(0) < 0$, as it should be. The possible points of inflection are given by setting $3x^2 - 1 = 0$ and thus are the points $(-1/\sqrt{3}, 3/4)$ and $(+1/\sqrt{3}, 3/4)$. Clearly, the graph is concave downward for $-1/\sqrt{3} < x < 1/\sqrt{3}$ and is concave upward for $x < -1/\sqrt{3}$ and for $x > +1/\sqrt{3}$.

Figure 4-14 presents a symbolic flowchart giving the procedures for finding, identifying, and evaluating relative extrema.

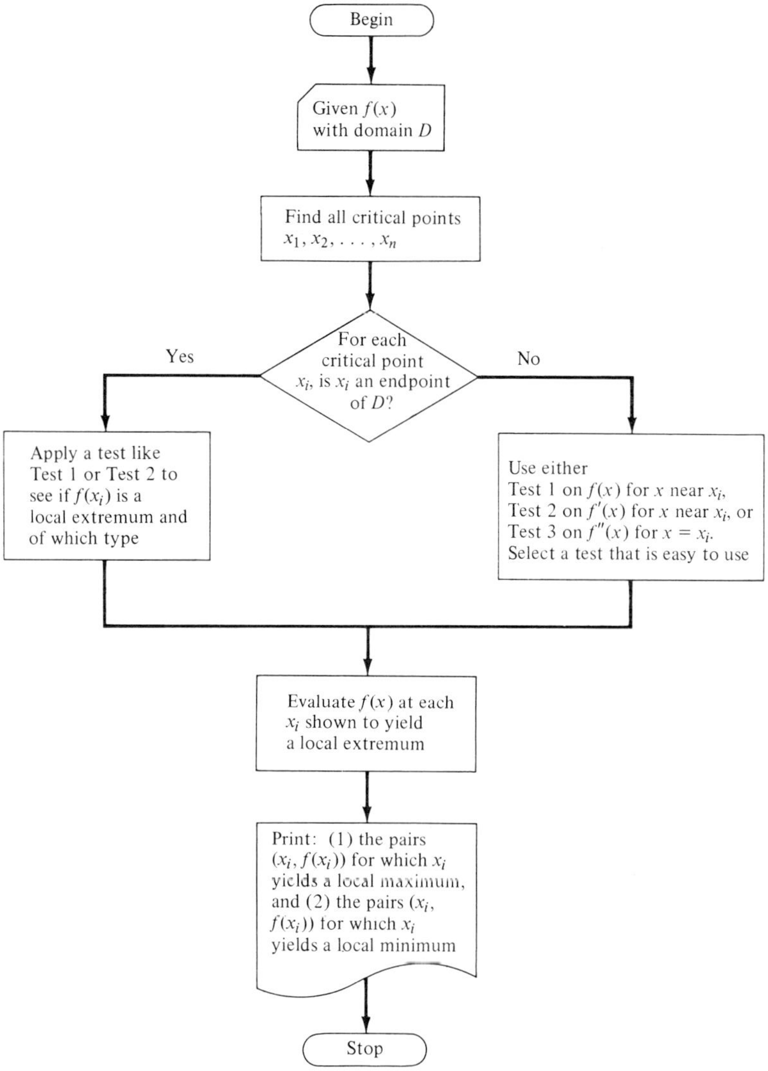

FIGURE 4-14

EXERCISES 4-3

In the following exercises, determine the critical points and use the second-derivative test to determine if the critical points yield local maxima or local minima.

1. $f(x) = x^2 - 6x + 1$

2. $f(x) = 1 + 2x - x^2$

3. $f(x) = 2x^3 + 3x^2 - 12x - 1$

4. $f(x) = 2 + 3x - x^3$

5. $f(x) = x^4 - 2x^2 - 5$

6. $f(x) = 1 + 12x^2 - 4x^3 - 3x^4$

HINT: Factor an x from $f'(x)$

7. $f(x) = \dfrac{1}{x^2 + 4}$

8. $f(x) = \dfrac{1}{x^2 - 1}$

9. $f(x) = \dfrac{x + 2}{x^2}$

10. $f(x) = \dfrac{x^2 + 1}{x^2 - 1}$

For each of the functions in the following exercises, (**a**) state where $f(x)$ is defined, (**b**) state where $f(x)$ is increasing or decreasing, (**c**) determine the critical points, (**d**) determine where $f(x)$ is concave upward or concave downward, (**e**) locate the inflection points, and (**f**) determine relative maxima and minima.

11. $f(x) = x^2 + 4x - 3$

12. $f(x) = 1 + 8x - x^2$

13. $f(x) = \dfrac{x^3}{3} + x^2 - 3x + 4$

14. $f(x) = 3 + 3x^2 - x^3$

15. $f(x) = x^4 - 2x^2 + 3$

HINT: Factor an x from $f'(x)$

4-4. ABSOLUTE EXTREMA

In Sections 4-2 and 4-3 we have studied ways of finding relative or local extrema of a function $f(x)$. To review, we first find all critical points, then determine which of these critical points yield relative extrema (perhaps using the second derivative test to help identify them), and finally evaluate $f(x)$ at these critical points that yield relative extrema.

Now we ask how we find *absolute* maxima and *absolute* minima. The guiding principle is the following: *Any absolute maximum is also a relative maximum, and any absolute minimum is also a relative minimum.*

This principle gives us a straightforward method of *starting* the problem of locating absolute extrema. We can list *all candidates* for points that yield absolute extrema by listing all points that yield relative extrema. Then, by evaluating the function at these points, we automatically list all candidates for absolute extrema.

If a function has an absolute extremum of either type, we now have a routine way of locating it—find all critical points and evaluate the function at all of them. The largest of these functional values is the absolute maximum (*if there is one*) and the smallest is the absolute minimum (*if there is one*.) For many practical problems, there will only be finitely many critical points and it is possible to find all of them. Thus we can locate the absolute extrema if there are any.

How do we know when a function has an absolute maximum or absolute minimum? For many practical problems, the following property guarantees that there will be an absolute extremum of the type we seek.

Property 4-2 If $f(x)$ is continuous over the *closed* interval $[a, b]$, then $f(x)$ has an absolute maximum and an absolute minimum over $[a, b]$.

The proof of this property depends on concepts not covered in this book. We simply accept the property as a fact. The property asserts that there exist numbers c and d in $[a, b]$ such that $f(c) \geq f(x)$ and $f(d) \leq f(x)$ for every x in $[a, b]$. The numbers $f(c)$ and $f(d)$ are the absolute maximum and absolute minimum, respectively, and c and d are points that yield them. Recall from Section 4-1 that there may be more than one point yielding either the absolute maximum or the absolute minimum.

Figure 4-15 presents a symbolic flowchart telling how to find the absolute maximum of a function *if it exists.* A similar flowchart can be

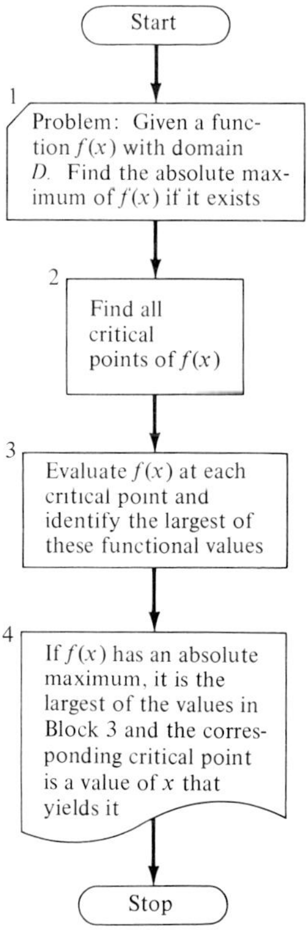

FIGURE 4-15

prepared for the absolute minimum. The procedure for identifying an absolute extremum and the points that yield it is, therefore, clearcut *if* we know or can find out that the given function has an absolute extremum of the type sought.

Example 1. Find the absolute maximum and absolute minimum of $f(x) = 3x^4 + 4x^3 - 12x^2 + 24$ over the interval $[-3, 3]$. Since $f(x)$ is a polynomial, it is continuous for all x. Since $f(x)$ is continuous and the interval $[-3, 3]$ is closed, then an absolute maximum and an absolute minimum must exist.

To determine the critical points, we determine $f'(x)$ and factor the result:

$$\begin{aligned} f'(x) &= \frac{d}{dx}(3x^4 + 4x^3 - 12x^2 + 24) \\ &= 12x^3 + 12x^2 - 24x \\ &= 12x(x^2 + x - 2) \\ &= 12x(x + 2)(x - 1) \end{aligned}$$

The zeros of $f'(x)$ are $x = 0$, $x = -2$, and $x = 1$. The critical points of $f(x)$ are these zeros plus the end points, $x = -3$ and $x = 3$.

Next, we evaluate $f(x)$ at each of the critical points, as shown in Table 4-1. The absolute minimum is -8 and occurs at $x = -2$; the absolute maximum is 267 and occurs at $x = 3$.

TABLE 4-1

Critical point	Functional value
$x = -3$	$f(-3) = 51$
$x = -2$	$f(-2) = -8$
$x = 0$	$f(0) = 24$
$x = 1$	$f(1) = 19$
$x = 3$	$f(3) = 267$

The conditions that the interval $[a, b]$ be closed and that the function $f(x)$ be continuous are vital to Property 4-2, as the examples in Figure 4-16 illustrate. In the graph on the left we consider $y = x^2 + 1$ over the *open* interval $(0, 1)$. There can be *no* c or d on $(0, 1)$ yielding a maximum or a minimum. For example, for any x_0 on $(0, 1)$, the point $\frac{1}{2}x_0$, which is also on $(0, 1)$, has a lesser functional value.

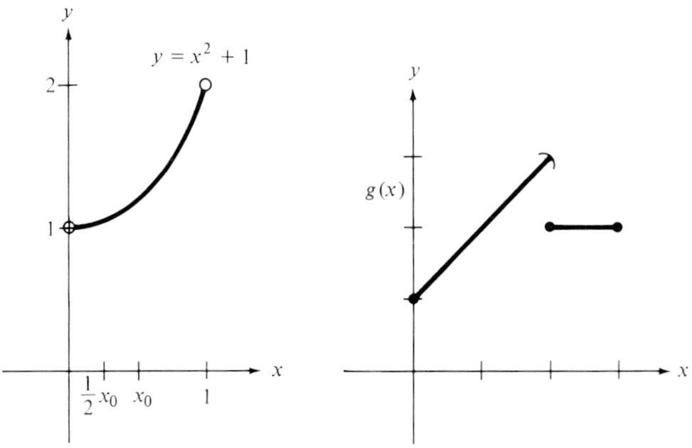

FIGURE 4-16

For the noncontinuous function on the right defined by

$$g(x) = \begin{cases} x + 1 & \text{whenever } 0 \le x < 2 \\ 2 & \text{whenever } 2 \le x \le 3 \end{cases}$$

we note that $g(0) = 1$ is the minimum but $g(x)$ has no maximum. There is no point on [0, 3] with functional value 3, but there are points with functional values as close to 3 as we wish.

In many practical problems, the function $f(x)$ that we wish to maximize or minimize can be readily recognized to be continuous and may have a specified domain D. Thus we concentrate on analyzing extrema for continuous functions. Frequently, the domain is for all $x \ge 0$ or perhaps for all x in some closed interval. The practical problem customarily restricts this domain.

We have the following cases for $f(x)$ continuous. We assume that $f(x)$ is differentiable except possibly at finitely many points.

Case 1

The domain D of $f(x)$ is a closed interval. Then $f(x)$ has an absolute maximum and an absolute minimum, which may be found by the critical point analysis as outlined in the symbolic flowchart in Figure 4-15. This was the case in Example 1 presented earlier. This procedure also works if the domain is a finite union of closed intervals.

Case 2

The domain D of $f(x)$ includes all large positive x or all large negative x, or both. In this case, $f(x)$ may or may not have an absolute maximum or an absolute minimum. We may evaluate (or think about evaluating) $f(x)$ at large positive or negative values of x to get relevant information. For example, let the domain D be the set of all $x \geq 0$. If for large x, $f(x)$ is large, then we can conclude from Property 4-1 that $f(x)$ must have a minimum on D. As illustrated in Figure 4-17, we regard D as the union of two domains D_1 and D_2, with D_1 being the interval $[0, h]$ and D_2 the set of all points for $x > h$ with h being large. Since D_1 is closed and $f(x)$ is continuous, $f(x)$ must have a minimum on D_1. Observe that for all $x \in D_2$, we have $f(x) > f(h)$. This says that all $f(x)$ with x on D_2 exceed one of the functional values, namely $f(h)$, on D_1. Therefore, the minimum over D_1 must be less than all functional values over D_2, and thus the minimum on D_1 must be the minimum on D.

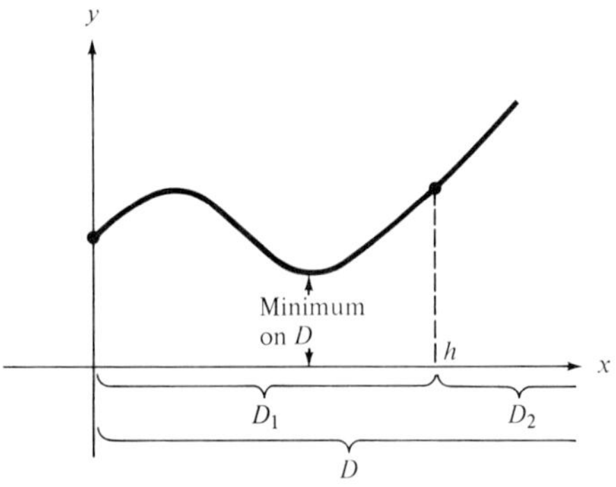

FIGURE 4-17

Observe that there would be no maximum on D. Note also that if $\lim_{x \to \infty} f(x) = +\infty$, then for large x, $f(x)$ is large. But generally we can observe the latter fact without formally taking limits.

Example 2. Let $f(x) = 3x^4 + 4x^3 - 12x^2 + 24$, but let the domain be all $x \geq -3$. The critical points are now $x = -3$, $x = -2$, $x = 0$, and $x = 1$. Recall that $f(-3) = 51$, $f(-2) = -8$, $f(0) = 24$, and $f(1) = 19$. *If* there is an absolute minimum, it is $f(-2) = -8$; *if* there is an absolute maximum, it is $f(-3) = 51$. However, we observe that for x large, $f(x)$ is large. Therefore, there is no absolute maximum. There is an absolute minimum, namely, $f(-2) = -8$.

Case 3

The domain D of $f(x)$ contains an open interval but not the corresponding closed interval. In this case, $f(x)$ may or may not have a maximum or a minimum on D. We may test $f(x)$ near an end point of such an interval, where the end point is not in D. This procedure frequently reveals whether the candidates for absolute maxima or minima are really maxima or minima on D.

Example 3. Consider $f(x) = (x - 1)/(x - 2)$ with specified domain D the set of all points for $2 < x \leq 5$ as illustrated in Figure 4-18. Since for $x > 2$ and near 2, $(x - 1)/(x - 2)$ is very large, we see immediately that $(x - 1)/(x - 2)$ can have no absolute maximum on D. We compute

$$f'(x) = \frac{(x-2)-(x-1)}{(x-2)^2} = \frac{-1}{(x-2)^2}$$

and see that $f'(x)$ is never zero (on D or any place). The only critical point on D is the end point 5. But $f(5) = \frac{4}{3}$. This is the minimum on D. We know this, since $f'(x)$ is negative on D and hence decreasing on D. Or we know D may be regarded as the union of D_1 and D_2, with D_1 being the set of all x for which $2 < x \leq h$, and D_2 being the set of all x for which $h \leq x \leq 5$. Then $f(5) = \frac{4}{3}$ is the minimum on D_2 and, for any $x \in D_1, f(x) > f(5)$, so we have $f(5) = \frac{4}{3}$, a minimum on D.

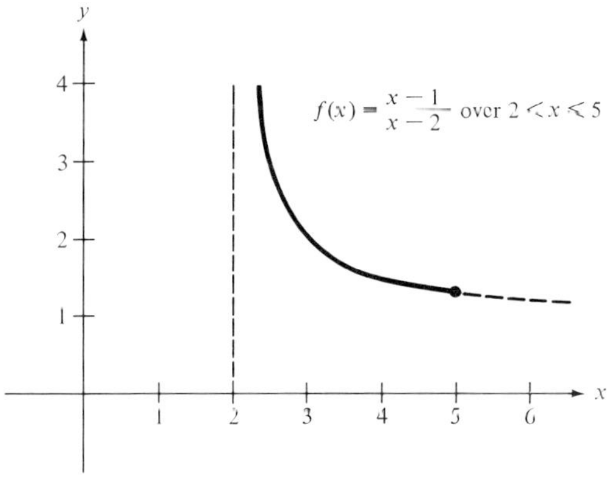

FIGURE 4-18

Example 4. If the function in Example 3 is altered to have the specified domain D, the set of all $x > 2$, then $(x - 1)/(x - 2)$ has neither a maximum nor a minimum on D, since on D there are *no* critical points. There can be no maximum or minimum except at a critical point.

The symbolic flowchart in Figure 4-19 is an extension of the flowchart in Figure 4-15 and encompasses the decision process to determine if the continuous function $f(x)$ has an absolute maximum on D.

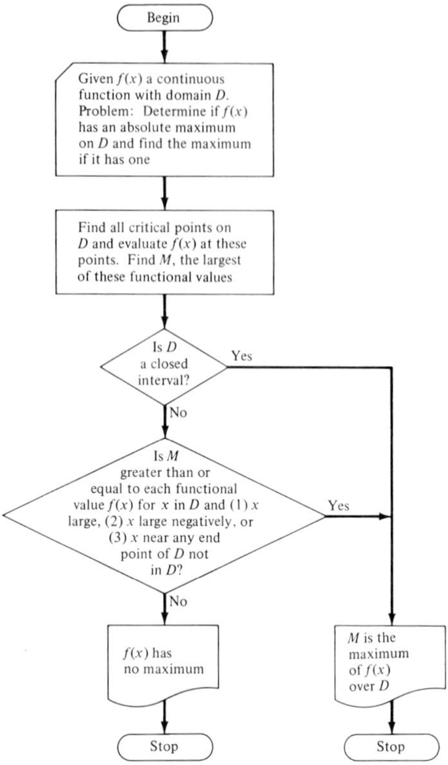

FIGURE 4-19

EXERCISES 4-4

For each function $f(x)$, state whether $f(x)$ has an absolute minimum or an absolute maximum or both on the specified domain D. Use an analysis like that of the text. Do *not* try to find the maximum or the minimum.

	$f(x)$	D		$f(x)$	D
1.	$x^4 - x^3 + x^2 + 10x$	$x \geq 2$	**2.**	$\dfrac{1}{x-1}$	$x > 1$
3.	$\dfrac{1}{x^2+10}$	all x	**4.**	$\dfrac{x}{x^2+10}$	all x
5.	$\dfrac{x+2}{x^2-9}$	$0 \leq x \leq 1$	**6.**	$x^6 - 1$	$-5 \leq x \leq 0$
7.	$\dfrac{x^3+1}{x^2+5}$	all x	**8.**	$\dfrac{1}{x+1} + \dfrac{2}{x+2}$	$x \geq 0$

	$f(x)$	D		$f(x)$	D
9.	$x + x^{100}$	$x \geq 0$	**10.**	$\frac{1}{x} + x$	$x \geq 1$

For each of the following functions $f(x)$ with domain D, find the absolute maximum and the absolute minimum *if they exist*. Otherwise, state that they do not exist. First, find all critical points and the values of the function at these points.

	$f(x)$	D		$f(x)$	D
11.	$(x-1)^4$	$0 \leq x \leq 3$	**12.**	$\frac{2}{(x-1)^2}$	$0 \leq x \leq 3$
13.	$x^3 + 8x$	$x \geq 1$	**14.**	$\frac{1}{x+1}$	$x \geq 1$
15.	$x^3 + 12x + 100$	all x	**16.**	$x^4 - 4x^3$	$-2 \leq x \leq 0$
17.	$\frac{x+1}{x-1}$	$x > 2$	**18.**	$\frac{x+1}{x-1}$	$x \geq 2$

For the following functions over the specified domains, find the absolute maximum, the absolute minimum, and the points that yield them.

	$f(x)$	D		$f(x)$	D
19.	$x^3 - 3x$	$-10 \leq x \leq 10$	**20.**	$x^2 - 4x + 6$	$0 \leq x \leq 2$
21.	$\frac{1}{x+1}$	$0 \leq x \leq 2$	**22.**	$2x^3 - 24x - 24$	$-3 \leq x \leq 3$
23.	$3x^2 - x - 1$	$0 \leq x \leq 1$	**24.**	$\frac{1}{x^2+1}$	$-2 \leq x \leq 2$
25.	$x^2 + 1$	$-2 \leq x \leq 2$	**26.**	$x^2 + 4$	$0 \leq x \leq 4$

4-5. APPLICATIONS OF MAXIMUM AND MINIMUM CONCEPTS

In this section we wish to apply the concepts in the previous sections to problems in the field of business. A wide variety of problems call for determining the value of one or more variables that optimize (i.e., maximize or minimize) some function (called the criterion function). In business, the classical problem is that of maximizing profits, but a variety of optimization problems exist. In this section, we shall consider problems in which the function to be optimized can be written as a function of only one variable (called the independent variable). In business applications, we are almost always concerned with absolute extrema and the point or points that yield them. We regard the study of relative extrema as a means to the end of identifying absolute extrema.

In solving optimization problems in business, the first step is to relate the criterion function (which we wish to optimize) to the variable or variables whose values may be specified. This entails an analysis of the problem and in practice is often the most difficult aspect of optimization problems. The ability to successfully write the proper equations or functions only comes with practice. One must first attempt to understand the problem at hand and then express the relationships involved in formulas or equations.

Once the relationships have been written, the maximum or minimum may often be identified using the procedures in the previous sections. The function to be optimized is expressed as a function of the independent variable. The derivative is obtained and the critical points located. Each critical point is then examined for a possible local maximum or minimum. The second derivative test or other convenient approaches may be used.

In problems with a business interpretation, the very nature of the problem frequently is sufficient to conclude that an absolute maximum or a minimum must exist. In solving such a problem, only one critical point may be obtained. The tendency to assume immediately that this critical point corresponds to the desired maximum or minimum should be avoided, but instead it should be verified that (1) the critical point does result in a maximum or minimum as expected, and that (2) the value of the independent variable corresponding to the critical point is reasonable (e.g., a price cannot be negative or so large as to be ridiculous). Should either or both of these "reasonableness tests" not hold, an error has been made, perhaps in writing the relationships describing the process.

As an example, suppose that in working a problem of maximizing profits only one critical point is found. Using the second derivative, a positive value is obtained, indicating a minimum. Something may be wrong. Alternatively, suppose that the critical point corresponds to a maximum, but the value of the selling price (the independent variable) is negative. Then either (a) an error has been made or (b) no goods should be produced and sold.

Example 1. A rancher purchases 600 meters* of fence and wants to enclose a rectangular region of maximum area. If we let x be one length (in meters) of the rectangle and y the other dimension, the area A (in square meters) is given by

$$A = xy$$

* A meter is a measure of length in the metric system, and is slightly longer than a yard (3 feet).

But since the rancher has only 600 meters of fence, x and y must satisfy the equation

$$2x + 2y = 600$$

Thus the value of y is determined if x is determined. We solve for y, obtaining $y = 300 - x$, and substitute into the expression for A:

$$A = x(300 - x) = 300x - x^2$$

We now have A as a function of the single independent variable x, and A is the desired criterion function. We wish to find the value of x, if any, that maximizes A.

Figure 4-20 presents the graph of A as a function of x. When $x = 0$, the area A is zero. When $x = 300$, then $y = 300 - x = 0$ and consequently A is zero. Since negative values for x, y, or A would be meaningless, we conclude that $0 \leq x \leq 300$.

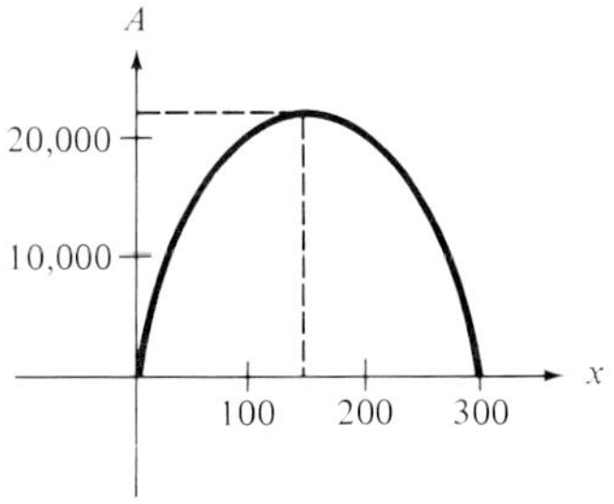

FIGURE 4-20

The end points $x = 0$ and $x = 300$ produce $A = 0$, and thus produce local minima, not maxima. To find all other critical points, we differentiate A with respect to x:

$$\frac{dA}{dx} = 300 - 2x$$

The only critical point (given by $dA/dx = 0$) is $x = 150$. Since the second derivative $[(d^2A/dx^2) = -2]$ is negative everywhere, this critical point yields a relative maximum. The value $x = 150$ is within the bounds $0 \leq x \leq 300$ established by the nature of the problem. Since $y = 300 - x$ and $x = 150$, then $y = 150$, and the rectangle that maximizes the area is a square, 150 meters on each side. The maximum area is given by $A = xy = (150)^2$ or 22,500 square meters.

Example 2. In some (but not all) business operations, we may wish to set the price so as to maximize the return from sales. Such a situation would exist when the profit is a fixed percentage of the sales (e.g., a salesman working on commission), or in a new business venture where evidence of a large sales volume would make the stock more attractive to potential investors.

The return R from the sale of q units of a product at price p per unit clearly must be given by

$$R = qp$$

However, the quantity sold, q, normally depends on the selling price p. Suppose that a businessman believes or knows that the following quantity-price relationship applies:

$$q = 10{,}000 - 10p$$

Then R as a function of p is

$$R = (10{,}000 - 10p)p = 10{,}000p - 10p^2$$

We wish to find the value of p that maximizes R.

Clearly, the selling price p must be positive. In the quantity-price relationship, observe that if p exceeds 1000, then q will be negative, which is meaningless. Consequently, we conclude that $0 \leq p \leq 1000$. The graph of R as a function of p is given in Figure 4-21.

As in Example 1, the end points are critical points that yield minima, not maxima. To locate the other critical points, we examine the derivative

$$\frac{dR}{dp} = 10{,}000 - 20p$$

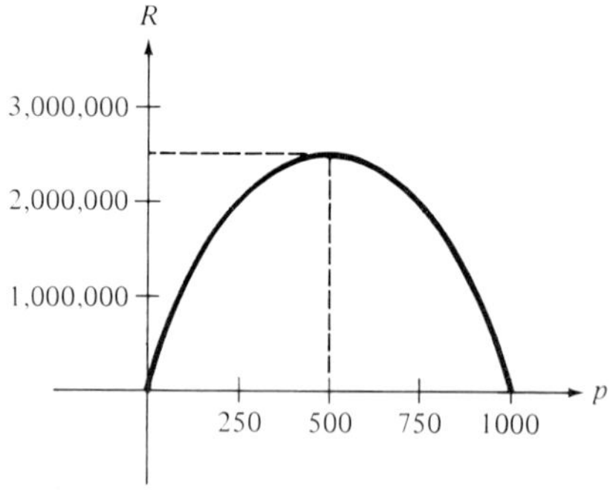

FIGURE 4-21

The only critical point (given by $dR/dp = 0$) is $p = 500$, which is in the specified domain for p. Since the second derivative, $d^2R/dp^2 = -20$, is negative everywhere, this must yield a maximum. The corresponding values of q and R are given by $q = 10{,}000 - 10p = 5000$ and

$$R = qp = (5000)(500) = 2{,}500{,}000$$

Example 3. The classical maximization problem in business is that of maximizing profit. For product x, suppose the revenue R (in millions of dollars) as a function of quantity sold q (in thousands of units) is

$$R = 3 \cdot (\sqrt{q})$$

Furthermore, suppose that the production cost C (in millions) as a function of quantity (in thousands) is

$$C = \tfrac{1}{4}q^2 + 1$$

The profit P is given by

$$P = R - C = 3 \cdot (\sqrt{q}) - \tfrac{1}{4}q^2 - 1$$

We wish to find the value of q that maximizes P.

Here we know that $q > 0$ and that q must be such that $P > 0$. Hence q cannot be as much as 5, since $3\sqrt{5} - \frac{1}{4} \cdot 25 - 1 < 0$. In practical problems, we usually do not need precise values specifying the domain. Approximate ones will do. The graph of P as a function of q is given in Figure 4-22.

To obtain the critical points, we determine the derivative:

$$\frac{dP}{dq} = \frac{3}{2\sqrt{q}} - \frac{q}{2}$$

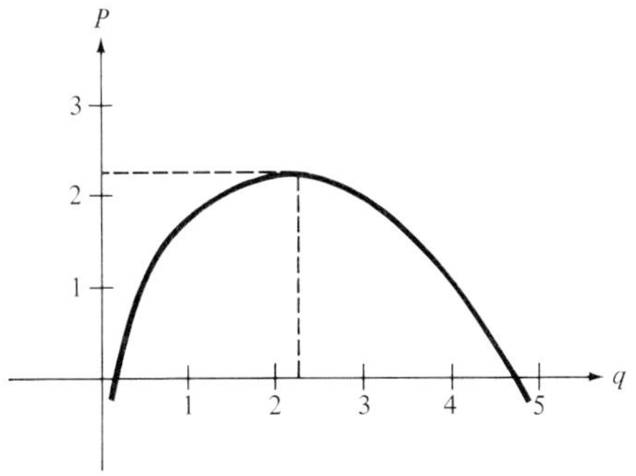

FIGURE 4-22

The derivative equals zero when

$$\frac{3}{2\sqrt{q}} = \frac{q}{2}$$

Thus we get $q^{3/2} = 3$, or $q^3 = 9$, or $q = \sqrt[3]{9} \approx 2.08$. To be certain that this value yields a maximum, we investigate the second derivative:

$$\frac{d^2P}{dq^2} = -\frac{3}{4q^{3/2}} - \frac{1}{2}$$

We observe that for $q > 0$, both terms are negative, and consequently $d^2P/dq^2 < 0$ at $q \approx 2.08$. Therefore, a maximum occurs at $q \approx 2.08$, and the corresponding value of P is $3\sqrt{2.08} - \frac{1}{4}(2.08)^2 - 1 \approx 2.245$. Since P is positive, the value of q is plausible. Translated into production figures, if we produce 2080 units we should make a profit of \$2,245,000.

EXERCISES 4-5

---

1. The sum of two numbers x and y is 30. Find the values of x and y that maximize their product, xy.

2. The sum of two numbers x and y is 20 and each is positive or zero. Find the values of x and y that *minimize* their sum of squares, $x^2 + y^2$. Find the values of x and y that maximize $x^2 + y^2$.

3. Rework Exercise 1, except use the product of their squares, x^2y^2, and require that $x \geq 0$ and $y \geq 0$.

4. Modify Example 1 in the text by assuming that the rancher will use an existing fence as one side of the enclosure.

5. In Example 2 in the text, we expressed R as a function of p and determined where $dR/dp = 0$. An alternate approach would be to express R as a function of q and determine the value of q at which $dR/dq = 0$. Solve the problem this way. Your answers should be consistent with those obtained in Example 2.

6. Solve an exercise like Example 3 in the text using the following functions for R and C:

$$R = \frac{6q}{q + 2}$$

$$C = \frac{q^2}{q + 2}$$

Use the same approach.

7. An alternative approach to the problem in Example 3 is via the marginal return and marginal cost as discussed in Section 3-4. The profit P is the revenue R less cost C:

$$P = R - C$$

When R and C are functions of the quantity q, the value of q that maximizes (or minimizes) P satisfies $dP/dq = 0$. Since $P = R - \bar{C}$, then

$$\frac{dP}{dq} = \frac{dR}{dq} - \frac{dC}{dq}$$

Therefore, $dP/dq = 0$ when

$$\frac{dR}{dq} = \frac{dC}{dq}$$

Since dR/dq is the marginal revenue and dC/dq is the marginal cost, then, at the maximum profit, the marginal revenue equals the marginal cost. For Example 3 in the text, determine the marginal revenue and the marginal cost. Equate these and solve for q. Your answer should check that obtained in the text.

8. Solve Exercise 6 using the marginal revenue and marginal cost approach as described in Exercise 8.

9. Let the total manufacturing cost to produce x units be given by

$$C = 900 + 20x + x^2$$

The average cost of producing a unit is given by C/x (the total cost divided by the number of units produced). Determine the value of x that *minimizes* the *average cost*. What is the minimum average cost?

10. Let the revenue generated from selling x units of a product be given by

$$R = 400x - 900 - x^2$$

The average revenue per unit is R/x (the total revenue divided by the number of units produced). Determine the value of x that *maximizes* the *average revenue*. What is the maximum average revenue?

11. The cost of operating a vehicle traveling at speed v (kilometers/hour*) is $20 + 0.2v$ in cents per kilometer. Suppose that the driver is paid \$9 per hour. At what speed should the vehicle be driven to minimize the total cost of a 600-kilometer trip, assuming that the driver observes the 90 kilometer per hour speed limit?

* A kilometer is a metric unit of length. The English mile equals approximately 1.6 kilometers.

12. A contractor has been requested to build a road from a mine to a processing plant, as illustrated in Figure LL. One approach is to widen an existing road for 5000 meters at a cost of \$75 per meter, and then construct a new 3000 meter road to the mine perpendicular to the existing road at a cost of \$150 per meter. Alternatively, he may widen only part of the existing road, say x meters, and then construct a new road of length $\sqrt{3000^2 + (5000 - x)^2}$. Determine the value of x that would produce the road at the least cost, assuming that the new road also costs \$150 per meter.

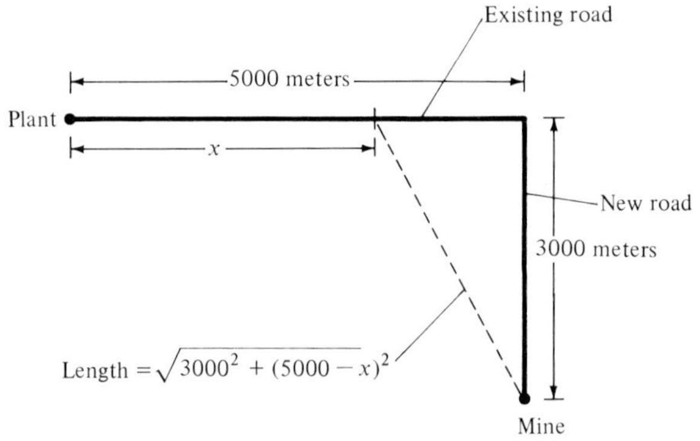

FIGURE LL

13. Refer to Example 1 in the text. Suppose that the rancher prefers a triangular-shaped pen with at least two sides of equal length. As illustrated in Figure MM, let x be the length of each of the two equal sides and let b be the length of the third side. Since the fence must still total 600 meters, then $2x + b = 600$. The area of the triangle is $\frac{1}{2}bh$, where $h = \sqrt{x^2 - (b^2/4)}$. Find the value of b or x that maximizes the area. Note that when either x or b is determined, the other is also (since $2x + b = 600$).

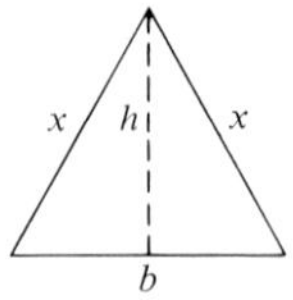

FIGURE MM

14. The unit cost to manufacture product Z is given by

$$C = \frac{16 + 4x + x^2}{x}$$

where x is the number of units produced in thousands. Because of competition, we shall be forced to sell product Z for \$1 per unit. How much, if any, of Z should we produce, and what will be our profit? If we cannot make a profit, we shall not produce any.

15. The revenue from the manufacture and sale of product Y is given by

$$R = 10 + 8x - 0.01x^2$$

where x is the number of units produced. If the maximum production capacity of our factory is 250 units, how many units of Y should we produce and what is the corresponding revenue?

*4-6. USING DERIVATIVES TO GRAPH POLYNOMIAL FUNCTIONS

The first and second derivatives can be very useful in deciding what the graph of a polynomial function must look like and in actually sketching the graph. As defined earlier, a polynomial in x is a function that can be written as a sum of terms each of which is a constant times a positive integral power of x. A polynomial is described by its highest-degree term having a nonzero coefficient. Thus $5x^3 - 6x^2 + 2$ is a third-degree polynomial, whereas $2x^{19} - 8x^{14} + 6x^5 - 97$ is a 19th-degree polynomial. As noted in Chapter 1, first-degree polynomials like $2x + 3$ or $-5x + 17$ have graphs that are straight lines. Zero degree polynomials like 6 or -2 also have graphs that are straight lines, but they are horizontal.

Indeed, if $f(x)$ is *any* polynomial in x, then $f(x)$ is defined and continuous for all values of x. The graph of $f(x)$ must cross each vertical line exactly once. Also, the first derivative $f'(x)$ of $f(x)$ exists for each value of x and may be routinely computed by the elementary formulas for differentiation. The first derivative of $f(x)$ is itself a polynomial of degree one less than the degree of $f(x)$. Thus the second derivative can also be routinely computed and is also a polynomial. For example, if

$$f(x) = x^5 - 6x^3 + 12x^2 - 5$$

then

$$f'(x) = 5x^4 - 18x^2 + 24x$$

and

$$f''(x) = 20x^3 - 36x + 24$$

In graphing $f(x)$, there is an important distinction between the graph of $f(x)$ if the highest-degree term in $f(x)$ is of odd degree or of even degree. An *odd*-degree polynomial like $3x + 2$ or $5x^3 - 6x^2 + 2x$ or $-x^{19} - x^{16} + 5$ has the property that its graph crosses every horizontal line at least once and thus $f(x)$ has no absolute extrema. Indeed, for large x the highest-degree term dominates; thus, if the coefficient of the highest-degree term is positive, then $f(x)$ must be large when x is large and must be large negatively when x is large negatively. Just the opposite is true if the coefficient of the highest-degree term is negative. These graphs are illustrated in Figure 4-23.

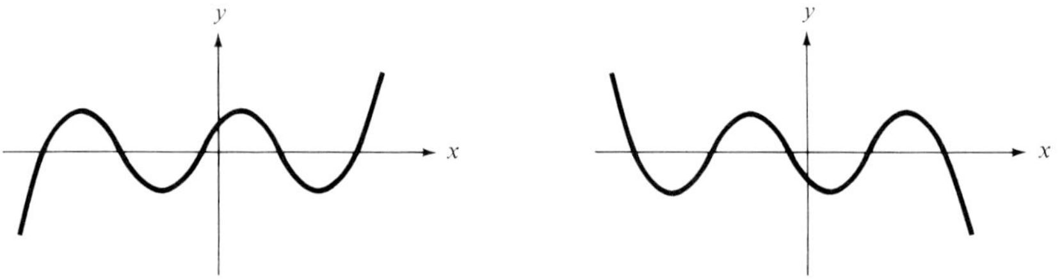

Highest-degree term is positive   Highest-degree term is negative

FIGURE 4-23

For any *even*-degree polynomial such as $x^{100} - x^{87} + 6x - 22$, the highest-degree term also dominates for large x. Thus, for the example above, the function is large positively when x is large positively *or* negatively. By the type of analysis in Section 4-4, such a function must have an absolute minimum but can have no absolute maximum. If the leading coefficient were negative, then for all large positive or negative x, the function is large negatively, and $f(x)$ has an absolute maximum but no absolute minimum. These graphs are illustrated in Figure 4-24.

So much for the general characteristics of the graph. How do we see what it looks like? In many cases, it is inconvenient to find accurately the

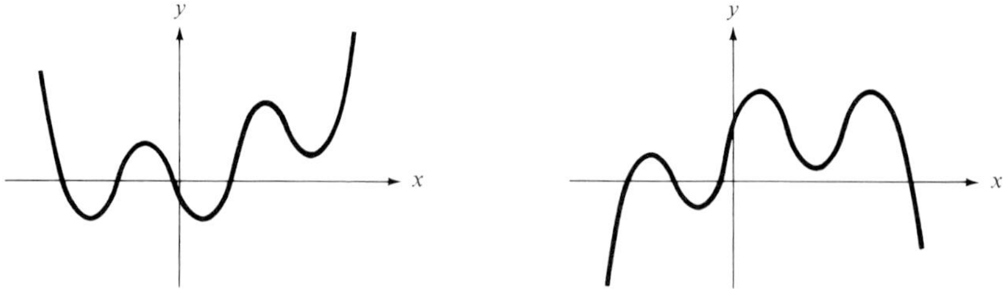

Highest-degree term is positive   Highest-degree term is negative

FIGURE 4-24

zeros of the function, i.e., the x coordinates of the points where the graph crosses the x axis. But since the degree of the derivative is one less than that of the function itself, it may be possible, or even easy, to find where the derivative is zero. If so, we can find all critical points and hence all local maxima and minima. Computing the second derivative lets us distinguish the maxima from the minima (in many cases) and also tells us where the curve is concave downward or concave upward.

Certainly, if our original polynomial is a cubic (or third-degree) polynomial, we can always find the critical points, since we can find the roots of any quadratic equation. The second derivative is a linear, first-degree polynomial, and we can always tell when a linear function is positive or negative and thus where the original function is concave upward and where concave downward. By the use of derivatives we can rather accurately graph any third-degree polynomial showing all local maxima and minima and the inflection point.

Example 1. Sketch the graph of $f(x) = x^3 - 3x^2 - 9x + 3$. Since $f(x)$ is a polynomial, it is defined for all x and is continuous everywhere. The function is large positively for large x and is large negatively for large negative x. The zeros of $f(x)$ are not easily determined, so we ignore them.

The first derivative is

$$f'(x) = 3x^2 - 6x - 9 = 3(x^2 - 2x - 3) = 3(x + 1)(x - 3)$$

The critical points occur at $x = -1$ and $x = 3$.

Finally, we determine the second derivative:

$$f''(x) = \frac{d}{dx}(3x^2 - 6x - 9) = 6x - 6$$

Since $f''(-1) < 0$, then a relative maximum occurs at $x = -1$. Since $f''(3) > 0$, a relative minimum occurs at $x = 3$. Since $f''(x) = 0$ at $x = 1$, a point of inflection occurs there. For $x > 1, f''(x) > 0$, and the graph is concave upward. For $x < 1, f''(x) < 0$, and the graph is concave downward.

In sketching the graph in Figure 4-25, the two most useful points are those identified with relative extrema, namely, $(-1, 8)$ and $(3, -24)$. The next most useful point is the point of inflection, namely, $(1, -8)$. Knowing these three points with their significance and knowing the concavity, the graph in Figure 4-25 is readily constructed. A fourth point that is easy to determine is that at which $x = 0$, namely the point $(0, 3)$.

Example 2. Sketch the graph of $f(x) = 2x^4 - x^2 - 1$. Since $f(x)$ is of even degree and the leading coefficient is positive, we note that for x large positively or negatively, $f(x)$ is large. We compute $f'(x)$.

$$f'(x) = 8x^3 - 2x = 2(4x^3 - x) = 2x(4x^2 - 1)$$

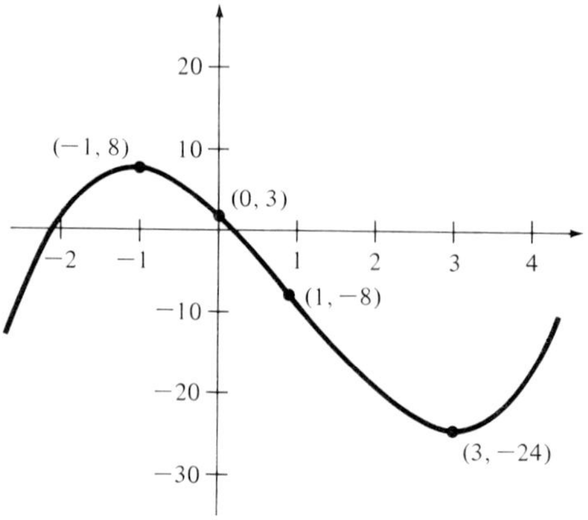

FIGURE 4-25

Equating $f'(x)$ to 0, we see that the zeros of $f'(x)$ are $x = 0$ and $x = \pm\frac{1}{2}$. Thus the critical points of $f(x)$ are $-\frac{1}{2}$, 0, $\frac{1}{2}$. We have

$$f''(x) = 24x^2 - 2 = 2(12x^2 - 1)$$

For $x = \pm\frac{1}{2}$, $f''(x) > 0$. For $x = 0$, $f''(x) < 0$. Thus $x = \pm\frac{1}{2}$ yield local minima and $x = 0$ yields a local maximum. We get $f(-\frac{1}{2}) = f(\frac{1}{2}) = -\frac{9}{8}$ and $f(0) = -1$. The graph is concave upward for $f''(x) > 0$, i.e., for $12x^2 - 1 > 0$ or $x^2 > \frac{1}{12}$ or $x < -1/(2\sqrt{3})$ and $x > +1/(2\sqrt{3})$. For $-1/(2\sqrt{3}) < x < 1/(2\sqrt{3})$ the curve is concave downward. The inflection points occur at $\pm 1/(2\sqrt{3})$. Plotting the local maxima and minima, we get a graph as indicated in Figure 4-26. It should also be noted that ± 1 are zeros of $f(x)$. Hence the graph crosses the x axis at ± 1.

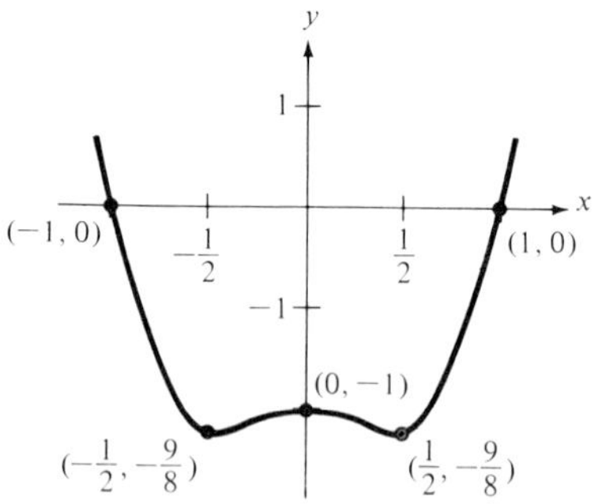

FIGURE 4-26

EXERCISES 4-6

For each of the following polynomials (**a**) find all critical points (if any). (**b**) Find all local maxima and minima. (**c**) Find where the graph is concave upward and where concave downward. (**d**) Graph the function.

1. $x^3 - 3x + 1$

2. $x^3 + 3x - 1$

3. $x^3 + 12x$

4. $x^3 - 12x$

5. $x^4 + 2x^2 - 1$

6. $x^4 - 2x^2 - 1$

7. $-x^4 - 8x^2$

8. $-x^4 + 8x^2$

9. $3x^3 - 9x^2 + 9x - 2$

10. $x^3 + 3x^2 + 3x + 1$

11. $-x^3 + x^2 + x$

12. $-x^3 - x^2 + x + 1$

*4-7. NEWTON'S METHOD

As indicated earlier in this chapter, finding the location of relative maxima and minima customarily involves determining the critical points, which are the end points of the domain and the points where the first derivative equals zero or is undefined. Locating the critical points where the first derivative equals zero means solving the equation

$$f'(x) = 0$$

Thus, in finding extrema, we are again led to the problem of finding the roots of an equation.

In our previous examples the problems were carefully selected so that the roots could be readily determined. Unfortunately, this is not the case for problems in general, especially many of those of practical interest. In Section 2-8 we presented a technique involving bracketing the root and interval halving that could be used to solve for such roots to any desired level of accuracy. In this section we present another technique, known as Newton's method, that lets us approximate the roots of a function by using the derivatives of that function. The method is named for the noted mathematician and physicist Sir Isaac Newton (1642–1727), one of the developers of the calculus and considered by some scholars to be as illustrious as any scientist who ever lived.

The idea of Newton's method is illustrated in Figure 4-27.

A tangent line may be regarded as approximating the graph of a function near the point of tangency. Suppose we know that there is a root of $f(x) = 0$ near a point a_1. Then find the tangent line to the graph at $(a_1, f(a_1))$ and locate the point a_2 where this tangent line crosses the x axis. (This is easy to do, since the tangent line has a linear equation.) Then find the tangent line to the graph at $(a_2, f(a_2))$ and locate the point a_3 where this

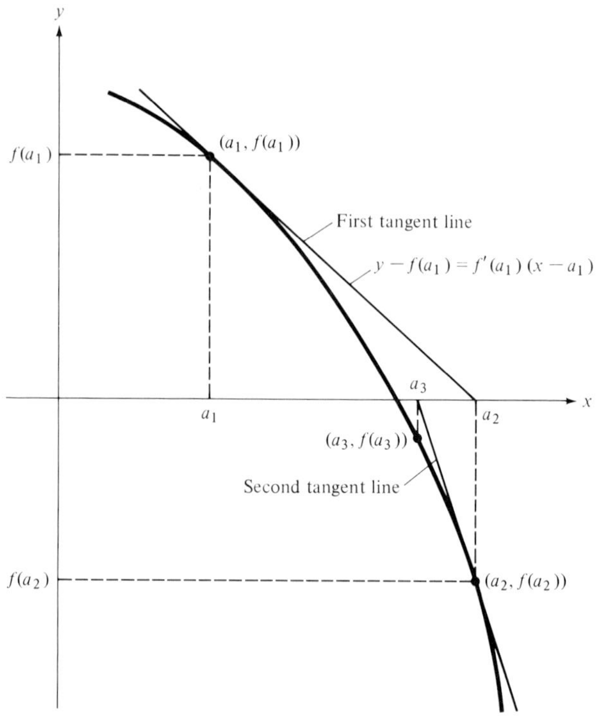

FIGURE 4-27

tangent line crosses the x axis. Repeating the process, we hope to get a sequence a_1, a_2, a_3, ... that converges to the desired root of the equation $f(x) = 0$. Each term of the sequence is defined recursively from the preceding term.

We can easily computerize the process. We shall stop the process at the nth step when $f(a_n)$ is as close to zero as we have specified. Thus we shall find a number a_n that is almost a root, in the sense that $f(a_n)$ is almost zero. In practice, $f(a_n)$ is compared to a small number ε (the Greek letter epsilon) called the tolerance, and if $|f(a_n)| < \varepsilon$, we take a_n to be the root.

Now let us analyze the algebra. The point-slope form of the equation of the line tangent to the graph of $f(x)$ at $(a_1, f(a_1))$ is

$$y - f(a_1) = f'(a_1)(x - a_1)$$

The x coordinate of the point where this line crosses the x axis is given by setting $y = 0$ and solving for x

$$-f(a_1) = f'(a_1)(x - a_1).$$

$$x - a_1 = -\frac{f(a_1)}{f'(a_1)}$$

$$x = a_1 - \frac{f(a_1)}{f'(a_1)}$$

This value of x is the desired a_2.

$$a_2 = a_1 - \frac{f(a_1)}{f'(a_1)}$$

The tangent line to the graph at $(a_2, f(a_2))$ is similarly

$$y - f(a_2) = f'(a_2) \cdot (x - a_2)$$

Setting $y = 0$ and solving for x gives a_3

$$a_3 = a_2 - \frac{f(a_2)}{f'(a_2)}$$

Except for the subscripts, the algebra is the same as before. Indeed, we have the recursion relation

$$a_{n+1} = a_n - \frac{f(a_n)}{f'(a_n)}$$

Thus we have a sequence $a_1, a_2, a_3, \ldots$, which hopefully converges to the desired root and will certainly converge to it if the conditions of Property 4-4 below are satisfied.

Newton's method is usually quite efficient, taking fewer steps (and thus less computer time) than the interval halving method. We have to be cautious how we use it, however, since we do not automatically know that the sequence $\langle a_n \rangle$ will converge and, even if it does converge, that it will converge to the desired root. The following property tells us conditions under which it is guaranteed to work.

Property 4-4 Let $f(x)$ be a function defined on $[a, b]$ such that

1. $f'(x)$ exists and is continuous on $[a, b]$.
2. $f(a)$ and $f(b)$ have opposite signs.
3. Neither $f'(x)$ nor $f''(x)$ is zero at any point on $[a, b]$.

Then there exists exactly one root c of $f(x) = 0$ on (a, b) and the sequence defined by the recursion relation

$$a_{n+1} = a_n - \frac{f(a_n)}{f'(a_n)}$$

converges to c, provided that a_1 is chosen so that $a \leq a_2 \leq b$.

These conditions will be satisfied for a great many cases of the type considered in this book, provided that the starting point a_1 is chosen fairly close to the desired root.

To illustrate with a specific numerical example, we shall determine the real value of x satisfying the equation

$$x^3 - 5 = 0$$

First, we observe that $f(x) = x^3 - 5$ and $f'(x) = 3x^2$. Using 2 as the starting point, the steps are as follows (continue to top of next page).

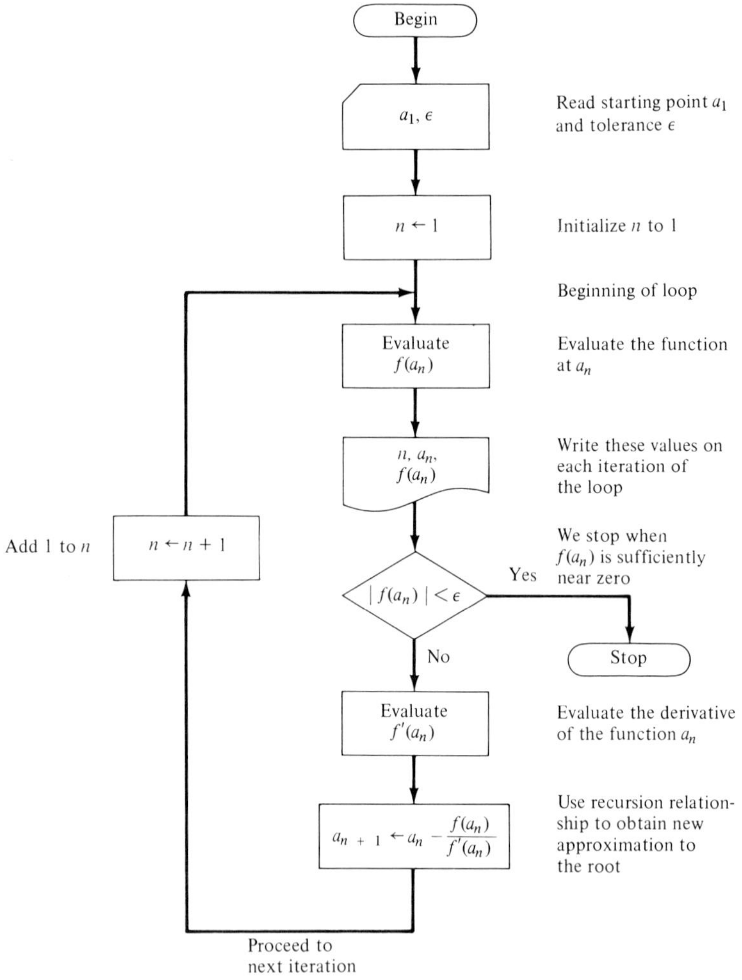

FIGURE 4-28

Iteration 1

$$\text{Evaluate } f(a_1) = f(2) = 2^3 - 5 = 3$$

$$\text{Evaluate } f'(a_1) = f'(2) = 3 \cdot 2^2 = 12$$

$$\text{Evaluate } a_2 = a_1 - \frac{f(a_1)}{f'(a_1)} = 2 - \frac{3}{12} = 1.75$$

Iteration 2

$$\text{Evaluate } f(a_2) = f(1.75) = 1.75^3 - 5 = 0.359375$$

$$\text{Evaluate } f'(a_2) = f'(1.75) = 3 \cdot 1.75^2 = 9.1875$$

$$\text{Evaluate } a_3 = a_2 - \frac{f(a_2)}{f'(a_2)} = 1.75 - \frac{0.359375}{9.1875} = 1.710884$$

Continuing this process, we obtain the sequence 2, 1.75, 1.710884, 1.709976, ..., which converges to $\sqrt[3]{5}$.

A flowchart representing Newton's method is shown in Figure 4-28. The program first reads a value for the initial starting point a_1 and the tolerance ε. The program initializes n to 1 and then enters a loop. In the loop, the first steps are to evaluate $f(a_n)$, print n, a_n, and $f(a_n)$, and then compare the absolute value of $f(a_n)$ to ε. If $|f(a_n)| < \varepsilon$, the program has located the root to the desired degree of accuracy. If not, the next steps are to calculate $f'(a_n)$ and determine a value for a_{n+1}. The value of n is increased by 1 and the process continued. As before, the execution of all blocks in the loop in Figure 4-28 comprises an iteration.

EXERCISES 4-7

In Exercises 1–5, you are to find a zero of $f(x)$. Use Newton's method and start at the indicated value of a_1. Stop when $|f(a_n)| < 0.2$.

1. $f(x) = x^2 - 2x - 4; \quad a_1 = 4$

2. $f(x) = x^2 - 2x - 4; \quad a_1 = -2$

3. $f(x) = x^3 - 2x^2 - 3x + 1; \; a_1 = 0$

4. $f(x) = x^3 - 2x^2 - 3x + 1; \; a_1 = 4$

5. $f(x) = x^3 - 2x^2 - 3x + 1; \; a_1 = -2$

In Exercises 6–10, use Newton's method to approximate the result to three decimal places. Unless a calculator or computer is available, you will have a good bit of arithmetic to do. Try to choose your first estimate close to the desired root.

6. $\sqrt[3]{9}$ HINT: Find the real root of $x^3 - 9 = 0$.

7. $\sqrt{5}$ **8.** $\sqrt[4]{7}$

9. $\sqrt[5]{2}$ **10.** $3^{2/5}$

*4-8. COMPUTING MAXIMA AND MINIMA WITHOUT CALCULUS

In the examples of the previous section, the derivatives could be easily obtained, and consequently so could the maximum or minimum. Many practical problems are not so easy to solve, and the maximum or minimum must be sought in other ways. Conceptually, the simplest approach is to perform an *exhaustive search*, where the function to be maximized or minimized is calculated at all possible points applicable to the particular practical problem. The most attractive of these is taken as the answer. Although this approach is not computationally very efficient, it can be applied to simple problems. For large problems, more efficient techniques are available, but these are beyond the scope of this text.

We shall illustrate the approach using the problem of Example 3 in Section 4-5. The revenue and cost functions were

$$R = 3\sqrt{q}$$

$$C = \frac{1}{4}q^2 + 1$$

where q is in units of a thousand. As posed, q could be any positive number. In reality, certain constraints on q exist. For most factories, it is not feasible to operate below some production rate or above another. For our factory suppose that we would shut down if production fell below 1000, and that plant capacity is such that no more than 5000 units can be produced. Therefore, q must satisfy $1 \leq q \leq 5$. Furthermore, suppose that we produce in units of 100. Now the only acceptable values of q are 1, 1.1, 1.2, 1.3, ..., 4.8, 4.9, and 5.

Using a computer, it would be easy to calculate R, C, and P for these values of q. A flowchart for this purpose is illustrated in Figure 4-29. In this flowchart, the statements to compute and print the values of R, C, and P for a particular value of q are placed inside a loop. Before entering this loop, q is assigned the value 1. Before computing values of R, C, and P, the value of q is compared to 5, and if $q > 5$, we stop. After printing the values of q, R, C, and P, we add 0.1 to q and proceed to compare q to 5. An iteration of the loop consists of testing q (comparing to 5); calculating R, C, and P; printing q, R, C, and P; and adding 0.1 to q.

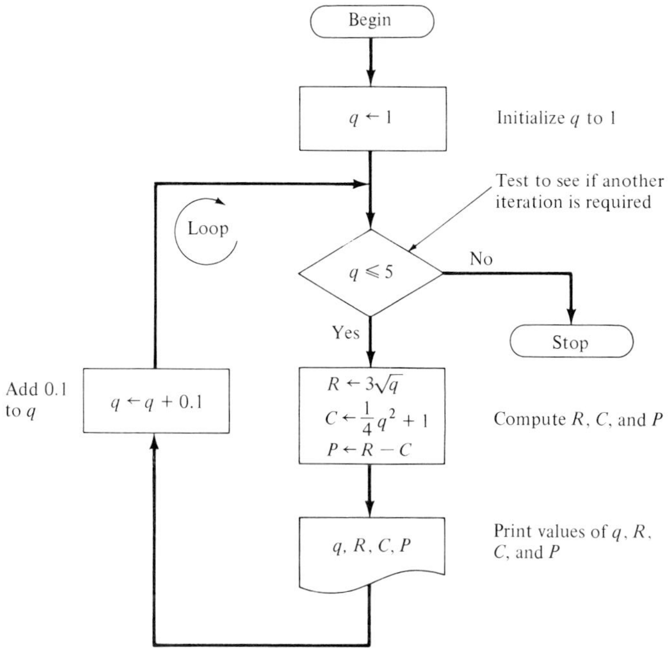

FIGURE 4-29

By suitable programming, the printout from the flowchart in Figure 4-29 can be arranged in the form of a table. For small problems, one can visually scan the values of P and select the largest value. But why not let the computer do it?

The flowchart to accomplish this is illustrated in Figure 4-30. This flowchart utilizes two variables, PX and QX, not found in the flowchart in Figure 4-30. At the beginning of the calculations for a particular value of q, the value of PX is the largest value of P that was encountered in previous iterations. The value of QX is the value of q corresponding to PX. For example, suppose that we are examining the calculations beginning with q equal to 2.4. The value of PX is the largest value of P calculated for $q = 1$, 1.1, 1.2, ..., 2.1, 2.2, or 2.3.

After calculating each value of P, it must be compared to the current value of PX. If P exceeds PX, we have found a larger profit, so the value of PX is changed to p and the value of QX is changed to q. If P does not exceed PX, the values of PX and QX are not changed. Blocks 6 and 7 in the flowchart in Figure 4-30 accomplish this.

Before entering the loop in Figure 4-30 values must be assigned to PX and QX. For this particular problem, one alternative is not to operate the plant at all. This would certainly result in no production and presumably

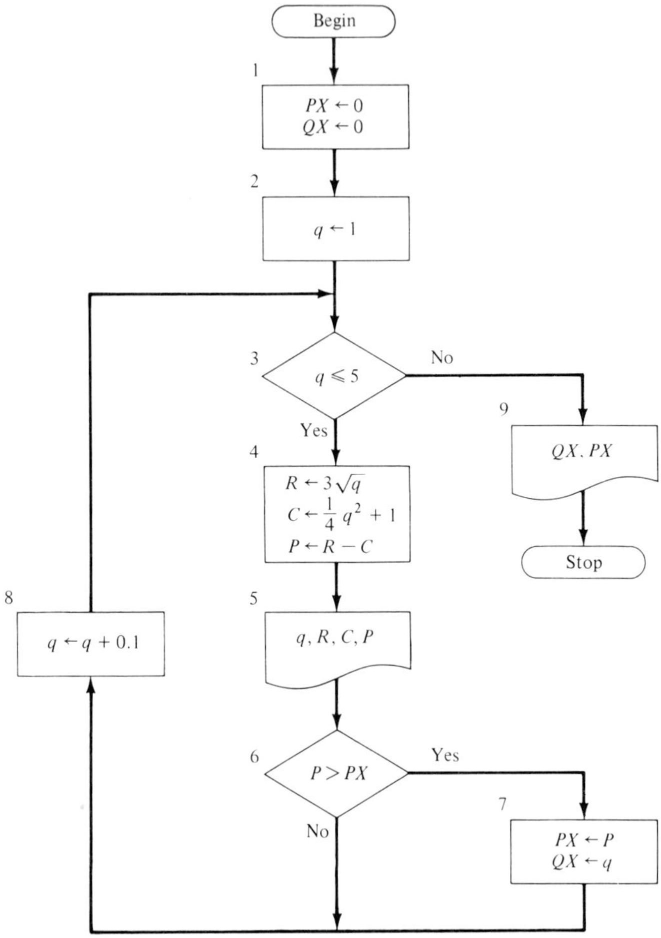

FIGURE 4-30

no profit. Therefore, we might choose to initially assign zero to both PX and QX. If no value of P greater than zero is encountered within the loop, the program would conclude that the "best" values of P and q are both zero. It should be emphasized that while assigning zero to both PX and QX is reasonable for this problem, the use of zero as initial value is not always appropriate.

In the flowchart in Figure 4-30, observe that immediately after exiting from the loop, the values of QX and PX are printed (Block 9). This block is executed only once, and only the final values of QX and PX are printed. For many problems this is sufficient output, and the output statement in Block 5 can be omitted, although it is particularly useful during program development.

EXERCISES 4-8

For each exercise in this section, you are to prepare a flowchart for a computer program to solve the problem.

1. Rework Example 1 of Section 4-5 assuming that the fence comes in 10-meter sections.
2. Rework Example 1 of Section 4-5 assuming that the fence comes in 20-meter sections.
3. Rework Exercise 1 of Section 4-5 under the restriction that x is an even integer.
4. Rework Exercise 2 of Section 4-5 under the restriction that x is an integer.
5. Rework Exercise 3 of Section 4-5 under the restriction that x is an odd integer.
6. Rework Exercise 9 of Section 4-5 assuming that x is produced in units of 12 and that our plant can produce no more than 1200 units.
7. Rework Exercise 10 of Section 4-5 assuming that x is sold in units of 12 and that we can sell no more than 600 units.
8. Rework Exercise 11 of Section 4-5 under the restriction that we will consider only speeds that are multiples of 5 but will allow speeds up to 120 kilometers per hour.

REVIEW EXERCISES: CHAPTER 4

For Exercises 1–10, determine **(a)** whether $f(x)$ is increasing or decreasing at the indicated point; **(b)** whether $f(x)$ is concave upward or downward at the indicated point; and **(c)** the inflection points, if any, of $f(x)$.

1. $f(x) = (x + 1)(x + 4)$ at $x = 2$
2. $f(x) = 1 + 2x - 2x^2$ at $x = -1$
3. $f(x) = x^3 - 2x - 5$ at $x = 3$
4. $f(x) = 2 - x - x^2$ at $x = 1$
5. $f(x) = 1 + x^2 - x^4$ at $x = 1$
6. $f(x) = x^{1/3}$ at $x = 1$
7. $f(x) = 3x\sqrt{x - 1}$ at $x = 2$
8. $f(x) = \sqrt{x^2 + 5}$ at $x = 2$
9. $f(x) = (x - 1)^4$ at $x = 2$
10. $f(x) = \dfrac{1}{x} - \dfrac{1}{x^2}$ at $x = 1$

In Exercises 11–20, the function $f(x)$ is defined for the indicated values of x. (**a**) determine all critical points; (**b**) for each critical point, indicate whether a relative maximum or minimum exists; and (**c**) determine the absolute maximum and minimum if they exist; if not, so state.

11. $f(x) = x^3 + 3x^2 - 9x + 5; [-5, 5]$

12. $f(x) = 3x^4 + 4x^3; [-2, 2]$

13. $f(x) = (x - 1)^3$; all x

14. $f(x) = 2x^3 = x^2 + 4; [-1, 4]$

15. $f(x) = 2x^3 + 3x^2 - 5; x \geq -2$

16. $f(x) = x^3 + 6x^2 - 19; x \leq 5$

17. $f(x) = \sqrt{x^2 + 3}; [-1, 1]$

18. $f(x) = \dfrac{x^2}{x - 1}; [-1, 1)$

19. $f(x) = \dfrac{2x}{x^2 - 1}; (-1, 1)$

20. $f(x) = \dfrac{x}{(x - 2)^2}$; [0, 2) and (2, 4]

In Exercises 21–26, refer to Figure NN. Give (**a**) all local maxima and minima and the values of x that yield them, and (**b**) all absolute maxima and minima and the values of x that yield them or state that they do not exist.

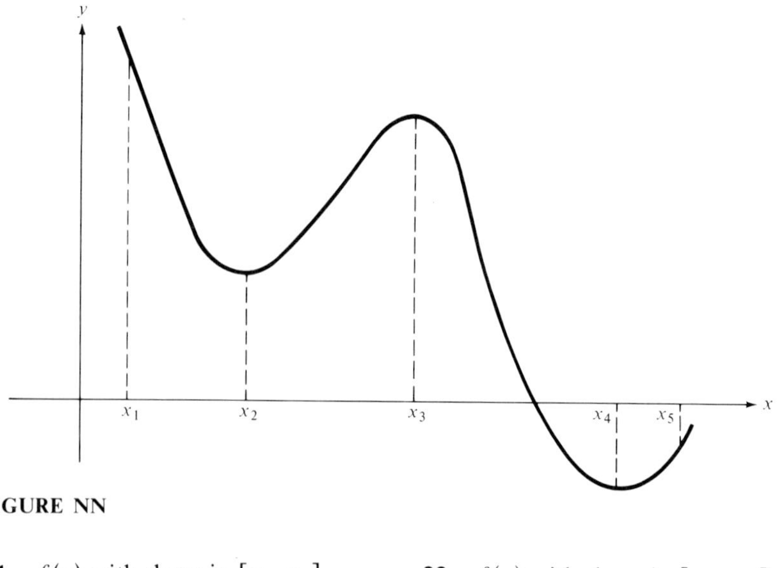

FIGURE NN

21. $f(x)$ with domain $[x_1, x_5]$

22. $f(x)$ with domain $[x_2, x_4]$

23. $f(x)$ with domain (x_2, x_4)

24. $f(x)$ with domain (x_1, x_3)

25. $f(x)$ with domain (x_3, x_4)

26. $f(x)$ with domain (x_1, x_5)

In Exercises 27–34, graph the polynomial functions using information from the first and second derivatives.

27. $x^2 + 4x - 1$

28. $1 - x - x^2$

29. $x^3 + 9$

30. $3 - x^3$

31. $2x^3 - 3x^2 - 36x + 10$

32. $x^4 - x^2$

33. $5 - 12x - 9x^2 - 2x^3$

34. $x^3 - 3x^4$

CHAPTER FIVE

EXPONENTIAL AND LOGARITHMIC FUNCTIONS

5-1. GROWTH AND DECAY

In our daily lives we are frequently exposed to growth processes and to decay processes many of which have a direct impact upon our personal welfare. As examples of growth processes, we hear such statements as "The rate of inflation is 9% per year"; "The world's population is increasing at 2% per year"; "Our use of energy is growing at 3% per year"; "Farm production is growing at 1% per year"; and so on. Examples of decay processes include the decreasing purchasing power of the dollar, the declining church attendance, the decreasing number of Americans living on farms, and so on.

Since these growth and decay processes are so common and so important, we must attempt to comprehend their impact upon our daily lives.

A growth or decay process that involves a fixed percentage change per unit time is called *exponential growth or decay*. For example, an inflation rate of 10% per year means that on the average from a given fixed time it takes 10% more dollars to purchase a given quantity of goods 1 year later. If i is the inflation rate expressed as a fraction and P is a given amount of money at our fixed starting point, then after 1 year we need $P + Pi = P(1 + i)$ dollars for the same purchases. After 2 years we need

$$[P(1 + i)] + i[P(1 + i)] = [P(1 + i)](1 + i) = P(1 + i)^2$$

dollars and, indeed, after n years we need

$$P(1 + i)^n$$

dollars. Since P is a constant, we have a formula $P \cdot a^x$, where $a = 1 + i$ and x represents the number of years. In Section 5-2, we shall identify such a function as an exponential function. We note that $P(1 + i)^n$ is also the value at n years of an investment of P dollars invested at an interest rate of i. The same type of formula comes up in any exponential growth problem.

In Table I of the Appendix we give a short table of exponential growth factors for various rates i and various times n. The top row lists the rates i and the left-hand column the number n of periods (e.g., years) concerned. The entries in the table are the numbers $(1 + i)^n$, which represent compound interest factors or exponential growth factors.

A natural question to ask is "At a given investment rate i, how many periods does it take for money to double?" In more general terms the same question is "For a given value of i, for what value of n is $(1 + i)^n \approx 2$?"

We can go to Table I for answers to such questions, but first let us observe that if $i = 0.1$, for example, an investment will certainly have doubled within 10 years, $(1 + 0.1)^{10} > 2$. If the earned interest each year were not reinvested, then at the end of 10 years, the money would have earned interest of exactly $(10) \cdot (0.1) = 1$ times the original investment. We call such investing, investing at *simple* interest. With earned interest being reinvested, the time for doubling your investment would be shorter. Indeed, from Table I we get $(1 + 0.1)^7 = 1.949$ and $(1 + 0.1)^8 = 2.144$ so that money is doubled in a little over 7 years at 10% compounded annually.

Table 5-1 gives the time in years required to double money at various interest rates.

TABLE 5-1

Interest rate	1%	2%	4%	6%	8%	10%
Years to double	69.7	35.0	17.7	11.9	9.0	7.3

Observe that the years to double at compound interest are about 0.7 times the years to double at simple interest. For example, at 4% simple interest, 25 $(= 1/0.04)$ years would be required to double your money. Using our rule, we have $0.7 \cdot 25 = 17.5$ years to double at compound interest. This "rule of thumb" value compares to the value 17.7 from Table 5-1.

To double our money in a little over 7 years sounds very good if we have money and can invest it at 10% per annum. But, on the other hand, at 10% inflation per year we see that in only a little over 7 years money is halved in value, and it will take twice as much money for the same standard of living. No wonder we are concerned with inflation rates of 8.8% in 1973 and 12.2% in 1974. At an average of 10% inflation per year, money is halved in value every 7+ years. If it is halved in value every 7+ years, then after 14+ years money has only $\frac{1}{4}$, that is $\frac{1}{2} \cdot \frac{1}{2}$, its earlier value, and after about 22 years it has only $\frac{1}{8}$ its earlier value. The graph in Figure 5-1 illustrates the rapidity of the decreasing value of the dollar at a 10% inflation rate. From Figure 5-1, we read that, at a continuing 10% inflation, in 20 years the dollar will be worth only about 15 cents. This means that about 15 cents today will purchase the same goods as will \$1 in 20 years. If inflation goes on at 10%

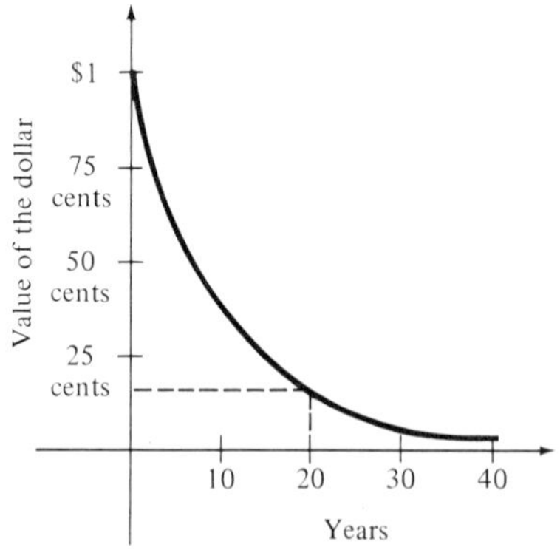

FIGURE 5-1

per year for the next 44 years, the dollar will be worth only $\frac{1}{8} \cdot \frac{1}{8} = \frac{1}{64}$ of its present value, about a cent and a half at present values. An 18-year-old college freshman will be 62 in 44 years. At a continuing 10% annual inflation rate the value of the dollar will be drastically changed in his or her lifetime.

Of course, salaries and wages will also go up so that in terms of the individual worker's real income, there may be not be such a striking change. But think of a person who retires at age 60 at a comfortable but fixed income of $25,000 per year. At 10% inflation per year, he will be a virtual pauper at age 82, and by age 90 his income would be less than $\frac{1}{16}$ of 25,000, or only about $1500 in terms of present dollars. The effect of long-term inflation at a rate like 10% on those with fixed incomes is frightening to contemplate.

Of course, inflation rates are not fixed at 10% or any other rate. But if inflation hovers around 10% for very long, the estimates made above are all too close to reality.

The world population is increasing at about 2% per year. From Table I we note that $(1 + 0.02)^{18} \approx 1.428$ and thus $(1.02)^{36} = [(1.02)^{18}]^2 \approx (1.428)^2 \approx 2.040$. Indeed, $(1.02)^{35} \approx 2$, as indicated in Table 5-1. At a 2% increase a year, the world's population will double every 35 years. It is close to 4 billion in 1975. If the 2% rate of increase continues, it will be 8 billion in 2010 and 16 billion in 2045. If feeding 4 billion people now is difficult, what will be the situation in the year 2000 or in the year 2050? The graph in Figure 5-2 illustrates the exponential growth of the world's population for a growth rate of 2% per year. Clearly, mankind is headed for great problems if the present 2% per year growth rate continues.

The previous discussion shows how Table I can be used for values of $n > 20$. We note that $(1 + i)^{m+n} = (1 + i)^m(1 + i)^n$. Thus, for example,

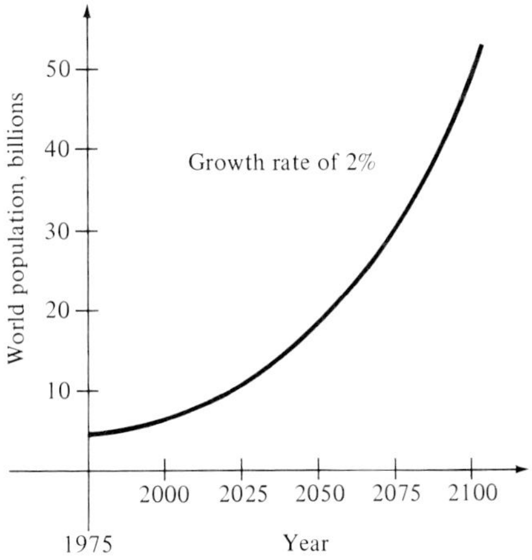

FIGURE 5-2

$(1+i)^{32} = (1+i)^{20}(1+i)^{12} = (1+i)^{16}(1+i)^{16}$. We can choose whichever is more convenient in using Table I.

Exponential decay is expressed in terms of a fixed percentage decrease every so often. Thus after 1 year of a percentage decrease corresponding to a rate i, there is $P(1-i)$ left of an original amount P. After 2 years we have $P(1-i)^2$ left, after n years, $P(1-i)^n$ left. In Table II in the Appendix we give values of $(1-i)^n$ for various values of i and n.

It is interesting that a 10% increase followed by a 10% decrease results in a net 1% decrease. The same result follows for a 10% decrease followed by a 10% increase, since

$$(1+i)(1-i) = (1-i)(1+i) = 1^2 - i^2 = 1 - i^2$$

for $i = 0.1$, $1 - i^2 = 0.99$. Starting from 100, we get 110 following a 10% increase and then 99 following a 10% decrease from 110. In general, a percentage decrease is more pronounced than the same percentage increase. A 50% decrease followed by a 50% increase gives only 75% of the original value.

The time it takes for a quantity in an exponential growth or decay process to double or halve in value is called the *doubling life* or *half-life* of the quantity. The half-life of a radioactive substance is the time it takes for radioactive decay to reduce the quantity by a factor of 2.

In social and economic phenomena, we should not require that the numbers used to describe exponential growth or decay be too precise. We should "round off" in our computations where appropriate. Thus inflation rates change year by year (or even month by month), and even interest rates on money are subject to frequent changes. We customarily want figures in the right "ballpark," not precise ones. But the fact that rates are subject to change does not mean that we should ignore the long-term effects of exponential growth or decay. Indeed, it is the longer-term effects that are properly the cause of concern.

Finally, let us analyze the effect of a pay increase that is only partially tied to inflation. Suppose that inflation is at 8% per year while a person's income increases at 4% per year. Then his real income decreases at a bit less than 4% per year, the computation for the net inflationary effect of inflation and salary increase being $(1+0.08)/(1+0.04)$. From Table I of the Appendix we can see that if the pattern continues for 17 or 18 years, his real income will be cut in half.

EXERCISES 5-1

In Exercises 1–18, use Tables I or II of the Appendix to find the indicated values.

1. $(1+0.04)^{10}$ **2.** $(1+0.06)^{12}$

3. $(1 + 0.08)^{20}$ 4. $(1 + 0.02)^{15}$

5. $(1 + 0.10)^{25}$ 6. $(1 - 0.02)^{20}$

7. $(1 - 0.04)^{10}$ 8. $(1 - 0.06)^{15}$

9. $(1 - 0.08)^{8}$ 10. $(1 - 0.10)^{30}$

In Exercises 11–18, find by bracketing between consecutive integers the doubling life or half-life of quantities at the following annual exponential growth or decay rates. For example, at a 10% growth rate we have 1.95 after 7 years and 2.14 after 8. Therefore, money doubles in 7 to 8 years.

11. Growth 4% 12. Growth 6%

13. Growth 8% 14. Growth 2%

15. Decay 4% 16. Decay 6%

17. Decay 8% 18. Decay 10%

In Exercises 19–22 find the net growth or decay of (**a**) the indicated growth followed by the indicated decay, and (**b**) the indicated decay followed by the indicated growth. Use a base figure of 100 if you wish.

19. Growth 20%, decay 20% 20. Growth 30%, decay 30%

21. Growth 60%, decay 60% 22. Growth 40%, decay 40%

23. The following problem, "Heads or Tails," appeared in Henry E. Dudeney's *Amusements in Mathematics*:*

> Crooks, an inveterate gambler, at Goodwood recently said to a friend, "I'll bet you half the money in my pocket on the toss of a coin—heads I win, tails I lose." The coin was tossed and the money handed over. He repeated the offer again and again, each time betting half of the money then in his possession. We are not told how long the game went on, or how many times the coin was tossed, but we are told, that *the number of times that Crooks lost was exactly equal to the number of times that he won.* Now, did he gain or lose by this little venture?

Use what you learned from Exercises 19 through 22 to obtain the answer.

24. The percentage of the U.S. population "on the farm" generally follows exponential decay and has had a half-life of about 20 years. If 60% of the U.S. population was "on the farm" in 1900, find the percentage on the farm in 1960. In 1980. Is it likely to continue to follow the same exponential decay pattern for another 100 years?

* New York: Dover, 1917.

25. The number of students in college in the United States generally followed the exponential growth law from 1950 to 1970, quadrupling to 8 million in that time. What was the doubling life of the number of college students? If the exponential growth were to have continued, what would have been the number of U.S. college students in 1990? In 2020? In 2030? If the U.S. population also followed the exponential growth law and was 200 million in 1970 with an approximate 1% per year increase (a 2% increase every 2 years), what would the total U.S. population have been in 2020? In 2030? When would the college student population have exceeded the total population? Clearly, projections of long-term exponential growth are subject to many limitations.

26. If the world population continues to grow at 2% a year and is 4 billion in 1975, when will the population be so large that each human being will have only 2 square feet of land area? (Assume that the total land area of the world is about 60 million square miles or $1.7 \cdot 10^{15}$ square feet.) Use the doubling life of 35 years for your estimate.

27. Suppose that the Indians had taken the $24 they got for Manhattan in about 1628 and had invested it at 8% interest compounded annually. What would the investment be worth now?

NOTE: Approximate by finding the "doubling life" at 8% per year from Table I of the Appendix and divide into 347 (the number of years since 1628). Then raise 2 to that power (noting that $2^{10} \approx 1000$ [$2^{10} = 1024$]). In a hypothetical question like this, we obviously want only an approximate answer.

5-2. a^x AND THE NATURAL NUMBER e

In this section, we shall begin by defining the exponential function a^x. Here we assume throughout that a is a fixed positive number. We shall eventually explain how the exponent x can be any real number.

If n is a positive integer, a^n is defined as the product of a by itself for a total of n factors:

$$a^n = \underbrace{a \cdot a \cdots a}_{n \text{ factors}}$$

Two properties immediately follow from this definition of a^n.

Property 5-1 For any two positive integers m and n,

$$a^m \cdot a^n = a^{m+n}$$

This is so, since

$$\underbrace{(a \cdot a \cdots a)}_{m \text{ factors}} \cdot \underbrace{(a \cdot a \cdots a)}_{n \text{ factors}} = \underbrace{a \cdot a \cdots a}_{m+n \text{ factors}}$$

Property 5-2 For any two positive integers m and p,

$$(a^m)^p = a^{m \cdot p}$$

This is so, since m factors p times yields $m \cdot p$ factors.

We now agree that we want these two properties to be true for any real numbers used as exponents. Then $a^m \cdot a^0 = a^{m+0} = a^m$. Hence a^0 must be 1, since it is the number which when multiplied by a^m gives a^m. Similarly, $a^m \cdot a^{-m} = a^{m-m} = a^0 = 1$. Hence $a^{-m} = 1/a^m$.

Next we define $a^{1/m}$ for m, an integer. We want $(a^{1/m})^m = a^{m/m} = a$ if Property 5-2 is to be true. So $a^{1/m}$ must be a number whose mth power is a. Therefore, $a^{1/m}$ must be $\sqrt[m]{a}$, the mth root of a. For example, $8^{1/3} = 2$. Indeed, $(8^{1/3})^3 = 8^{3/3} = 8$ and $2^3 = 8$. In general, we know that there must be an mth root of a positive number a, since the polynomial function $x^m - a$ is negative for $x = 0$ and is positive for large x. Since every polynomial is continuous, the intermediate value theorem guarantees the existence of a positive number x for which $x^m - a = 0$, i.e., for which $x^m = a$.

For any rational number m/n we define $a^{m/n} = (a^{1/n})^m = (\sqrt[n]{a})^m$, which agrees with Property 5-2. We could alternatively have used $a^{m/n} = (a^m)^{1/n} = \sqrt[n]{a^m}$, also agreeing with Property 5-2.

Finally, we explain how we can define a^x for x, a real number that need not be rational. We consider a decimal expansion for x. For example, $\pi = 3.141592 \ldots$ and we seek a^π. Consider the sequence of numbers

$$a^3, a^{3.1}, a^{3.14}, a^{3.141}, \ldots$$

observing, for example, that $a^{3.1}$ is really $a^{31/10}$ and, therefore, $a^{3.1} = \sqrt[10]{a^{31}}$. For any positive number a, it can be proved that the sequence $a^3, a^{3.1}, a^{3.14}, \ldots$ converges to a limit. We call the limit a^π. By a similar procedure, using decimal expansions, we could define a^x for any real number x and it is true that with this definition of a^x, the properties $a^m \cdot a^n = a^{m+n}$ and $(a^m)^p = a^{mp}$ hold for all real numbers m, n, and p. Of course, we also have $a^0 = 1$ and $a^{-m} = 1/a^m$. Indeed $a^x = 1$ if and only if either $x = 0$, $a = 1$, or both.

In Figure 5-3 we have graphed the exponential function 2^x. This graph is typical of the graph of any exponential function a^x when $a > 1$. The following are some important properties of any exponential function a^x for $a > 1$. We accept these without further proof.

1. The domain is the set of all real numbers.
2. a^x is positive for all x.

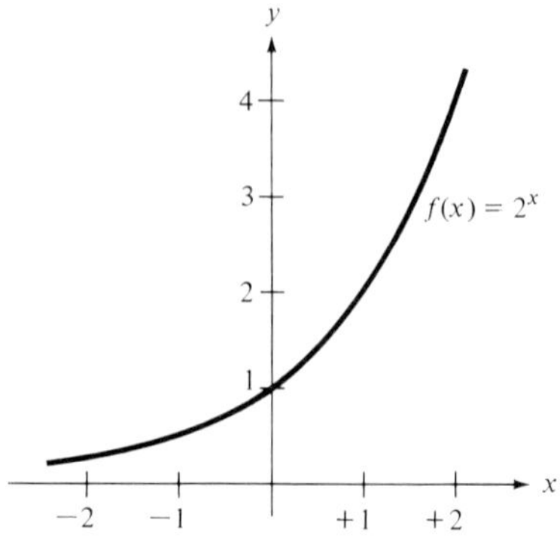

FIGURE 5-3

3. a^x is increasing for all x, a^x is near zero for x large negatively, and a^x is very large for x large positively.
4. a^x is continuous for all x.
5. For each positive number y, there is one and only one x such that $a^x = y$. That there is one such x follows from Properties 3 and 4 and the intermediate value theorem. That there is only one x, follows from the fact that if $a^{x_1} = a^{x_2}$, then $a^{x_1}/a^{x_2} = 1$ and $a^{x_1 - x_2} = 1$ and thus $x_1 - x_2 = 0$ or $x_1 = x_2$. This is reflected in the following property.

Property 5-3 For $a \neq 1$, $a^x = a^y$ if and only if $x = y$.

In Figure 5-4, the graphs of several exponential functions are drawn. Notice that $(\frac{1}{2})^x = 1/2^x = 2^{-x}$. The graphs of 2^x and 2^{-x} are mirror images of each other across the y axis; for example, the point (1, 2) is on the graph of 2^x; hence the point $(-1, 2)$ is on the graph of 2^{-x}.

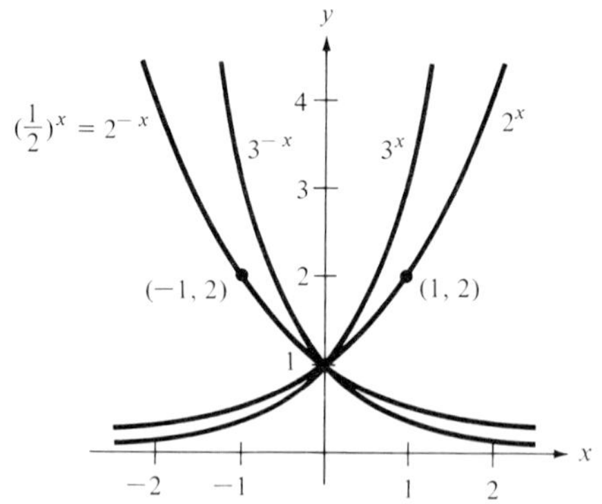

FIGURE 5-4

NOTE: The reader should carefully distinguish between the *power* function x^n and the *exponential* function a^x. Examples of power functions are x^2, x^{30}, and x^{100}. Examples of exponential functions are 2^x, 4^x, and 10^x. In a power function, x is raised to a constant power. In an exponential function, x is the exponent to which a constant is raised.

There is one special and very useful limit that arises in connection with exponential functions. It is $\lim_{t\to 0} (1 + t)^{1/t}$. The limit exists and is denoted by e. As we shall see, the number e in the form of $\lim_{t\to 0} (1 + t)^{1/t}$ arises naturally. There is no way to avoid it in the calculus. Along with the number π, e plays a very important role in mathematics and its applications. Table 5-2 gives the values of the function $(1 + t)^{1/t}$ for various values of t.

TABLE 5-2

t	$(1 + t)^{1/t}$
0.1	2.59374246
0.01	2.70481383
0.001	2.71692393
0.0001	2.71814593
−0.0001	2.71841775
−0.001	2.71964222
−0.01	2.73199903
−0.1	2.86797199

Indeed, to 15 decimal places the decimal expansion of e is

$$e = 2.718281828459045 \ldots$$

An alternative form for e as a limit is to let $t = 1/M$ and thus $(1 + t)^{1/t} = [1 + (1/M)]^M$. As $M \to \infty$, $t \to 0$. Hence if $\lim_{t\to 0} (1 + t)^{1/t} = e$, then $\lim_{M\to\infty} [1 + (1/M)]^M = e$. In other words, for the sequence $(1 + \frac{1}{2})^2$, $(1 + \frac{1}{3})^3$, $(1 + \frac{1}{4})^4, \ldots, [1 + (1/M)]^M, \ldots$ we have $e = \lim_{M\to\infty} [1 + (1/M)]^M$.

EXERCISES 5-2

Find the values of the following.

1. 2^{10}

2. 3^5

3. $2^{21/2}$ given that $2^{1/2} = \sqrt{2} \approx 1.414$ HINT: $2^{21/2} = 2^{10} \cdot 2^{1/2}$

4. $3^{11/2}$ given that $\sqrt{3} \approx 1.732$ HINT: $3^{11/2} = 3^5 \cdot 3^{1/2}$

5. $(0.1)^{10}$

6. $(0.01)^5$

In Exercises 7–10, graph the two given exponential functions. Determine approximate functional values for $x = -2, -1, 0, 1, 2$ at least.

7. e^x and e^{-x} HINT: Use Table IV in the Appendix

8. 4^x and 4^{-x}

9. 5^x and 5^{-x}

10. 10^x and 10^{-x}

11. For what values of $a > 0$ is a^x an increasing function? A decreasing function? A constant function?

12. For $x > 0$, is the graph of 4^x above or below the graph of 3^x? For $x < 0$, is the graph of 4^x above or below the graph of 3^x?

13. Compute an approximation of e using the average of the values of

$$\left(1 + \frac{1}{1000}\right)^{1000} \quad \text{and} \quad \left(1 - \frac{1}{1000}\right)^{-1000}$$

as given in the text. Do the same for the average of

$$\left(1 + \frac{1}{10{,}000}\right)^{10{,}000} \quad \text{and} \quad \left(1 - \frac{1}{10{,}000}\right)^{-10{,}000}.$$

Since $2^{x+1} = 2^x \cdot 2^1 = 2 \cdot 2^x$, we see that the functional values of 2^x double (increase by a factor of 2) when x is increased by one unit. In Exercises 14 through 21, determine by what factor the functional values increase or decrease when x is increased by one unit. Verify by evaluating at $x = 1$ and $x = 2$.

14. 3^x

15. 4^x

16. e^x

17. 10^x

18. 3^{-x}

19. 4^{-x}

20. e^{-x}

21. 10^{-x}

In Exercises 22–25, find the necessary increase in x for which the functional values will double. (The problems have been selected to have easily computable answers.)

22. 4^x

23. 8^x

24. $(\sqrt{2})^x$

25. $(\sqrt[4]{2})^x$

5-3. a^x, $\log_a x$, AND THEIR PROPERTIES

In this section we introduce the concept and notation of logarithms. We shall use logs (as they are commonly called) only in connection with exponential functions and elementary formulas in calculus. We are not concerned with the usual computational aspect of logs. Before discussing logs specifically, we investigate properties of inverse functions.

Some functions have inverses. If a function $f(x)$ has the property that for each functional value there is *only one* value of x which yields that value, then $f(x)$ has an inverse. The inverse of such a function $f(x)$ is a function called "f inverse of x" and written $f^{-1}(x)$ [in general, *not* the reciprocal of $f(x)$], which interchanges the role of the domain and the set of functional values of $f(x)$. Thus, if $f(2) = 4$, then $f^{-1}(4) = 2$. If $f(0) = -5$, then $f^{-1}(-5) = 0$.

Here is a way of representing an inverse function. If $y = x^3$, then we can solve for x in terms of y, getting $x = \sqrt[3]{y}$. But since we want the values previously identified with y to be the domain and the values previously identified with x to be the set of functional values, we can or should reverse the roles of x and y and write $y = \sqrt[3]{x}$ as the inverse function of x^3.

This convention makes sense in terms of "function of a function" notation. Thus both $f^{-1}(f(x))$ and $f(f^{-1}(x))$ should be the identity function, x. The former carries a point x to $f(x)$ and then the point $f(x)$ back to x. The latter carries a point $f(x)$ to x and then back to $f(x)$ where it started. Thus each carries any point to itself. We can apply first the function and then its inverse or first the inverse and then the function. For $y(x) = x^3$ we have $y^{-1}(x) = \sqrt[3]{x}$. We note that $\sqrt[3]{x^3} = x$ and $(\sqrt[3]{x})^3 = x$, as should be the case.

Here is another example. Let $y = x^2 + 1$ for $x \geq 0$. The domain is as specified. A graph of the function is shown in Figure 5-5. The domain of y is the set of all numbers $x \geq 0$. The set of functional values is the set of all numbers ≥ 1. The inverse function is found by solving for x in terms of y and then reversing the roles of x and y. Thus the domain of the inverse function

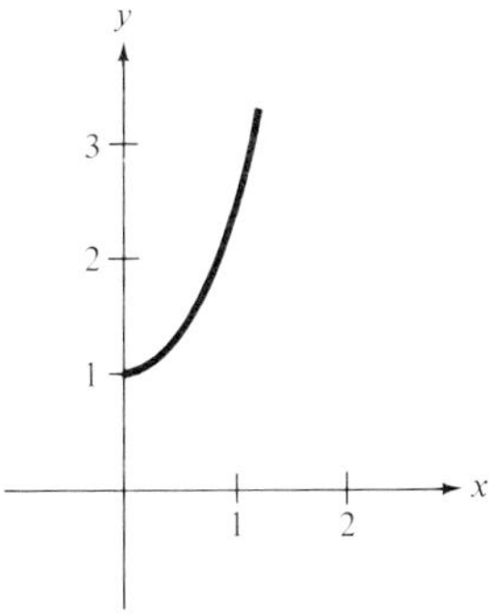

FIGURE 5-5

$y^{-1}(x)$ is the set of all numbers ≥ 1 and the set of functional values will be the set of all positive or zero numbers. Algebraically, we have $y = x^2 + 1$; since on solving for x, we get $x = \sqrt{y - 1}$, we have, on reversing the variables, the inverse function as $y = \sqrt{x - 1}$. Note that $(\sqrt{x - 1})^2 + 1 = (x - 1) + 1 = x$ and $\sqrt{(x^2 + 1) - 1} = \sqrt{x^2} = x$ for $x \geq 0$. Thus $y(y^{-1}(x)) = x$ and $y^{-1}(y(x)) = x$.

In the previous section we remarked that for $a > 1$ the exponential function a^x is a continuous function defined for all values of x but having only positive functional values, i.e., $a^x > 0$ for all x. Also, for any number $y > 0$ it is possible to find a value of x that will satisfy the equation $y = a^x$, provided that $a > 1$. Furthermore, for any $y > 0$ there will be *only* one such value of x. [The same observations concerning y are true for any a between 0 and 1 $(0 < a < 1)$, but this fact is not used in this text.]

The above observations are sufficient to establish that the exponential function has an inverse. The domain of the exponential function is the set of all real numbers. The domain of its inverse is the set of all positive numbers. In Figure 5-6, graphs for the exponential function e^x and its inverse

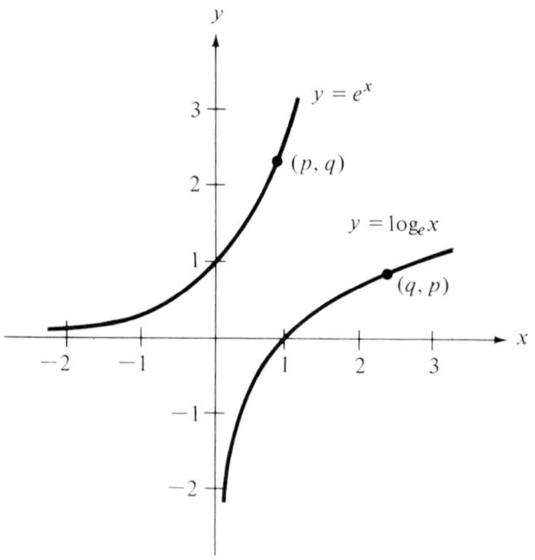

FIGURE 5-6

are given. The inverse of the exponential function is called the logarithm function. For the inverse of the exponential function a^x, we write $\log_a x$, where a is called the base, and we read "logarithm to the base a of x" or "log of x to the base a." The lower graph in Figure 5-6 represents the function $y = \log_e x$ which is the inverse of the function $y = e^x$. The reader should verify that for any point (p, q) on the graph of e^x, the point (q, p) is on the graph of its inverse.

Although the base a of the logarithm as shown in Figure 5-6 may be any number greater than 1, the two bases for logarithms most frequently encountered are 10 and e. When performing arithmetic calculations, base 10 logs are more convenient, but we use them only occasionally in this book. In later sections we shall see that the number e arises naturally as a base for logs in many problems, particularly in the calculus. In fact, e is used so frequently as a base that we call $\log_e x$ the *natural log* of x and write

$$\ln x \quad \text{in place of} \quad \log_e x$$

and we shall henceforth use this notation.

Now we ask how we can interpret $\log_a M$ or $\ln M$ as numbers given $a > 1$ and $M > 0$. We think of $\log_a M$ as an exponent. *It is the exponent to which the number a must be raised to give M*. In symbols, $a^{\log_a M} = M$, since we are raising a to the power needed to get M. Thus the following identity is a direct consequence of the definition of $\log_a M$:

Identity 5-1 For $a > 1$ and $M > 0$, $a^{\log_a M} = M$.

By the same token we see that $\log_a (a^p) = p$, since p is the power to which you must raise a to get a^p. This results in the following identity:

Identity 5-2 For $a > 1$, $\log_a (a^p) = p$.

The two equations $a^x = M$ and $x = \log_a M$ are equivalent. They are just different ways of saying the same thing. Any statement about exponents can be translated into a statement about logs, and vice versa.

In the previous section of this chapter we presented some fundamental properties of exponents. We list them here opposite the corresponding properties of logarithms. For these properties, $a > 1$, $M > 0$, and $N > 0$.

Properties of Logarithms	**Properties of Exponents**
5-4. $\log_a \left(\frac{1}{M}\right) = -\log_a M$	$\frac{1}{a^x} = a^{-x}$
5-5. $\log_a (MN) = \log_a M + \log_a N$	$a^x a^y = a^{x+y}$
5-6. $p \log_a M = \log_a (M^p)$	$(a^x)^p = a^{px}$
5-7. $\log_a M = \log_a N$ if and only if $M = N$	$a^x = a^y$ if and only if $x = y$

All of these properties of logs follow from the definition of a log as an exponent and the properties of exponents. We show only Properties 5-6 and 5-7, leaving the others to the exercises.

To show Property 5-6, we first write M^p in several different ways. By Identity 5-1, we can write $M^p = a^{\log_a (M^p)}$. From this same identity, we observe that $M = a^{\log_a M}$; therefore, we can write $M^p = (a^{\log_a M})^p$. Using Property 5-2 for exponents, as cited in the previous section, we note that $(a^{\log_a M})^p = a^{p \log_a M}$. Combining all of the equalities, we have

$$a^{p \log_a M} = (a^{\log_a M})^p = M^p = a^{\log_a (M^p)}$$

Since $a^{p \log_a M} = a^{\log_a (M^p)}$, Property 5-3 for exponents as cited in Section 5-2 enables us to conclude that $p \log_a M = \log_a (M^p)$.

To show Property 5-7 that $\log_a M = \log_a N$ if and only if $M = N$, we write $M = a^{\log_a M}$ and $N = a^{\log_a N}$. But by Property 5-3 of exponents, $a^{\log_a M} = a^{\log_a N}$ if and only if $\log_a M = \log_a N$. Therefore, $M = N$ if and only if $\log_a M = \log_a N$.

Example 1. $\ln \left(\frac{1}{e^2}\right) = -\ln e^2 = -2 \ln e = -2$

This example first uses Property 5-4 that, for natural logs, reads $\ln (1/M) = -\ln M$, and then applies Property 5-6 that, for natural logs, reads $\ln M^p = p \ln M$. Finally, the definition of the log is used to conclude that $\ln e = 1$.

Example 2. $\ln (e^{x^2+5}) = (x^2 + 5) \ln e = x^2 + 5$

This example uses Property 5-6 for the logarithm of a number raised to a power, and finally the definition of the log to conclude that $\ln e = 1$. We can also get the result directly from the definition. That is, $\ln (e^{x^2+5})$ is the power to which e must be raised to give (e^{x^2+5}). But $x^2 + 5$ is obviously such a number.

Example 3. Here is a third technique for getting a solution of the previous example.

$$\ln (e^{x^2+5}) = \ln (e^{x^2}e^5) = \ln (e^{x^2}) + \ln (e^5) = x^2 + 5$$

We first use Property 5-1 of exponents to write $e^{x^2+5} = e^{x^2}e^5$, and then use Property 5-5 which, for natural logs, reads $\ln (MN) = \ln M + \ln N$. We then use Identity 5-2, which for natural logs reads $\ln e^x = x$.

Example 4. $\log_{10} \sqrt{10} = \frac{1}{2} \log_{10} 10 = \frac{1}{2}$

In this example, we use Property 5-6 for the logarithm of a number raised to a power. We note that $\sqrt{10} = 10^{1/2}$, and thus

$$\log_{10} \sqrt{10} = \log_{10} [10^{1/2}] = \frac{1}{2} \log_{10} 10$$

Finally, from the definition of the log, we conclude that $\log_{10} 10 = 1$.

For certain types of computation, the use of logarithms is vital. For example, how would you compute $2.1^{1.6}$? We have already shown that a^x can be written as $(e^{\ln a})^x = e^{x \ln a}$. Therefore, $2.1^{1.6} = e^{1.6 \ln 2.1}$. First, we need to evaluate ln 2.1. For this purpose, a short table of natural logarithms is provided in Table VI in the Appendix. This table gives the values of natural logarithms by tenths. To find ln 2.1, we look down the left column to locate the row corresponding to 2.0, and then look across to the column with heading 0.1. The number 0.7419, which is located in this row and column, is the value of ln 2.1. We now compute $1.6 \ln 2.1 \approx 1.187$ by multiplying 1.6 by 0.7419.

We must now look up $e^{1.187}$. Table IV in the appendix gives values of the exponential function e^x by tenths. We round off 1.187 to 1.2 and get $e^{1.2} \approx 3.32$ from Table IV (which is used in the same way as Table VI). Therefore, $2.1^{1.6} \approx 3.32$.

Example 5. $1.2^{10} = e^{10 \ln 1.2} = e^{10(0.1823)} = e^{1.823} \approx e^{1.8} \approx 6.05$

In these examples, we have "rounded off" so that we could use Tables IV and VI directly and without interpolation. By using more complete tables and/or by interpolation, we could get more accurate results. However, in this book we are concerned primarily with the ideas of calculus and of computation and thus ask only for answers obtainable from the techniques and the short tables available.

EXERCISES 5-3

Simplify the formulas in Exercises 1–10.

1. $e^{\ln (e^2)}$ **2.** $e^{\ln 5}$

3. $e^{\ln (x^2+1)}$ **4.** $e^{\ln y}$

5. $5^{\log_5 17}$ **6.** $3^{\log_3 (x+1)}$

7. $\log_2 (2^{1+x^2})$ **8.** $\log_{10} (10^{\sqrt{x}})$

9. $\ln (e^{2x-5})$ **10.** $\ln e^{(1-x^2)}$

In Exercises 11–17, write the given expression as e to some power, e.g., $3^2 = (e^{\ln 3})^2 = e^{2 \ln 3}$.

11. 10^3 **12.** 2^{π}

13. 3^e **14.** $10^{\sqrt{x}}$

15. 4^{x+1} **16.** 5^{1+x^2}

17. x^2

In Exercises 18–23, solve the equation for x. HINT: Write in exponential form, use Property 5-7, and/or use common numerical sense.

18. $\log_4 x = 1$

19. $\log_x 81 = 4$

20. $\log_e x - \log_e (2x - 3) = 0$

21. $\log_2 x^2 - \log_2 x = 1$

22. $e^x e^{x-2} = 1$

23. $e^{2x} = \log_e (e)$

In Exercises 24–27, graph the functions.

24. $\log_2 x \quad x > 0$

25. $\log_4 x \quad x > 0$

26. $y = \ln (x + 1) \quad x > -1$

27. $2 \ln x \quad x > 0$ HINT: Note for $x > 0$, $2 \ln x$ is just twice $\ln x$. We just double the y coordinates for $\ln x$.

In Exercises 28–31, use Tables IV and VI in the appendix to evaluate the expressions (approximately, as in the text).

28. $8.6^{0.1}$

29. $5.2^{0.7}$

30. $(0.5)^{3.1}$ HINT: $(0.5)^{3.1} = (\frac{1}{2})^{3.1} = 1/2^{3.1}$.

31. $2.6^{-1.2}$ HINT: $2.6^{-1.2} = 1/2.6^{1.2}$.

In Exercises 32–37, (**a**) find the inverse function for the given function, (**b**) state the domain and set of all functional values for the inverse function, and (**c**) compute $y(y^{-1}(x)) = x$ and $y^{-1}(y(x)) = x$ to verify your results.

32. $y = x^3 + 1$ with domain all x

33. $y = \sqrt{x}$ with domain $x \geq 0$

34. $y = \sqrt{1 - x^2}$ with domain $0 \leq x \leq 1$

35. $y = \sqrt{4 - x^2}$ with domain $0 \leq x \leq 2$

36. $y = x^5 + 3$ with domain all x

37. $y = x^5 - 3$ with domain all x

38. Show Property 5-4 of logs.

39. Show Property 5-5 of logs.

40. In order to use Table VI in the Appendix to obtain logs to any base, we may use the relationship

$$\log_a b = \frac{\ln b}{\ln a}$$

Show this, by first observing that

$$a^{\log_a b} = b$$

and then taking the natural logarithm of both sides. Apply Property 5-6, and then solve for $\log_a b$ to obtain the desired formula.

5-4. DERIVATIVES OF $\ln x$ AND $\log_a x$

We assume that $a > 1$. We also assume that $x > 0$, since $\log_a x$ is defined only for $x > 0$.

To differentiate $y = \log_a x$, we begin by using the definition of the derivative:

$$y' = \lim_{h \to 0} \frac{\log_a (x + h) - \log_a (x)}{h}$$

if this limit exists. Recalling that $\log_a M - \log_a N = \log_a (M/N)$, we have

$$\frac{\log_a (x + h) - \log_a x}{h} = \frac{1}{h} \log_a \frac{(x + h)}{x} = \frac{1}{h} \log_a \left(1 + \frac{h}{x}\right)$$

To simplify the formulas, we now substitute t for h/x; therefore, $h = tx$.

$$\frac{\log_a (x + h) - \log_a x}{h} = \frac{1}{tx} \log_a (1 + t) = \frac{1}{x}\left(\frac{1}{t} \log_a (1 + t)\right)$$

$$= \frac{1}{x} \log_a (1 + t)^{1/t}$$

since $p \log_a M = \log_a M^p$. We observe that for $h/x = t$, $t \to 0$ if and only if $h \to 0$, since x is fixed throughout this development. Thus

$$y' = \lim_{h \to 0} \frac{\log_a (x + h) - \log_a x}{h} = \lim_{t \to 0} \left\{\frac{1}{x} \log_a [(1 + t)^{1/t}]\right\}$$

$$= \frac{1}{x} \lim_{t \to 0} \{\log_a [(1 + t)^{1/t}]\}$$

$$= \frac{1}{x} \log_a \left\{\lim_{t \to 0} [(1 + t)^{1/t}]\right\}$$

Since $\lim_{t\to 0} [(1 + t)^{1/t}] = e$, we have

$$y' = \frac{d \log_a x}{dx} = \frac{1}{x} \log_a e$$

Here we used the fact that the limit of the log of a function is the log of the limit of the function. This follows from the continuity of the log function.

This development shows why the number e is so important. The $\lim_{t\to 0} (1 + t)^{1/t}$ naturally arises when we take the derivative of a log function. Also, we see why logs to the base e arise. Noting that $\log_e e = \ln e = 1$, we have

$$\frac{d \ln x}{dx} = \frac{1}{x}$$

which is a simple formula. For any base other than e, the formula is not so simple.

The term $\log_a e$ in the formula

$$\frac{d}{dx}(\log_a x) = \frac{1}{x} \log_a e$$

deserves special attention, since Table VI in the Appendix is for the natural (base e) logarithm. To obtain an alternate expression for $\log_a e$, we first write the identity $a^{\log_a e} = e$. Since $\ln e = 1$, we have $\ln (a^{\log_a e}) = 1$. However, by Property 5-5, $\ln (a^{\log_a e}) = (\log_a e) \cdot (\ln a)$. Consequently, we conclude that

$$\log_a e = \frac{1}{\ln a}$$

Thus the derivative of $\log_a x$ can be written as

$$\frac{d}{dx}(\log_a x) = \frac{1}{x} \log_a e = \frac{1}{x} \cdot \frac{1}{\ln a}$$

In Figure 5-7, we show a graph of $\ln x$ and observe that $1/x$ is a reasonable expression for $(d \ln x)/dx$. Note that if x is large, the slope of the graph appears to be small and indeed for x large, $1/x$ is small. Also for x positive and x near zero, the slope of the graph appears large and indeed $1/x$ is large for such x. At $x = 1$ the slope is 1, as the formula shows.

Of course, the formulas for the derivative of a sum, product, difference, or quotient of two functions also apply to logs. For example,

$$\frac{d(x + 5 + \ln x)}{dx} = 1 + 0 + \frac{1}{x} = 1 + \frac{1}{x}$$

$$\frac{d(x^3 + \ln x)}{dx} = \frac{d(x^3)}{dx} + \frac{d \ln x}{dx} = 3x^2 + \frac{1}{x}$$

Using the chain rule, we can easily find a formula for the derivative of $\ln f(x)$ if $f(x) > 0$:

$$\frac{d}{dx}[\ln f(x)] = \frac{d}{df(x)}[\ln f(x)] \cdot \frac{df(x)}{dx} = \frac{1}{f(x)} \cdot f'(x) = \frac{f'(x)}{f(x)}$$

For logarithms to bases other than e, we have

$$\frac{d}{dx}[\log_a f(x)] = \frac{f'(x)}{f(x)} \cdot \log_a e = \frac{f'(x)}{f(x)} \cdot \frac{1}{\ln a}$$

These formulas are exactly what one expects from the chain rule. If one knows the chain rule and the formulas for

$$\frac{d \ln x}{dx} \quad \text{and} \quad \frac{d \log_a x}{dx}$$

one need not memorize these new formulas.

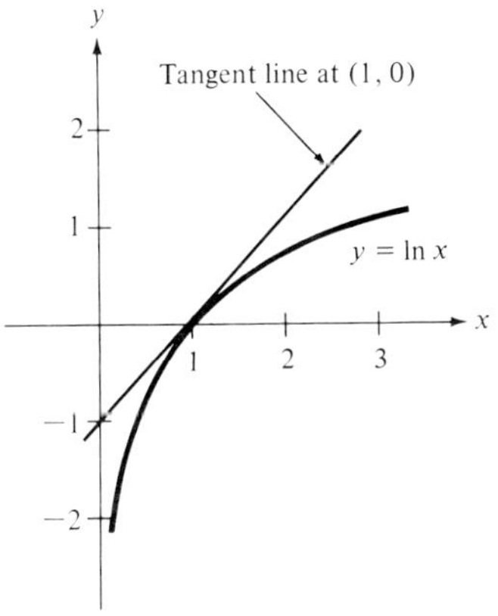

FIGURE 5-7

To illustrate the use of the chain rule, we shall consider differentiating $\ln(x^2 + 1)$. We choose $u = x^2 + 1$ and write

$$\frac{d}{dx}[\ln(x^2 + 1)] = \frac{d}{du}[\ln u]\frac{du}{dx} = \frac{1}{u}\frac{d}{dx}(x^2 + 1)$$

$$= \frac{1}{x^2 + 1} \cdot 2x = \frac{2x}{x^2 + 1}$$

With a little practice, one can write the answer down directly, as illustrated by the following example involving a product of two functions.

$$\frac{d[x \ln (1 + x^4)]}{dx} = x \cdot \frac{d \ln (1 + x^4)}{dx} + \ln (1 + x^4) \cdot \frac{d(x)}{dx}$$

$$= x \cdot \frac{1}{1 + x^4} \cdot 4x^3 + \ln (1 + x^4) \cdot 1$$

$$= \frac{4x^4}{1 + x^4} + \ln (1 + x^4)$$

EXERCISES 5-4

In Exercises 1–10, calculate the derivatives of the given functions.

1. $\log_3 x$

2. $1 + \log_{10} x$

3. $x + \ln x$

4. $x^2 - \ln x$

5. $1 + x^3 - \log_2 x$

6. $\log_3 x - \log_2 x$

7. $x \ln x$

8. $x^2 \log_{10} x$

9. $(1 + x) \ln x$

10. $\dfrac{\ln x}{x}$

In Exercises 11–20, calculate the derivatives using the chain rule.

11. $\ln (2x + 3)$

12. $\ln (x^2 - x - 1)$

13. $\ln \left(\dfrac{1}{x}\right)$ NOTE: This can be done regarding $1/x$ as $f(x)$ or by writing $\ln 1/x = -\ln x$. The answers should be the same.

14. $\ln (\ln x)$

15. $(\ln x)^3$ NOTE: This is different from $\ln (x^3)$.

16. $(\log_{10} x)^2$

17. $\log_{10} (x^2)$

18. $\log_5 \left(\dfrac{x + 2}{x^2}\right)$ HINT: $\log_5 [(x + 2)/x^2] = \log_5 (x + 2) - \log_5 x^2$
$= \log_5 (x + 2) - 2 \log_5 x$

19. $\ln ((x + 1)^2 \sqrt{x + 4})$ HINT: Note that $\ln ((x + 1)^2 \sqrt{x + 4}) = 2 \ln(x + 1) + \frac{1}{2} \ln (x + 4)$

20. $\ln \dfrac{x^3(x+1)^{3/2}}{(x+2)^4}$ HINT: Follow the approach used in Exercises 18 and 19.

For Exercises 21–24, (**a**) determine all local maxima and minima (using the second derivative test for deciding which), and (**b**) graph the function.

21. $f(x) = x \ln x \quad x > 0$

22. $f(x) = (\ln x)^2 \quad x > 0$

23. $f(x) = \ln (x^2 + 1) \quad \text{all } x$

24. $f(x) = x - \ln x \quad x > 0$

25. Find an equation of the line tangent to the graph of $y = (\ln x)^2$ at $x = 1$.

26. Find an equation of the line tangent to the graph of $y = x^2 \ln x$ at $x = 1$.

27. Best-Guess Market Research has informed Doughy Biscuits that the total revenue r generated by selling x units of biscuits is given by the formula

$$r = \ln (x + 1)$$

where r is in millions of dollars and x is in millions of cases. The manufacturing cost c in millions of dollars is given by

$$c - \frac{x}{4}$$

How many units should Doughy sell to maximize profits?

28. The profits from a proposed chain store venture are related to the number x of stores opened by the formula

$$p = 8 - x \ln \left(\frac{x}{8e}\right)$$

Determine the number of stores that maximizes profits. HINT: Recall that

$$\ln (x/8e) = \ln x - \ln (8e) = \ln x - \ln 8 - \ln e = (\ln x - \ln 8) - 1$$

5-5. DERIVATIVES OF e^x AND a^x

We shall first obtain the derivative of e^x and then extend to the more general a^x.

To obtain the derivative of e^x, we let $y = e^x$ and observe that $dy/dx = y' = d(e^x)/dx$. Now we write $y = e^x$ in log form:

$$\ln y = x$$

Next we differentiate with respect to x.

$$\frac{d}{dx}(\ln y) = \frac{d}{dx}(x)$$

Of course, $d(x)/dx = 1$. We use the chain rule to obtain $d(\ln y)/dx$:

$$\frac{d}{dx}(\ln y) = \frac{d}{dy}(\ln y)\frac{dy}{dx} = \frac{1}{y} \cdot y'$$

By substituting for the derivatives, the earlier equation becomes

$$\frac{1}{y}y' = 1 \qquad \text{or} \qquad y' = y$$

Since $y = e^x$ and $y' = d(e^x)/dx$, we have

$$\frac{d}{dx}(e^x) = e^x$$

Observe that the derivative of the natural exponential function is the function itself. The slope at each point of the graph is the y coordinate of that point.

In Figure 5-8 we may observe the tangent lines to the graph of $y = e^x$ at $x = -1$, 0, and 1. Of course the slope is positive at each point.

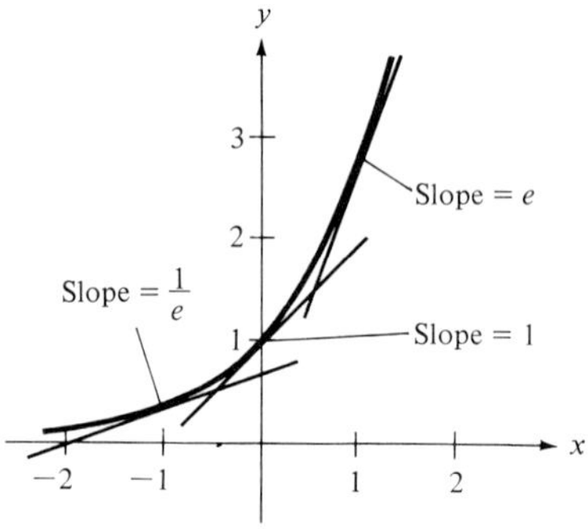

FIGURE 5-8

Next we wish to consider functions of the type $e^{f(x)}$. Applying the chain rule, we choose $u = f(x)$ and write

$$\frac{d}{dx}[e^{f(x)}] = \frac{d}{du}[e^u]\frac{du}{dx} = e^u\frac{df(x)}{dx} = e^{f(x)} \cdot f'(x)$$

We need not memorize this formula. We simply apply the chain rule. To illustrate, consider differentiating e^{x^2+1}. The derivative will be e^{x^2+1} times the derivative of $(x^2 + 1)$ or $2x \cdot e^{x^2+1}$. More formally, we choose $u = x^2 + 1$ and write

$$\frac{d}{dx}(e^{x^2+1}) = \frac{d}{du}[e^u]\frac{du}{dx} = e^u \frac{d}{dx}(x^2 + 1) = e^{x^2+1} \cdot (2x) = 2xe^{x^2+1}$$

With a little practice, you can write such answers directly.

Finally, we shall consider the derivative of a^x for $a > 0$ and $a \neq 1$. From the definition of ln a, we know that ln a is that power to which e must be raised to give a. Therefore, $e^{\ln a} = a$. We then write a^x as

$$a^x = (e^{\ln a})^x = e^{x \ln a}$$

Thus we can differentiate a^x as follows (we choose $u = x \cdot \ln a$ in applying the chain rule):

$$\frac{d}{dx}(a^x) = \frac{d}{dx}(e^{x \ln a}) = \frac{d}{du}[e^u]\frac{du}{dx} = e^u \frac{d}{dx}[x \ln a]$$

$$= e^{x \ln a} \cdot \ln a = a^x \ln a$$

Thus we have the formula $d(a^x)/dx = a^x \ln a$. However, for many people it is easier to remember the procedure of writing $a^x = e^{x \ln a}$ rather than to remember the formula.

The chain rule can be used to differentiate $a^{f(x)}$. We have

$$\frac{d}{dx}(a^{f(x)}) = a^{f(x)}f'(x) \ln a$$

For example,

$$a^{(x^2+1)} = e^{(x^2+1) \ln a}$$

$$\frac{d[a^{(x^2+1)}]}{dx} = \frac{d[e^{(x^2+1)\ln a}]}{dx} = e^{[(x^2+1) \ln a]} \cdot \frac{d[(x^2 + 1) \cdot \ln a]}{dx}$$

$$= a^{x^2+1}(2x) \ln a = 2x\, a^{x^2+1} \ln a$$

In this example, we could have written

$$a^{x2+1} = a^{x2} \cdot a = a \cdot a^{x2},$$

regarding the first factor as a constant. We would eventually have gotten the same formula for $d(a^{x^2+1})/dx$.

In computing work, as well as in some mathematics texts, the symbol exp (x) is used to denote the exponential function of x, that is e^x. Thus exp $(x^2) = e^{x^2}$, exp $(1 - x + x^3) = e^{1-x+x^3}$, etc. One of the advantages of using exp (x) in place of e^x is that formulas can be written without superscripts (the raised x) or in more complicated formulas, without as many superscripts, this being easier to handle, especially on computer printouts. In this notation, we have

$$\frac{d[\exp(x)]}{dx} = \exp(x)$$

EXERCISES 5-5

In Exercises 1–20, calculate the derivatives and evaluate at the indicated point if given. (You may use properties of exponents to change forms if you wish.) In most problems, you will need to use the chain rule.

1. e^{-x} **2.** e^{x-1}

3. 2^x **4.** 3^x

5. $3\exp\left(\frac{1}{x}\right)$ **6.** $\exp\left(2x+\frac{1}{x}\right)$

7. $e^x + \ln x$ **8.** $e^x \ln x \quad x = 1$

9. $xe^x \quad x = -1$ **10.** $(e^x)^2 \quad x = \ln 2$

11. $e^{\ln(x+1)} \quad x = e - 1$ **12.** $e^{x^2-x-1} \quad x = 2$

13. $10^{(x^2)} \quad x = 0$ **14.** $2^{-x} \quad x = -1$

15. $3^{x^2-x} \quad x = 0$ **16.** $e^x - \ln(x+1) + x^2$

17. $\frac{1}{4}\exp(x)$ **18.** $\frac{1}{x+1}\exp(2x)$

19. $(x+1)^x$ HINT: $(x+1)^x = e^{x\ln(x+1)}$

20. $(2x+1)^{1/x}$ See Hint for 19.

For Exercises 21–24, **(a)** determine all local maxima and minima (using the second-derivative test for deciding which), and **(b)** graph the function.

21. $f(x) = xe^{-x} \quad$ all x **22.** $f(x) = (xe^{-x})^2 \quad$ all x

23. $f(x) = e^{-x^2} \quad$ all x **24.** $f(x) = 2^x - x \quad$ all x

25. Find an equation of the line tangent to the graph of $y = xe^x$ at $x = 0$.

26. Find an equation of the line tangent to the graph of $y = e^x/x$ at $x = 1$.

27. From past experience, Widgets, International has developed the following formula relating its return r from sales to the quantity q that is sold:

$$r = 1.2 - e^{-q/4}$$

where r is in units of a million and q is in units of a thousand. Determine the marginal return, dr/dq.

28. Suppose company A has a monopoly on commodity B. If A decides to sell x units of B, past data indicate that the price must be $e^{-x/1000}$. What value of x maximizes the total revenue from sales (i.e., sales price times units sold)?

5-6. CONTINUOUS COMPOUNDING

Whenever banks advertise their savings accounts, two facts are usually stated: (1) the annual interest rate, and (2) the compounding period. That the interest rate must be stated should be obvious. We shall now see why stating the compounding period is also necessary.

Suppose that two banks offer 5% annual interest to savings depositors with the option that a depositor can withdraw his money at any time and get interest at that rate for the time his money has been in the bank. So Mr. Sharp figures he has a way to make some extra money. He puts his money (P dollars) in Bank A and withdraws it after 6 months, immediately depositing it in Bank B at the same rate. In this way he gets 6 months interest on the interest he earned from Bank A. Indeed, he has

$$P\left(1 + \frac{0.05}{2}\right) = 1.025P$$

after the first 6 months and

$$P\left(1 + \frac{0.05}{2}\right)^2 \approx 1.0506P$$

after a year.

If Mr. Sharp had left his money in Bank A for an entire year, he would have $P(1 + 0.05) = 1.05P$. By changing in midyear, he has $1.0506P$, which is slightly more. If it is worth changing banks every 6 months, why not do it every quarter? If he did this, alternating between banks A and B, he would have $P[1 + (0.05/4)]$ after the first quarter, $P[1 + (0.05/4)]^2$ after the second quarter, $P[1 + (0.05/4)]^3$ after the third quarter, and finally

$$P\left(1 + \frac{0.05}{4}\right)^4 \approx 1.0509P$$

at year end.

If Mr. Sharp had the time and energy, he could change banks every day (or at least every banking day) and get even more interest on his interest. But bankers understand this process, so they state and use a daily compounding period. In this way, banks pay interest on accounts on deposit effectively adding interest every day. If the interest is not withdrawn, it becomes principal for the next compounding period and draws interest. Thus Mr. Sharp gets no more by switching banks than by keeping his money in one bank.

In order to compare various interest rates and various compounding periods, it is convenient to define the *effective annual interest rate* as the annual interest rate compounded annually that would yield an equivalent return. As pointed out above, a 5% annual interest rate with semiannual compounding yields a total value of $[1 + (0.05/2)]^2 \approx 1.0506$ at year end so the effective annual interest rate is 5.06%. If quarterly compounding is used, the total value is $[1 + (0.05/4)]^4 \approx 1.0509$, and the effective annual interest rate is 5.09%.

What is the effective annual interest rate equivalent to a specified annual rate compounded daily? With m compounding periods per year at interest rate i, we have $P[1 + (i/m)]^m$ as the value of the deposit P at the end of the year. At 5% annual interest, the effective annual interest rate for various numbers of compounding periods per year is given by Table 5-3.

TABLE 5-3

Compounding periods	Effective annual interest rate
1	5.0000
2	5.0625
4	5.0945
6	5.1053
12	5.1162
52	5.1246
360	5.1267

The larger the number of compounding periods, the higher the effective interest rate, but note how slowly it rises toward the bottom of the table.

What, if anything, is $\lim_{m \to \infty} [1 + (i/m)]^m$? As we shall see, it is related to our old friend e; indeed it is e^i. We have

$$\left(1 + \frac{i}{m}\right)^m = \left[\left(1 + \frac{i}{m}\right)^{m/i}\right]^i.$$

Letting $i/m = t$, we have

$$\left[\left(1 + \frac{i}{m}\right)^{m/i}\right]^i = [(1 + t)^{1/t}]^i$$

and as $m \to \infty$, $t \to 0$. Thus

$$\lim_{m \to \infty} \left(1 + \frac{i}{m}\right)^m = \lim_{t \to 0} [(1 + t)^{1/t}]^i = e^i$$

since the ith power does not affect the limit process in this case.

An annual interest rate of i compounded *continuously* has an effective annual rate that may be computed from e^i. Indeed, if r is the effective annual rate, then $e^i = 1 + r$ or $r = e^i - 1$. In Table III in the Appendix, we give values of e^x for small values of x; these values are listed here:

x	0.00	0.01	0.02	0.03	0.04	0.05	0.06	0.07	0.08	0.09
0.0	1.0000	1.0100	1.0202	1.0305	1.0408	1.0513	1.0618	1.0725	1.0833	1.0942
0.1	1.1052	1.1163	1.1275	1.1388	1.1503	1.1618	1.1735	1.1853	1.1972	1.2092

The left-hand column lists values of x to tenths and the top row the hundredth's values. To find $e^{0.12}$, we find the entry in the row with 0.1 on the left and in the column with 0.02 at the top. The value is 1.1275.

What is the effective annual interest for 5% compounded continuously? From the table, the value of $e^{0.05} = 1.0513$. This means that 5% compounded continuously is equivalent to an annual rate of 5.13% compounded once a year. At usual interest rates, compounding continuously increases the effective rate but not so much as to be ridiculous. Compounding daily (360 times per year) changes an 8% annual rate to an 8.30% effective annual rate. Compounding continuously changes it to an 8.33% effective annual rate.

The converse of the problem discussed in the above paragraph is to determine the annual interest rate i compounded continuously that is equivalent to an effective annual rate of r. Since i and r are related by $e^i = 1 + r$, we can obtain an explicit equation by taking natural logarithms in the equation $e^i = 1 + r$, obtaining $i = \ln(1 + r)$. For example, if the effective annual interest rate is 8%, then the equivalent annual rate compounded continuously is given by $\ln(1 + 0.08) \approx 0.0770$ or 7.70%. The natural logarithm of numbers close to one can be obtained from Table V in the Appendix.

Next, we might ask how long it takes to double your money if the bank pays 5% compounded continuously. A rate of interest i compounded continuously is equivalent to a rate $r = e^i - 1$ compounded annually. Hence $(1 + r) = e^i$ and the value of the investment after t years is $(1 + r)^t$ or $(e^i)^t$ or e^{it} times the original investment. The original investment has doubled when

$e^{it} = 2$, or when $i \cdot t = \ln 2 \approx 0.693$ (from Table VI in the Appendix). For 5% interest, we have $t \approx 0.693/i = 0.693/0.05 \approx 13.86$ years.

Problems involving interest on money are somewhat unique in that they may naturally involve either continuous compounding or stated compounding periods. For many processes, the growth or decay is fundamentally occurring continuously. For example, the growth of the world's population is occurring essentially on a continuous basis. Such processes are described by the equation

$$Q(t) = Q_0 e^{it}$$

where $Q(t)$ is the quantity at time t; Q_0 the quantity at some instant of time, say t_0; t the time past t_0; and i the rate of growth or decay on an annual basis compounded continuously. For growth processes, i is positive. For decay processes, i is negative.

For a stated rate of i compounded or converted continuously, we sometimes call i the *instantaneous* growth (or decay) rate. This makes sense in the formula e^{it}, since

$$\frac{de^{it}}{dt} = ie^{it}$$

and for $t = 0$,

$$\left.\frac{d(e^{it})}{dt}\right|_{t=0} = i$$

For any other value of t, i is the factor by which we multiply e^{it} to determine the rate of change.

When expressing the growth (or decay) rate for a continuous process, it should be noted whether the rate is an effective annual rate or an instantaneous rate. For example, when we read in the newspaper that inflation is 10% per annum, it is usually understood that this is the effective annual rate. If we let r be the effective annual growth rate and i the instantaneous growth rate, these are related by

$$1 + r = e^i$$

as noted earlier. Using this relationship and the tables in the Appendix, we can convert from continuous or instantaneous growth rates to effective annual rates, and vice versa.

EXERCISES 5-6

1. For an annual interest rate of 10%, determine the following:
 (a) The effective annual interest rate if interest is compounded semiannually.

(**b**) The doubling life for semiannual compounding (to the nearest half-year).
Use Table I in the Appendix.

2. Rework Exercise 1 for an interest rate of 12% compounded semiannually.

3. For an annual interest rate of 8%, determine the following:
(**a**) The effective annual interest rate if the interest is compounded semiannually.
(**b**) The effective annual interest rate if the interest is compounded quarterly.

4. Many credit card companies charge interest of $1\frac{1}{2}$% per month on the unpaid balance (equivalent to an annual rate of 18% compounded monthly). If $(1.015)^{12} = 1.1956$, what is the effective annual interest rate?

For Exercises 5–8, determine the effective annual growth rate corresponding to the given instantaneous rate, i.e., find e^i for the specified rate i and compute $e^i - 1$, writing your answer as a percent.

5. 2% **6.** 6%

7. 9% **8.** 12%

For Exercises 9–12, determine the instantaneous annual rate corresponding to the given effective annual rate, find $\ln(1 + r)$ and write your answer as a percent.

9. 2% **10.** 6%

11. 9% **12.** 12%

13. Suppose, in a given process, that the instantaneous decay rate is 4%. Then, after a year, the quantity remaining is $e^{-0.04} = 1/e^{0.04}$. From Table III in the Appendix, we determine that $e^{0.04} \approx 1.0408$, and, therefore,

$$e^{-0.04} = \frac{1}{1.0408} \approx 0.9608$$

The effective annual decay rate is computed from $(1 - 0.9608)$ and is 3.92%, which is less than the quoted rate. Can you explain why the effective annual decay rate is less than the instantaneous decay rate, whereas the effective annual growth rate is greater than the instantaneous growth rate?

For Exercises 14–16, determine the effective annual decay rate for the given instantaneous decay rate. Refer to Exercise 13 for further explanation.

14. 6% **15.** 8% **16.** 12%

17. In Exercise 4, it was concluded that $1\frac{1}{2}\%$ per month was equivalent to the effective annual interest rate of 19.56%. Observe that $1\frac{1}{2}\%$ per month is 18% per year compounded monthly. Use Table III in the Appendix to determine the effective annual interest rate for 18% compounded continuously.

18. In 1960, a city's population was 150,000. In 1970, its population was 210,000. Assuming that the growth rate remains constant, what will be its population in 1980? In 1990? In 2000? About when will its population first exceed 1 million? 2 million? Assuming that the city's rate of growth was constant during the period 1960–1970, what was the instantaneous growth rate?

19. On August 1, 1973, a 5-pound bag of sugar cost about $1. On January 1, 1975, a 5-pound bag of sugar cost $3.20. Assuming the price increased at a constant rate, what was the instantaneous growth rate?

20. A company purchased a piece of equipment in 1964 for $820,000. In 1974, it sold that piece of equipment for $100,000. The decline in value of the equipment with use is known as depreciation. Assuming the depreciation rate was constant, what was the depreciation rate expressed as a percentage per year on a continuous basis?

21. By earning a college diploma, one generally feels that he or she has made a permanent investment in his or her future. But is a college education really that permanent? The utility of knowledge changes, which results in changes in the subjects studied in college. One approach to assessing this change in the college curricula is to list for each course the specific topic covered each class meeting. Examples could be "The Exponential Function," "The Natural Logarithm," etc. Suppose that we first list all the topics studied by you in seeking your degree. Next we shall list the topics studied by the typical graduate in your field who earned his degree 10 years ago. We shall then determine what fraction of the topics studied by the graduate of 10 years ago you also study. Suppose that you study only half of the topics studied by him. Then the half-life of his education is 10 years. What is this expressed as a continuous decay rate? In fields such as engineering, the half-life of the degree is around 5 years. What is this expressed as a continuous decay rate? This reveals the need to continue to read and study even after receiving a degree.

22. Suppose that about 1%, i.e., 1 out of every 100, of the adult population in the United States has attended business school. Furthermore, suppose that the population is increasing at a continuously compounded rate of 1% (effective annual rate = 1.005%) and that the number of business graduates is increasing at a continuously compounded rate of 5% (effective annual rate = 5.127%). If these growths continue at these rates, in what year will the entire adult population have attended business school? To solve this problem, we consider that in year 0 there are 100 adults, one of whom is a business graduate. The number of business graduates in year t is $1 \cdot e^{0.05t}$. How many total adults are there in year t? When are these equal?

NOTE: $\ln 100 = 2 \ln 10$, and $\ln 10$ can be obtained from the tables in the Appendix.

REVIEW EXERCISES: CHAPTER 5

For these review exercises use the tables in the Appendix where appropriate. Find the doubling life (to the nearest year) at the following effective annual growth rates:

1. 3% **2.** 5% **3.** 7% **4.** 9%

Find the half-life (to the nearest year) at the following effective annual decay rates:

5. 3% **6.** 5% **7.** 7% **8.** 9%

Find the doubling life (to the nearest year) at the following *instantaneous* growth rates:

9. 2% **10.** 4% **11.** 7% **12.** 8%

Find the half-life (to the nearest year) at the following instantaneous decay rates:

13. 1% **14.** 3% **15.** 6% **16.** 10%

Find the instantaneous growth rate equivalent to an effective annual growth rate of:

17. 4% **18.** 5% **19.** 8% **20.** 9%

Find the effective annual growth rate equivalent to an instantaneous growth rate of:

21. 2% **22.** 3% **23.** 6% **24.** 7%

Find df/dx if $f(x)$ is the following function:

25. $\ln x + \ln (x^2 + 1)$ **26.** $\log_2 (1 - x^3)$

27. $\log_{10} (3x + 5)$ **28.** $\ln \left(\frac{x + 1}{x + 2}\right)$

29. $e^{2x} + e^{x^2}$

30. $e^{3x-1} + e^{\sqrt{x}}$

31. 10^{2x}

32. 4^{1-x-x^2}

33. Find an equation of the tangent line to the graph of e^{-3x} at the point for which $x = 2$.

34. Find the minimum of the function $\ln (x^2 + e)$ and any point or points that yield it.

35. Find the set of values of x for which the graph of $2xe^{-x}$ is concave downward.

36. Find dy/dx if y is defined implicitly as a function of x by $ye^x - \ln (x^2 + y^2 + 1) = 1$.

37. Graph the function $y = e^{[(x-1)^2]}$ showing all local maxima and minima.

In Exercises 38–43, simplify.

38. $e^{\ln(3x^2-1)}$

39. $\log_{10} (100^{1-x^2})$

40. $[\ln (e^2)]^4$

41. $\log_{10} (e^2) + \log_{10} \left(\frac{1}{e^2}\right)$

42. $\ln (x^3 - 3x) - \ln x$

43. $5^{\log_5 (x+1)}$

In Exercises 44–47, solve for x.

44. $e^{x-1} = e^{3x}$

45. $\ln (x^2 - 1) = \ln 3$

46. $5^x = e^{1-x}$

47. $\log_{10} (10^{x+3}) = \log_2 (2^{-x})$

48. If the population of the world is 4 billion in 1975 and the effective annual growth rate is 2% per year, find the first year at which the world population will exceed 5 billion.

49. If parents invest $10,000 at 8% compounded annually in a trust fund for a newborn child, at what age will the child first become a millionaire from this fund (assuming the money is left in the fund until that time)?

50. If a rubber ball bounces to two thirds its height when dropped, from how great a height should it be dropped so that after the sixth bounce it reaches a height of at least 1 centimeter?

51. If the daily decay rate of a load of vegetables being moved to market is 6%, what is the half-life of the load of vegetables?

52. If inflation continues at a 5% rate, at what age will a recent 65-year-old retiree on a fixed income of $12,000 have his income cut in half in terms of actual spending power?

CHAPTER SIX

THE INTEGRAL

6-1. AREAS UNDER GRAPHS

A very important problem in mathematics and in many applications of mathematics is that of finding the area of a given region in the plane. In elementary geometry, formulas for the areas of various special types of regions are found. For example, some elementary regions and their areas are shown in Figure 6-1.

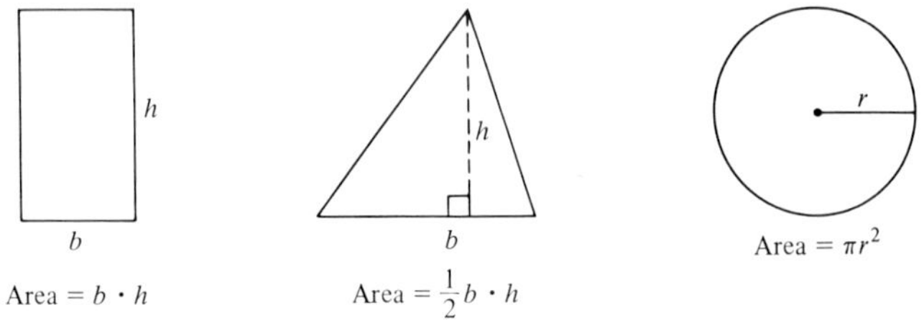

FIGURE 6-1

By elementary geometry, we also can compute the area of any region that is the union of finitely many such smaller nonoverlapping regions. We simply add the areas of the smaller regions.

For some general types of regions such as the region under the graph of a function $f(x)$ from a to b, as illustrated in Figure 6-2, we need to develop

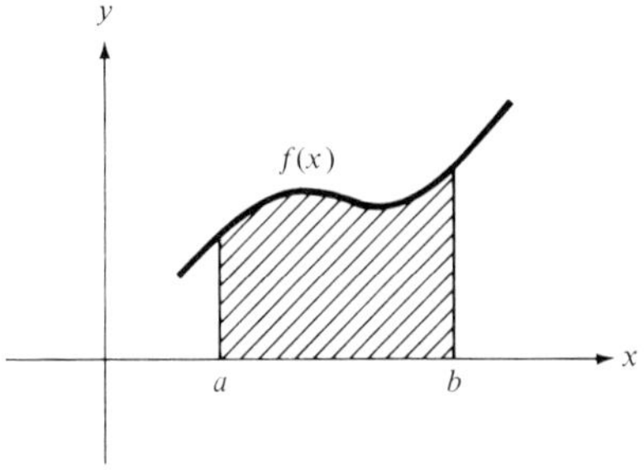

FIGURE 6-2

further concepts concerning area. We need a common-sense approach to area, one that is consistent with elementary uses of area, for example that used in determining the amount of paint needed to cover a particular surface or region. Therefore, we assume an intuitive understanding of area.

As illustrated in Figure 6-3, we may consider subdividing the interval $[a, b]$ into a finite number, say n, of nonoverlapping subintervals of equal length. The length of any of these subintervals is $(b - a)/n$, but for convenience we denote it by Δx. On the jth subinterval, we select some point x_j and consider a rectangle whose base is Δx and whose height is $f(x_j)$. Then the area of this jth rectangular region is $f(x_j) \cdot \Delta x$. We do this for each j from 1 to n. The sum of the areas of the n rectangular regions so formed is $f(x_1) \cdot \Delta x + f(x_2) \cdot \Delta x + \cdots + f(x_n) \cdot \Delta x$. In this notation, Δ is the Greek letter delta.

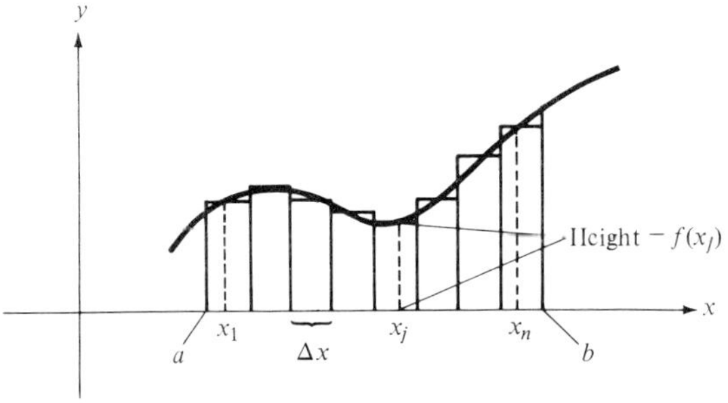

FIGURE 6-3

From Figure 6-3 it appears geometrically evident that $f(x_1) \cdot \Delta x + f(x_2) \cdot \Delta x + \cdots + f(x_n) \cdot \Delta x$, representing the sum of the areas of the rectangular regions, should be an approximation for the area of the region under the graph of $f(x)$ from a to b. Summarizing, we proceed as follows:

1. Take a subdivision of $[a, b]$ into nonoverlapping subintervals of small equal length, Δx.
2. Choose points, one in each of these subintervals.
3. Compute $f(x_1)\ \Delta x + f(x_2)\ \Delta x + \cdots + f(x_n)\ \Delta x$.

The result is a number that is approximately the area under the graph of $f(x)$. Indeed, by taking Δx sufficiently small or, equivalently, the number of subintervals sufficiently large, we are able to approximate the desired area as closely as we wish.

To illustrate, we shall calculate an approximation to the area under the graph of $f(x) = x^2$ over the interval $[1, 3]$. As in Figure 6-4 and as a first approximation, we shall use four subintervals. The subintervals are $[1, 1\frac{1}{2}]$, $[1\frac{1}{2}, 2]$, $[2, 2\frac{1}{2}]$, and $[2\frac{1}{2}, 3]$. We have $\Delta x = \frac{1}{2}$. Since we can choose for x_j any point in the jth subinterval, for convenience we shall choose the left-hand end point for x_j. Therefore, $x_1 = 1$, $x_2 = 1\frac{1}{2}$, $x_3 = 2$, and $x_4 = 2\frac{1}{2}$. In Figure 6-4, we show the graph of $f(x) = x^2$ with the four rectangular areas we are using to approximate the area under the graph of $f(x) = x^2$. The respective values of $f(x_j)\ \Delta x$ are computed as illustrated in Table 6-1.

The total area is

$$f(x_1) \cdot \Delta x + f(x_2) \cdot \Delta x + f(x_3) \cdot \Delta x + f(x_4) \cdot \Delta x = \tfrac{1}{2} + 1\tfrac{1}{8} + 2 + 3\tfrac{1}{8} = 6\tfrac{3}{4}$$

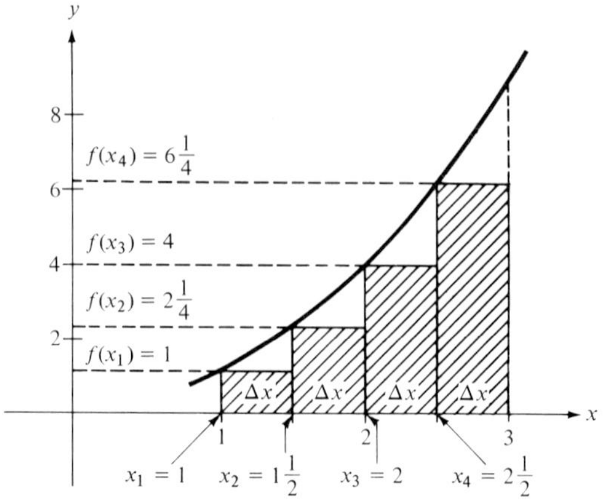

FIGURE 6-4

Therefore, our first approximation to the area is $6\frac{3}{4}$.

To obtain a better approximation, we could use more subintervals. Using 10 subintervals and left-hand end points, the approximation (to three decimal places) is 7.880. Similarly, using 100 subintervals, the approximation is 8.587. In subsequent sections, we shall present methods by which it can be determined that the area is $8\frac{2}{3}$. As common sense indicates, we could obtain a better approximation by using midpoints of the subintervals for the points $x_1, x_2, \ldots, x_n$. For 10 subintervals and using midpoints of these intervals, we get the value 8.660, which is very close to the actual area.

TABLE 6-1

Subinterval j	Length Δx	Left end point x_j	Functional value $f(x_j)$	Area of rectangle $f(x_j) \cdot \Delta x$
1	$\Delta x = \frac{1}{2}$	$x_1 = 1$	$f(x_1) = 1$	$f(x_1) \cdot \Delta x = \frac{1}{2}$
2	$\Delta x = \frac{1}{2}$	$x_2 = 1\frac{1}{2}$	$f(x_2) = 2\frac{1}{4}$	$f(x_2) \cdot \Delta x = 1\frac{1}{8}$
3	$\Delta x = \frac{1}{2}$	$x_3 = 2$	$f(x_3) = 4$	$f(x_3) \cdot \Delta x = 2$
4	$\Delta x = \frac{1}{2}$	$x_4 = 2\frac{1}{2}$	$f(x_4) = 6\frac{1}{4}$	$f(x_4) \cdot \Delta x = 3\frac{1}{8}$

Sigma Notation There is a shorthand notation called *sigma* or *summation* notation for sums such as those used above. We use $\sum_{j=1}^{n} f(x_j) \cdot \Delta x$ to represent $f(x_1) \cdot \Delta x + f(x_2) \cdot \Delta x + \cdots + f(x_n) \cdot \Delta x$. Here $\sum$ is the Greek capital letter "sigma," similar to our S, and is used to indicate a sum in mathematics. The symbol j (sometimes written i) is called the *index*. For example, $\sum_{j=1}^{n} y_j$ means the sum of the y_j's from $j = 1$ to $j = n$ or $y_1 + y_2 +$

$\cdots + y_n$. Similarly, $\sum_{j=1}^{n} f(x_j) \cdot \Delta x$ means the sum of terms of the form $f(x_j) \cdot \Delta x$ from $j = 1$ to $j = n$, and thus

$$\sum_{j=1}^{n} f(x_j) \cdot \Delta x = f(x_1) \cdot \Delta x + f(x_2) \cdot \Delta x + \cdots + f(x_n) \cdot \Delta x$$

as indicated.

The value of n in the $\sum$ notation specifies the index of the last term in the sum. When we write $\sum_{j=1}^{4} x_j$, we mean $x_1 + x_2 + x_3 + x_4$. If we write $\sum_{j=1}^{5}$, we mean $x_1 + x_2 + x_3 + x_4 + x_5$. Although in most cases we start with $j = 1$, this is not necessary. For example, $\sum_{j=3}^{5} x_j = x_3 + x_4 + x_5$. Here are some examples in which we have explicit formulas for x_j.

Example 1

$$\sum_{j=1}^{5} j = 1 + 2 + 3 + 4 + 5 = 15$$

Example 2

$$\sum_{j=1}^{3} j^2 = 1^2 + 2^2 + 3^2 = 1 + 4 + 9 = 14$$

Example 3

$$\sum_{j=1}^{4} \frac{j-1}{2} = \frac{1-1}{2} + \frac{2-1}{2} + \frac{3-1}{2} + \frac{4-1}{2} = 0 + \frac{1}{2} + 1 + \frac{3}{2} = 3$$

Since the $\sum$ notation is simply a shorthand representation of a sum of terms, any algebraic operation that is valid on a sum is valid on a formula written using the $\sum$ notation. For example, consider

$$\sum_{j=1}^{4} 3x_j = 3x_1 + 3x_2 + 3x_3 + 3x_4$$

Since $3x_1 + 3x_2 + 3x_3 + 3x_4 = 3(x_1 + x_2 + x_3 + x_4)$, we have

$$\sum_{j=1}^{4} 3x_j = 3(x_1 + x_2 + x_3 + x_4) = 3 \sum_{j=1}^{4} x_j$$

This illustrates that we can factor constant coefficients from the summation. In the exercises, we discuss the fact that

$$\sum_{j=1}^{n} (x_j + y_j) = \sum_{j=1}^{n} x_j + \sum_{j=1}^{n} y_j$$

However, it should be noted that the corresponding property does *not* hold for products. That is, in general,

$$\sum_{j=1}^{n} x_j y_j \neq \left(\sum_{j=1}^{n} x_j\right) \cdot \left(\sum_{j=1}^{n} y_j\right)$$

This is illustrated in Exercise 20.

EXERCISES 6-1

For Exercises 1–8, write the sum in $\sum$ notation.

1. $x_1^2 + x_2^2 + \cdots + x_n^2$ **2.** $x_1^3 + x_2^3 + \cdots + x_n^3$

3. $1^2 + 2^2 + \cdots + n^2$

4. $1^3 + 2^3 + \cdots + n^3$ HINT: In Exercises 3 and 4, compare with Exercises 1 and 2 and let $x_j = j$.

5. $[1 \cdot 2] + [2 \cdot 3] + [3 \cdot 4] \cdots + [n \cdot (n+1)]$ HINT: The jth term is $j \cdot (j+1)$.

6. $\frac{1}{2} + \frac{1}{4} + \frac{1}{8} + \cdots + \frac{1}{2^n}$ **7.** $1 + \frac{1}{2} + \frac{1}{3} + \cdots + \frac{1}{n}$

8. $\frac{2}{1} + \frac{3}{2} + \frac{4}{3} + \cdots + \frac{n+1}{n}$

For Exercises 9–13, write as an explicit sum of numbers.

9. $\sum_{j=1}^{6} j \cdot (2j+1)$ **10.** $\sum_{j=1}^{4} (j + j^2)$

11. $\sum_{j=1}^{5} \frac{j+1}{2j}$ **12.** $\sum_{j=1}^{5} \frac{2j}{j+1}$

13. $\sum_{j=1}^{10} 4$ HINT: This is equivalent to $\sum_{j=1}^{10} x_j$ where, for each j, $x_j = 4$.

In Exercises 14–17, determine what constant, if any, can be conveniently factored out of the sums. Note that technically any constant $\neq 0$ could be factored out but restrict yourself to obvious ones.

14. $\sum_{j=1}^{8} 3j$ **15.** $\sum_{j=1}^{5} \frac{j^2}{2}$

16. $\sum_{j=1}^{7} 8x_j$ **17.** $\sum_{j=1}^{20} \frac{x_j + 3}{2}$

18. Consider the sum $\sum_{j=1}^{4} (j + j^3)$. Evaluate the sum. Also evaluate $\sum_{j=1}^{4} j$ and $\sum_{j=1}^{4} j^3$. Do your results agree with the following relationship?

$$\sum_{j=1}^{4} (j + j^3) = \sum_{j=1}^{4} j + \sum_{j=1}^{4} j^3$$

19. Write $\sum_{j=1}^{6} x_j$ as $x_1 + x_2 + \cdots + x_6$. Write $\sum_{j=1}^{6} y_j$ and $\sum_{j=1}^{6} (x_j + y_j)$ similarly. Rearrange the terms in the expansion of $\sum_{j=1}^{6} (x_j + y_j)$ to show that

$$\sum_{j=1}^{6} (x_j + y_j) = \sum_{j=1}^{6} x_j + \sum_{j=1}^{6} y_j$$

20. Let $x_j = j - 1$ and let $y_j = 1/j$. Determine $\sum_{j=1}^{4} x_j$, $\sum_{j=1}^{4} y_j$, and $\sum_{j=1}^{4} (x_j y_j)$. Your results should show that, in general,

$$\sum_{j=1}^{n} (x_j y_j) \neq \left(\sum_{j=1}^{n} x_j\right) \cdot \left(\sum_{j=1}^{n} y_j\right)$$

In Exercises 21–30, evaluate $f(x_1) \cdot \Delta x + f(x_2) \cdot \Delta x + \cdots + f(x_n) \cdot \Delta x$ using 4 equal-lengthed subintervals and choosing the left-hand end points of the subintervals as the x_j's. Sketch the graph of the function over the given interval showing **(a)** the region under the graph, and **(b)** the union of the rectangular regions. For Exercises 25–28, use Tables IV and V in the Appendix.

21. $f(x) = x^2 + 1$ $[a, b] = [0, 2]$ **22.** $f(x) = \frac{1}{x}$ $[a, b] = [1, 2]$

23. $f(x) = x + 5$ $[a, b] = [-1, 1]$

24. $f(x) = \frac{1}{x + 3}$ $[a, b] = [-2, 2]$

25. $f(x) = e^x$ $[a, b] = [0, 2]$ **26.** $f(x) = \ln x$ $[a, b] = [1, 5]$

27. $f(x) = \ln (1 + x^2)$ $[a, b] = [0, 4]$

28. $f(x) = e^{1+x}$ $[a, b] = [-2, 2]$

29. In Exercise 21, the area under $x^2 + 1$ from $x = 0$ to $x = 2$ was approximated using the left-hand end points of the subintervals. **(a)** Use the right-hand end points for the evaluation, and **(b)** use the midpoints of the subintervals. Compare your answers. NOTE: For any interval $[c, d]$, the midpoint is $(c + d)/2$.

30. Repeat Exercise 29 for Exercise 22.

6-2. THE DEFINITE INTEGRAL

In the previous section, we approximated what appeared to be the area under the graph of a function by the sum

$$A \approx f(x_1) \cdot \Delta x + f(x_2) \cdot \Delta x + \cdots + f(x_n) \cdot \Delta x = \sum_{j=1}^{n} f(x_j) \cdot \Delta x$$

We also observed that the use of smaller subintervals should produce a better approximation.

We now consider the fundamental question of how we "define" the area under the graph of a function $f(x)$ from a to b. We assume that $f(x)$ is continuous and that $f(x) \geq 0$ over $[a, b]$.

Let

$$A_1 = f(xn_1) \cdot \Delta x + f(x_2) \cdot \Delta x + \cdots + f(xn_1) \cdot \Delta x$$

for a subdivision using n_1 subintervals of $[a, b]$. Now subdivide $[a, b]$ again into n_2 nonoverlapping subintervals by subdividing *each* of the subintervals used in the first subdivision and make a similar calculation. Let

$$A_2 = f(xn_2) \cdot \Delta x + f(x_2) \cdot \Delta x + \cdots + f(xn_2) \cdot \Delta x$$

for this second subdivision. Since we are using more subintervals, we should have more confidence in A_2 as an approximation of the "area" under the graph of $f(x)$ from a to b than we had in A_1. We may continue the process, getting a sequence A_1, A_2, A_3, ... of areas of finite unions of rectangular regions, which according to our intuition appear to approximate the area of the desired region. Indeed, for $f(x)$ continuous over $[a, b]$, we have the following property.

Property 6-1 If the sequence of the lengths of the subintervals converges to zero, then limit A_i exists. Furthermore, for any two such sequences $\langle A_i \rangle$ and $\langle A'_i \rangle$, $\lim_{i \to \infty} A_i = \lim_{i \to \infty} A'_i$.

Property 6-1 guarantees us that for any continuous function $f(x)$ over $[a, b]$ there is exactly one number, $A = \lim_{i \to \infty} A_i$, which is obtained by the procedure we have described. We may use any convenient process of subdivision and selection of points in the appropriate subintervals to find such a number. The unique number guaranteed by Property 6-1 is called the *area under the graph of $f(x)$ from a to b*, and we use the symbol $\int_a^b f(x)\,dx$ to denote this number.

The symbol $\int_a^b f(x)\,dx$ can be regarded as developed from $\sum_{j=1}^{n} f(x_j)\,\Delta x$ by "passing to the limit" with the Greek letter $\sum$ becoming an S written as $\int$, $f(x_j)$ becoming $f(x)$, and Δx becoming dx. We call $\int_a^b f(x)\,dx$ the *Riemann* (or *definite*) *integral of $f(x)$ from a to b* or just *the integral of $f(x)$ from a to b*. The expressions $\sum_{j=1}^{n} f(x_j)\,\Delta x$ used in the definition of the integral are called *Riemann sums*. Both the integral and the sum are named for Bernard Riemann (pronounced Ree′mann), a great German mathematician of the nineteenth century. A function is *Riemann-integrable* over an interval if all sequences of Riemann sums with length of the subintervals going to zero converge to the same number.

The numbers a and b are usually called the *lower limit* and the *upper limit*, respectively, of $\int_a^b f(x)\,dx$. They are just the end points of the interval. The use of the term "limit" in this sense is perhaps unfortunate, since a and b are fixed numbers in the process and it is the number $\int_a^b f(x)\,dx$ that is $\lim_{i\to\infty} A_i$ as explained above. The function $f(x)$ is called the *integrand.*

We have so far defined $\int_a^b f(x)\,dx$ for any continuous function $f(x)$ over $[a, b]$ with $f(x) \geq 0$. For such a function, $\int_a^b f(x)\,dx$ is a number that is the area under the graph of $f(x)$ over $[a, b]$. The restriction that $f(x) \geq 0$ is needed for our interpretation of area, but it is not needed for Property 6-1 to be true for continuous functions. Indeed, $\int_a^b f(x)\,dx$ is defined to be $\lim_{i\to\infty} A_i$ for any sequence $\langle A_i \rangle$, whether or not $f(x)$ is positive or zero over $[a, b]$. In Figure 6-5 we see a graph of a continuous function over $[a, b]$ for which $\int_a^b f(x)\,dx$ is approximately zero. The process of developing the integral as a Riemann sum tells us that the region below the curve and above the axis is counted positively and the region above the curve and below the axis is counted negatively, since in $\sum_{j=1}^n f(x_j)\,\Delta x$, the sign of $f(x_j)$ determines the sign of the term $f(x_j)\,\Delta x$.

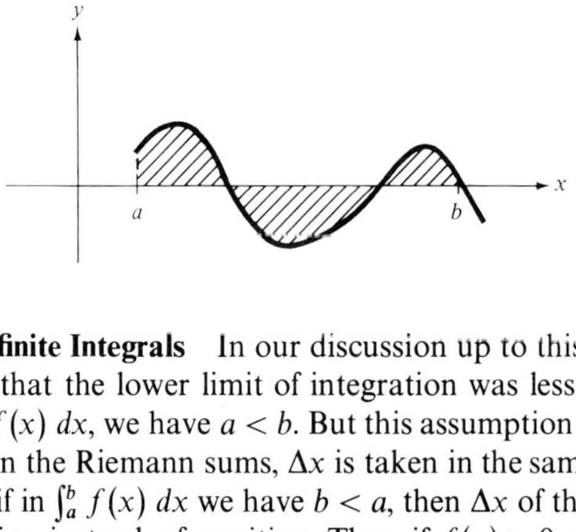

FIGURE 6-5

Properties of Definite Integrals In our discussion up to this point, we have tacitly assumed that the lower limit of integration was less than the upper limit, i.e., for $\int_a^b f(x)\,dx$, we have $a < b$. But this assumption is not necessary if we agree that in the Riemann sums, Δx is taken in the same sense as a and b. For example, if in $\int_a^b f(x)\,dx$ we have $b < a$, then Δx of the Riemann sum would be negative instead of positive. Thus if $f(x) \geq 0$ and $b < a$, then $(\int_a^b f(x)\,dx) \leq 0$, since for each term of the Riemann sum $f(x_j) \geq 0$ while $\Delta x < 0$. In fact, we have

Property 6-2 $\int_a^b f(x)\,dx = -\int_b^a f(x)\,dx.$

We also agree that if $b = a$, then the area is zero, since the base has length zero. Thus we have

Property 6-3 $\int_a^a f(x)\,dx = 0.$

Our next property is a natural one in terms of the definition of the definite integral.

Property 6-4 $\int_a^b f(x)\,dx = \int_a^c f(x)\,dx + \int_c^b f(x)\,dx$ provided that any two of the three integrals exist.

If $a < c < b$, and if we could use c as a point of subdivision in getting our subintervals for the Riemann sums, then we could write each Riemann sum for $\int_a^b f(x)\,dx$ as the sum of two Riemann sums, one for $\int_a^c f(x)$ and the other for $\int_c^b f(x)\,dx$. On passing to the limit, we get Property 6-4. Indeed, the property is valid for any c. For example, if c is not between a and b, then we can use Properties 6-2 and 6-3 to justify the more general form of Property 6-4.

The next two properties will be very useful in evaluating integrals of polynomials and other functions. We assume that the indicated functions are Riemann integrable over the indicated intervals.

Property 6-5 $\int_a^b kf(x)\,dx = k\int_a^b f(x)\,dx$ for any constant k.

This property is true, since in each Riemann sum $\sum_{j=1}^n kf(x_j)\,\Delta x$, we can factor out k getting $k\sum_{j=1}^n f(x_j)\,\Delta x$. On passing to the limit, we get Property 6-5.

Our final property is

Property 6-6 $\int_a^b [f(x) + g(x)]\,dx = \int_a^b f(x)\,dx + \int_a^b g(x)\,dx$.

For the Riemann sums, we have

$$\sum_{j=1}^{n} [f(x_j) + g(x_j)]\,\Delta x = \sum_{j=1}^{n} f(x_j)\,\Delta x + \sum_{j=1}^{n} g(x_j)\,\Delta x$$

On passing to the limit, we get Property 6-6, which is sometimes read "the integral of a sum is the sum of the integrals."

Observe that these properties of the definite integral parallel the properties of a sum. In Section 6-1, we noted that, in general, $\sum_{j=1}^n (x_j y_j) \neq (\sum_{j=1}^n x_j)\cdot(\sum_{j=1}^n y_j)$. The same observation holds for the definite integral: in general,

$$\int_a^b [f(x)\cdot g(x)]\,dx \neq \left[\int_a^b f(x)\,dx\right]\cdot\left[\int_a^b g(x)\,dx\right]$$

"The integral of a product is, in general, *not equal to* the product of the integrals." See Exercise 20 for an example.

Establishing Bounds for Integrals and Estimating Integrals In learning to evaluate definite integrals, we should try to ensure that the answers we get are reasonable, i.e., "fit the facts." For example, for the definite integral of a function that is positive, it is a very good idea to make a preliminary estimate of the answer by (mentally) calculating the areas of two rectangular regions over $[a, b]$, the smaller one contained in the specified region and the larger one containing it. The correct answer must be between these two numbers, which are called *lower* and *upper bounds* on the value of the integral.

Consider $\int_0^4 \sqrt{9 + x^2}\, dx$. The graph of $\sqrt{9 + x^2}$ over $[0, 4]$ is illustrated in Figure 6-6. The inner region has area 12 and the outer one has area 20. Therefore $12 < \int_0^4 \sqrt{9 + x^2}\, dx$ and $\int_0^4 \sqrt{9 + x^2}\, dx < 20$. Indeed 15 would be a good guess for $\int_0^4 \sqrt{9 + x^2}\, dx$, since the area under the dotted line joining (0, 3) and (4, 5) is $16 = 12 + 4$. In some of the exercises, the reader is asked to estimate some integrals.

For integrals of functions that are not always positive, one or both of the bounds one gets by the type of procedure indicated may be negative. In some problems it may be easy to get much better bounds (closer to the actual value) or much better estimates than that given by the specific procedure suggested above. For example, by breaking the interval $[a, b]$ up into smaller intervals and employing Property 6-4, we may be able to estimate the value quite accurately.

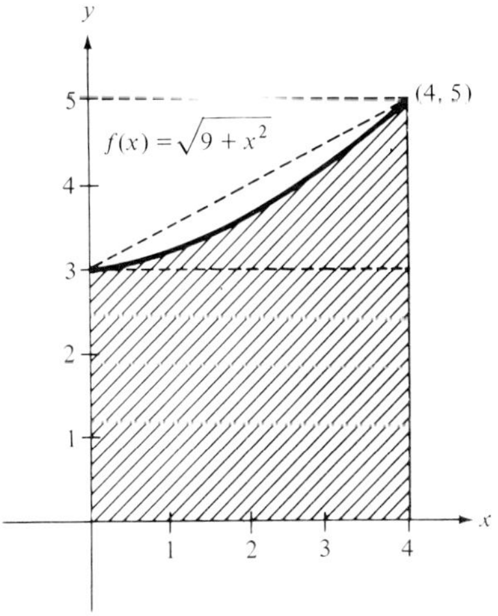

FIGURE 6-6

EXERCISES 6-2

The function $f(x)$ indicated in Figure OO is used for Exercises 1-8. For each integral, give (**a**) reasonable upper and lower bounds on the integral, and (**b**) a realistic estimate of the value of the integral accurate to the nearest integer.

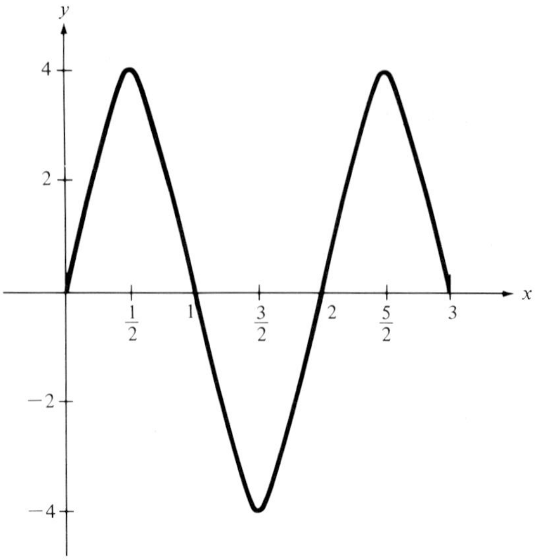

FIGURE OO

1. $\int_0^1 f(x)\,dx$ **2.** $\int_1^2 f(x)\,dx$

3. $\int_0^2 f(x)\,dx$ **4.** $\int_{1/2}^{5/2} f(x)\,dx$

5. $\int_0^3 f(x)\,dx$ **6.** $\int_0^{3/2} f(x)\,dx$

7. $\int_1^0 f(x)\,dx$ **8.** $\int_2^{3/2} f(x)\,dx$

In Exercises 9–18, (**a**) graph the given function over the specified interval, (**b**) find reasonable upper and lower bounds on the value of $\int_a^b f(x)\,dx$, and (**c**) by inspection of the graph estimate the value of $\int_a^b f(x)\,dx$.

9. $f(x) = x^2 + 1$ $[-1, 1]$ **10.** $f(x) = 4 - x^2$ $[-2, 2]$

11. $f(x) = e^x$ $[0, 2]$ **12.** $f(x) = \sqrt{9 - x^2}$ $[0, 3]$

13. $f(x) = \ln x$ $\left[\frac{1}{e}, e\right]$ **14.** $f(x) = x \cdot 2^x$ $[-1, 1]$

15. $f(x) = x^3 - 1$ $[-2, 2]$ **16.** $f(x) = 1 - x^4$ $[-1, 1]$

17. $f(x) = e^{-x}$ $[-1, 2]$ **18.** $f(x) = \ln(1 + x^2)$ $[0, 2]$

In Exercises 19–22, use Properties 6-5 and 6-6 to write each integral as a sum as indicated in the following example.

$$\int_a^b (2x^5 - 3x^2 + 2x - 5)\,dx = 2\int_a^b x^5\,dx - 3\int_a^b x^2\,dx + 2\int_a^b x\,dx - 5\int_a^b dx$$

19. $\int_0^1 (3x^4 - 2x^3 + 3x + 5)\,dx$ **20.** $\int_0^3 (4x^3 - 5x^2 - 2x + 4)\,dx$

21. $\int_{-1}^2 (4x^5 - 3x^3 - 2x + 6)\,dx$ **22.** $\int_6^8 (x^6 - 4x^4 + 8x^2 - 16)\,dx$

23. In the text, it was observed that, in general, $\int_a^b [f(x)g(x)]\,dx \neq (\int_a^b f(x)\,dx)(\int_a^b g(x)\,dx)$. Let $f(x) = x$ and $g(x) = x$; consequently, $f(x)g(x) = x^2$. Observe that $\int_{-1}^1 f(x)g(x)\,dx = \int_{-1}^1 x^2\,dx$ which must be positive, since $x^2 \geq 0$. But $\int_{-1}^1 f(x)\,dx = \int_{-1}^1 g(x)\,dx = \int_{-1}^1 x\,dx = 0$ by geometry. Therefore, $\int_{-1}^1 f(x)g(x)\,dx \neq (\int_{-1}^1 f(x)\,dx)(\int_{-1}^1 g(x)\,dx)$. Do your results support the observation in the text?

For Exercises 24–27, we are given that $\int_1^2 f(x)\,dx = 5$, $\int_2^3 f(x)\,dx = 6$, and $\int_1^2 g(x)\,dx = 2$. Find the integrals.

24. $\int_1^3 f(x)\,dx$ **25.** $\int_2^1 f(x)\,dx$

26. $\int_1^1 f(x)\,dx$ **27.** $\int_1^2 [f(x) + g(x)]\,dx$

6-3. THE FUNDAMENTAL THEOREM OF THE CALCULUS

In the expression for a definite integral $\int_a^b f(x)\,dx$, the symbol or variable x is sometimes called a "dummy variable," since the value of the integral depends only on the function concerned and on the upper and lower limits of integration. Thus we have $\int_a^b f(x)\,dx = \int_a^b f(t)\,dt = \int_a^b f(u)\,du$. Indeed, designating the function by the symbol f and not $f(x)$ as we have customarily done, some authors write $\int_a^b f$, omitting the dummy variable entirely.

If we regard the lower limit and the integrand function as given, then we may regard the definite integral as a function of the upper limit. In Figure 6-7, we have $F(x) = \int_a^x f(t)\,dt$. Whenever we assign x a specific value in an appropriate domain, we get a specific value for $F(x)$. In particular, $F(a) = 0$, since $\int_a^a f(t)\,dt = 0$. Note that we could have written $F(x) = \int_a^x f(x)\,dx$, where the upper limit denotes the variable of the function $F(x)$. The x of $f(x)\,dx$ is a dummy variable. For the function indicated in Figure 6-7, there will be some number x_1, for which $F(x_1) = \int_a^{x_1} f(t)\,dt = 1$, there will be some number x_2 for which $F(x_2) = 2$, etc.

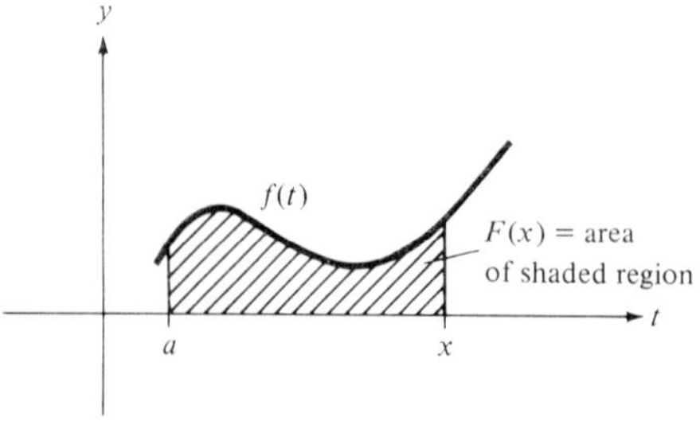

FIGURE 6-7

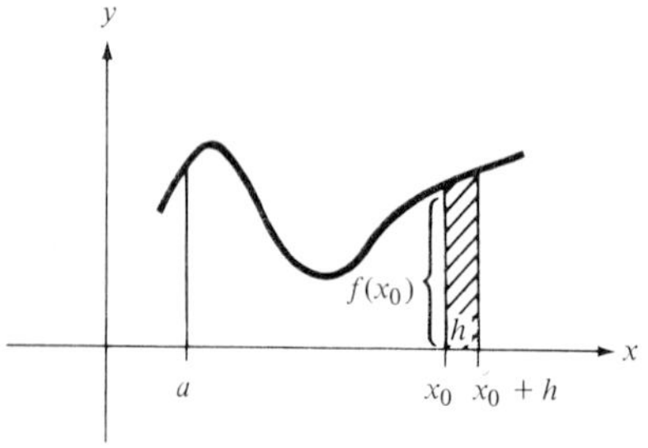

FIGURE 6-8

As illustrated in Figure 6-8, there is a very important property that relates the integral with the derivative. Let $F(x) = \int_a^x f(t)\,dt$ and assume that $f(t)$ is Riemann-integrable over each closed interval $[a, x_0]$ contained in a larger open interval (c, d).

Property 6-7 If x_0 is a point of (c, d) at which $f(t)$ is continuous, then

$$\left.\frac{dF(x)}{dx}\right|_{x_0} = f(x_0)$$

To see why this property is true, we first use the definition of the derivative to obtain

$$\left.\frac{dF(x)}{dx}\right|_{x_0} = \lim_{h\to 0} \frac{F(x_0 + h) - F(x_0)}{h}$$

Since $F(x) = \int_a^x f(t)\,dt$, then $F(x_0)$ and $F(x_0 + h)$ are given by

$$F(x_0) = \int_a^{x_0} f(t)\,dt$$

$$F(x_0 + h) = \int_a^{x_0+h} f(t)\,dt$$

By Property 6-4, $F(x_0 + h)$ can be written as follows:

$$F(x_0 + h) = \int_a^{x_0} f(t)\,dt + \int_{x_0}^{x_0+h} f(t)\,dt$$

Therefore, $F(x_0 + h) - F(x_0)$ is given by

$$F(x_0 + h) - F(x_0) = \int_a^{x_0} f(t)\,dt + \int_{x_0}^{x_0+h} f(t)\,dt - \int_a^{x_0} f(t)\,dt$$

$$= \int_{x_0}^{x_0+h} f(t)\,dt$$

Now $f(t)$ is given as continuous at x_0; so for each t near x_0, $f(t)$ is almost equal to $f(x_0)$. Therefore, for h near zero,

$$\int_{x_0}^{x_0+h} f(t)\,dt \approx f(x_0)\cdot h$$

since h is the length of the interval from x_0 to x_0+h and all functional values in $[x_0, x_0+h]$ are close to $f(x_0)$. We have

$$\frac{F(x_0+h)-F(x_0)}{h} = \frac{\int_{x_0}^{x_0+h} f(t)\,dt}{h} \approx \frac{f(x_0)\cdot h}{h} = f(x_0)$$

Indeed, it is not only plausible but true that

$$\left.\frac{dF(x)}{dx}\right|_{x_0} = \lim_{h\to 0}\frac{F(x_0+h)-F(x_0)}{h} = \lim_{h\to 0} f(x_0) = f(x_0)$$

given that x_0 is a point of continuity of $f(t)$. Since a function is continuous at any point at which its derivative exists, we also know *that* $F(x)$ *is continuous at every point at which* $f(t)$ *is continuous.*

Property 6-7 can be used to help us develop a very important technique for evaluating definite integrals of continuous functions over closed intervals. First, we note an important elementary property.

Property 6-8 If $F(x)$ and $G(x)$ are any two functions for which $F'(x) = G'(x)$ for each x in $[a, b]$, then for some constant C, $F(x) = G(x) + C$ over $[a, b]$.

In other words, over $[a, b]$ either of $F(x)$ or $G(x)$ can be written as a constant plus the other. Property 6-8 can be easily proved on the basis of an even more elementary property which states that if $H(x)$ is a function for which $H'(x) = 0$ over $[a, b]$, then $H(x)$ is a constant over $[a, b]$. However, the proof of this latter property requires material we have omitted from this book.

We now can state the other half of the very important property that relates the derivative and the integral and enables us to evaluate many integrals easily.

Property 6-9 Let $f(x)$ be continuous over an open interval (c, d) containing $[a, b]$ and let $G(x)$ be *any* function for which $G'(x) = f(x)$ for each x in $[a, b]$. Then $\int_a^b f(x)\,dx = G(b) - G(a)$.

To see that this is so, let $F(x) = \int_a^x f(t)\,dt$. Then

$$F(b) = \int_a^b f(t)\,dt \qquad \text{and} \qquad F(a) = \int_a^a f(t)\,dt = 0$$

Now, since $G'(x) = F'(x) = f(x)$ for each x in $[a, b]$, we may use Property 6-8 to get $G(x) = F(x) + C$ and hence $G(b) = F(b) + C$ and $G(a) = F(a) + C$.

Thus

$$G(b) - G(a) = F(b) + C - [F(a) + C] = F(b) - F(a) = F(b) - 0$$

But,

$$F(b) = \int_a^b f(x)\,dx$$

and thus

$$G(b) - G(a) = \int_a^b f(x)\,dx$$

which is what we wanted to show.

The two properties 6-7 and 6-9 taken together are sometimes referred to as the *Fundamental Theorem of the Calculus.* They relate the process of differentiation with the process of integration.

Example. Find $\int_1^3 x^2\,dx$.

Recalling our derivative formulas, we may observe that $d(\frac{1}{3}x^3)/dx = \frac{1}{3} \cdot 3x^2 = x^2$ for each x, and we know that the integrand x^2 is continuous for each x. Therefore, letting $G(x) = x^3/3$, we have $G'(x) = x^2$ and

$$\int_1^3 x^2\,dx = G(3) - G(1) = \frac{3^3}{3} - \frac{1^3}{3} = 9 - \frac{1}{9} = \frac{80}{9}$$

If we do our calculations correctly, we must get the right final answer no matter which function we use, as long as its derivative is the integrand x^2. For instance, $(x^3/3) + 10$ is also a function whose derivative is x^2. Thus

$$\int_1^3 x^2\,dx = \frac{3^3}{3} + 10 - \frac{1^3}{3} - 10 = 9 - \frac{1}{9} = \frac{80}{9}$$

as it should be.

REMARK: The conditions in Properties 6-7 and 6-9 that there be an open interval (c, d) larger than $[a, b]$ over which we have Riemann integrability or continuity are really unnecessary. Our earlier definitions technically required us to use such conditions. By suitably changing our earlier definitions, we could omit the references to the larger open interval, simply requiring Riemann integrability and continuity over $[a, b]$.

EXERCISES 6-3

In Exercises 1–8, (**a**) verify by recalling what you have known about continuous functions that the given function $f(x)$ is continuous over the prescribed interval, (**b**) verify that the derivative of the given function $G(x)$ is the function $f(x)$ over $[a, b]$, and (**c**) evaluate $\int_a^b f(x)\,dx$ as $G(b) - G(a)$.

Write your answers in the form $\int_a^b f(x)\,dx = G(b) - G(a)$ and simplify where convenient.

	$f(x)$	$G(x)$	$[a, b]$
1.	$3x^2$	$x^3 + 2$	$[-1, 1]$
2.	$-x$	$-\dfrac{x^2}{2} + 5$	$[0, 2]$
3.	$2x + 1$	$x^2 + x + 2$	$[-1, 2]$
4.	$4x^3 + x$	$x^4 + \dfrac{x^2}{2} - 3$	$[1, 3]$
5.	$\dfrac{1}{x^2} - \sqrt{x}$	$-\dfrac{1}{x} - \dfrac{2}{3}x^{3/2}$	$[1, 2]$
6.	$\dfrac{5}{2}x^{3/2} + \dfrac{3}{2}x^{1/2} + 1$	$x^{5/2} + x^{3/2} + x + 1$	$[1, 4]$
7.	xe^x	$xe^x - e^x$	$[0, 4]$
8.	$\dfrac{1}{x}$	$2 + \ln x$	$[1, 5]$

9. For each of Exercises 1, 3, 5, and 7 write a function different from the given $G(x)$ but whose derivative is the given $f(x)$.

10. For each of Exercises 2, 4, 6, and 8, write a function different from the given $G(x)$ but whose derivative is the given $f(x)$.

In Exercises 11–18, use Property 6-7 to find the value of $F'(x)$ at the indicated point.

11. $F(x) = \int_0^x (t^3 - 2t^2 + t)\,dt.$ Find $F'(3)$.

12. $F(x) = \int_0^x \dfrac{t-1}{t+1}\,dt.$ Find $F'(2)$.

13. $F(x) = \int_0^x \sqrt{\dfrac{t+1}{t^2+2}}\,dt.$ Find $F'(1)$.

14. $F(x) = \int_0^x \left(\dfrac{1}{t+2} - \dfrac{1}{t+1}\right)dt.$ Find $F'(e)$.

15. $F(x) = \int_0^x (t^6 - t^4 + t^2 - 1)\,dt.$ Find $F'(\sqrt{7})$.

16. $F(x) = \int_0^x (\sqrt{t^2 + 5} - 3t)\,dt.$ Find $F'(2)$.

17. $F(x) = \int_0^x t \ln (t+1)\, dt.$ Find $F'(e-1)$.

18. $F(x) = \int_0^x e^{(t^2)}\, dt.$ Find $F'(1)$.

6-4. FINDING THE ANTIDERIVATIVE

As the fundamental theorem of the calculus tells us, *if* we can find a function $G(x)$ whose derivative at each point of $[a, b]$ is $f(x)$, then we can evaluate $\int_a^b f(x)\, dx$ by the simple formula

$$\int_a^b f(x)\, dx = G(b) - G(a)$$

Given the function $f(x)$, how do we find $G(x)$?

First we introduce some notation. The symbol $\int f(x)\, dx$ written like $\int_a^b f(x)\, dx$ but without the a and b is called the *indefinite integral* or *antiderivative* of $f(x)$ and refers to any function whose derivative is $f(x)$. Recall that over any interval, if $F(x)$ and $G(x)$ are functions for which $F'(x) = G'(x) = f(x)$, we must have $F(x) = G(x) + C$. Indeed, by properly specifying C, any antiderivative of $f(x)$ over an interval is expressible in the form $G(x) + C$. Thus, if $G'(x) = f(x)$, we can write $\int f(x)\, dx = G(x) + C$, with the understanding that the domain of $G(x)$ is any subset of the x axis that is an interval, a ray, or line over which this formula holds. For example, in giving a formal expression for $\int x^2\, dx$ we should write $(x^3/3) + C$, since (a) for any value of C, $d[(x^3/3) + C]/dx = x^2$, and (b) any function whose domain is the x axis and whose derivative is x^2 can be obtained from $(x^3/3) + C$ by a suitable choice of C.

For the purposes of this chapter, we need only three basic integration formulas:

Formula 6-1 $\int x^m\, dx = \dfrac{x^{m+1}}{m+1} + C$ for all $m \neq -1$

Formula 6-2 $\int \dfrac{1}{x}\, dx = \ln x + C$ for $x > 0$

Formula 6-3 $\int e^{ax}\, dx = \dfrac{e^{ax}}{a} + C$ for $a \neq 0$

Each of these formulas can be readily verified by differentiating the right-hand side and showing that the result is the integrand. For example, to verify Formula 6-1, we differentiate $[(x^{m+1}/(m+1)] + C$:

$$\frac{d}{dx}\left[\frac{x^{m+1}}{m+1} + C\right] = \frac{1}{m+1}\frac{d}{dx}[x^{m+1}] + \frac{d}{dx}(C)$$

$$= \frac{1}{m+1} \cdot (m+1)x^m + 0$$

$$= x^m$$

We shall verify Formulas 6-2 and 6-3 later in this section.

To apply these formulas in many problems, we need the two properties of the indefinite integral or antiderivative that are the counterparts of Properties 6-5 and 6-6 for the definite integral. In stating these, we assume that the integrals on the right exist.

Property 6-10 $\int kf(x)\,dx = k\int f(x)\,dx$, where k is a constant.

Property 6-11 $\int [f_1(x) + f_2(x)]\,dx = \int f_1(x)\,dx + \int f_2(x)\,dx$.

To show that these formulas hold for indefinite integrals or antiderivatives, we may simply apply the corresponding formulas for derivatives:

$$\frac{d}{dx}[kF(x)] = k\frac{dF(x)}{dx}$$

$$\frac{d}{dx}[F_1(x) + F_2(x)] = \frac{dF_1(x)}{dx} + \frac{dF_2(x)}{dx}$$

where $F(x)$, $F_1(x)$, and $F_2(x)$ are the antiderivatives of $f(x)$, $f_1(x)$, and $f_2(x)$, respectively.

Formula 6-1 coupled with Properties 6-10 and 6-11 enables us to find the antiderivative (or indefinite integral) of any polynomial function, as illustrated in the following two examples:

Example 1

$$\int (x^3 - 2x^2 + x - 1)\,dx = \int x^3\,dx - 2\int x^2\,dx + \int x\,dx - \int dx$$

$$= \frac{x^4}{4} - 2\frac{x^3}{3} + \frac{x^2}{2} - x + C$$

After a little practice, it is customary to write the answer down directly without actually writing the intermediate step involving $\int x^3\,dx - 2\int x^2 + \int x\,dx - \int dx$.

Example 2

$$\int \left(3x^5 - 4x^4 + \frac{2}{3}x + 5\right) dx = 3\frac{x^6}{6} - 4\frac{x^5}{5} + \frac{2}{3} \cdot \frac{x^2}{2} + 5x + C$$

$$= \frac{x^6}{2} - \frac{4}{5}x^5 + \frac{1}{3}x^2 + 5x + C$$

Indeed, by using a termwise approach to the arithmetic, the final simplified form of the answer can be written down immediately. However, in learning to do problems like these, many students are well advised to write down the intermediate steps and then to omit them later when they are able to do so without making mistakes.

Formula 6-1 is also applicable to fractional values of m, as illustrated in the next example.

Example 3

$$\int x^{1/2}\, dx = \frac{2}{3}x^{3/2} + C$$

Here

$$\frac{d(\frac{2}{3}x^{3/2} + C)}{dx} = \frac{2}{3} \cdot \frac{3}{2}x^{1/2} + 0 = x^{1/2}$$

Although Formula 6-1 applies when m is any negative number except $m = -1$, some people prefer to put the function in the denominator instead of using a negative exponent. When this is done, the following formula is more convenient.

Formula 6-1(a) $\displaystyle\int \frac{1}{x^r}\, dx = \frac{1}{1-r} \cdot \frac{1}{x^{r-1}} + C \quad \begin{cases} \text{for } r > 0 \\ \text{and } r \neq 1 \end{cases}$

To see that Formulas 6-1 and 6-1(a) are really equivalent, we consider $\int x^{-50}\, dx$, which is representative of any case.

By Formula 6-1: $\displaystyle\int x^{-50}\, dx = \frac{x^{-50+1}}{-50+1} + C = \frac{x^{-49}}{-49} + C$

By Formula 6-1(a):

$$\int x^{-50}\,dx = \int \frac{dx}{x^{50}} = \frac{1}{-49}\cdot\frac{1}{x^{49}} + C = \frac{x^{-49}}{-49} + C$$

Some people prefer Formula 6-1 while others find Formula 6-1(a) easier to work with.

To verify Formula 6-2, we need only recall that

$$\frac{d}{dx}(\ln x) = \frac{1}{x} \qquad \text{and hence} \qquad \int \frac{dx}{x} = \ln x + c$$

One of the reasons why, in elementary calculus, we must deal with logs, in fact, with logs to the base e, is that any function whose derivative is $1/x$ for $x > 0$, can be regarded as being of the form $\ln x + C$. We cannot express such a function as a polynomial or even as a rational function of x.

As stated, Formula 6-2 applies only when x is positive. However, as illustrated in Figure 6-9, the definite integral $\int_{-2}^{-1} 1/x\,dx$ clearly exists. But since $x < 0$, Formula 6-2 is not applicable. Such cases can be handled using the following alternate form of Formula 6-2.

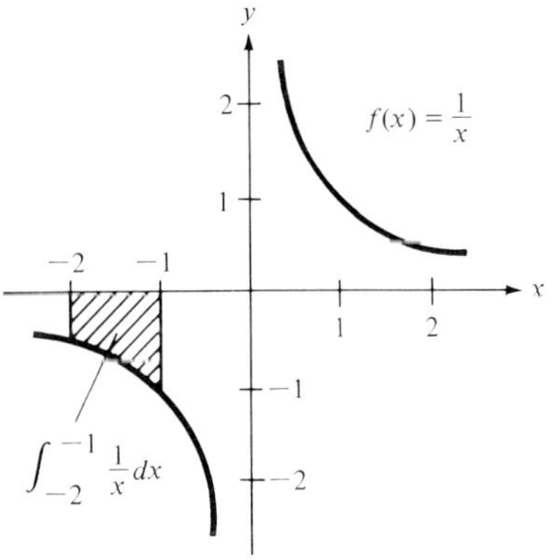

FIGURE 6-9

Formula 6-2(a) $\int \frac{1}{x}\,dx = \ln|x| + C$

This formula holds for x positive or for x negative, but not both.

Since $d(e^{ax})/dx = ae^{ax}$, Formula 6-3 follows directly. This formula can be easily used for finding $\int 10^x\,dx$, for example. As before, we note that

$10 = e^{\ln 10}$ (since ln 10 is the power to which e must be raised to give 10). Then

$$\int 10^x\,dx = \int e^{x \ln 10}\,dx = \frac{1}{\ln 10} e^{x \ln 10} + C$$

$$= \frac{1}{\ln 10} 10^x + C$$

since $e^{x \ln 10} = 10^x$ (reversing the process above).

Here are some other examples. You can verify the results by differentiating the answer.

Example 4. $\int e^{-5x}\,dx = \frac{1}{-5} e^{-5x} + C$

Example 5. $\int 6^x = \int e^{x \ln 6}\,dx = \frac{1}{\ln 6} e^{x \ln 6} + C = \frac{1}{\ln 6} 6^x + C$

Example 6. $\int \left(\frac{2}{x} - x^3 + e^{-2x}\right) dx = (2 \ln x) - \frac{x^4}{4} - \frac{1}{2} e^{-2x} + C$

There is another way of determining the coefficients of the terms in problems like these. If one knows the general form of the answer for a particular term, then multiply it by a constant k and differentiate. Determine k to make the derivative the desired function. This method can be used if you forget the formula or get confused about the coefficients. To illustrate, consider $\int e^{6x}\,dx$. The answer should be in the form $ke^{6x} + C$. Therefore, on differentiating we obtain:

$$\frac{d(ke^{6x} + C)}{dx} = k \cdot 6 \cdot e^{6x}$$

Hence $6ke^{6x}$ must be e^{6x} and, therefore, $6k = 1$ or $k = \frac{1}{6}$ yielding the answer $\frac{1}{6}e^{6x} + C$.

EXERCISES 6-4

Find the indefinite integral or antiderivative. Remember to add the constant C to each answer. Always verify your answer by taking the derivative.

1. $\int x^{10}\,dx$

2. $\int x^{3/2}\,dx$

3. $\int \dfrac{dx}{x^2}$

4. $\int x^{-1}\,dx$

5. $\int x^{-3/2}\,dx$

6. $\int x^{-0.1}\,dx$

7. $\int e^{3x}\,dx$

8. $\int e^{-2x}\,dx$

9. $\int (x^3 - 3x^2 + x^{1/2})\,dx$

10. $\int (e^{2x} - e^{-2x})\,dx$

11. $\int \left(\dfrac{3}{x} - \dfrac{2}{x^2} + \dfrac{5}{x^3}\right) dx$

12. $\int (2x^{1/2} + 3x^{1/3} + 4x^{1/4})\,dx$

13. $\int (x^{3/2} + x^{1/2} + x^{-1/2})\,dx$

14. $\int \left(\dfrac{1}{x^{1/2}} - \dfrac{1}{x} + \dfrac{3}{x^{3/2}}\right) dx$

15. $\int (2^x - 2^{3x})\,dx$

16. $\int \left(x^{10} - \dfrac{5}{x} + e^{-3x}\right) dx$

In Exercises 17–26, assume the right general form of the answer and determine the appropriate factor, for example, let $\int e^{3x}\,dx = ke^{3x} + C$ and determine k. Check your answers by differentiation.

17. $\int e^{-10x}\,dx$

18. $\int \dfrac{10}{x}\,dx$

19. $\int 8x^{43}\,dx$

20. $\int 3x^{5/4}\,dx$

21. $\int -\tfrac{2}{3}x^{1/5}\,dx$

22. $\int 3e^{4x}\,dx$

23. $\int \dfrac{-8}{x}\,dx$

24. $\int \tfrac{1}{3}x^{32}\,dx$

25. $\int -\tfrac{1}{2}\,x^{6/5}\,dx$

26. $\int -\tfrac{4}{3}x^{-1/7}\,dx$

In Exercises 27–30, you are asked to analyze some indefinite integrals in the following way: Find $\int (x + 1)^2\,dx$. Note that there are two ways of doing this problem. We can observe that $(x + 1)^3/3 + C$ is an antiderivative or we can multiply $(x + 1)^2$ to get $x^2 + 2x + 1$ and then

$$\int (x^2 + 2x + 1)\,dx = \frac{x^3}{3} + x^2 + x + C'$$

where the constant C' may well be (and in fact is) different from the constant C in the answer $(x + 1)^3/3 + C$. However, note that if $C' = \tfrac{1}{3} + C$, then the two answers agree. Thus there are two natural solutions that differ by a constant.

Find the indefinite integral by both techniques illustrated above.

27. $\int (1 + x)^3 \, dx$ **28.** $\int (2 + x)^2 \, dx$

29. $\int (x + 3)^2 \, dx$ **30.** $\int (x + 3)^3 \, dx$

6-5. USE OF ANTIDERIVATIVES

In this section we consider two applications of antiderivatives: (1) evaluating definite integrals, and (2) integrating rates of change.

Evaluating Definite Integrals In applying the fundamental theorem of the calculus, the following notation is common [$F(x)$ being an antiderivative of $f(x)$]:

$$\int_a^b f(x)\, dx = F(x)\Big|_a^b$$

where $F(x)\,|_a^b$ is understood to mean $F(b) - F(a)$. We use this in the following examples:

Example 1. $$\int_1^2 (x^2 - 1)\, dx = \left(\frac{x^3}{3} - x + C\right)\Bigg|_1^2$$
$$= \left(\frac{8}{3} - 2 + C\right) - \left(\frac{1}{3} - 1 + C\right) = \frac{4}{3}$$

Observe that the constant C will always be added and subtracted, and consequently does not appear in the final answer. In fact, it may be omitted entirely when the antiderivative is used to evaluate the definite integral, as we shall do in the following example.

Example 2. $$\int_0^1 4e^{-x}\, dx = 4\int_0^1 e^{-x}\, dx = -4e^{-x}\Big|_0^1$$
$$= -4(e^{-1} - e^0) = -4(e^{-1} - 1)$$
$$= 4\left(1 - \frac{1}{e}\right)$$

Remember that, as in Example 2, the evaluation of the antiderivative at the lower limit 0 may *not* be 0.

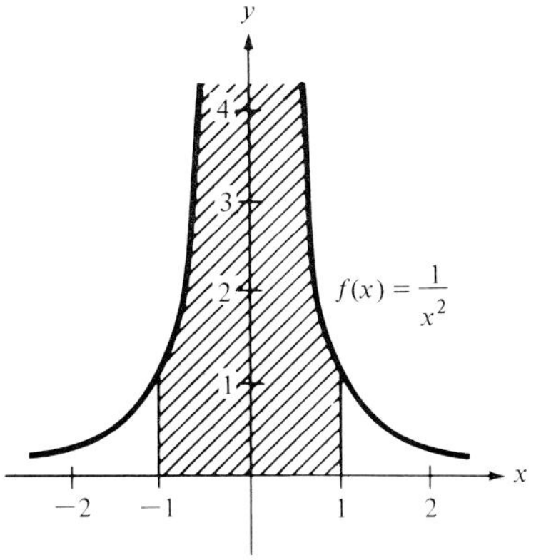

FIGURE 6-10

A word of caution must be issued regarding the fundamental theorem. The existence of $F(a)$ and $F(b)$ does not insure that $\int_a^b f(x)\,dx$ exists. For example, consider $\int_{-1}^{1} (1/x^2)\,dx$. By Formula 6-1, we have $\int (1/x^2)\,dx = -(1/x) + C$. Apparently, applying the Fundamental Theorem, we have

$$\int_{-1}^{1} \frac{1}{x^2}\,dx = \left(-\frac{1}{x}\right)\Bigg|_{-1}^{1} = -1 - 1 = -2$$

From Figure 6-10, it should be clear that $\int_{-1}^{1} (1/x^2)\,dx$ could not be negative (if it existed) and thus we must have made a mistake.

In applying $\int_a^b f(x)\,dx = F(b) - F(a)$ to evaluate $\int_a^b f(x)\,dx$, we *must* know that $F'(x) = f(x)$ at *every* point of $[a, b]$. Otherwise, the fundamental theorem simply does not apply. This condition is not satisfied in the above example. Specifically, the point 0 is contained in the interval $[-1, 1]$. At $x = 0$, neither $f(x) = 1/x^2$ nor $F(x) = -1/x$ is defined. For either reason, $f(x)$ cannot be the derivative of $F(x)$ at this point.

Integrating Rates of Change In a number of important problems, we know that two functions are related because one is the rate of change (or derivative) of the other. Let $F(x)$ and $f(x)$ be related as follows:

$$F'(x) = f(x)$$

over an interval. Therefore, we can obtain $F(x)$ by integrating $f(x)$ over an appropriate interval.

A practical difficulty in problems of this type is that there are many functions $F(x)$ that will satisfy the relationship $F'(x) = f(x)$ for a given $f(x)$. For example, suppose that $f(x) = 2x$. Then the functions $F(x) = x^2$, $F(x) = x^2 + 1$, $F(x) = x^2 - 8000$, and so on, all satisfy $F'(x) = f(x)$. We observe that all of these functions are of the type $F(x) = x^2 + C$.

These observations are consistent with the following statement: "When we integrate the continuous rate of change $f(x)$ of the function $F(x)$ over an interval, the integral is the function $F(x)$ plus or minus an appropriate constant."

How do we obtain a specific value of C to meet conditions given in a particular problem? It suffices to have a value of $F(x)$ at a particular value of x. For example, suppose we know that $F(1) = 5$. Then only the function $F(x) = x^2 + 4$ satisfies both $F'(x) = 2x$ and $F(1) = 5$.

In problems of this type, we are given three facts:

1. The function $f(x)$ is known; in the above example, $f(x) = 2x$.
2. The function $F(x)$ is related to $f(x)$ by $F'(x) = f(x)$, i.e., $f(x)$ is the rate of change of $F(x)$.
3. The value of $F(x)$ is known for $x = a$, for some a, that is, we know $F(a)$; in the above example, $F(1) = 5$.

This information permits us to solve for $F(x)$.

There are two formulations used to solve this problem. In one formulation, we use the indefinite integral and write

$$F(x) = \int f(x)\,dx$$

The constant C that appears in the antiderivative of $f(x)$ is determined from $F(a)$. For example, let $f(x) = 2x$ and $F(1) = 5$. Then $F(x) = \int 2x\,dx = x^2 + C$. Therefore, $F(1) = 1 + C$. Since we know that $F(1) = 5$, then $1 + C = 5$ or $C = 4$. Hence $F(x) = x^2 + 4$ is the desired function.

In the second formulation, we use the definite integral to integrate $F'(x) = f(x)$:

$$\int_a^x F'(t)\,dt = \int_a^x f(t)\,dt$$

By the fundamental theorem of the calculus, we have

$$\int_a^x F'(t)\,dt = F(x) - F(a).$$

Therefore,

$$F(x) - F(a) = \int_a^x f(t)\,dt$$

To illustrate, let $f(x) = 2x$ and $F(1) = 5$. Then we choose $a = 1$, and we have

$$\begin{aligned} F(x) - F(1) &= \int_1^x 2t\,dt \\ &= t^2 \Big|_1^x \\ &= x^2 - 1 \end{aligned}$$

Since $F(1) = 5$, we have

$$F(x) - 5 = x^2 - 1$$

or

$$F(x) = x^2 + 4$$

This is the same result as determined earlier. However, this formulation is frequently more natural for practical problems.

To illustrate, consider the situation where gas is pumped into the tank of your automobile. The meter on the pump basically measures the rate of flow $v(t)$ in units such as gallons per minute. The total volume of gas pumped into your tank from time 0 to time T is $\int_0^T v(t)\,dt$. Furthermore, $v(t)$ is the rate of change of the total volume of gasoline $V(t)$ in your tank, i.e., $v(t) = V'(t)$. Thus we have

$$\int_0^T v(t)\,dt = \int_0^T V'(t)\,dt$$

Since $V(t)$ is the antiderivative of $V'(t)$, we have

$$\int_0^T V'(t)\,dt = V(t)\Big|_0^T = V(T) - V(0)$$

Consequently,

$$V(T) - V(0) = \int_0^T v(t)\,dt$$

By integrating the flow $v(t)$ from 0 to T, we obtain the total volume added between time 0 and time T, which is $V(T) - V(0)$.

The meter at the gasoline pump works as a mechanical integrator; it converts the rate of flow into your tank into the total amount of gas pumped into your tank. Water meters, electric meters, and natural gas meters work on much the same principle; they are mechanical integrators converting the rate of flow into total consumption. The following statement applies to all of these meters: "When we integrate the rate of flow over a period of time, we get the total flow over that period."

In business, the relationship between cost and marginal cost, for example, is that the latter is the derivative of the former. Given the marginal cost, we find the cost by integrating. However, we must also have one point on the cost curve to obtain the cost function. Problems of this type are considered in the exercises.

EXERCISES 6-5

Evaluate the following integrals by use of the fundamental theorem of the calculus.

1. $\int_1^3 (3x^2 - 2x + 1)\,dx$ **2.** $\int_{-1}^1 (x^2 - x + 2)\,dx$

3. $\int_0^1 (e^x - e^{-x})\,dx$

4. $\int_1^e \left(\frac{1}{x} - \frac{1}{x^2}\right) dx$

5. $\int_{-2}^{-1} (x^4 + x^2 + 1)\,dx$

6. $\int_{-1}^{0} (x + e^x)\,dx$

7. $\int_0^1 (e^{-2x} + e^{-3x})\,dx$

8. $\int_{-1}^{2} (x^4 - x)\,dx$

9. $\int_1^2 (3x^2 - 4x^3 + 5x^4)\,dx$

10. $\int_{-1}^{0} (e^x + e^{3x} + 1)\,dx$

11. $\int_1^2 \frac{1}{x}\,dx$

12. $\int_{-1}^{1} (3e^{2x} - 5e^{-x} - 1)\,dx$

State why the integrals in Exercises 13–16 could not be evaluated using the fundamental theorem of the calculus even if the integrals existed.

13. $\int_{-1}^{1} \frac{1}{x^{1/3}}\,dx$

14. $\int_{-3}^{1} \frac{1}{x^{1/5}}\,dx$

15. $\int_{-1}^{3} \left(\frac{1}{x^{1/5}} + x + 1\right) dx$

16. $\int_0^2 \frac{1}{\sqrt[3]{x-1}}\,dx$ NOTE: $\frac{d[(3/2)(x-1)^{2/3}]}{dx} = \frac{1}{\sqrt[3]{x-1}}$ wherever both functions are defined.

In Exercises 17–20, we are told that the given function $f(x)$ is the rate of change of a function $F(x)$, for which the value at a specific value of x is given. In these exercises, **(a)** obtain $F(x)$ using the indefinite integral, **(b)** obtain $F(x)$ using the definite integral, and **(c)** evaluate $F(2)$.

17. $f(x) = 3x^2 - x + 1$; $F(0) = 4$

18. $f(x) = -e^x$; $F(3) = 10$

19. $f(x) = x + e^x$; $F(1) = -5$

20. $f(x) = -5 - 2e^{-x}$; $F(0) = 200$

21. Let the marginal cost be related to the quantity x that is produced by the relationship

$$\text{M.C.} = 2 + 0.1x$$

If the fixed cost (i.e., the cost to produce zero units) is \$20,000, determine the cost function, i.e., the cost as a function of x.

22. Let the marginal return be related to the quantity q that is sold by the relationship

$$\text{M.R.} = 2 - e^{-q/100}$$

If no goods are sold, the return is $-\$80{,}000$, i.e., we lost \$80,000. Determine the return as a function of q.

23. At time zero, an airplane's fuel tanks have 5000 gallons of fuel. For the next 2 hours, the fuel is consumed at a rate of $3000e^{-t/2}$ gallons per hour. (**a**) How much fuel is consumed during these 2 hours? (**b**) How much remains in the tanks after 2 hours? (**c**) Obtain a formula giving the gallons of fuel in the tanks as a function of time. (**d**) At what time is all of the fuel consumed?

24. Upon retirement, a couple has \$20,000 in savings. They withdraw money from their savings at a rate of $\$1000\sqrt{t}$, t being years since retirement. (**a**) How much money do they withdraw during the first 4 years? (**b**) How much is left after 4 years? (**c**) Obtain a formula giving the money left in their savings as a function of time. (**d**) When are their savings exhausted?

6-6. APPLICATIONS OF THE INTEGRAL

In this section we shall discuss the use of the integral to solve two representative business problems.

Total Profit A variety of problems fall under the general category of finding the total profit from an operation. Some involve time; others involve increasing production capacity, advertising effort, or the like.

As our first example, suppose that our company is considering a project involving the purchase of a large automatic handling system for the commodity it produces. We shall presume that the rate of return (in millions of dollars per year) as a function of time is given by

$$R'(t) = 8 - 0.02t^2$$

and is plotted in Figure 6-11, where t is time in years. The decrease in the rate of return with time is due to such factors as increased maintenance and shutdown costs as the system gets older. The operating costs (expressed in millions of dollars per year, which is also a rate) are assumed to be

$$C'(t) = 0.6t$$

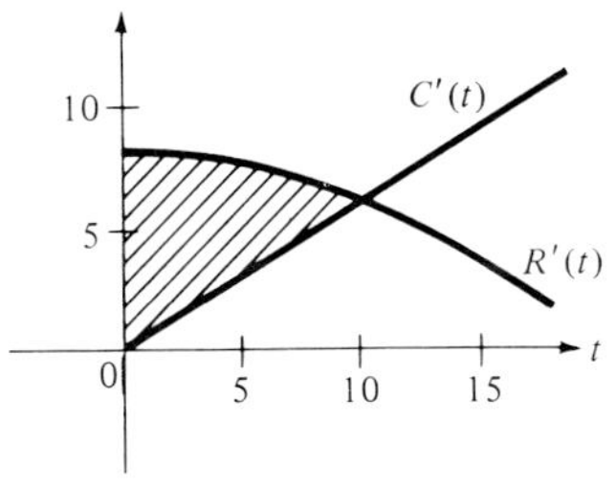

FIGURE 6-11

and $C'(t)$ is also plotted in Figure 6-11. When there is no salvage value of the equipment, the optimum time to dispose of the equipment is when the rate of return equals the rate of operating costs, which is given by

$$R'(t) = C'(t)$$

$$8 - 0.02t^2 = 0.6t$$

which gives

$$t = 10$$

The total profit realized from the operation of this equipment is the total return from operating the equipment for 10 years (which is $\int_0^{10} R'(t)\,dt$) less the total cost of operating the equipment for 10 years (which is $\int_0^{10} C'(t)\,dt$):

$$\begin{aligned}\int_0^{10} R'(t)\,dt - \int_0^{10} C'(t)\,dt &= \int_0^{10} [R'(t) - C'(t)]\,dt \\ &= \int_0^{10} (8 - 0.02t^2 - 0.6t)\,dt \\ &= (8t - 0.00667t^3 - 0.3t^2)\Big|_0^{10} \\ &= 80 - 6.67 - 30 = 43.33\end{aligned}$$

Geometrically, this is the area of the cross-hatched region in Figure 6-11.

The profit from time 0 to any time T is given by

$$\begin{aligned}P(T) &= \int_0^{T} [R'(t) - C'(t)]\,dt \\ &= \int_0^{T} (8 - 0.02t^2 - 0.6t)\,dt \\ &= (8t - 0.00667t^3 - 0.3t^2)\Big|_0^{T} \\ &= 8T - 0.3T^2 - 0.00667T^3\end{aligned}$$

This function is plotted in Figure 6-12. It has its maximum at $T = 10$.

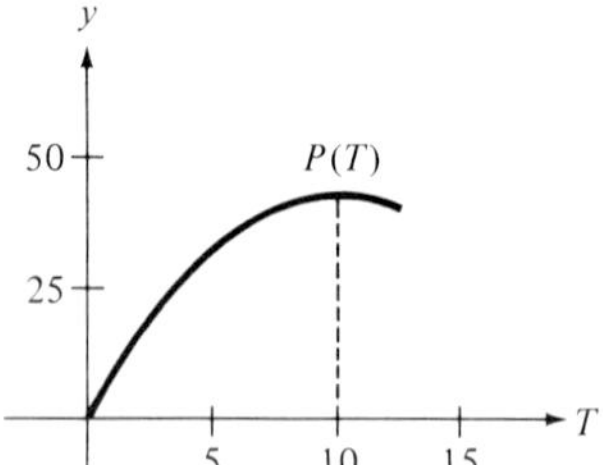

FIGURE 6-12

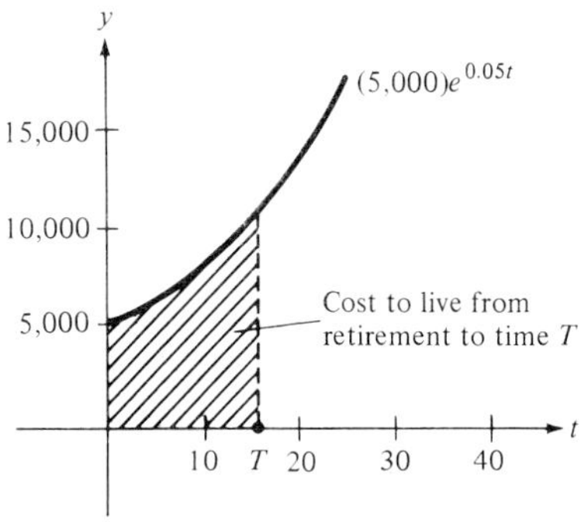

FIGURE 6-13

A Retirement Problem Suppose that at age 65, a miser retires with \$200,000 buried in a jar in his backyard. Not being accustomed to the finer things of life, he feels that he can live on \$5000 a year. Therefore, he believes that he can live comfortably for up to 40 years (40 = 200,000/5000).

Were it not for inflation, he would be correct. However, some inflation seems to be inevitable nowadays, and our miser friend will feel its effects. Let us assume that the inflation rate is 5% per year, compounded continuously. To maintain his \$5000 living standard at which he retired, his cost of living must follow the exponential growth pattern given by \$5000 $e^{0.05t}$, where t is the years since retirement. This is graphed in Figure 6-13.

How much money is required to maintain his standard of living for T years past retirement? This is the accumulated expenditures from retirement to time T, which is the shaded area in Figure 6-13. Mathematically, this is given by the integral

$$\int_0^T (5000)e^{0.05t}\,dt = \frac{(5000)}{(0.05)}e^{0.05t}\Big|_0^T$$

$$= 100{,}000(e^{0.05T} - 1)$$

The cumulative cost to live from retirement to T years is graphed in Figure 6-14.

Treating inflation as a continuous growth process is probably quite realistic in this case, since our miser friend probably spends his money at various times over the year, as opposed to making the entire expenditure on the first day of the year. To maintain a constant living standard throughout the year, the expenditure for the first year is given by

$$\int_0^1 (5000)e^{0.05t}\,dt = \frac{5000}{0.05}e^{0.05t}\Big|_0^1$$

$$= (100{,}000)(e^{0.05} - 1) \approx \$5127$$

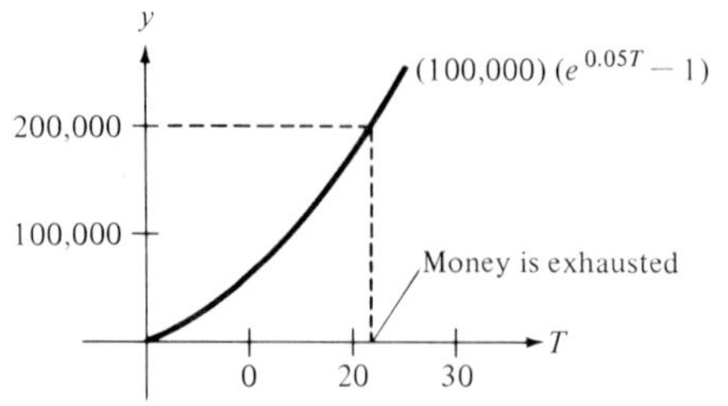

FIGURE 6-14

Thus, because of inflation that occurred during the first year, his expected expenditures were \$5127 rather than \$5000.

What is of utmost concern to our miser is how long his money will last. This is the value of T for which $\int_0^T 5000e^{0.05t}\,dt$ equals \$200,000. Since $\int_0^T 5000e^{0.05t}dt = 100{,}000(e^{0.05T} - 1)$, we solve for T as follows:

$$(100{,}000)(e^{0.05T} - 1) = 200{,}000$$

$$e^{0.05T} - 1 = \frac{200{,}000}{100{,}000} = 2$$

$$e^{0.05T} = 3$$

$$0.05T = \ln 3 \approx 1.1$$

(from Table VI in the Appendix)

$$T \approx \frac{1.1}{0.05} = 22$$

Therefore, our miser's funds last only about 22 years, or only slightly over half of what he expected.

EXERCISES 6-6

1. In Figure 6-12, we graphed the profit realized to date as a function of t. Another useful graph is that of the profit yet to be realized. Assuming a 10-year life for the equipment, the profit yet to be realized is given by $P_1(T) = \int_T^{10} [R'(t) - C'(t)]\,dt$. Evaluate this integral for the rate of return and rate of cost functions of the text and graph $P_1(T)$.

2. In practice, problems involving total profit usually involve consideration of the salvage value of the equipment. In terms of the first example of this section, the total profit realized through time T is $P(T)$. Let the salvage value of the equipment at time T be given by $S(T)$. If the

equipment is operated to time T and then sold immediately, the total monetary receipts would be $P(T) + S(T)$, which we shall call $P_M(T)$. We would like to determine the time T at which $P_M(T)$ is maximized, which is given by

$$\frac{dP_M(T)}{dT} = \frac{dP(T)}{dT} + \frac{dS(T)}{dT} = 0$$

(**a**) In the first example, we determined that $P(T) = 8T - 0.3T^2 - 0.00667T^3$. Determine $dP(T)/dT$.

(**b**) In the first example, we showed that $P(T) = \int_0^T [R'(t) - C'(t)]\, dt$. Express $dP(T)/dT$ in terms of $R'(T)$ and $C'(T)$. Substitute $R'(t) = 8 - 0.02t^2$ and $C'(t) = 0.6t$. Your answer should agree with that obtained in part (a).

(**c**) Let $S(T) - 5 - 0.2T$. Graph $dP(T)/dT$ and $dS(T)/dT$, and determine the time T at which the equipment should be sold, that is, the value of T for which $d[P_M(T)]/dT = 0$.

3. In the example of the miser's retirement, we calculated that his expenditure during his first year of retirement is \$5127. What is his expenditure during the second year? During the tenth? During the twentieth?

NOTE: $e^{0.45} = (e^{0.4}) \cdot (e^{0.05})$, both of which are directly available from the tables in the Appendix.

4. Suppose that in the example of the miser's retirement, inflation was specified as 5% on an effective annual basis. The effect of inflation on his living standard in year t would be given by $(\$5000)(1.05)^t$, and the accumulated expenditures from retirement to year t would be given by $\int_0^T (5000)(1.05)^t\, dt$. Evaluate this integral, and determine how long his funds last. NOTE: $\ln 1.05$ can be obtained from Table V in the Appendix.

5. A distributor is considering opening more stores. The operating cost per store (in millions) of adding more stores is given by

$$C(x) = 1 + \frac{x}{2}$$

where x is the number of stores added. The additional revenue per store generated by these stores is given by R (in millions).

$$R(x) = 3\tfrac{1}{6}\sqrt{x}$$

In effect, $C(x)$ and $R(x)$ are rates per store in this problem. The optimum number of stores is where $R = C$, or $x = 36$. The profit from adding these stores would be given by $\int_0^{36} [R(x) - C(x)]\, dx$. Evaluate this integral.

6. The average or mean value of a function $f(x)$ over interval $[a, b]$ is given by

$$\frac{1}{b-a}\int_a^b f(x)\,dx$$

Determine the average value of $f(x) = x^2$ over the interval $[3, 5]$.

7. The rate of return from an investment for the next 10 years is expected to be approximated by the formula $r(t) = \$100{,}000(1 - e^{-t/2})$. Determine the total return and the average return over the next 10 years. (See Exercise 6.)

8. A truck starts a trip with a gross weight of 50,000 kilograms.* The truck makes deliveries along the way, such that the truck's weight is approximately given by $w(t) = 50{,}000 - 5000t$ for t in hours. The trip requires 8 hours. The truck consumes fuel at a rate of 0.1 liters† per hour per kilogram gross weight. That is, the consumption of fuel in liters per hour is given by $(0.1)(50{,}000 - 5000t)$. How much fuel is consumed on the trip?

9. Suppose that our country's known oil and gas reserves were to be depleted in 40 years at the current rate of energy consumption. If the rate of consumption is increasing at the continuous rate of 2% per year, how long will our reserves last? To solve this problem, we first observe that we are using up $\frac{1}{40}$ or 0.025 of our reserves per year at the current rate. If the rate is increasing at 2%, the rate of consumption at year t is $\frac{1}{40}e^{0.02t}$. The energy consumed through year T is $\int_0^T (e^{0.02t}/40)\,dt$. For what value of T is the value of this integral equal to 1 (i.e., when will all of the known energy reserves have been consumed)?

10. Rework the above exercise if the effective annual rate is 2%. Now the rate of energy consumption in year t is $(1.02)^t/40$, and the integral is $\int_0^T (1.02)^t/40\,dt$. NOTE: $\ln 1.02 \approx 0.0198$.

*6-7. PIECEWISE-CONTINUOUS FUNCTIONS AND THE RIEMANN INTEGRAL

The concept of the Riemann or definite integral also applies to certain noncontinuous functions. In particular, as in Figure 6-15, we may consider a *piecewise-continuous* function $f(x)$ over $[a, b]$.

* A kilogram is a metric unit of weight, and is equal approximately to 2.2 pounds.

† A liter is a metric unit of volume; a gallon equals approximately 3.8 liters, or in other words, a liter is slightly larger than a quart.

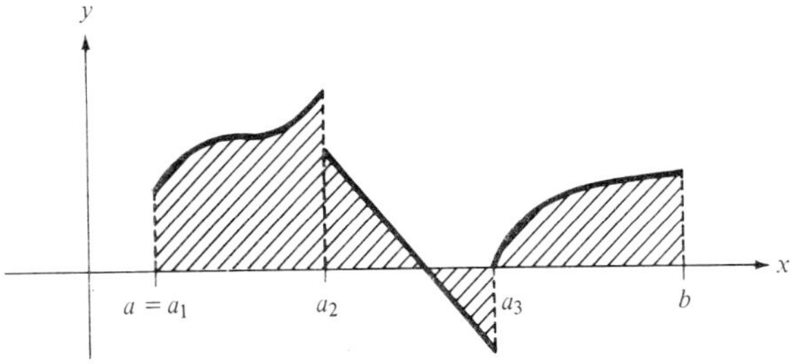

FIGURE 6-15

Definition A function $f(x)$ is *piecewise continuous* over $[a, b]$ if (i) $f(x)$ is defined over $[a, b]$ except possibly at finitely many points, and (ii) $[a, b]$ is the union of finitely many nonoverlapping, closed subintervals $[a_1, a_2], [a_2, a_3], \ldots [a_{m-1}, a_m]$, where $a_1 = a$ and $a_m = b$ and, for each one of these closed subintervals, there is a continuous function $g(x)$ defined over it such that $g(x) = f(x)$ for each x in the corresponding open subinterval.

The values of $f(x)$ at the end points of the subintervals are not important in this context, but we do require that the function $g(x)$ be continuous over the whole, closed subinterval. In general, we say that a function $f(x)$ is *Riemann*-integrable over $[a, b]$ if $\int_a^b f(x)\,dx$ exists, i.e., if Property 6-1 holds for such a function. *Any piecewise-continuous function over* $[a, b]$ *is Riemann-integrable over* $[a, b]$. For such a function there are at most only finitely many points of discontinuity. If we choose Δx small enough, the contribution of those rectangular regions involving the discontinuities to the Riemann sum is small. Thus the situation is almost as if we had a continuous function.

Example. Perhaps the best known piecewise-continuous function is the so-called postage stamp function. We will use as our example the postal rates before the changes in late 1975. The cost in cents of a first class letter of x ounces was 10 times the least integer greater than or equal to x (up to some maximum weight). For example, a letter of $2\frac{1}{2}$ ounces cost $3 \cdot 10$ or 30 cents, since 3 is the least integer greater than or equal to $2\frac{1}{2}$. The graph of the postage stamp function for $0 < x \leq 4$ is shown in Figure 6-16.

In a strict sense, most cost or price functions in business are piecewise continuous but not continuous, since costs or prices go up by units of 1 cent or in some cases (as for most cars) by units of \$1. However, we normally approximate such cost or price functions by continuous functions that are usually easier to handle mathematically. In dealing with derivatives in Chapters 3 and 4, we have done just that. Functions like the postage stamp

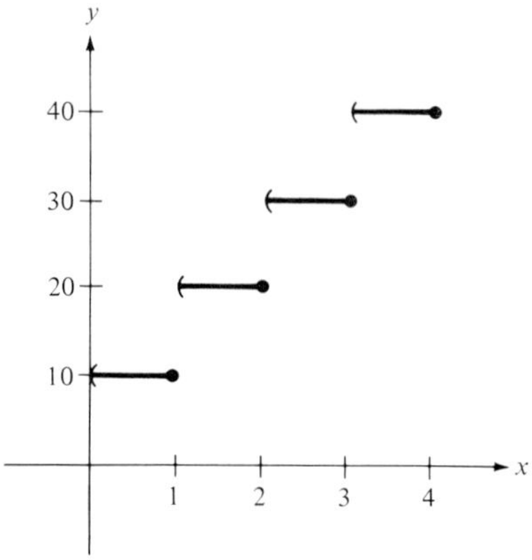

FIGURE 6-16

function are not conveniently approximated by continuous functions, since the jump in price at each integer weight in ounces is too big relative to the other numbers considered.

Another example of a piecewise-continuous function is illustrated in Figure 6-17. The function $f(x)$ shown in the figure may be defined as

$$f(x) = \begin{cases} 1 & \text{whenever } 0 \le x < 2 \\ x & \text{whenever } 2 < x < 4 \\ 3 & \text{whenever } 4 < x \le 6 \end{cases}$$

The function $f(x)$ is piecewise continuous over [0, 6], since the functions 1, x, and 3 are continuous over the appropriate closed subintervals. Hence $\int_0^6 f(x)\,dx$ exists. We can compute the value of the integral using

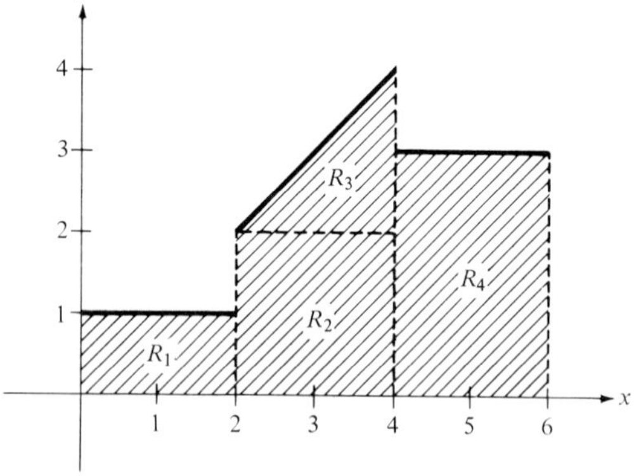

FIGURE 6-17

elementary geometry and our knowledge that $\int_0^6 f(x)\,dx$ is the area of the region under the graph of $f(x)$. This region is the union of four nonoverlapping regions, three rectangular and one triangular, as illustrated in Figure 6-17.

$$\int_0^6 f(x)\,dx = \text{area } R_1 + \text{area } R_2 + \text{area } R_3 + \text{area } R_4$$

$$= 2 + 4 + 2 + 6$$

$$= 14$$

We can also compute the value of $\int_0^6 f(x)\,dx$ by writing

$$\int_0^6 f(x)\,dx = \int_0^2 1 \cdot dx + \int_2^4 x\,dx + \int_4^6 3\,dx = x\Big|_0^2 + \frac{x^2}{2}\Big|_2^4 + 3x\Big|_4^6$$

$$= 2 + 8 - 2 + 18 - 12 = 14$$

Property 6-4 tells us how we can reduce the evaluation of any integral of a piecewise-continuous function to the evaluation of integrals of continuous functions as illustrated in Figure 6-18:

$$\int_a^d f(x)\,dx = \int_a^b f(x)\,dx + \int_b^c f(x)\,dx + \int_c^d f(x)\,dx$$

Over each one of the subintervals $[a, b]$, $[b, c]$, and $[c, d]$, $f(x)$ can be regarded as equivalent to a continuous function.

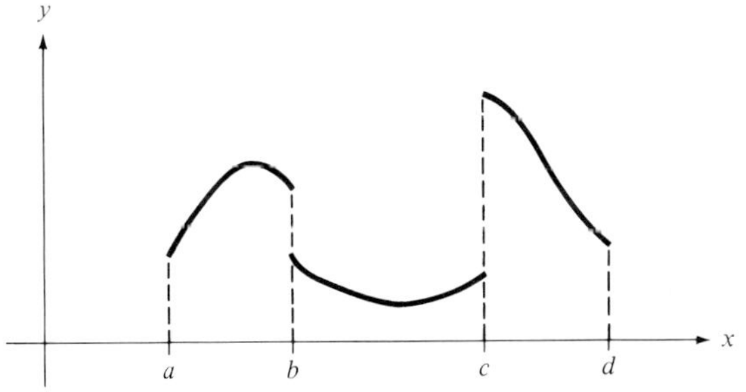

FIGURE 6-18

In this chapter, we have so far defined the Riemann integral over a specified interval for any function that is continuous or piecewise continuous. In so doing, we have defined Riemann sums as $\sum_{j=1}^{n} f(x_j)\,\Delta x$ with the understanding that the interval $[a, b]$ of the integral is subdivided into n equal subintervals of length $(b - a)/n = \Delta x$. On each subinterval the function $f(x)$ is to be evaluated at a point x_j of that subinterval. In a broader context, it is not necessary that the subintervals be of the same length. We can designate their lengths as $\Delta x_1, \Delta x_2, \ldots, \Delta x_n$, and our Riemann sum then becomes $\sum_{j=1}^{n} f(x_j)\,\Delta x_j$. For dealing with more general functions that are to be Riemann-integrable, this latter definition is sometimes considered preferable. With such an understanding, Property 6-1 would specify that the maximal length of the Δx_j's goes to zero rather than that the length Δx goes to zero.

EXERCISES 6-7

---

1. Graph the 1975 overseas airmail postage stamp function. (Postage was 26 cents per $\frac{1}{2}$ ounce or fraction thereof.)

2. Graph the "greatest integer function" for $0 \le x \le 4$, where for any positive number x, the greatest integer in x is the whole number part of the usual decimal representation of x.

In Exercises 3–12, evaluate the integrals by first writing them as the sum of two or more integrals over shorter subintervals.

3. $\displaystyle\int_0^2 f(x)\,dx \qquad$ where $f(x) = \begin{cases} x & \text{whenever } x \le 1 \\ x^2 & \text{whenever } x > 1 \end{cases}$

4. $\displaystyle\int_0^3 f(x)\,dx \qquad$ where $f(x) = \begin{cases} 3 & \text{whenever } x < 1 \\ x + 2 & \text{whenever } 1 \le x \le 2 \\ x^2 & \text{whenever } x > 2 \end{cases}$

5. $\displaystyle\int_{-1}^1 f(x)\,dx \qquad$ where $f(x) = \begin{cases} \sqrt{1 - x} & \text{whenever } -1 \le x \le 0 \\ -\sqrt{1 - x} & \text{whenever } \quad 0 < x < 1 \end{cases}$

HINT: $d[(1 - x)^{3/2}]/dx = -\frac{3}{2}\sqrt{1 - x}$.

6. $\displaystyle\int_{-2}^2 f(x)\,dx \qquad$ where $f(x) = \begin{cases} 2 + x & \text{whenever } -2 \le x \le 0 \\ 2 - x & \text{whenever } \quad 0 < x \le 2 \end{cases}$

7. $\displaystyle\int_0^6 f(x)\,dx \qquad$ where $f(x) = \begin{cases} x^2 & \text{whenever } 0 \le x < 2 \\ 3 - x & \text{whenever } 2 \le x < 5 \\ x^3 & \text{whenever } 5 \le x < 6 \end{cases}$

8. $\displaystyle\int_0^4 f(x)\,dx \qquad \text{where } f(x) = \begin{cases} e^x & \text{whenever } 0 \le x \le 1 \\ e^{-x} & \text{whenever } 1 < x < 2 \\ 1/x & \text{whenever } 2 \le x \le 4 \end{cases}$

9. $\displaystyle\int_0^e f(x)\,dx \qquad \text{where } f(x) = \begin{cases} e^x & \text{whenever } 0 \le x \le 1 \\ 1 & \text{whenever } 1 < x < 2 \\ 1/x & \text{whenever } 2 \le x \le e \end{cases}$

10. $\displaystyle\int_0^4 f(x)\,dx \qquad \text{where } f(x) = \begin{cases} xe^{-x^2} & \text{whenever } 0 \le x \le 2 \\ -1 & \text{whenever } 2 < x \le 4 \end{cases}$

HINT: $d(e^{-x^2})/dx = -2xe^{-x^2}$

11. $\displaystyle\int_{-1}^1 f(x)\,dx \qquad \text{where } f(x) = \begin{cases} -e^x & \text{whenever } -1 \le x < 0 \\ e^{-x} & \text{whenever } 0 \le x \le 1 \end{cases}$

12. $\displaystyle\int_0^2 f(x)\,dx \qquad \text{where } f(x) = \begin{cases} \dfrac{x}{1+x^2} & \text{whenever } 0 \le x \le 1 \\ -\sqrt{x-1} & \text{whenever } 1 < x \le 2 \end{cases}$

HINT:

$d[\ln(1+x^2)]/dx = 2x/(1+x^2)$ and $d[(x-1)^{3/2}]/dx = \frac{3}{2}(x-1)^{1/2}$.

*6-8. NUMERICAL COMPUTATION OF INTEGRALS

In developing the concept of the integral for a continuous function in Chapter 6, we first introduced the sum

$$\sum_{j=1}^{n} f(x_j)\,\Delta x = f(x_1)\,\Delta x + f(x_2)\,\Delta x + \cdots + f(x_n)\,\Delta x$$

We observed that as the length of Δx became small, the value of this sum approached a number that we called the definite integral.

One approach to computing approximations to the definite integral is to use this sum. To this end, we shall develop a flowchart for computing an approximation to $\int_a^b f(x)\,dx$.

We regard Δx as given. For x_j, we may use any point within the appropriate subinterval. Let us arbitrarily take the right end point. For the first (leftmost) subinterval, we have $x_1 = a + \Delta x$; for the second, we have $x_2 = a + 2\,\Delta x$; for the jth subinterval, we have $x_j = a + j\,\Delta x$.

Next, we need to address the question of how to organize the computation of sums such as $\sum_{j=1}^{n} f(x_j)\,\Delta x$. Suppose that we let S_k denote the sum of the first k terms. That is, $S_k = \sum_{j=1}^{k} f(x_j)\,\Delta x$, where $k \le n$. Thus S_{k-1} is $\sum_{j=1}^{k-1} f(x_j)\,\Delta x$. Now we ask, "What is the difference between S_k and S_{k-1}?" This is best seen by writing down the expansions for S_k and S_{k-1}:

$$S_{k-1} = f(x_1)\,\Delta x + f(x_2)\,\Delta x + \cdots + f(x_{k-1})\,\Delta x$$

$$S_k = f(x_1)\,\Delta x + f(x_2)\,\Delta x + \cdots + f(x_{k-1})\,\Delta x + f(x_k)\,\Delta x$$

The distinction is that S_k contains one more term than S_{k-1}, that term being $f(x_k)\,\Delta x$. Therefore, we conclude that

$$S_k = S_{k-1} + f(x_k)\,\Delta x$$

This equation constitutes a recursion relationship relating S_k to S_{k-1}. For completeness, we must specify the first term in the sequence. We may use either $S_0 = 0$ or $S_1 = f(x_1)\,\Delta x$.

The S_k's are called partial sums. With $S_1 = f(x_1)\,\Delta x$, S_k is the sum of the first k terms of $\sum_{j=1}^{n} f(x_j)\,\Delta x$. Therefore, S_1 consists of only the first term, which is $f(x_1)\,\Delta x$. If we were to use S_0, we could think of S_0 as the sum of no terms, and thus it seems reasonable that S_0 equals zero.

We shall first consider the computation of the sum $\sum_{j=1}^{20} j^2$. This computation is illustrated by the flowchart in Figure 6-19. For $\sum_{j=1}^{20} j^2$, the partial sums are related by the recursion relationship

$$S_j = S_{j-1} + j^2$$

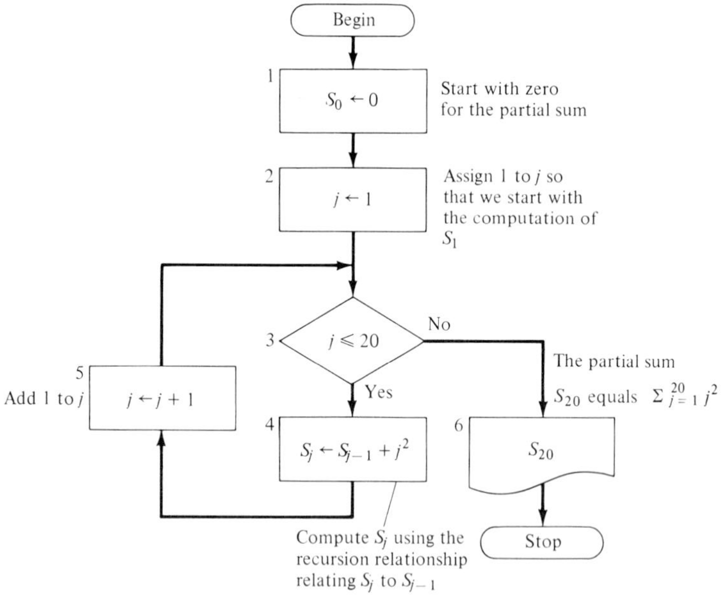

FIGURE 6-19

where $S_0 = 0$. In the flowchart in Figure 6-19, S_0 is assigned the value zero in Block 1, and in Block 4 the recursion relationship is used to compute S_j from S_{j-1}. Blocks 2, 3, and 5 control the execution of the loop. In Block 6, we write the last partial sum, namely, S_{20}, which equals $\sum_{j=1}^{20} j^2$. Note that we use the S_j's for $j < 20$ only as computational aids; we do not print them.

A better insight to the problem can often be gained from a "trace" of the flowchart in Figure 6-19. The flow of execution is as follows:

Block 1. Assign 0 to S_0.
Block 2. Assign 1 to j.
Block 3. The current value assigned to j is 1. This is less than 20, so we go to Block 4.
Block 4. Since j is currently 1, the statement $S_j \leftarrow S_{j-1} + j^2$ is actually $S_1 \leftarrow S_0 + 1^2$. Since the value of S_0 is 0, $S_1 \leftarrow 0 + 1 = 1$.
Block 5. We add 1 to j, so j now has the value 2.
Block 3. Since j is now 2, which is not more than 20, we proceed to Block 4.
Block 4. Since j is now 2, the statement $S_j \leftarrow S_{j-1} + j^2$ is actually $S_2 \leftarrow S_1 + 2^2$. Since the value of S_1 is 1, $S_2 \leftarrow 1 + 2^2 = 5$. Observe that $S_2 = \sum_{j=1}^{2} j^2 = 1^2 + 2^2 = 5$.
Block 5. We add 1 to j, so j is now 3.
Block 3. Since j is now 3, which is not more than 20, we proceed to Block 4.
Block 4. Since j is now 3, the statement $S_j \leftarrow S_{j-1} + j^2$ is actually $S_3 \leftarrow S_2 + 3^2$. Since the value of S_2 is 5, then $S_3 \leftarrow 5 + 3^2 = 14$. Observe that $S_3 = \sum_{j=1}^{3} j^2 = 1^2 + 2^2 + 3^2 = 14$.

For the flowchart in Figure 6-19, the sequence of calculations in Blocks 3, 4, and 5 is repeated until we have computed S_{20}.

We now proceed from the computation of a sum to the computation of an integral by approximation.

The computation of an approximation to the integral $\int_a^b f(x)\,dx$ is given by the flowchart in Figure 6-20. In Block 1, we enter the end points a and b for the interval of integration, along with n, the number of subintervals to be used. We can then compute Δx as $(b - a)/n$ as in Block 2. Block 3 starts the partial sums at zero. Blocks 4, 5, and 8 control the execution of the loop. On each iteration of the loop, we compute the area $f(x_j)\,\Delta x$ of one of the rectangular areas used in the approximation and add to the previous partial sum (S_{j-1}) to get the current partial sum (S_j). To do this, we must first compute x_j as the right end point of the present subinterval (Block 6). In Block 7, we add $f(x_j)\,\Delta x$ to the previous partial sum to get the current partial sum. The output from the program is only the last partial sum, namely, S_n.

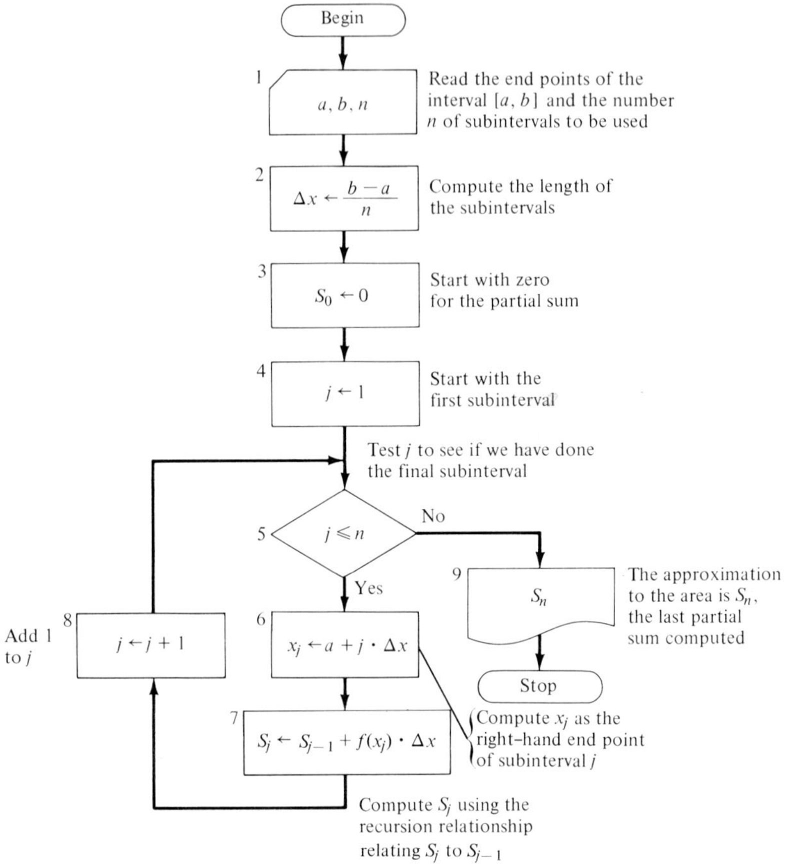

FIGURE 6-20

The selection of the number of subintervals (n in the flowchart in Figure 6-20) entails a compromise between the accuracy of the approximation and the computational effort involved. As n increases, the subinterval size Δx decreases, and the accuracy should improve. But associated with an increase in n is an increase in the computational effort, due to the fact that we have more subintervals.

For a typical problem such as $\int_0^2 x^2\,dx$, a reasonable value of Δx would be 0.01. Since the interval for integration is [0, 2], the value of n would be 200. Since we have to use such large values for n, these calculations are only feasible using a computer. There exist other techniques that can be used to calculate approximations to integrals. Some of these employ better approximations within each subinterval, and thus are able to attain equivalent accuracy using fewer subintervals and thus less work. However, not even these techniques reduce the number of calculations sufficiently to eliminate the need for the computer. Although the number of calculations is large, the computer program is simple, and consequently numerical techniques are frequently used to calculate approximations to integrals (such as $\int_0^2 e^{-x^2}\,dx$) that are difficult or even impossible to integrate using techniques of indefinite integration.

EXERCISES 6-8

In Exercises 1–4, do not simplify your answers.

1. For $\sum_{j=1}^{6} j \cdot (2j+1)$, determine the partial sums $S_1, S_2, S_3, \ldots, S_6$.

2. For $\sum_{j=1}^{4} (1+j^2)$, determine the partial sums S_1, S_2, S_3, and S_4.

3. For $\sum_{j=1}^{5} \frac{j+1}{2j}$, determine the partial sums $S_1, S_2, \ldots, S_5$.

4. For $\sum_{j=1}^{5} \frac{2j}{j+1}$, determine the partial sums $S_1, S_2, \ldots, S_5$.

For Exercises 5–10, prepare a flowchart to compute the sum. Use the one in Figure 6-19 as your model.

5. $\sum_{j=1}^{6} j \cdot (2j+1)$ **6.** $\sum_{j=1}^{4} (1+j^2)$

7. $\sum_{j=1}^{5} \frac{j+1}{2j}$ **8.** $\sum_{j=1}^{5} \frac{2j}{j+1}$

9. $\sum_{j=3}^{7} \frac{j^2}{j+1}$ **10.** $\sum_{j=0}^{9} (j+1)^2$

11. State how the flowchart in Figure 6-20 can be modified so that the left-hand end point is used for x_j. This can be accomplished by changing the arithmetic assignment statement in one of the blocks.

12. State how the flowchart in Figure 6-20 can be modified so that the midpoint of the subinterval is used for x_j. This can be accomplished by changing the arithmetic assignment statement in one of the blocks. Recall that the midpoint of $[c, d]$ is $(c+d)/2$.

For Exercises 13–18, prepare a flowchart similar to the one in Figure 6-20 to compute an approximation to the given definite integral, using the indicated points within each interval to compute the rectangular areas (see

Exercises 11 and 12). The limits of integration and the number n of subintervals should be entered as in the flowchart in Figure 6-20.

13. $\int_1^2 \frac{1}{x}\,dx$; right end points

14. $\int_2^8 x^3\,dx$; left end points

15. $\int_{-1}^2 (x-1)^2\,dx$; midpoints See Exercise 12

16. $\int_{-2}^{-1} \frac{1}{(x-1)^2}\,dx$; right end points

17. $\int_0^4 e^x\,dx$; midpoints See Exercise 12

18. $\int_1^2 \ln x\,dx$; left end points

REVIEW EXERCISES: CHAPTER 6

In each of the following exercises, evaluate by the fundamental theorem of the calculus.

1. $\int_0^1 2e^{-2t}\,dt$

2. $\int_1^3 \frac{2}{x}\,dx$

3. $\int_{-1}^1 x^3\,dx$

4. $\int_1^4 (1-x^2)\,dx$

5. $\int_1^e (1-x^2)\,dx$

6. $\int_0^{1/2} (e^{2x}+e^{-2x})\,dx$

7. $\int_{-1}^1 (3x-x^2)\,dx$

8. $\int_{-1}^0 4e^{4x}\,dx$

9. $\int_{-2}^{-1} (2x^2-3x)\,dx$

10. $\int_0^2 e^{-(x/2)}\,dx$

11. $\int_0^1 2^x\,dx$ HINT: $2^x = e^{x\ln 2}$

12. $\int_1^2 (2-e^{-x})\,dx$

13. $\int_1^2 \left(\frac{1}{x}-\frac{1}{x^2}\right) dx$

14. $\int_{-2}^{-1} \left(1-\frac{2}{x^2}-\frac{1}{x^3}\right) dx$

15. $\int_1^3 \left(\frac{1}{x^3}+\frac{1}{x^2}+\frac{1}{x}+1\right) dx$

16. $\int_1^2 \left(\frac{2}{x^3} - \frac{1}{x^2} + \frac{3}{x} - 4\right) dx$

For Exercises 17–24 refer to the function $f(x)$ as graphed in Figure PP. Estimate the value of the integral to the nearest integer.

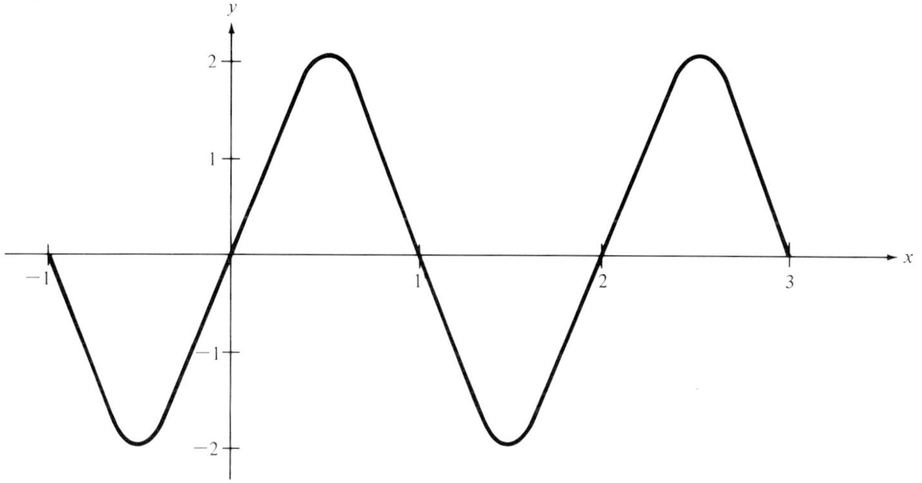

FIGURE PP

17. $\int_{-1}^{2} f(x)\,dx$

18. $\int_0^2 f(x)\,dx$

19. $\int_2^1 f(x)\,dx$

20. $\int_3^0 f(x)\,dx$

21. $\int_{-1}^{-1} f(x)\,dx$

22. $\int_0^1 \frac{1}{10} f(x)\,dx$

23. $\int_2^3 (f(x) + 1) dx$

24. $\int_{-1}^{3} (2 + f(x))\,dx$

For Exercises 25–28, refer to Figure QQ. Determine reasonable upper and lower bounds for the following integrals.

25. $\int_1^3 f(x)\,dx$

26. $\int_3^7 f(x)\,dx$

27. $\int_1^3 4f(x)\,dx$

28. $\int_1^3 [f(x) + 2]\,dx$

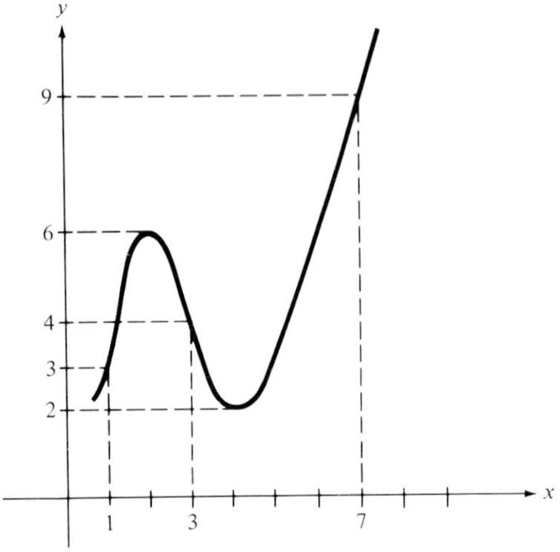

FIGURE QQ

In Exercises 29–37, we are given that $\int_1^2 f(x)\,dx = 3$, $\int_0^2 f(x)\,dx = 1$, $\int_2^5 f(x)dx = -1$, and $\int_0^2 g(x)\,dx = 5$. For each of these exercises, determine the values of the indicated integrals if possible, using the information given. If the information is insufficient, state why.

29. $\displaystyle\int_1^5 f(x)\,dx$

30. $\displaystyle\int_1^2 5f(x)\,dx$

31. $\displaystyle\int_0^2 [f(x) - 1]\,dx$

32. $\displaystyle\int_0^1 f(x)\,dx$

33. $\displaystyle\int_1^1 f(x)\,dx$

34. $\displaystyle\int_2^0 f(x)\,dx$

35. $\displaystyle\int_0^2 [f(x) + g(x)]\,dx$

36. $\displaystyle\int_1^2 xf(x)\,dx$

37. $\displaystyle\int_0^2 [f(x)g(x)]\,dx$

38. Using a Riemann sum with $n = 4$ and the x_j's chosen at the left end points of the subintervals, find the approximate area under the graph of $x^2 - x + 1$ from $x = 2$ to $x = 4$.

39. Using the tables in the Appendix, find the approximate value of $\int_1^3 e^{x/2}\,dx$ by computing a Riemann sum with 10 subintervals and using right-hand end points for the x_j's.

Evaluate the following:

40. $\sum_{j=1}^{4} 2(j-1)$

41. $\sum_{j=3}^{6} j^2$

42. $\sum_{j=1}^{10} (j-5)$

43. $\sum_{j=1}^{5} j(j+1)$

44. $\sum_{j=1}^{4} \frac{j}{j^2+2}$

Write in $\sum$ notation:

45. $1 \cdot 2 + 2 \cdot 3 + \cdots + n(n+1)$

46. $2^2 + 3^2 + 4^2 + \cdots + 10^2$

47. $1 + \frac{1}{2} + \frac{1}{3} + \cdots + \frac{1}{n}$

48. $1 - 2 + 4 - 8 \cdots - 1024$. HINT: $1024 = 2^{10}$

49. $\frac{2}{3} + \frac{4}{5} + \frac{8}{7} + \frac{16}{9} + \cdots + \frac{2^n}{2n+1}$

50. A man invests \$5000 in a venture whose rate of return in dollars per year is expected to be given by the following function:

$$R(t) = 1 - e^{-t}$$

(**a**) Determine the return in the first year.
(**b**) Determine the return for the first 10 years.
(**c**) Determine the return for the tenth year.

51. A store's rate of sales at the beginning of the year is \$10,000 per week. If the sales increase at a continuous rate of 1% per week, what is the total sales for the year? Assume that the year has 52 weeks.

HINT: $e^{0.52} = (e^{0.5}) \cdot (e^{0.02})$

Determine the average values of the following functions over the indicated intervals.

52. $f(t) = e^{-t}$, over $[1, 5]$

53. $f(x) = \frac{1}{x}$, over $[1, 3]$

CHAPTER SEVEN

MORE ON INTEGRATION

7-1. CHANGING VARIABLES

To find the antiderivatives of given functions, we have used the following three formulas:

Formula 6-1 $\int x^m\,dx = \dfrac{x^{m+1}}{m+1} + C \quad \text{for } m \neq -1$

Formula 6-2 $\int \dfrac{1}{x}\,dx = \ln x + C \quad \text{for } x > 0$

Formula 6-3 $\int e^{ax}\,dx = \dfrac{1}{a}e^{ax} + C \quad \text{for } a \neq 0$

Along with Property 6-10 for the integral of a constant times a function and Property 6-11 for the integral of the sum of two functions, these formulas were used in Chapter 6 to obtain the integrals of a variety of functions.

The utility of these formulas is further enhanced by the fact that each of these formulas can also be used in conjunction with a process known as "changing variables," or "substitution," which is based on the chain rule for differentiation.

The chain rule supposes that we have y as a function of u and u as a function of x. We may consider y as a function of x, and we have

$$\frac{dy}{dx} = \frac{dy}{du} \cdot \frac{du}{dx}$$

provided that the derivatives dy/du and du/dx exist.

Let us consider the example $y = (x^2 + x)^3$. To apply the chain rule, we let $y = u^3$ and $u = x^2 + x$. Since $dy/du = 3u^2$ and $du/dx = 2x + 1$, we have

$$\frac{dy}{dx} = \frac{dy}{du} \cdot \frac{du}{dx} = 3u^2 \cdot (2x + 1) = 3(x^2 + x)^2(2x + 1)$$

In the last step, $u = x^2 + x$ was substituted to obtain dy/dx as an explicit function of x.

Suppose that we want to find $\int 3(x^2 + x)^2(2x + 1)\,dx$. We know the answer already. It is $(x^2 + x)^3 + C$, since for any specific value of C, the derivative of this function is $3(x^2 + x)^2(2x + 1)$. But suppose that we were not given the function $(x^2 + x)^3$ in advance. Then how could we find it?

We could have noticed that the integrand could be put in quite a special form. If $u = x^2 + x$, then $du/dx = 2x + 1$ and the integrand is in the form $3u^2 \cdot (du/dx)$. But we recognize that $3u^2 = d(u^3)/du$. Thinking of y as u^3, we have

$$\int 3(x^2 + x)^2(2x + 1)\, dx = \int \frac{d(u^3)}{du}\frac{du}{dx}\, dx$$

$$= \int \frac{dy}{du}\frac{du}{dx}\, dx$$

The integrand can now be recognized as dy/dx, since by the chain rule

$$\frac{dy}{dx} = \frac{dy}{du} \cdot \frac{du}{dx}$$

Thus we have $\int (dy/dx)\, dx$, which is simply

$$y + C \qquad \text{or} \qquad u^3 + C \qquad \text{or} \qquad (x^2 + x)^3 + C$$

when we replace y by u^3 and u by $x^2 + x$.

We can use a notational device that sometimes makes the process easier. In the formula

$$\int \frac{dy}{du} \cdot \frac{du}{dx}\, dx$$

we interpret $(du/dx)\, dx$ as du, getting $\int (dy/du)\, du$. In the example above,

$$\int \frac{dy}{du}\, du = u^3 + C$$

since y stood for u^3. Since $u = (x^2 + x)$, we get the same answer in terms of x, namely, $(x^2 + x)^3 + C$.

We call dx the *differential* of x and, similarly, du the *differential* of u. In any expression, $\int g(t)\, dt$, the dt tells us to integrate with respect to t. When we can write the integrand $g(t)$ as $h(u) \cdot (du/dt)$, then we can write $\int g(t)\, dt$ as

$$\int h(u) \cdot \frac{du}{dt}\, dt = \int h(u)\, du$$

We have used a substitution of variables to change the form of the integral.

The process we have outlined can be tried even if we do not see our way through to the answer. Sometimes, in making intelligent guesses we are led to answers.

Example 1. Find $\int xe^{-x^2}\, dx$. So far, we have no formulas for handling e^{-x^2} in an integrand. We may try the substitution of $u = x^2$, since we do know something about e^{-u} as an integrand. For $u = x^2$, we have $du/dx = 2x$ or $du = (du/dx)\,dx = 2x\,dx$. Thus writing

$$\int xe^{-x^2}\,dx = \tfrac{1}{2}\int e^{-x^2}\cdot 2x\,dx$$

we have

$$\begin{aligned}\int xe^{-x^2}\,dx &= \frac{1}{2}\int e^{-u}\cdot\frac{du}{dx}\,dx\\ &= \tfrac{1}{2}\int e^{-u}\,du\\ &= -\tfrac{1}{2}e^{-u} + C\\ &= -\tfrac{1}{2}e^{-x^2} + C\end{aligned}$$

In this case, our guess of substituting u for x^2 helped. In some other cases even our best guesses will not help.

In this or any other indefinite integration problem, any answer obtained is valid if it works, i.e., if the derivative of the answer is the integrand. We check the result obtained in Example 1.

$$\begin{aligned}\frac{d(-e^{-x^2}/2 + C)}{dx} &= -\frac{1}{2}(-2x)e^{-x^2} + 0\\ &= xe^{-x^2}\end{aligned}$$

"Changing variables" or "substitution" helped us get the answer.

Example 2. $\int (x^3 - 4x + 1)^4(3x^2 - 4)\,dx$. Here it is natural to let $u = x^3 - 4x + 1$, since $du/dx = 3x^2 - 4$ exists as a factor in the integrand. Thus we have

$$\begin{aligned}\int (x^3 - 4x + 1)^4(3x^2 - 4)\,dx &= \int u^4\cdot\frac{du}{dx}\,dx\\ &= \int u^4\,du\\ &= \frac{u^5}{5} + C\\ &= \frac{(x^3 - 4x + 1)^5}{5} + C\end{aligned}$$

It is easy to verify by differentiation that the derivative of the answer is the integrand as it should be.

Example 3. Find

$$\int \frac{x}{\sqrt{1+x^2}}\,dx$$

We try $u = 1 + x^2$. Then

$$du = \frac{d}{dx}(1 + x^2)\,dx = 2x\,dx$$

We have

$$\int \frac{x}{\sqrt{1+x^2}}\,dx = \frac{1}{2}\int \frac{(2x\,dx)}{\sqrt{1+x^2}} = \frac{1}{2}\int \frac{du}{\sqrt{u}} = \sqrt{u} + C = \sqrt{1+x^2} + C$$

In the first step of this process, we multiply and divide by 2 so as to explicitly obtain $du = 2x\,dx$, which makes substitution for u and du quite easy in the next step.

After some practice, we can often do the substitution process mentally, in much the same fashion that we use the chain rule for differentiation without explicitly substituting u as a function of x. In examples like Example 3, one is able, after some practice, to recognize immediately that the answer should be of the form $\sqrt{1 + x^2} + C$.

There are no hard and fast rules that tell you what substitution to make. It is perfectly proper to try a substitution hoping that it may simplify the process. If it does not succeed, you are free to try something else.

How do we know that a substitution is working? The symbolic flowchart in Figure 7-1 should help to clarify this question. As the flowchart indicates, we first guess a formula for u as a function of x, determine du, and substitute into $\int f(x)\,dx$. At this step, we hope that x does not appear in the result. In other words, we hope that we obtain $f(x)\,dx$ as a function of u times du, i.e., $g(u)\,du$. If this is the case, we then see if we recognize $\int g(u)\,du$. If so, then substitution is working, and we simply write the antiderivative of $g(u)$ and replace u by the function of x to obtain $\int f(x)\,dx$.

If either of these two conditions is not true, then we must try again. We reexamine $f(x)$ and see if another substitution appears promising. If so, we try it. If not, we have no choice but to abandon the substitution approach for our particular problem. There are many integrals for which substitution does not work. Unfortunately, there is no clearcut way to distinguish between failures due to (1) the problem's being insoluble via the substitution process, and (2) our inability to find the proper substitution.

When applying substitutions to evaluate *definite* integrals by the fundamental theorem of calculus, we must note that the limits of integration are in the original variable, not the new one. We can either change back to the

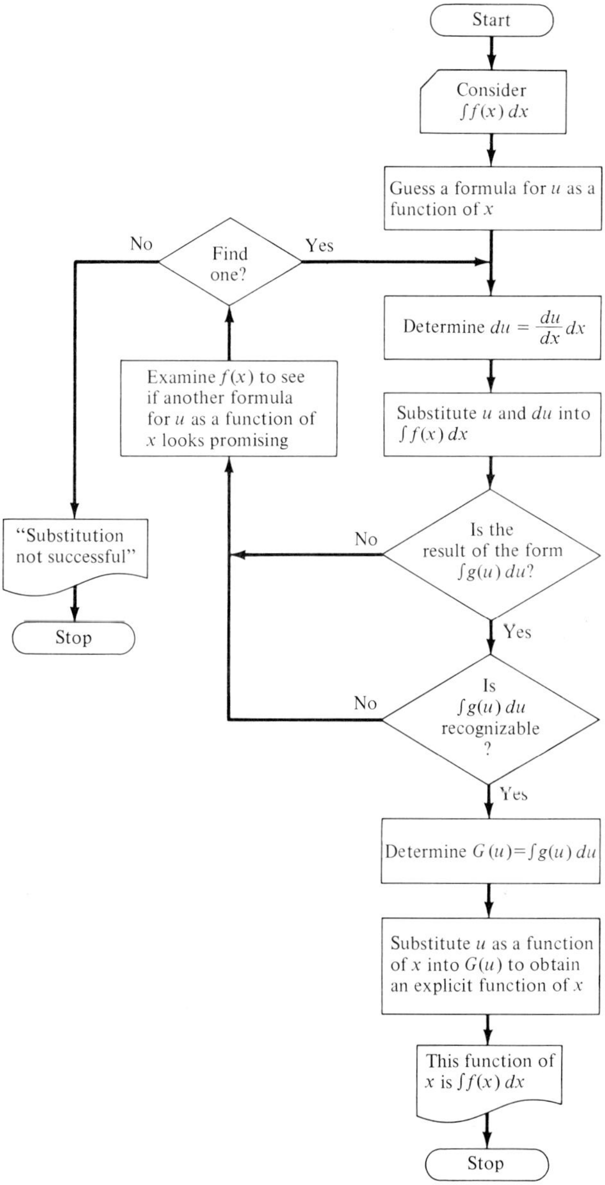
Start
Consider
∫f(x) dx
Guess a formula for u as a function of x
Determine du = du/dx dx
Substitute u and du into ∫f(x) dx
Is the result of the form ∫g(u) du?
Yes
No
Is ∫g(u) du recognizable ?
Yes
No
Determine G(u)=∫g(u) du
Substitute u as a function of x into G(u) to obtain an explicit function of x
This function of x is ∫f(x) dx
Stop
Examine f(x) to see if another formula for u as a function of x looks promising
Find one?
Yes
No
"Substitution not successful"
Stop

FIGURE 7-1

original variable or find the values of the new variable corresponding to the limits for the old. The latter process works if, between the original limits of integration, the substitute function has an inverse and is continuous. Consider the following examples.

Example 4. $\int_0^1 (2x+3)(x^2+3x+5)^3\,dx$. Here we let $u = x^2 + 3x + 5$, and then $du = (2x+3)\,dx$. Thus we have

$$\int_0^1 (2x+3)(x^2+3x+5)^3\,dx = \int_{x=0}^{x=1} u^3\,du = \frac{u^4}{4}\Big|_{x=0}^{x=1} = \frac{(x^2+3x+5)^4}{4}\Big|_0^1$$

$$= \frac{9^4}{4} - \frac{5^4}{4} = \frac{1}{4}(6561 - 625)$$

$$= \frac{1}{4}(5936)$$

$$= 1484$$

We could have observed that when x is 0, u is 5, and when x is 1, u is 9, getting

$$\int_{x=0}^{x=1} (2x+3)(x^2+3x+5)^3\,dx = \int_{u=5}^{u=9} u^3\,du = \frac{u^4}{4}\Big|_5^9 = \frac{9^4}{4} - \frac{5^4}{4} = 1484$$

as before.

The observant reader may have noted that for this example, it would have been possible (though tedious) to have multiplied the factors together to get a single polynomial to integrate. There is often more than one way to skin a cat.

Example 5

$$\int_1^2 \frac{(2x+2)}{(x^2+2x)}\,dx$$

Let $u = x^2 + 2x$ and then $du = (2x+2)\,dx$. When $x = 1$, then $u = 3$; when $x = 2$, then $u = 8$. Therefore,

$$\int_{x=1}^{x=2} \frac{2x+2}{x^2+2x}\,dx = \int_{u=3}^{u=8} \frac{du}{u} = \ln u\Big|_3^8 = \ln 8 - \ln 3 = \ln\frac{8}{3}$$

Some students prefer formulas to the methods of changing variables that we have used. The following three formulas are the modified versions of those at the beginning of this section.

Formula 6-1′ $\int [f(x)]^m f'(x)\,dx = \dfrac{[f(x)]^{m+1}}{m+1} + C$

Formula 6-2′ $\int \dfrac{f'(x)}{f(x)}\,dx = \ln\,[f(x)] + C$

Formula 6-3′ $\int e^{f(x)} f'(x)\,dx = e^{f(x)} + C$

In each case the formula can be justified by either letting $f(x) = u$ and using the methods of the text or directly by the use of the chain rule in differentiating the right sides of the three equations. These formulas may be used in working many of the exercises of this section.

EXERCISES 7-1

In Exercises 1–8, find the indefinite integral by change of variable using the suggested substitution.

1. $\int xe^{x^2-1}\,dx.$ Let $u = x^2 - 1.$

2. $\int \dfrac{3+8x}{3x+4x^2}\,dx.$ Let $u = 3x + 4x^2.$

3. $\int (2x-1)(x^2-x+2)^4\,dx.$ Let $u = x^2 - x + 2.$

4. $\int x(x^2+3)^{5/2}\,dx.$ Let $u = x^2 + 3.$

5. $\int \dfrac{1}{x}\ln x\,dx.$ Let $u = \ln x.$

6. $\int \dfrac{3x^2-1}{(x^3-x+5)^2}\,dx.$ Let $u = x^3 - x + 5.$

7. $\int \dfrac{x^2}{\sqrt{x^3-2}}\,dx.$ Let $u = x^3 - 2.$

8. $\int (3x^2+1)e^{x^3+x}\,dx.$ Let $u = x^3 + x.$

In Exercises 9–16, find a suitable change of variable and integrate.

9. $\int 3xe^{x^2}\,dx$

10. $\int \dfrac{x}{x^2+1}\,dx$

11. $\int (3x^2 + 1)(x^3 + x + 2)^{3/2}\, dx$

12. $\int \frac{x}{(x^2 + 1)^5}\, dx$

13. $\int (1 - 3x^2)(x - x^3)^5\, dx$

14. $\int (5 + 2x)(1 + 5x + x^2)^{-1/2}\, dx$

15. $\int (3x^2 + 2)e^{x^3 + 2x}\, dx$

16. $\int \frac{1 - 2x + 3x^2}{1 + x - x^2 + x^3}\, dx$

In Exercises 17–24, evaluate the definite integral.

17. $\int_0^{\sqrt{2}} x(x^2 + 2)^{1/2}\, dx$

18. $\int_0^1 xe^{-x^2}\, dx$

19. $\int_0^1 (1 - x^2)^4 x\, dx$

20. $\int_0^2 \sqrt{4 - x^2}\, x\, dx$

21. $\int_0^1 \frac{x}{1 + 3x^2}\, dx$

22. $\int_0^4 \frac{x}{\sqrt{9 + x^2}}\, dx$

23. $\int_3^4 (2x - 1)(x^2 - x - 6)^3\, dx$

24. $\int_0^1 \frac{1 + 2x}{(1 + x + x^2)^4}\, dx$

7-2. INTEGRATION BY PARTS

There is a special device called "integration by parts" for finding some antiderivatives. We explain the method by starting with the formula for the derivative of a product. If $u(x)$ and $v(x)$ are differentiable functions of x, we have

$$\frac{d}{dx}[u(x)v(x)] = v(x)u'(x) + u(x)v'(x)$$

Solving for the last term, we get

$$u(x)v'(x) = \frac{d}{dx}[u(x)v(x)] - v(x)u'(x)$$

We regard each of the three terms as a function of x, which is integrable. Therefore, when we integrate the functions on both sides of the equation with respect to x, we get

$$\int u(x)v'(x)\, dx = \int \frac{d}{dx}[u(x)v(x)]\, dx - \int v(x)u'(x)\, dx$$

or

$$\int u(x)v'(x)\,dx = u(x)v(x) - \int v(x)u'(x)\,dx$$

We have evaluated the middle integral using the fact that the integral of the derivative of a function is the function itself; hence $\int (d[u(x)v(x)]/dx)dx$ is simply $u(x)v(x)$. We could include a constant for that integral, but such a constant can be absorbed into the constant for $\int v(x)u'(x)\,dx$. Consequently, it is customarily omitted.

The objective in applying integration by parts is to write the integrand as a product of two functions $u(x)$ and $v'(x)$ for which the following conditions hold:

1. The antiderivative of $v'(x)$, namely, $v(x) = \int v'(x)\,dx$, is recognizable.
2. The integral $\int u'(x)v(x)\,dx$ can be determined.

If either of these do not hold for the chosen $u(x)$ and $v'(x)$, then another choice must be attempted. Technically, we should include in our conditions that the derivative $u'(x) = d[u(x)]/dx$ is recognizable, but this is almost never a problem.

Like the substitution method presented in the previous section, integration by parts is not guaranteed to work. Starting with an integrand $f(x)$, we choose two functions $u(x)$ and $v'(x)$ such that $f(x) = u(x)v'(x)$. We then determine $u'(x) = du(x)/dx$ and $v(x) = \int v'(x)\,dx$. We then examine the integral $\int u'(x)v(x)\,dx$. If we can recognize this integral, we have succeeded. If not, we must choose another $u(x)$ and $v'(x)$ and try again. This process is illustrated by the symbolic flowchart in Figure 7-2 on page 290.

Example 1. Find $\int xe^x\,dx$. Here we interpret x as $u(x)$ and e^x as $v'(x)$, i.e., we let $u(x) = x$ and $v'(x) = e^x$. Then $u'(x) = d(x)/dx = 1$ and $v(x) = \int v'(x)\,dx = \int e^x\,dx = e^x$. Thus we have

$$\int u(x) \cdot v'(x)\,dx = u(x) \cdot v(x) - \int v(x) \cdot u'(x)\,dx$$

$$\int x \cdot e^x\,dx = x \cdot e^x - \int e^x \cdot (1)\,dx$$

But $\int e^x\,dx = e^x + C$ and, therefore, $\int xe^x\,dx = xe^x - e^x + C$. We note that the process gives a valid answer in this case, since the derivative of the function we have found as the answer is the integrand:

$$\frac{d(xe^x - e^x + C)}{dx} = 1 \cdot e^x + x \cdot e^x - e^x + 0 = xe^x$$

It is always advisable to verify antiderivatives in this way, and we shall consistently do so in our examples.

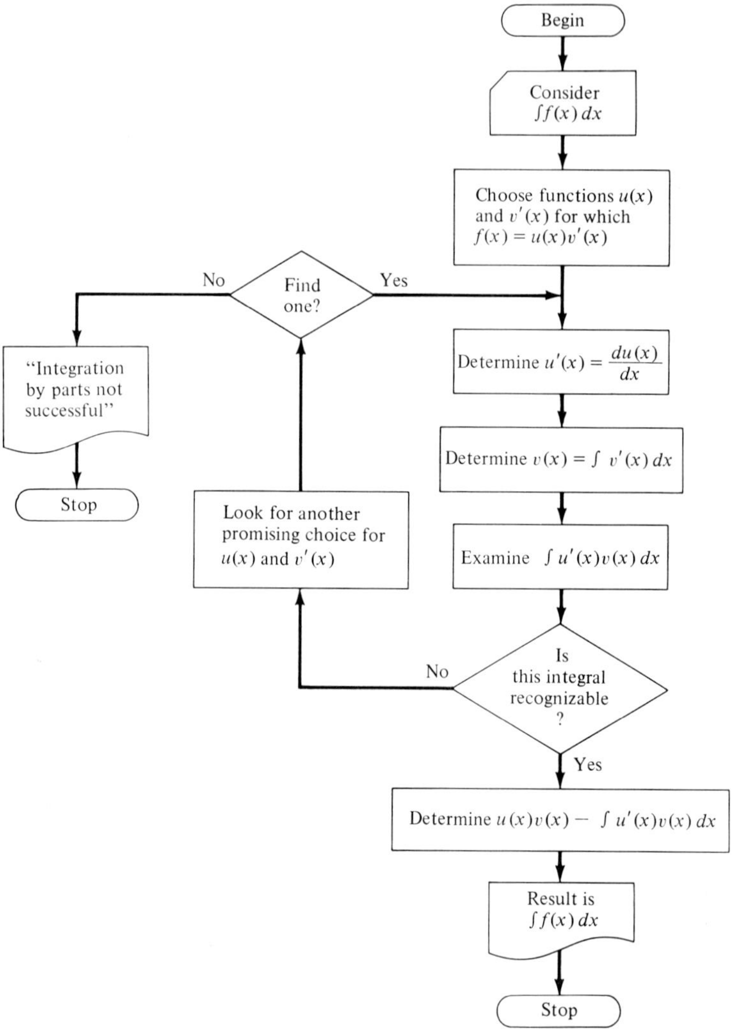

FIGURE 7-2

In a sense, the strategy of integrating by parts is to take an integrand for which we do not recognize an indefinite integral and, using the formula, write it in terms of an integrand for which we can (or at least hope to) recognize an indefinite integral. Thus we took the integrand xe^x and reduced it to the integrand e^x for which we recognized the integral. We let $x = u(x)$ so that du/dx would be 1 and thus our new integrand would not involve an x factor.

Let us note how *not* to do the problem. Reversing the roles of x and e^x, we could let $u(x) = e^x$ and $x = v'(x)$. Then $u'(x) = d(e^x)/dx = e^x$ and $v(x) = \int v'(x)\, dx = \int x\, dx = x^2/2$, and we write

$$\int v'(x) \cdot u(x)\, dx = u(x) \cdot v(x) - \int v(x) \cdot u'(x)\, dx$$

$$\int x \cdot e^x\, dx = e^x \cdot \frac{x^2}{2} - \int \frac{x^2}{2} \cdot e^x\, dx$$

But here the new integral $\int (x^2/2)e^x\, dx$ is more complicated than the integral $\int xe^x\, dx$ with which we started. We are going in the wrong direction.

Example 2. Find $\int x \ln x\, dx$. Here we let $u(x) = \ln x$ and $v'(x) = x$. Then $u'(x) = d(\ln x)/dx = 1/x$ and $v(x) = \int v'(x)\, dx = \int x\, dx = x^2/2$. Thus we write

$$\int u(x) \cdot v'(x)\, dx = u(x) \cdot v(x) - \int u'(x) \cdot v(x)\, dx$$

$$\int (\ln x) \cdot x\, dx = (\ln x) \cdot \frac{x^2}{2} - \int \frac{1}{x} \cdot \frac{x^2}{2}\, dx$$

$$= \frac{x^2}{2} \ln x - \frac{x^2}{4} + C$$

We check the answer

$$\frac{d[(x^2/2) \cdot \ln x - (x^2/4) + C]}{dx} = x \ln x + \frac{x^2}{2} \cdot \frac{1}{x} - \frac{2x}{4} + 0 = x \ln x$$

In Example 2, we assigned $u(x) = \ln x$ and $v'(x) = x$ in spite of the fact that $v(x) = x^2/2$ would be of higher degree than x itself. If we had assigned $\ln x = v'(x)$, then $v(x) = \int v'(x)\, dx = \int (\ln x)\, dx$, which is not a function that we know how to integrate. By letting $\ln x = u(x)$, then $u'(x) = 1/x$, and the logarithm function does not appear in the integrand of the new integral.

In any event, the proof of the pudding is in the eating. When the process produces a function whose derivative is the given integrand, we know we have succeeded.

In integrating by parts, it is appropriate, and may sometimes be necessary, to go through the process two or more times as illustrated in the following example. Obviously, one does not want to undo the effect of the first integration by parts when doing the second one.

Example 3. Find $\int x^2e^{-x}\,dx$. We let $u(x) = x^2$ and $v'(x) = e^{-x}$. Then $u'(x) = d(x^2)/dx = 2x$ and $v(x) = \int v'(x)\,dx = \int e^{-x}\,dx = -e^{-x}$. Then we write

$$\int u(x) \cdot v'(x)\,dx = u(x) \cdot v(x) - \int u'(x) \cdot v(x)\,dx$$

$$\int x^2 \cdot e^{-x}\,dx = x^2 \cdot (-e^{-x}) - \int 2x \cdot (-e^{-x})\,dx$$

$$= -x^2e^{-x} + 2\int xe^{-x}\,dx$$

In the first application of integration by parts (Step 1), we have gone from an x^2 factor to a $2x$ factor, so we try it again on $\int xe^{-x}\,dx$.

To evaluate $\int xe^{-x}\,dx$, we let $u(x) = x$ and $v'(x) = e^{-x}$. Then $u'(x) = d(x)/dx = 1$ and $v(x) = \int v'(x)\,dx = \int e^{-x}\,dx = -e^{-x}$. We now write

$$\int u(x) \cdot v'(x)\,dx = u(x) \cdot v(x) - \int u'(x) \cdot v(x)\,dx$$

$$\int x \cdot e^{-x}\,dx = x \cdot (-e^{-x}) - \int (1) \cdot (-e^{-x})\,dx$$

$$= -xe^{-x} + \int e^{-x}\,dx$$

Since $\int e^{-x}\,dx = -e^{-x} + C$, we have

$$\int xe^{-x}\,dx = -xe^{-x} - e^{-x} + C$$

For this problem, the integral resulting from the second application (Step 2) of integration by parts could easily be evaluated.

The final result is obtained by combining the two steps:

$$\int x^2e^{-x}\,dx = -x^2e^{-x} + 2\int xe^{-x}\,dx \qquad \text{(Step 1)}$$

$$= -x^2e^{-x} + 2(-xe^{-x} - e^{-x} + C) \qquad \text{(Step 2)}$$

$$= -x^2e^{-x} - 2xe^{-x} - 2e^{-x} + C$$

In the algebraic operations used in the last step, we have replaced $2C$ by C, since C is an arbitrary constant.

As always, when learning the technique we should verify the answer by differentiating:

$$\frac{d(-x^2e^{-x} - 2xe^{-x} - 2e^{-x} + C)}{dx} = -2xe^{-x} + x^2e^{-x} - 2e^{-x} + 2xe^{-x} + 2e^{-x} + 0 = x^2e^{-x}$$

Since the method of integration by parts is designed to produce an antiderivative of the integrand, it can be applied to evaluate definite integrals by use of the fundamental theorem of the calculus, as illustrated by the following example.

Example 4. Find $\int_1^e \ln x\, dx$. We must first find $\int \ln x\, dx$. We let $u(x) = \ln x$ and $v'(x) = 1$. Then $u'(x) = d(\ln x)/dx = 1/x$ and $v(x) = \int v'(x)\, dx = \int (1)\, dx = x$. We write

$$\int u(x) \cdot v'(x)\, dx = u(x) \cdot v(x) - \int u'(x) \cdot v(x)\, dx$$

$$\int (\ln x) \cdot (1)\, dx = (\ln x) \cdot x - \int \frac{1}{x} \cdot x\, dx$$

$$= x \ln x - \int dx = x \ln x - x + C$$

As always, we should check the answer by differentiating:

$$\frac{d(x \ln x - x + C)}{dx} = \ln x + x \cdot \frac{1}{x} - 1 = \ln x$$

Since we have found the antiderivative, we use it to evaluate the definite integral:

$$\int_1^e \ln x\, dx = (x \ln x - x)\Big|_1^e = e \cdot \ln e - e - 1 \cdot \ln 1 + 1$$

$$= e - e - 0 + 1$$

$$= 1$$

Some authors prefer to indicate the definite integral as they go along. Thus they write

$$\int_1^e \ln x\, dx = x \ln x\Big|_1^e - \int_1^e dx = (e - 0) - x\Big|_1^e = e - 0 - e + 1 = 1$$

Some authors use the symbolism of differentials as introduced at the end of Section 7-1 in integrating by parts. Thus $u'(x)\, dx$ is called du and $v'(x)\, dx$ is called dv. The formula $\int u(x)v'(x)\, dx = u(x)v(x) - \int v(x)u'(x)\, dx$ becomes $\int u(x)\, dv = u(x)v(x) - \int v(x)\, du$. The procedures are essentially identical, just the notation being different. We illustrate by re-solving Example 1 using differentials.

Example 1(a). Find $\int xe^x\,dx$. Let $u(x) = x$ and $dv = e^x\,dx$. Then

$$du = u'(x)\,dx = \frac{d}{dx}(x) \cdot dx = 1 \cdot dx = dx$$

and

$$v(x) = \int dv = \int e^x\,dx = e^x$$

Then we write

$$\begin{aligned}\int u(x) \cdot dv &= u(x) \cdot v(x) - \int v(x) \cdot du \\ \int x \cdot e^x\,dx &= x \cdot e^x - \int e^x \cdot dx \\ &= xe^x - e^x + C\end{aligned}$$

This is the same result as obtained previously.

Some students find it convenient to organize their work (and their thoughts) by using the following format illustrated for the preceding example. Write u and dv and immediately below them write du and v.

$$\int xe^x\,dx \qquad\qquad u = x \qquad dv = e^x\,dx$$

$$\int u\,dv = uv - \int v\,du \qquad\qquad du = dx \qquad v = e^x$$

$$\begin{aligned}\int xe^x\,dx &= xe^x - \int e^x\,dx \\ &= xe^x - e^x + C\end{aligned}$$

EXERCISES 7-2

In Exercises 1–8, find the indefinite integrals by integrating by parts. Use the indicated functions for $u(x)$ and $v'(x)$. You may need to use the results or techniques of Section 7-1. Check your answers by differentiation.

1. $\int xe^{2x}\,dx \qquad\qquad u(x) = x,\ v'(x) = e^{2x}$

2. $\int x^2 \ln x \, dx$ $\quad u(x) = \ln x,\ v'(x) = x^2$

3. $\int x^3 e^{x^2} \, dx$ $\quad u(x) = x^2,\ v'(x) = xe^{x^2}$

4. $\int \frac{1}{x^2} \ln x \, dx$ $\quad u(x) = \ln x,\ v'(x) = \frac{1}{x^2}$

5. $\int (x + 3)e^{-x} \, dx$ $\quad u(x) = x + 3,\ v'(x) = e^{-x}$

6. $\int (2x - 1)e^{-2x} \, dx$ $\quad u(x) = 2x - 1,\ v'(x) = e^{-2x}$

7. $\int (x + 1) \ln (x + 1) \, dx$ $\quad u(x) = \ln (x + 1),\ v'(x) = (x + 1)$

8. $\int x^2 e^{3x} \, dx$ $\quad u(x) = x^2,\ v'(x) = e^{3x}$
and use two steps.

In Exercises 9–14, find the definite integrals. You will probably want to use integration by parts.

9. $\int_0^{e-1} \ln (1 + x) \, dx$ HINT: First change variables, let $1 + x = y$.

10. $\int_{-1}^{1} xe^{-x} \, dx$

11. $\int_0^2 (x - 1)e^x \, dx$

12. $\int_1^e x \ln x \, dx$

13. $\int_0^1 x^2 e^{-x} \, dx$ HINT: Use 2 steps.

14. $\int_0^1 (1 + x^2)e^{2x} \, dx$ HINT. Use 2 steps.

15. The profit from a business venture is anticipated to be given by

$$p(t) = \ln t \qquad \text{for } t > 1$$

with t expressed in time in years. Determine the profit (**a**) realized from the end of the fourth year to the end of the fifth year, i.e., $\int_4^5 p(t) \, dt$, and (**b**) from the end of the first year to the end of the sixth year, a total of 5 years.

16. The rate of sales of fuel oil in millions of gallons as a function of time is given by the formula

$$s(t) = 1 - e^{-t}$$

The price of fuel oil is given by

$$p(t) = 1 + 0.1t$$

Determine the total revenue from sales for the first 5 years of operation.

HINT: The total revenue is $\int_0^5 s(t)p(t)\,dt$.

7-3. IMPROPER INTEGRALS

One of the important applications of definite integrals to problems of business and economics involves probability and statistics. In this connection, we are particularly concerned with integrals whose value is 1. If the integrand is nonnegative, such an integral is called a *probability integral* and a function $f(x)$ for which

1. $f(x) \geq 0$ for all x in the interval $[a, b]$.
2. $\int_a^b f(x)\,dx = 1$.

is called a *probability function*. Notice that if $g(x) \geq 0$ for all x in the interval $[a, b]$ and $\int_a^b g(x)\,dx = k \neq 0$, $g(x)/k$ becomes a probability function, since

1. $\dfrac{1}{k}g(x) \geq 0$ for all x in $[a, b]$.
2. $\displaystyle\int_a^b \frac{1}{k}g(x)\,dx = \frac{1}{k}\int_a^b g(x)\,dx = \frac{1}{k}k = 1.$

Some naturally arising functions that are important to consider as probability functions have domains which are the whole x axis. An example is the important function $e^{-x^2}/\sqrt{\pi}$, which is known as the normal curve function. The graph of $e^{-x^2}/\sqrt{\pi}$ is given in Figure 7-3.

The concept of the Riemann integral is not directly applicable for a function $e^{-x^2}/\sqrt{\pi}$ whose domain is the whole x axis, since Riemann integrals are defined only over finite intervals. We need an extended notion of inte-

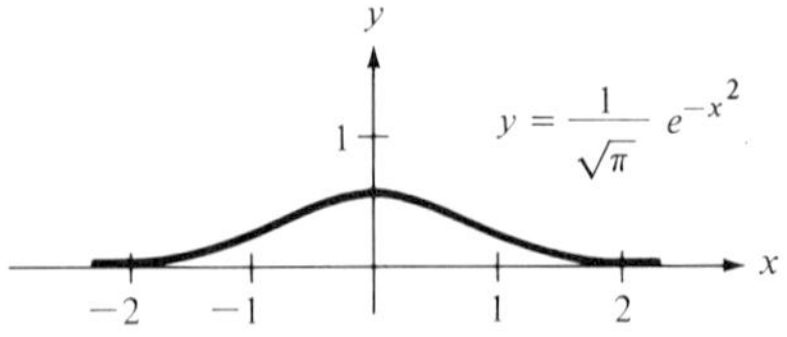

FIGURE 7-3

grals to cover such cases, and thus we define what are called *improper integrals.* In fact, improper integrals arise in many applications other than for probability integrals; our treatment is valid in quite a general context.

Let $f(x)$ be a function that is Riemann integrable over each interval $[a, b]$. For example, the function $e^{-x^2}/\sqrt{\pi}$ is continuous for all x and thus is Riemann integrable over every finite interval. For such a function $f(x)$, we know that $\int_a^h f(x)\,dx$ exists for each value of h and, in fact, is a function of h. We ask whether $\lim_{h\to\infty} \int_a^h f(x)\,dx$ exists? If it does, then we define $\int_a^\infty f(x)\,dx$ as $\lim_{h\to\infty} \int_a^h f(x)\,dx$. If it does not, we say that $\int_a^\infty f(x)\,dx$ does not exist. The limit will exist and be equal to L if, for each sufficiently large value of h, $\int_a^h f(x)\,dx$ is a number as close to L as we wish. Figure 7-4 gives a symbolic flowchart for considering improper integrals of the type $\int_a^\infty f(x)\,dx$.

On page 298 we give an example (Example 1) of an improper integral that exists.

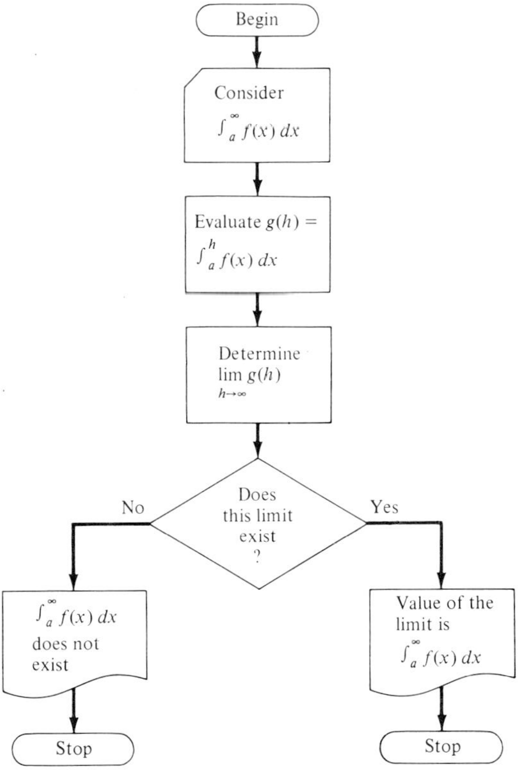

FIGURE 7-4

Example 1. Find $\int_1^\infty (1/x^2)\,dx$. We first determine $\int_1^h (1/x^2)\,dx$:

$$\int_1^h \frac{1}{x^2}\,dx = -\frac{1}{x}\Big|_1^h = \left(1 - \frac{1}{h}\right)$$

Next we take the limit:

$$\int_1^\infty \frac{1}{x^2}\,dx = \lim_{h\to\infty} \int_1^h \frac{1}{x^2}\,dx = \lim_{h\to\infty} \left(1 - \frac{1}{h}\right) = 1 - 0 = 1$$

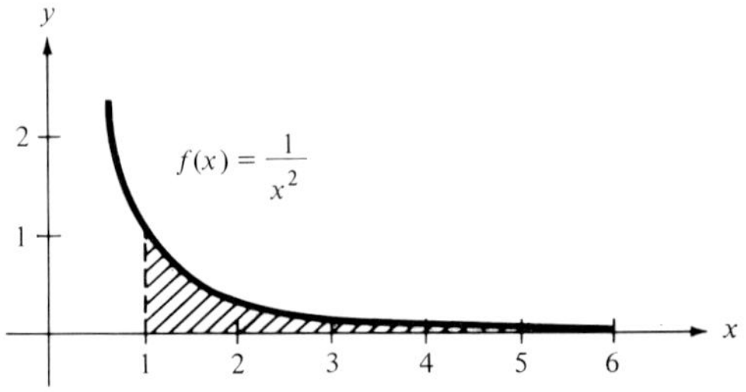

FIGURE 7-5

We refer to Figure 7-5 to visualize both the area represented by $\int_1^h (1/x^2)\,dx$ and also the area represented by $\int_1^\infty (1/x^2)\,dx$. Since $\int_1^h (1/x^2)\,dx = 1 - (1/h)$ as determined above, the area under the graph of $1/x^2$ from 1 to h is less than 1 for each value of $h > 1$. Thus $\int_1^{10} (1/x^2)\,dx = 1 - \frac{1}{10} = 0.9$, $\int_1^{100} (1/x^2)\,dx = 1 - \frac{1}{100} = 0.99$, etc. It makes sense to talk about the area under the graph of $1/x^2$, above the x axis, and to the right of $x = 1$, and to say that the area equals 1.

Not every function whose graph looks somewhat like that shown in Figure 7-5 has an integral from 1 to ∞. We illustrate with the following example.

Example 2. Consider $\int_1^\infty (1/x)\,dx$. We first determine $\int_1^h (1/x)\,dx$:

$$\int_1^h \frac{1}{x}\,dx = \ln x\Big|_1^h = \ln h - \ln 1 = \ln h$$

By choosing h sufficiently large, $\int_1^h (1/x)\,dx$ can be made as large as we want; for $h = e^{100}$, $\ln h = 100$, for example. Since

$$\lim_{h\to\infty} \int_1^h \frac{1}{x}\,dx = \lim_{h\to\infty} (\ln h)$$

does not exist, we say that $\int_1^\infty (1/x)\,dx$ does not exist.

The improper integral $\int_a^\infty f(x)\,dx$ exists if and only if $\lim_{h\to\infty}\int_a^h f(x)\,dx$ exists, and if it exists, this limit is called the value of the improper integral. We can similarly define $\int_{-\infty}^a f(x)\,dx$ as $\lim_{h\to-\infty}\int_h^a f(x)\,dx$ if this limit exists. We also define

$$\int_{-\infty}^{\infty} f(x)\,dx = \int_{-\infty}^{0} f(x)\,dx + \int_{0}^{\infty} f(x)\,dx$$

if the two improper integrals on the right exist. We shall later explain why it does not make sense to define $\int_{-\infty}^{\infty} f(x)\,dx$ as $\lim_{h\to\infty}\int_{-h}^{h} f(x)\,dx$, as some people might be tempted to do.

We shall illustrate these facts with the probability function $e^{-x^2}/\sqrt{\pi}$ graphed in Figure 7-3. We write

$$\int_{-\infty}^{\infty}\frac{1}{\sqrt{\pi}}e^{-x^2}\,dx = \int_{-\infty}^{0}\frac{1}{\sqrt{\pi}}e^{-x^2}\,dx + \int_{0}^{\infty}\frac{1}{\sqrt{\pi}}e^{-x^2}\,dx$$

By methods beyond the scope of this text, it can be shown that $\int_0^\infty (e^{-x^2}/\sqrt{\pi})\,dx = \frac{1}{2}$. See Exercise 14 for a plausibility argument that $\int_0^\infty (e^{-x^2}/\sqrt{\pi})\,dx$ must exist. By the symmetry of the graph of $e^{-x^2}/\sqrt{\pi}$ with respect to the y axis (see Figure 7-3), we expect that

$$\int_{-\infty}^{0}\frac{1}{\sqrt{\pi}}e^{-x^2}\,dx = \int_{0}^{\infty}\frac{1}{\sqrt{\pi}}e^{-x^2}\,dx$$

Therefore, we have

$$\int_{-\infty}^{\infty}\frac{1}{\sqrt{\pi}}e^{x^2}\,dx = \frac{1}{2}+\frac{1}{2} = 1$$

It is in the sense that the improper integral $\int_{-\infty}^{\infty}(e^{-x^2}/\sqrt{\pi})\,dx = 1$ that we call $e^{-x^2}/\sqrt{\pi}$ a probability function. Indeed, the coefficient $1/\sqrt{\pi}$ was chosen to make the improper integral have the value 1.

We have defined improper integrals from $-\infty$ to $+\infty$ only in terms of the two improper integrals from $-\infty$ to 0 and from 0 to ∞. We must do so, since if we erroneously took $\int_{-\infty}^{\infty} f(x)\,dx = \lim_{h\to\infty}\int_{-h}^{h} f(x)\,dx$, we would have improper integrals existing that should not exist. We want improper integrals from $-\infty$ to ∞ to behave as much as possible like ordinary integrals; for example, it should be that $\int_{-\infty}^{\infty} f(x)\,dx = \int_{-\infty}^{c} f(x)\,dx + \int_c^\infty f(x)\,dx$ for any number c.

To illustrate, consider $\int_{-\infty}^{\infty} x\,dx$. If we *erroneously* take $\int_{-\infty}^{\infty} x\,dx = \lim_{h\to\infty}\int_{-h}^{h} x\,dx$, the first step would be to evaluate $\int_{-h}^{h} x\,dx$:

$$\int_{-h}^{h} x\,dx = \left.\frac{x^2}{2}\right|_{-h}^{h} = \frac{h^2}{2} - \frac{h^2}{2} = 0$$

Since this integral is zero for each value of h, we would erroneously conclude that $\int_{-\infty}^{\infty} x\,dx = 0$.

If we *correctly* define $\int_{-\infty}^{\infty} x\,dx = \int_{-\infty}^{0} x\,dx + \int_{0}^{\infty} x\,dx$, we first observe that $\int_0^\infty x\,dx = \lim_{h\to\infty} \int_0^h x\,dx$ and evaluate $\int_0^h x\,dx$:

$$\int_0^h x\,dx = \frac{x^2}{2}\Big|_0^h = \frac{h^2}{2}$$

Since

$$\int_0^\infty x\,dx = \lim_{h\to\infty} \int_0^h x\,dx = \lim_{h\to\infty} \frac{h^2}{2} = \infty$$

we conclude that $\int_0^\infty x\,dx$ does not exist. Similar reasoning leads to the conclusion that $\int_{-\infty}^{0} x\,dx$ also does not exist. We conclude that $\int_{-\infty}^{\infty} x\,dx$ does not exist since, if it did, each of $\int_0^\infty x\,dx$ and $\int_{-\infty}^{0} x\,dx$ would have to exist and neither of them does.

EXERCISES 7-3

In Exercises 1–8, determine whether the indicated improper integrals exist, and if they do, evaluate them.

1. $\int_0^\infty e^{-x}\,dx$

2. $\int_1^\infty \frac{dx}{x^3}$

3. $\int_1^\infty \frac{dx}{\sqrt{x}}$

4. $\int_{-\infty}^{0} e^x\,dx$

5. $\int_0^\infty xe^{-x^2}\,dx$. Let $t = x^2$.

6. $\int_1^\infty xe^{-x}\,dx$. Use integration by parts. Assume the fact that

$$\lim_{h\to\infty} he^{-h} = 0.$$

7. $\int_{-\infty}^{0} xe^{x}\,dx$. Use integration by parts. Assume the fact that

$$\lim_{h\to\infty} he^{-h} = 0.$$

8. $\int_0^\infty \frac{x}{(1+x^2)^2}\,dx$. Let $1 + x^2 = u$.

9. Find $\lim_{h\to\infty} \int_{-h}^{h} x^3\,dx$ and show that $\lim_{h\to\infty} \int_0^h x^3\,dx$ does not exist.

10. Find properties of the graph of $f(x) = e^{-x^2}/\sqrt{\pi}$ by finding the critical value(s) of the function and by locating where the graph is concave upward and where concave downward. Use this information to verify that the graph in Figure 7-3 is what you expect.

11. Show that for $m > 1$, $\int_1^\infty (1/x^m)\,dx$ exists.

12. Show that for $0 \le m \le 1$, $\int_1^\infty (1/x^m)\,dx$ does not exist.

13. Suppose $\int_1^\infty f(x)\,dx$ does not exist because $\int_1^h f(x)\,dx$ gets arbitrarily large as h gets large. Also suppose $g(x)$ is a continuous function with $g(x) \ge f(x)$ and $f(x) \ge 0$ for every $x > 1$. Determine whether $\int_1^\infty g(x)\,dx$ can exist. Note that $\int_1^h g(x)\,dx$ must be greater than or equal to $\int_1^h f(x)\,dx$.

14. Here is a plausibility argument showing that $\int_1^\infty (e^{-x^2}/\sqrt{\pi})\,dx$ exists. Since the integrand $e^{-x^2}/\sqrt{\pi}$ is positive for all x, then $\int_1^h (e^{-x^2}/\sqrt{\pi})\,dx$ is an increasing function of h and if $\lim_{h\to\infty} \int_1^h (e^{-x^2}/\sqrt{\pi})\,dx$ does not exist, we must have

$$\lim_{h\to\infty} \int_1^h \frac{1}{\sqrt{\pi}} e^{-x^2}\,dx = \infty.$$

But if we compare the integrand $e^{-x^2}/\sqrt{\pi}$ and the function $e^{-x}/\sqrt{\pi}$, we note that for every $x > 1$, $e^{-x^2} < e^{-x}$, since x^2 must be greater than x. For example, for $x = 10$, e^{-100} is much less than e^{-10}. But $\int_1^\infty (e^{-x}/\sqrt{\pi})\,dx$ can be shown to exist and thus $\int_1^\infty (e^{-x^2}/\sqrt{\pi})\,dx$ should also exist, since the region under the graph of $e^{-x^2}/\sqrt{\pi}$ is totally contained in the region under the graph of $e^{-x}/\sqrt{\pi}$ for $x > 1$. Show that $\int_1^\infty (e^{-x}/\sqrt{\pi})\,dx$ does exist and evaluate it.

Finally we note that

$$\int_0^\infty \frac{1}{\sqrt{\pi}} e^{-x^2}\,dx = \int_0^1 \frac{1}{\sqrt{\pi}} e^{-x^2}\,dx + \int_1^\infty \frac{1}{\sqrt{\pi}} e^{-x^2}\,dx$$

if the latter integral exists, since the middle integral is the integral of a continuous function over the closed interval $[0, 1]$.

*7-4. ONE-SIDED LIMITS AND IMPROPER INTEGRALS

In our treatment of limits of functions in Chapter 2, we considered what are sometimes called *two-sided limits*. That is, for $\lim_{x\to a} f(x)$, we did not restrict x to being either necessarily greater than a or necessarily less than a. We said that $\lim_{x\to a} f(x)$ exists if there is a number L such that by choosing x sufficiently close to a but different from a, $f(x)$ can be required to be as close to L as we wish. If we choose to consider $f(x)$ only for values of $x > a$, it still makes sense to ask whether $f(x)$ is close to some one number whenever x is close to a *and* greater than a.

We use the notation $\lim_{x\to a^+} f(x)$ to denote the limit of $f(x)$ as x approaches a from above. Note the plus sign near a. The limit exists and is the number L if whenever x is taken sufficiently close to *a and is greater than*

$a, f(x)$ is as close to L as we wish. Such a limit is called a *right-hand* limit. As with ordinary two-sided limit questions, we also use notation like $\lim_{x\to a^+} f(x) = +\infty$ if the right-hand limit does not exist but $f(x)$ is arbitrarily large whenever x is sufficiently close to a *and* greater than a.

We can similarly define the concept of a left-hand limit, $\lim_{x\to a^-} f(x)$, or a limit as x approaches a from below. Consider the example of $y = 1/x$, as shown in Figure 7-6.

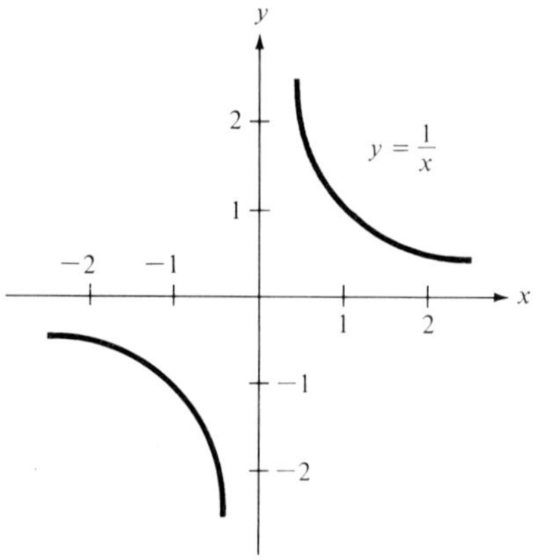

FIGURE 7-6

We have $\lim_{x\to 0^+} (1/x) = +\infty$, $\lim_{x\to 0^-} (1/x) = -\infty$, and $\lim_{x\to 0} (1/x)$ does not exist and is neither $+\infty$ nor $-\infty$. Writing $+\infty$ or $-\infty$ is a special way of saying "does not exist," but such a symbol actually says more. Saying $\lim_{x\to 0^+} (1/x) = +\infty$ says that $1/x$ is arbitrarily large when x is positive and sufficiently close to zero; for example, for the function $1/x$, whenever $x > 0$ and $x < 1/1{,}000{,}000$, then $1/x > 1{,}000{,}000$.

One-sided limits can also be regarded as applying to functions whose domains exist only on one side of a number, such as $\lim_{x\to 1^-} \sqrt{1 - x^2} = 0$. For $x > 1$, $\sqrt{1 - x^2}$ is not a real number at all. As in Chapter 2, we can also write $\lim_{x\to 1} \sqrt{1 - x^2} = 0$, where x is necessarily restricted to the domain of $\sqrt{1 - x^2}$.

We now apply the notions of one-sided limits to questions about certain types of improper integrals.

In addition to integrals like $\int_1^\infty (1/x^2)\,dx$ or $\int_{-\infty}^0 e^x\,dx$, we may also consider the integral of certain functions over an interval where the functions are not continuous (or even piecewise continuous). As illustrated in Figure 7-7, we consider functions like $f(x) = 1/\sqrt[3]{x}$ for which the $\lim_{x\to 0^+} f(x)$ does not exist. We ask if there is an area under the graph from 0 to 1, for example. The question as to whether there is a number that

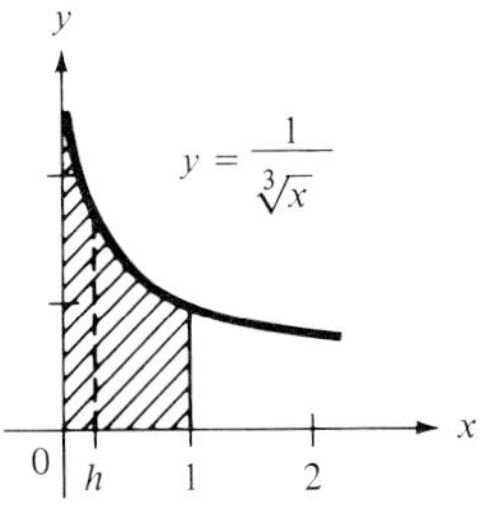

FIGURE 7-7

represents the area of the shaded region is very similar to the questions we considered in the last section. There the cross-hatched regions "went to infinity" horizontally. Here they "go to infinity" vertically.

We ask what happens when the lower limit of integration is slightly more than zero rather than zero itself. For any $h > 0$, $\int_h^1 (1/\sqrt[3]{x})\,dx$ exists, since the function $1/\sqrt[3]{x}$ is continuous on the interval $[h, 1]$. If we can find an antiderivative of $1/\sqrt[3]{x}$, we can evaluate $\int_h^1 (1/\sqrt[3]{x})\,dx$ by the fundamental theorem of calculus. But by formula or observation we note that $(\frac{3}{2}x^{2/3}$ is a function whose derivative is $1/x^{1/3} = 1/\sqrt[3]{x}$, since $\frac{3}{2} \cdot \frac{2}{3} \cdot x^{2/3-1} = x^{-1/3} = 1/x^{1/3}$. Therefore,

$$\int_h^1 \frac{1}{\sqrt[3]{x}}\,dx = \frac{3}{2}x^{2/3}\bigg|_h^1 = \frac{3}{2} - \frac{3}{2}h^{2/3} \qquad \text{for } h > 0$$

Since

$$\lim_{h\to 0^+} \int_h^1 \frac{1}{\sqrt[3]{x}}\,dx = \lim_{h\to 0^+} \left(\frac{3}{2} - \frac{3}{2}h^{2/3}\right) = \frac{3}{2}$$

then $\frac{3}{2}$ should be the area under the graph from 0 to 1. Observe that from the geometric description, we are totally unconcerned about what happens for $h < 0$. Therefore, we really need only the right-hand limit at zero.

An observant reader may note that the two-sided $\lim_{h\to 0} (\frac{3}{2} - \frac{3}{2}h^{2/3})$ is also $\frac{3}{2}$ and wonder why we worried about one-sided limits. As applied to this problem, we had to know about one-sided limits because h had to be positive for $\int_h^1 (1/\sqrt[3]{x})\,dx$ to be Riemann integrable. The nature of the discontinuity of $1/\sqrt[3]{x}$ at $x = 0$ prevents the Riemann integral from existing if $h \leq 0$. It also prevents us from evaluating the integral by the fundamental theorem (even if it did exist).

Observe that we are making a distinction between the Riemann integral and an improper integral. The *Riemann* integral of $1/\sqrt[3]{x}$ does not exist over $[0, 1]$, whereas the improper integral does exist. In a sense, we use improper integrals to evaluate certain areas in which the Riemann integral itself is inappropriate.

The basic limit theorems and evaluation techniques of chapter 2 on limits are just as valid for one-sided limits as for two-sided limits. Thus to handle one-sided limits we do not need to learn new techniques—we just need to restrict our attention to the appropriate domain.

Let us return to the example of $f(x) = 1/x$ as given in Figure 7-6. Here

$$\int_h^1 \frac{1}{x}\,dx = \ln x \Big|_h^1 = \ln 1 - \ln h = -\ln h$$

For h between 0 and 1, it may appear as if $-\ln h$ is negative, which we know it cannot be. But recall that $\ln h < 0$ for $0 < h < 1$, so $-\ln h$ will indeed be positive. However

$$\lim_{h \to 0+} \int_h^1 \frac{1}{x}\,dx = \lim_{h \to 0+} (-\ln h)$$

does not exist, since $\lim_{h \to 0^+}(-\ln h) = -\infty$. Therefore, the area under the graph of $1/x$ from 0 to 1 does not exist; by choosing h small enough, $\int_h^1 (1/x)\,dx$ can be made as large as we please.

EXERCISES 7-4

In Exercises 1–8, evaluate the one-sided limits or show nonexistence and indicate the nature of the nonexistence.

1. $\lim_{x \to 1+} \dfrac{1}{x - 1}$

2. $\lim_{x \to 1-} \sqrt{1 - x}$

3. $\lim_{x \to 2-} \sqrt{4 - x^2}$

4. $\lim_{x \to 0+} \dfrac{1}{x^2 - x}$

5. $\lim_{x \to 1+} \dfrac{1 + x}{1 - x}$

6. $\lim_{x \to 1+} \dfrac{1}{x^2 - 1}$

7. $\lim_{x \to 1-} \dfrac{1}{x^2 - 1}$

8. $\lim_{x \to 0+} \dfrac{x}{|x|}$

In Exercises 9–16, give the value of x for which the integrand is not continuous, and evaluate the indicated improper integrals by use of one-sided limits or show that the integrals do not exist. Substitute $u = 1 - x^2$ in Problems 11 and 12; substitute $u = 4 - x^2$ in Problems 13 and 14.

9. $\displaystyle\int_0^1 \frac{dx}{\sqrt{x}}$

10. $\displaystyle\int_0^e \frac{3\,dx}{x}$

11. $\displaystyle\int_0^1 \frac{x\,dx}{(1-x^2)^2}$

12. $\displaystyle\int_0^1 \frac{x\,dx}{\sqrt{1-x^2}}$

13. $\displaystyle\int_{-2}^0 \frac{x\,dx}{\sqrt{4-x^2}}$

14. $\displaystyle\int_0^2 \frac{x\,dx}{(4-x^2)^3}$

15. $\displaystyle\int_0^1 \frac{dx}{x^m} \qquad m > 1$

16. $\displaystyle\int_0^1 \frac{dx}{x^m} \qquad 0 < m < 1$

17. By reference to Exercises 15 and 16, and the example in the text, state conditions under which $\int_0^1 (1/x^m)\,dx$ for $m > 0$, **(a)** exists or **(b)** does not exist.

18. By reference to Exercise 17 and Exercises 11 and 12 in Section 7-3, compare $\int_0^1 (1/x^m)\,dx$ and $\int_1^\infty (1/x^m)\,dx$ for **(a)** $0 < m < 1$ and **(b)** $1 < m$. Note that for $m = 1$, neither improper integral exists.

Identify points of discontinuity of the integrand and evaluate the following improper integrals or show that they do not exist. You may have to use techniques from Sections 7-1 or 7-2 to find the appropriate indefinite integrals.

19. $\displaystyle\int_0^2 \frac{e^x\,dx}{1-e^x}$

20. $\displaystyle\int_0^1 \frac{e^x}{\sqrt{e^x-1}}\,dx$

21. $\displaystyle\int_0^1 \ln x\,dx.$ Assume the fact that $\lim_{h\to 0} (h \ln h) = 0$.

22. $\displaystyle\int_0^1 \frac{1}{x} \ln x\,dx$ HINT: Try the substitution $u = \ln x$.

*7-5. INTEGRALS FOR OTHER PURPOSES

In Chapter 6 we introduced the concept of the integral as the area under a graph, and so far we have considered only examples and problems that can naturally be interpreted in this way. One of the real powers of the calculus is that the integral can also be used to solve problems that appear quite divorced from the concept of an integral as the area under a graph.

A classical example of such a problem is that of the work required to pump a liquid out of a tank. Although such problems have previously been primarily of concern only to engineers, the major interest these days with the consumption of various forms of energy is an incentive for a wider consideration of problems of this type.

Our first example in this section concerns the use of the integral to solve a typical work problem. Problems of a similar type occur in business, and our second example will be one of these.

Figure 7-8 illustrates a buried tank that must be emptied by the pump, as shown. The tank is conical, being 6 feet in diameter at the top and 12 feet high with the vertex of the cone at the bottom. It is initially filled with water to a level of 8 feet. We wish to know how much work is required for the pump to empty the tank, i.e., to pump the water to ground level.

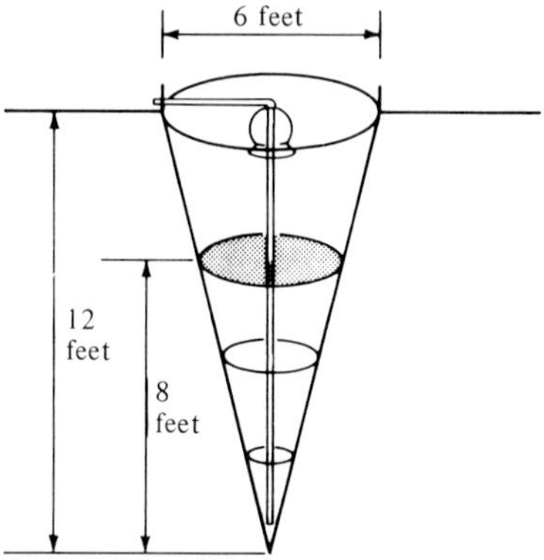

FIGURE 7-8

The measure of work is the foot-pound, which is simply the work required to lift 1 pound of material a distance of 1 foot. Referring to Figure 7-8, the pump must initially raise the water from its initial level (8 feet above the bottom) to ground level, or a total of 4 feet.

Although the pump delivers water continuously, it must do work analogous to what a man would have to do with a bucket. Suppose, for example, that we have a bucket (of negligible weight) that holds exactly 1 pound of water. For us to raise this bucket from the surface of the water in the tank to ground level would require 4 foot-pounds of work.

The pump does exactly the same amount of work in raising water from the liquid surface to ground level. Although the pump suction is at the bottom of the tank, water seeks its own level, and thus the pump has to do no work to raise the water from the bottom of the tank to the surface of the water.

If the liquid level were to remain constant at 8 feet, then the pump must perform 4 foot-pounds of work for every pound of water removed from the tank. However, as the pump removes water, the surface falls, and the pump must do increasingly more work to remove the same amount of water. If the surface of the water is x feet from the bottom of the tank, then to remove a pound of water, the pump must do $(12 - x)$ foot-pounds of work.

One approach to formulating such problems is to consider the water to be divided into thin slices of thickness Δx, as illustrated in Figure 7-9. It is

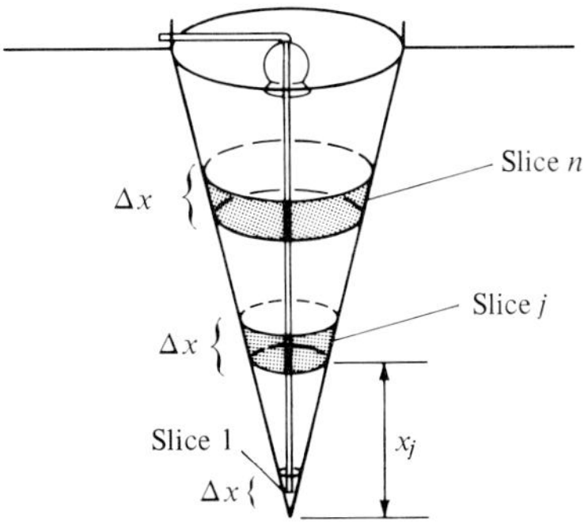

FIGURE 7-9

convenient to regard the slices as being of equal thickness. From the geometry as illustrated in Figure 7-10, the diameter of the slice j at distance x_j from the bottom of the tank is $x_j/2$. Next we want to know how much water is in this slice. Since we have a circular cone, the cross-sectional area is πr^2. The volume of the slice is approximately equal to the product of the cross-sectional area of the slice and its thickness Δx. Since the cross section is circular, the cross-sectional area is $\pi d^2/4$, where d is the diameter of the slice. Since the diameter of slice j is $x_j/2$, the cross-sectional area is

$$\frac{\pi}{4}\left(\frac{x_j}{2}\right)^2$$

To obtain the volume, we multiply by the thickness Δx, obtaining

$$\frac{\pi}{4}\left(\frac{x_j}{2}\right)^2 \cdot (\Delta x)$$

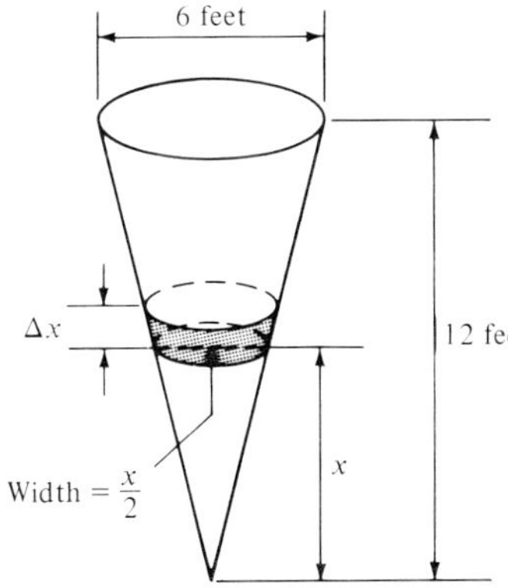

FIGURE 7-10

REMARK: Even though it does not make sense to slice water in a physical sense, it does in a mathematical sense. And the process does produce valid answers in the physical sense.

Since water is known to weigh 62.4 pounds per cubic foot, the weight of water in this slice is $(62.4)(\pi/4)(x_j/2)^2\,\Delta x$. The work involved in raising this water to the ground level is the product of its weight (in pounds) and the distance, namely, $12 - x_j$, that it must be raised (in feet). This product is $(62.4)(\pi/4)(x_j/2)^2(\Delta x)(12 - x_j)$.

To empty the tank, we must raise all of the slices which occur from $x = 0$ to $x = 8$. Therefore, the total work is approximately given by the following sum:

$$\text{work} \approx \sum_{j=1}^{n} \left[(62.4)\left(\frac{\pi}{4}\right)\left(\frac{x_j}{2}\right)^2(12 - x_j)(\Delta x)\right]$$

where n is the number of slices. This sum is essentially a Riemann sum as considered in Chapter 6. By letting the thickness Δx of the slices approach zero, we can replace the sum by the integral

$$\text{work} = \int_0^8 (62.4)\left(\frac{\pi}{4}\right)\left(\frac{x}{2}\right)^2(12 - x)\,dx$$

In this limit process, x_j becomes x, Δx becomes dx, and the finite summation is replaced by the integral.

In this example, the integral is evaluated as follows:

$$\begin{aligned}\text{work} &= (62.4)\left(\frac{\pi}{16}\right)\int_0^8 x^2(12 - x)\,dx \\ &= \frac{62.4}{16}\pi\int_0^8 (12x^2 - x^3)\,dx \\ &= \frac{(62.4)\pi}{16}\left(4x^3 - \frac{x^4}{4}\right)\Bigg|_0^8 = \frac{(62.4)\pi}{16}\cdot 1024 \approx 12{,}546 \text{ foot-pounds}\end{aligned}$$

If desired, we could express this in terms of the electrical energy required to do the pumping, but we shall not do so.

In this example, we do not directly use the concept of the integral as the area under a curve to obtain the solution to the problem as originally stated. Instead, we wrote an expression for the work done in lifting the water in a small horizontal slice of the system, summed over all the slices, and then observed that as the slices became very small, the sum could be replaced by a definite integral. In the above problem this definite integral was a constant times $\int_0^8 (12x^2 - x^3)\,dx$. Of course, this latter integral is the area under the graph of $f(x) = 12x^2 - x^3$ between $x = 0$ and $x = 8$. On the one hand, "passing to the limit" gives the integral, as we have noted. On the other hand, "passing to the limit" gives the total work to be performed since using very thin slices gives a very good approximation to the total work. Thus the integral does represent the work performed.

Problems of a similar nature can arise in a business context. As an example, consider an importer who receives goods in shipload quantities and then distributes them to retail stores. Suppose that on each unit he receives he must pay \$5, which includes purchase costs, shipping charges, and customs duties to the port of entry. He must distribute the goods to retail stores, which costs him \$1 per unit per each 100 miles the goods are shipped (or 1 cent per mile). Because of domestic competition, he must sell each unit for no more than \$10. Therefore, he cannot afford to ship units more than 500 miles. On the units sold at the port of entry, the importer realizes \$5 profit, but his profit decreases as the stores get further away from the port of entry.

Assuming that the port of entry is on a reasonably straight section of coast line, the sales region is the semicircular area illustrated in Figure 7-11. Let us further assume that the stores are uniformly distributed with, on the average, one store every 20 square miles. We shall also assume that each store sells 50 units each month.

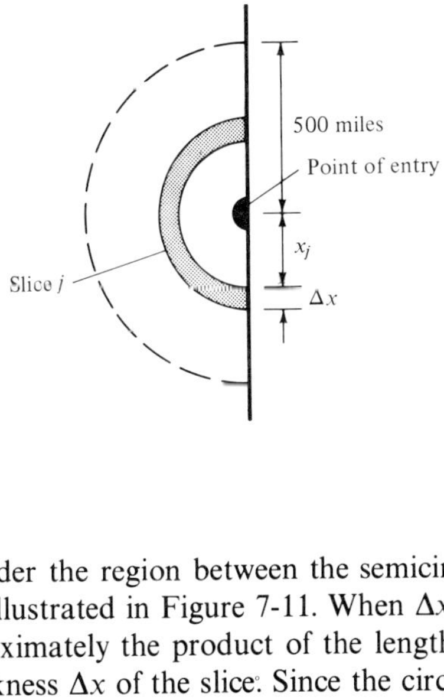

FIGURE 7-11

Let us consider the region between the semicircles of radius x_j and radius $x_j + \Delta x$ as illustrated in Figure 7-11. When Δx is small, the area of this region is approximately the product of the length of one of the semicircles and the thickness Δx of the slice. Since the circumference of a circle of radius x_j is $2\pi x_j$, the length of the semicircular arc of radius x_j is πx_j, i.e., half of the circumference of the circle. To obtain the area, we multiply by the thickness Δx, getting $\pi x_j \, \Delta x$. In effect, we have treated the area of the shaded region in Figure 7-11 as if the region were rectangular, which is acceptable when Δx is small.

Since there is one store for every 20 square miles, the number of stores in the shaded region in Figure 7-11 is $\pi x_j\ \Delta x/20$. Since each store sells 50 units per month, the number of units sold each month by stores in the shaded region in Figure 7-11 is

$$50 \cdot \left(\frac{\pi x_j\ \Delta x}{20}\right) \approx \frac{5\pi x_j\ \Delta x}{2}$$

The cost to ship a unit of the commodity from the port of entry to the shaded region in Figure 7-11 is $0.01x_j$. Therefore, the profit realized is $5 - 0.01x_j$ on each unit sold. The profit realized from all sales in the shaded region is

$$\left(\frac{5\pi x_j\ \Delta x}{2}\right)(5 - 0.01x_j)$$

The total profit realized from sales throughout the sales region is given by the sum

$$\text{profit} = \sum_{j=1}^{n} \cdot \left(\frac{5\pi x_j\ \Delta x}{2}\right)(5 - 0.01x_j)$$

This sum is essentially a Riemann sum as discussed in Chapter 6. By letting Δx approach zero, we can replace the sum by the integral

$$\text{profit} = \int_0^{500} \left(\frac{5\pi x}{2}\right)(5 - 0.01x)\, dx$$

As in the work problem, x_j becomes x, Δx becomes dx, and the finite summation is replaced by the integral.

This integral is evaluated as follows:

$$\begin{aligned} \text{profit} &= \frac{5\pi}{2}\int_0^{500} (5x - 0.01x^2)\, dx \\ &= \left(\frac{5\pi}{2}\right)\left(\frac{5}{2}x^2 - \frac{0.01}{3}x^3\right)\Bigg|_0^{500} \\ &\approx \left(\frac{5\pi}{2}\right)(208{,}333) \approx \$1{,}636{,}250 \end{aligned}$$

EXERCISES 7-5

1. Rework the first example in the section except let the orientation of the cone be as illustrated in Figure RR (pump is not shown).

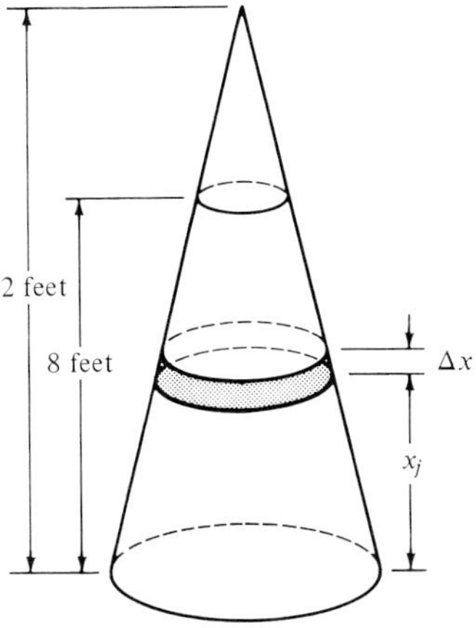

FIGURE RR

HINT: The diameter of the slice at a distance x_j from the bottom is $(12 - x_j)/2$. You may use similar triangles to verify this if you wish.

2. Rework the first example in the section except let the tank be shaped like an inverted pyramid as illustrated in Figure SS (pumping system not shown). The tank is initially filled with water to a depth of 8 feet from the bottom. The top of the tank is 6 square feet on a side.

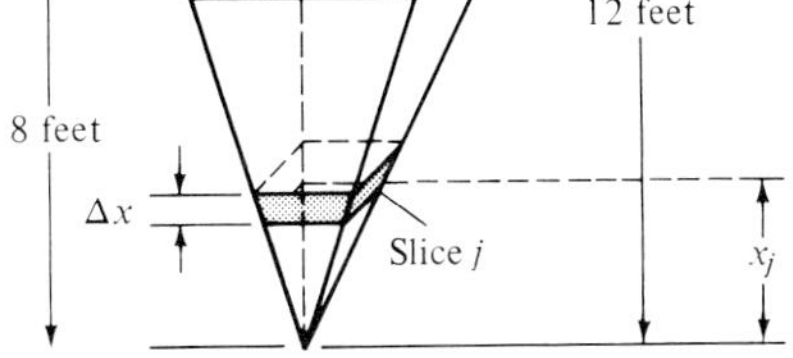

FIGURE SS

HINT: The length of a side of slice j is $x_j/2$. You may use similar triangles to verify this if you wish. Since the slice is square, the cross-sectional area is $(x_j/2)^2$.

3. Suppose we are pumping water out of a swimming pool of the shape illustrated in Figure TT (the pump is not shown). How much work is required?

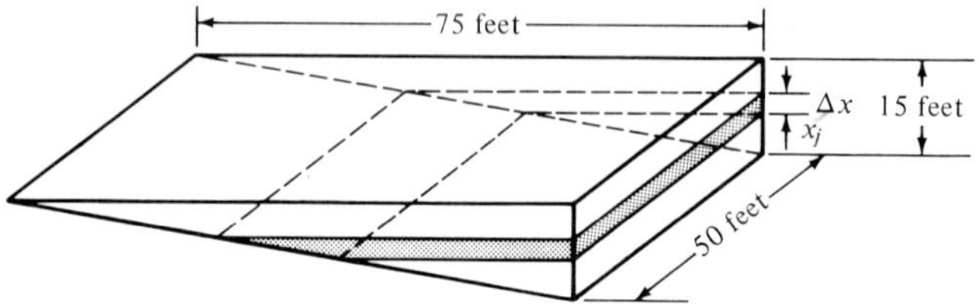

FIGURE TT

HINT: The length of the side of slice j (see Figure TT) is $x_j \cdot \frac{75}{15} = 5x_j$. The cross-sectional area is $50 \cdot (5x_j)$. The volume is $50 \cdot (5x_j) \cdot \Delta x$.

4. Rework the second example in the text except let the number of stores per square mile be given by $e^{-0.01x}$.

5. Suppose that your company is operating a marble quarry, as illustrated in Figure UU. Let the quarry be roughly circular with a diameter of

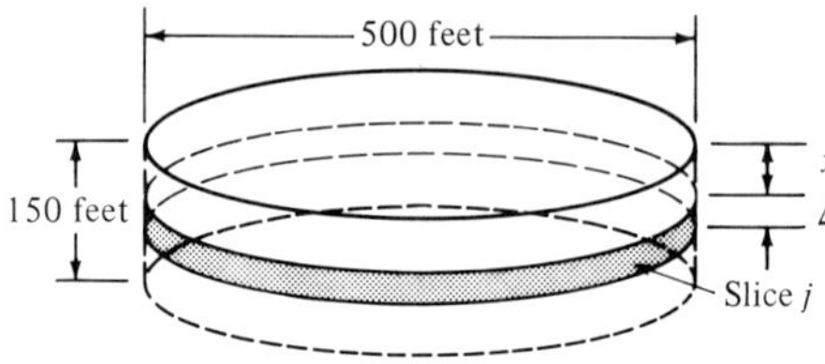

FIGURE UU

500 feet, and suppose that marble sells for \$100 per 100 cubic feet. Furthermore, suppose that the cost of mining (and bringing to the top) 100 cubic feet of marble varies with the depth x, being given by $20 + 0.002x^2$ with x measured in feet below the surface. If we choose to remove the marble to a depth of 150 feet, how much profit would be made?

HINT: The cross-sectional area of slice j is $(\pi/4) \cdot (500)^2$. Its volume is $(\pi/4) \cdot (500)^2 \cdot \Delta x$.

The profit per 100 cubic feet of marble is $100 - (20 + 0.002x_j^2) = 80 - 0.002x_j^2$. The profit realized by mining slice j is $(\pi/4) \cdot (500)^2 \cdot (\Delta x) \cdot (80 - 0.002x_j^2)/100$.

6. In Exercise 5, suppose that the quarry is shaped as in Figure VV. How much profit would be made?

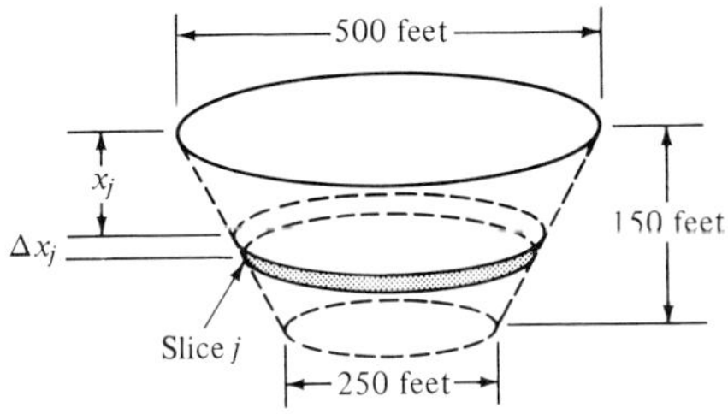

FIGURE VV

HINT: The diameter of the quarry at a distance of x feet below the surface is $(300 - x) \cdot \frac{500}{300} = 500 - 5x/3$. The volume of slice j is $(\pi/4) \cdot (500 - 5x_j/3)^2 \, \Delta x$.

7. Suppose that your company has been retained to dig a tunnel through the mountain illustrated in Figure WW. The cost of digging a foot of the tunnel is given by $20{,}000 \cdot e^{0.01x}$, where x is the distance from the inlet. In digging such tunnels, the practice is to start at both sides of the mountain and dig (hopefully) toward the center. Therefore, two

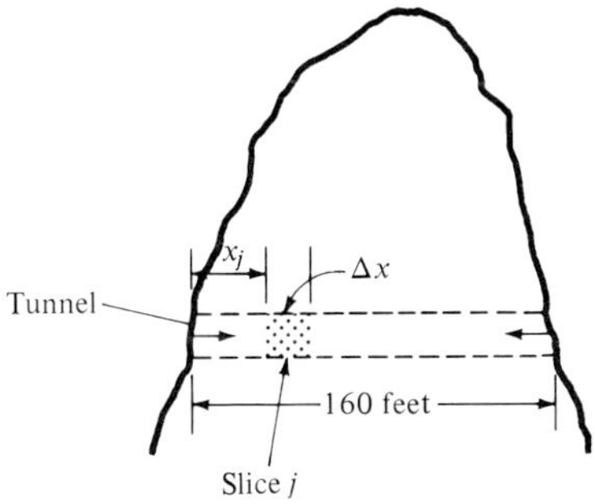

FIGURE WW

80-foot tunnels will be constructed instead of one 160-foot tunnel. How much does it cost to tunnel through the mountain?

HINT: The cost to dig 1 foot of the tunnel at a distance x_j from the inlet is $20{,}000 \cdot e^{0.01x_j}$. The cost to dig slice j is $20{,}000 \cdot (e^{0.01x_j}) \cdot \Delta x$.

8. In the construction of a multistory building, the cost of construction increases exponentially as the distance from the ground increases. For example, suppose that the cost of constructing a 1-foot height of the building at ground level is \$30,000. The cost of construction of 1 foot of the building at a distance x from the ground is given by $\$30{,}000 \cdot e^{0.02x}$, or in essence, the construction costs increase by about 2% for every additional foot of height. How much does it cost to construct a building that is 100 feet in height?

HINT: To solve this problem, we consider a slice of the building of thickness Δx measured in feet at a distance x_j from the ground, as illustrated in Figure XX. The cost of constructing 1 foot of the building at a distance x_j from the ground is $30{,}000 \cdot e^{0.02x_j}$. The cost of constructing slice j is $30{,}000 \cdot (e^{0.02x_j}) \cdot \Delta x$.

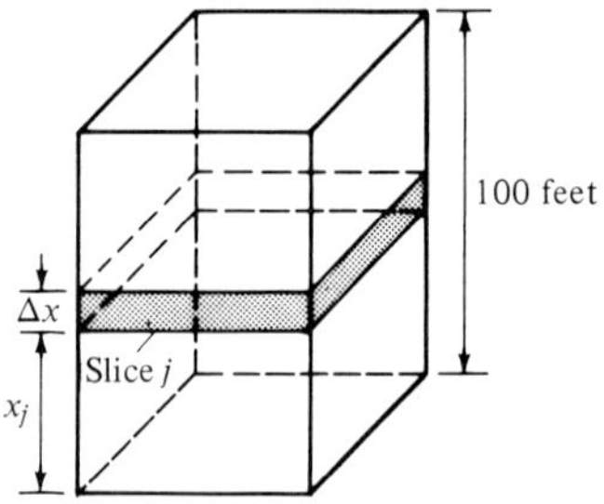

FIGURE XX

REVIEW EXERCISES: CHAPTER 7

In Exercises 1–12, obtain the indefinite integrals:

1. $\displaystyle\int \frac{x}{(x^2+2)^2}\,dx$

2. $\displaystyle\int 4x^2 e^{-x^3}\,dx$

3. $\displaystyle\int xe^{x-1}\,dx$

4. $\displaystyle\int (x+1)e^{-x}\,dx$

5. $\int 3x^2e^{-x}\,dx$

6. $\int (x+1)(4-2x-x^2)^4\,dx$

7. $\int 4x\sqrt{x^2+4}\,dx$

8. $\int \frac{x}{1-4x^2}\,dx$

9. $\int \frac{\ln x^2}{x^2}\,dx$ HINT: $\ln x^2 = 2\ln x$.

10. $\int (x-1)\ln x\,dx$

11. $\int x^2e^{x+1}\,dx$

12. $\int x^3e^{x^2}\,dx$ HINT: Choose $u = x^2$ and $dv = xe^{x^2}\,dx$.

In Exercises 13–18, evaluate the definite integral.

13. $\int_0^1 \frac{\ln(x+1)}{x+1}\,dx$

14. $\int_1^e \frac{\ln x^2}{x}\,dx$

15. $\int_1^e (x^2+x)\ln x\,dx$

16. $\int_0^1 \frac{x^3}{\sqrt{x^4+1}}\,dx$

17. $\int_{-1}^1 \frac{x}{x^2+1}\,dx$

18. $\int_0^4 \frac{1}{(\sqrt{7}+\sqrt{x})\sqrt{x}}\,dx$

19. $\int_1^3 \frac{e^{1/x}}{x^2}\,dx$

20. $\int_1^e \left(\frac{1}{x^2}+\frac{1}{x}\right)\ln x\,dx$

21. $\int_{-3}^3 (x+3)e^{-(x+3)}\,dx$

22. $\int_0^2 (x^2+x)e^{-2x}\,dx$

In Exercises 23–35, determine whether the indicated improper integrals exist, and if they do, evaluate them.

23. $\int_1^\infty \frac{1}{x^5}\,dx$

24. $\int_1^\infty \frac{1}{\sqrt[5]{x}}\,dx$

25. $\int_0^1 \frac{1}{x^5}\,dx$

26. $\int_0^1 \frac{1}{\sqrt[5]{x}}\,dx$

27. $\int_0^\infty e^{-(3x+1)}\,dx$

28. $\int_2^\infty \frac{x^2}{(1-x^3)^2}\,dx$

29. $\int_{-\infty}^{-1} \frac{1}{\sqrt{-x}}\,dx$

30. $\int_{-\infty}^0 e^{3x}\,dx$

31. $\int_{-1}^0 \frac{x}{(1-x^2)^2}\,dx$

32. $\int_{-1}^0 \frac{x}{\sqrt{1-x^2}}\,dx$

33. $\displaystyle\int_0^1 \frac{1}{(x-1)^3}\,dx$ **34.** $\displaystyle\int_3^4 \frac{1}{\sqrt{x-3}}\,dx$

35. $\displaystyle\int_1^2 \frac{1+x}{1-x}\,dx$ HINT: Substitute $u = 1 - x$.

36. A man invests $5000 in a venture whose rate of return in dollars per year per dollar invested is expected to be given by

$$R(t) = e^{-t}$$

Suppose that he leaves his money there forever.

(**a**) Devise an improper integral that would represent the total money that would be returned by this "infinite" investment.

(**b**) Does this integral exist (i.e., is it a real number), and if so, what is it?

CHAPTER EIGHT

PARTIAL DERIVATIVES

8-1. FUNCTIONS OF MORE THAN ONE VARIABLE

We consider a three-dimensional coordinate system as a simple generalization of a two-dimensional system, the coordinate plane. Thus every ordered triple (x, y, z) of real numbers corresponds to a unique point of three-dimensional space, and every point of three-dimensional space has a unique ordered triple of real numbers corresponding to it. As in Figure 8-1, we consider three mutually perpendicular lines which we call the x, y, and z axes.

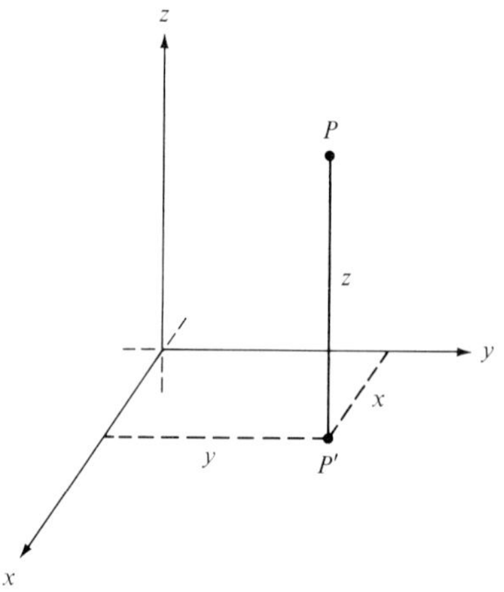

FIGURE 8-1

It is customary to consider the y axis as being from left to right with the positive direction to the right. The positive half of the x axis extends out front. A plane containing two of the axes is designated by the letters for the axes, as the xy plane. We regard the x and y axes as lying in a horizontal plane and the z axis as being vertical with the positive direction upward.

We consider each axis as coordinatized by the real numbers. Thus, for any point P in space, we assign an ordered triple of numbers to P, the x coordinate, the y coordinate, and the z coordinate of P. As in Figure 8-1, the z-coordinate of P measures the height of P above or below the xy plane. Similarly, the x coordinate measures the distance of P in front of or behind the yz plane, and the y coordinate the distance to the right or left of the xz plane. We also can compute the x, y, and z coordinates of P by dropping a perpendicular from P to the xy plane. As before, the z coordinate of P is the height of P above or below the xy plane. The x and y coordinates of P are simply the x and y coordinates in the xy plane of the foot of the perpendicular from P to the xy plane.

If we are given coordinates (a, b, c) of a point, we can find the point by first finding the point P' in the xy plane with x and y coordinates a and b and then locating the point P to be c units above or below P' (according as c is positive or negative).

Example 1. The point $(0, 0, 0)$ is the origin, the point $(1, 0, 0)$ is the point on the x axis one unit in front of the yz plane, the point $(1, 1, 0)$ is the point P' in the xy plane whose x and y coordinates are both 1, and the point $(1, 1, 1)$ is the point P that is 1 unit above P'. These points are illustrated in Figure 8-2.

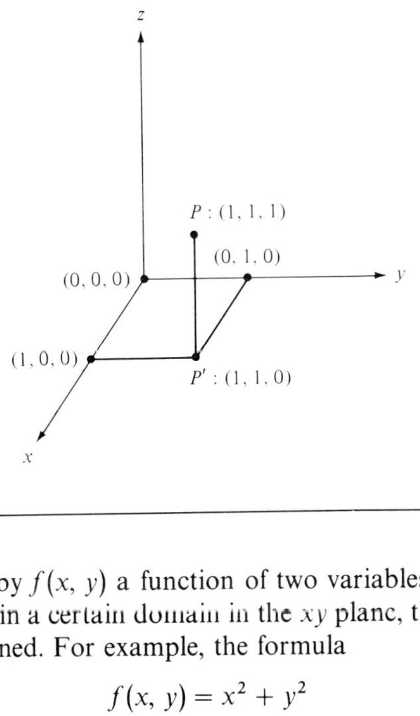

FIGURE 8-2

We denote by $f(x, y)$ a function of two variables, x and y. When x and y are specified in a certain domain in the xy plane, then the value of the function is determined. For example, the formula

$$f(x, y) = x^2 + y^2$$

gives such a function defined for all points in the xy plane. If $x = 1$ and $y = 2$, then $f(x, y) = 1^2 + 2^2 = 5$. If $x = -2$ and $y = 0$, then $f(x, y) = (-2)^2 + 0 = 4$. We can analyze a function of two variables very much as we analyze a function of one variable. The *graph* of $f(x, y)$ is the set of all points (x, y, z) in three-dimensional space whose coordinates satisfy the functional relationship $z = f(x, y)$. We interpret the height z above or below the point $(x, y, 0)$ in the xy plane as the value of $f(x, y)$.

In Figure 8-3, we show a graph of a function $f(x, y)$ whose domain is the set D in the xy plane. The graph is a surface defined over D. The domain of any function $f(x, y)$ is thus regarded as a set in the xy plane.

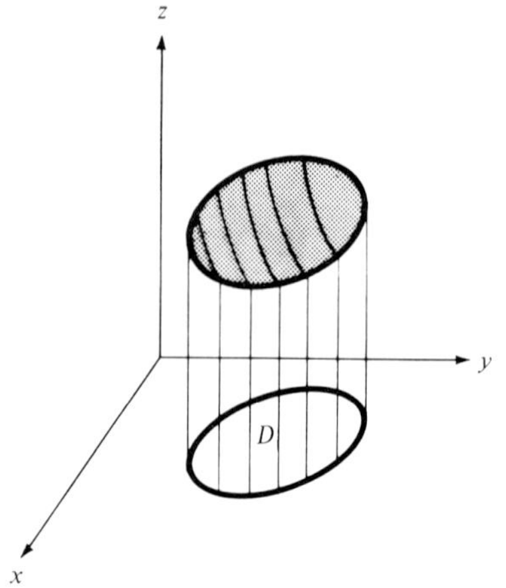

FIGURE 8-3

From a viewpoint of graphing $f(x, y)$ we regard the points of the xy plane as given by their x and y coordinates alone. Thus we can interpret any point of D either as (x, y) from the point of view of graphing $f(x, y)$ or as $(x, y, 0)$ from the point of view of the three-dimensional coordinate system.

Example 2. Let $f(x, y) = \sqrt{1 - x^2 - y^2}$. Since $f(x, y)$ must be a real number, the domain of $f(x, y)$ will be the set of all points (x, y) for which $1 - x^2 - y^2 \geq 0$, hence for which $1 \geq x^2 + y^2$. This set is the set of all points of the xy plane on or in the interior of the circle of radius 1 with center at the origin. We note that $f(x, y)$ is always positive or zero and can never be more than 1 (since $x^2 + y^2$ must be positive or zero). In fact, it can be shown that the graph is a hemisphere of radius 1 and center at (0, 0, 0), as illustrated in Figure 8-4. It lies on or above the xy plane. The graph intersects the xy plane

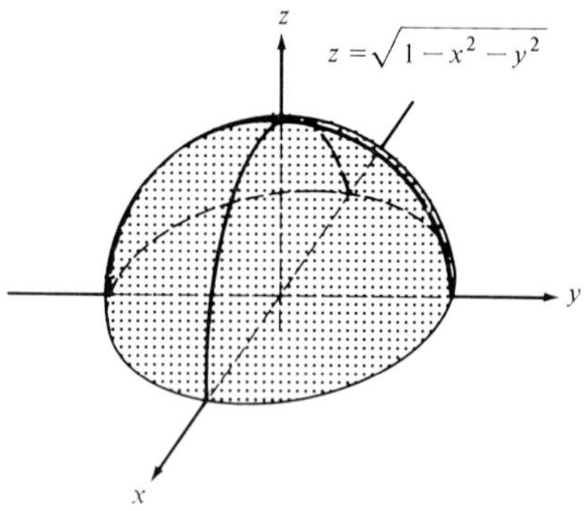

FIGURE 8-4

at those points for which $f(x, y) = 0$, i.e., for which $x^2 + y^2 = 1$. Indeed, the circle $x^2 + y^2 = 1$ in the xy plane serves as the equator of the sphere for which the graph is the upper or northern hemisphere.

Example 3. Let $f(x, y) = x^2 + y^2 + 1$. Here the function is defined for all values of x and y. Also $f(x, y)$ is always ≥ 1. Furthermore, $f(x, y) = 1$ only when $x^2 + y^2 = 0$, i.e., when $x = 0$ and $y = 0$.

As illustrated in Figure 8-5, the graph lies on or above the plane parallel to and one unit above the xy plane. It touches that plane only at the

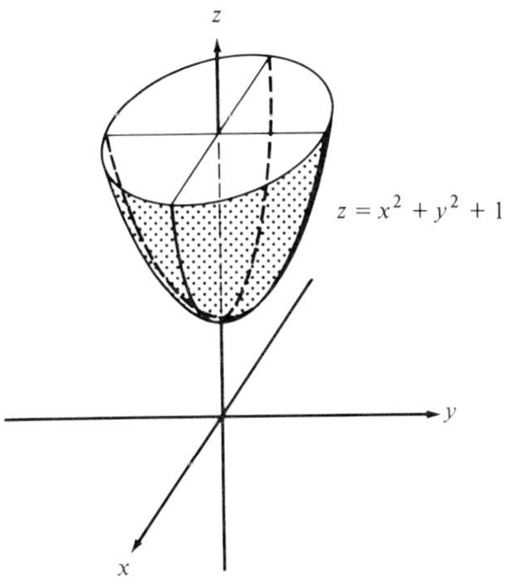

FIGURE 8-5

point (0, 0, 1). For large x or y, $f(x, y)$ is also large. The graph rises as we get farther from the z axis.

Example 4. Let $f(x, y) = ax + by + c$ for any constants a, b, and c. It can be shown that the graph is a plane. The plane crosses every vertical line, since the domain of $f(x, y)$ is the whole xy plane. If a and b are zero, then the plane is given by $z = c$ and will be perpendicular to the z axis, i.e., parallel to the xy plane unless $c = 0$, in which case it will be the xy plane. For any a and b, the plane $z = ax + by + c$ illustrated in Figure 8-6 crosses the z axis in the point $(0, 0, c)$. The plane intersects the xy plane in the line in that plane given by the equation $ax + by + c = 0$, since such points are given by setting z or $f(x, y)$ equal to 0.

In chapter 2 we defined what was meant by a function $g(x)$ being continuous. The concept of continuity is also applicable to functions of more

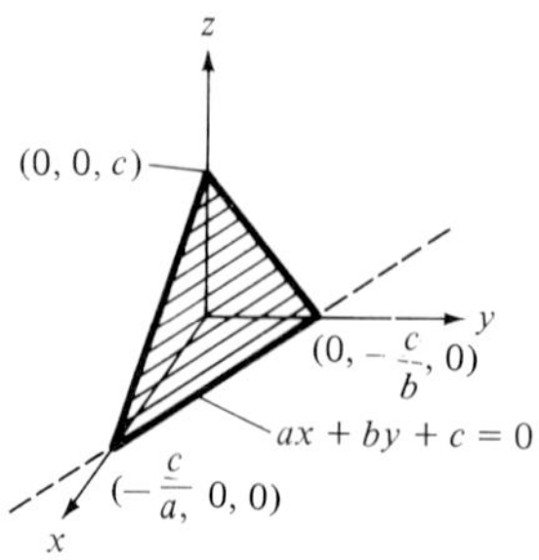

FIGURE 8-6

than one variable. Intuitively speaking, a function $f(x, y)$ is *continuous at a point* (x_0, y_0) in its domain if for each point (x_1, y_1) which is in its domain and is sufficiently close to (x_0, y_0), the functional value $f(x_1, y_1)$ is close to $f(x_0, y_0)$. A function is *continuous* if it is continuous at each point of its domain.

The functions cited in Examples 1 to 4 are all continuous as will be all other functions with which we are explicitly concerned in this chapter. A fuller discussion of continuity is omitted.

EXERCISES 8-1

In Exercises 1–8, (**a**) state the domain of the given functions and (**b**) identify the sets in the xy plane for which the functions are zero. The exercises have been chosen to make these sets easily recognizable.

1. $x^2 + y^2$ **2.** $2x - y - 4$

3. e^{x+y} **4.** xy

5. $\sqrt{4 - x^2 - y^2}$ **6.** $\dfrac{1}{x^2 + y^2}$

7. $\ln(1 + x^2 + y^2)$ **8.** $2^{x(y+1)}$

In Exercises 9–12, find the coordinates of the points where the given planes cross the coordinate axes. Note that any point on the x axis, for example, has both y and z coordinates zero.

9. $z = 3x - 2y - 6$ **10.** $z = x + 3y + 3$

11. $z = 2x - 5y - 10$ **12.** $z = x + 4y - 12$

In each of Exercises 13–16, draw the coordinate axes in the form shown in Figure YY and locate the points on the axes where the plane crosses these axes. Sketch the plane by drawing the triangular region through these points as in Figure 8-4.

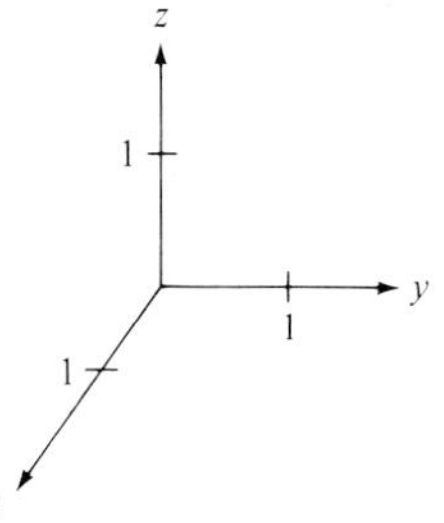

FIGURE YY

13. $z = -2x - 3y + 6$

14. $z = -x - y + 2$

15. $z = -x - 2y + 4$

16. $z = -3x - y + 3$

In each of Exercises 17–20, draw the x, y, and z axes and locate the indicated point by first locating the foot of the perpendicular from the point to the xy plane, as in Figure ZZ.

17. $(1, 4, 2)$

18. $(3, 1, 3)$

19. $(1, 2, 4)$

20. $(2, 3, 1)$

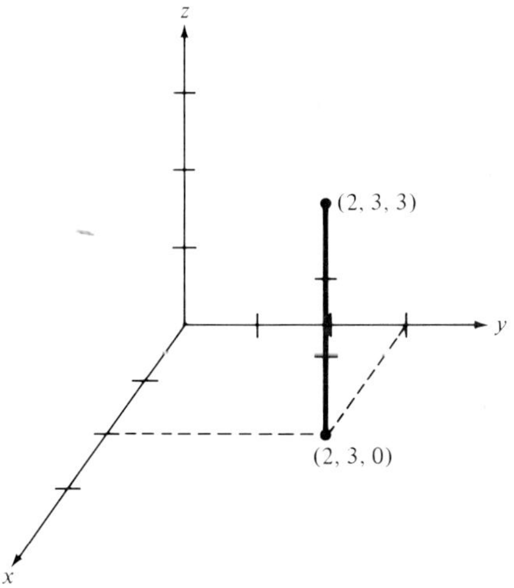

FIGURE ZZ

8-2. PARTIAL DERIVATIVES

If we are given a function $f(x, y)$, we might be tempted to ask if there is something analogous to the derivative for such a function. Can we talk about the rate of change of the function with respect to x and y? Unless we

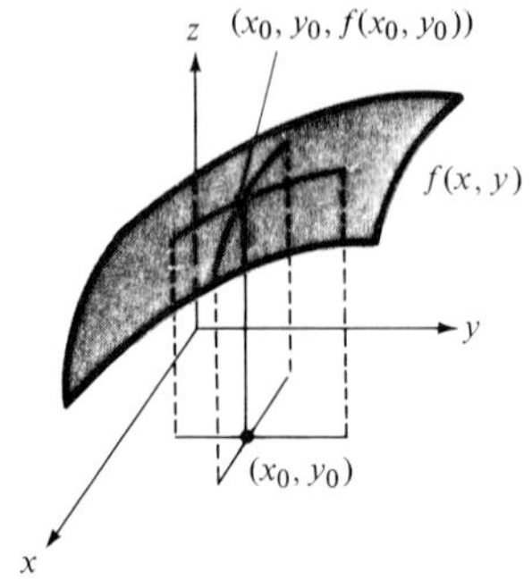

FIGURE 8-7

introduce more concepts, such a question does not have an easy answer. However, we can easily answer some related questions. What is the rate of change of $f(x, y)$ with respect to x alone? Similarly, what is the rate of change of $f(x, y)$ with respect to y alone? In Figure 8-7 we look at these questions graphically.

We interpret the question "What is the rate of change of $f(x, y)$ with respect to x alone?" by regarding y as a fixed quantity. Then we may simply take the derivative of $f(x, y)$ with respect to x, with y held constant. Thus we want

$$\lim_{h \to 0} \frac{f(x+h, y) - f(x, y)}{h}$$

If this limit exists, we call it the (first) *partial derivative of* $f(x, y)$ *with respect to* x or sometimes just the *partial of* f *with respect to* x. We denote it by $\partial f(x, y)/\partial x$, the symbol ∂ being something like a letter d. Other widely used symbols for $\partial f(x, y)/\partial x$ are $\partial f/\partial x$ and $f_x(x, y)$. We similarly define the partial derivative of f with respect to y, using a symbol like $\partial f/\partial y$ or $\partial f(x, y)/\partial y$ or $f_y(x, y)$. Thus

$$\frac{\partial f(x, y)}{\partial y} = \lim_{h \to 0} \frac{f(x, y+h) - f(x, y)}{h}$$

provided that the limit exists. We use

$$\left. \frac{\partial f(x, y)}{\partial y} \right|_{(x_0, y_0)}$$

as the value of $\partial f(x, y)/\partial y$ at the particular point (x_0, y_0). Sometimes, we write $z = f(x, y)$, and in such cases we may use symbols such as $\partial z/\partial x$ and $\partial z/\partial y$ for the partial derivatives.

In deciding what $\partial f/\partial x$ means, we note that with y held constant, $f(x, y)$ is simply a function of x. For example, let $f(x, y) = x^2y + y^3$. If y is fixed, then $x^2y + y^3$ is in the form of a constant times x^2 plus the constant cubed. For $y = 2$, we have $2x^2 + 8$; for $y = 3$, we have $3x^2 + 27$, and so on.

For each of these functions, we know how to take the derivative with respect to x, getting $4x$ as the derivative of $2x^2 + 8$, for example.

To find $\partial(x^2y + y^3)/\partial x$ at an arbitrary point (x, y), we regard y as a constant and get $2xy + 0$. Writing it explicitly as a function of x, we have $(2y) \cdot x$. For $y = 2$, the derivative is $4x$, as it should be. For $y = 3$, the derivative is $6x$, as it should be. In a sense, getting $\partial f/\partial x$ in terms of y as a constant lets us get a formula for the derivative with respect to x, which is applicable for each point for which the derivative exists. Alternatively, for any particular y we could assign y a numerical value first and then take the derivative with respect to x.

In Figure 8-8, we illustrate the graph of a function $f(x, y)$. We note that $\partial f(x, y)/\partial x|_{(x_0, y_0)}$ is the slope at the point (x_0, y_0) of the graph lying

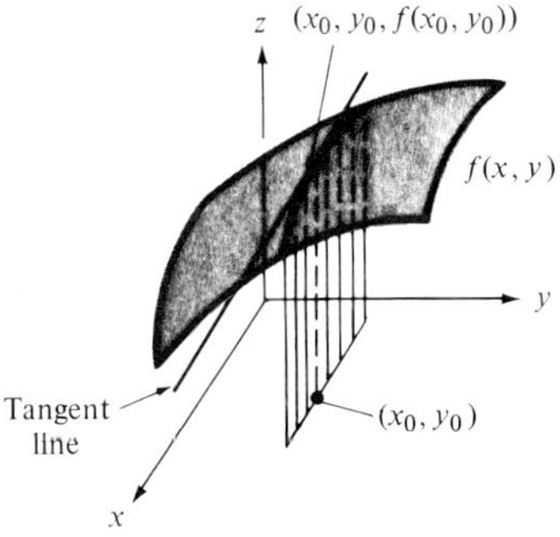

FIGURE 8-8

over the line in the xy plane through (x_0, y_0) and perpendicular to the y axis. With y_0 regarded as fixed, for each value of x near x_0, $f(x, y_0)$ exists and is the value of the function at the point x. If one knows formulas for ordinary derivatives, the actual computation of partial derivatives is usually rather easy as illustrated in the following examples.

Example 1. Let $f(x, y) = x^2 - xy + y^3 - 2$. Then

$$\frac{\partial f}{\partial x} = 2x - y \qquad \text{and} \qquad \frac{\partial f}{\partial y} = -x + 3y^2$$

At the particular point (2, 3), we have

$$\left.\frac{\partial f}{\partial x}\right|_{(2,3)} = 1 \qquad \text{and} \qquad \left.\frac{\partial f}{\partial y}\right|_{(2,3)} = 25$$

Example 2. Let $f(x, y) = e^{-(x^2+y^2)}$. Then

$$\frac{\partial f}{\partial x} = -2xe^{-(x^2+y^2)} \qquad \text{and} \qquad \frac{\partial f}{\partial y} = -2ye^{-(x^2+y^2)}$$

At the particular point (0, 1), we have

$$\left.\frac{\partial f}{\partial x}\right|_{(0,\,1)} = 0 \qquad \text{and} \qquad \left.\frac{\partial f}{\partial y}\right|_{(0,\,1)} = -\frac{2}{e}$$

Example 3. Let $f(x, y) = x \cdot \ln(1 + y^2)$. Then

$$\frac{\partial f}{\partial x} = \ln(1 + y^2) \qquad \text{and} \qquad \frac{\partial f}{\partial y} = \frac{2x \cdot y}{1 + y^2}$$

At the particular point (3, 1), we have

$$\left.\frac{\partial f}{\partial x}\right|_{(3,\,1)} = \ln 2 \qquad \text{and} \qquad \left.\frac{\partial f}{\partial y}\right|_{(3,\,1)} = 3$$

In general, we can interpret both $\partial f(x, y)/\partial x$ and $\partial f(x, y)/\partial y$ as functions of x and y. The domain of $\partial f(x, y)/\partial x$ is the set of all points in the domain of $f(x, y)$ for which $\partial f(x, y)/\partial x$ exists. For the function $f(x, y) = y(\ln x) - xy^2$, we have $\partial f/\partial x = (y/x) - y^2$. In this case, the domain of the function $\partial f/\partial x$ is the domain of $f(x, y)$, the set of all ordered pairs of numbers (x, y) for which $x > 0$. Geometrically, it is the set of all points in front of the y axis in the xy plane. Note that since $f(x, y)$ is not defined for $x \leq 0$, then neither is $\partial f/\partial x$.

Since $\partial f/\partial x$ and $\partial f/\partial y$ can be regarded as functions of x and y, we can take their partial derivatives with respect to x and with respect to y, called their *second partial derivatives*. We use the notation

$$\frac{\partial^2 f}{\partial x^2} = f_{xx} = \frac{\partial}{\partial x}\left[\frac{\partial f}{\partial x}\right], \qquad \frac{\partial^2 f}{\partial y^2} = f_{yy} = \frac{\partial}{\partial y}\left[\frac{\partial f}{\partial y}\right],$$

$$\frac{\partial^2 f}{\partial y\,\partial x} = f_{xy} = \frac{\partial}{\partial y}\left[\frac{\partial f}{\partial x}\right], \qquad \frac{\partial^2 f}{\partial x\,\partial y} = f_{yx} = \frac{\partial}{\partial x}\left[\frac{\partial f}{\partial y}\right]$$

We call the latter two the second *cross* partial derivatives. Note that the variable of the first partial derivative is written on the right.

A natural question that arises is whether the two second cross partial derivatives are necessarily equal to each other, i.e., must $\partial^2 f/\partial x\,\partial y = \partial^2 f/\partial y\,\partial x$? The answer, in general, is "No," but for almost all the functions we deal with in this book, the answer is "Yes." Specifically, at any point (x_0, y_0) for which each of $\partial^2 f/\partial x\,\partial y$ and $\partial^2 f/\partial y\,\partial x$ is continuous, then

$$\left.\frac{\partial^2 f}{\partial x\,\partial y}\right|_{(x_0,\,y_0)} = \left.\frac{\partial^2 f}{\partial y\,\partial x}\right|_{(x_0,\,y_0)}.$$

For functions of more than two variables, we use similar notation.

Thus, if $f(x, y, z)$ is a function of x, y, and z, then $\partial f/\partial x$, $\partial f/\partial y$, and $\partial f/\partial z$ are defined in each case by regarding the other two of the three variables as constants. In Section 8-4, we shall consider more general questions of functions of several variables.

EXERCISES 8-2

In Exercises 1–10, find $\partial f/\partial x$ and $\partial f/\partial y$ and evaluate each at the indicated point.

1. $xy^2 \quad (1, -2)$

2. $(x + 2y)^2 \quad (0, 1)$

3. $\ln (2x + y^2) \quad (1, 1)$

4. $e^{x^2 - 3y} \quad (0, 2)$

5. $xy - 3^{xy} \quad (-1, 1)$

6. $\sqrt{1 - x^2 - y^2} \quad (0, 0)$

7. $x^2y - 4x(1 + y^2) \quad (-2, 2)$

8. $(\ln x) - (x \cdot \ln y) \quad (1, 1)$

9. $(x + y)e^{2x+3y} \quad (-1, 1)$

10. $x^2 \ln (1 + xy) \quad (1, 0)$

In Exercises 11–16, find $\partial^2 f/\partial x^2$, $\partial^2 f/\partial y^2$, $\partial^2 f/\partial x\, \partial y$, and $\partial^2 f/\partial y\, \partial x$. (The latter two should agree; verify that they do.)

11. $2xy - y^2$

12. $\dfrac{3y}{x}$

13. $y \cdot e^x$

14. $\ln (xy)$

15. $\ln (x + y)$

16. $e^{-(x2+y2)}$

In Exercises 17–20, find $\partial f/\partial x$ and $\partial f/\partial y$ and explicitly state the domain of each. Each domain must be a subset of the doman of $f(x, y)$.

17. $\dfrac{x + 2y}{x - 2y}$

18. $y \ln (1 + x)$

19. $\sqrt{4 - x^2 - y^2}$

20. $ye^x - xe^y$

In Exercises 21–24, find $\partial f/\partial x$, $\partial f/\partial y$, and $\partial f/\partial z$ and explicitly state the domain of each. Note that if a variable is missing in a formula, then we understand that the function may be defined for all values of that variable.

21. $x^2 + y^2 + z^2$

22. xyz

23. $\sqrt{4 - x^2 - y^2 - z^2}$

24. $ye^x + ze^x + xe^y + ze^y + xe^z + ye^z$

8-3. MAXIMA AND MINIMA OF FUNCTIONS OF MORE THAN ONE VARIABLE

Consider a function of two variables, such as $x^2 + y^2 + 1$. We can define extrema and relative extrema in much the same way that we did for functions of one variable.

Definition An *(absolute) minimum* of a function $f(x, y)$ is a number m such that for some (x_0, y_0) in the domain of f, $f(x_0, y_0) = m$ and $f(x_0, y_0) \leq f(x_1, y_1)$ for all other points (x_1, y_1) in the domain.

The point (x_0, y_0) *yields* the minimum. We may similarly define an *(absolute) maximum* of the function $f(x, y)$ and the point that yields it.

For the function $x^2 + y^2 + 1$, the absolute minimum is 1 and occurs at the point (0, 0), since $0^2 + 0^2 + 1 = 1$ and for any other pair of real numbers for x and y, we have $x^2 + y^2 + 1 > 1$. There is clearly no absolute maximum of the function $x^2 + y^2 + 1$; for example, for any (x, y) in the domain, with x, y not both zero, the point $(2x, 2y)$ produces a functional value $4x^2 + 4y^2 + 1$, which is greater than $x^2 + y^2 + 1$.

In order to define *relative* extrema, we must first extend the concept of an open interval to domains of functions of two variables. In Figure 8-9, we show how this can be done. The domain of $f(x, y)$ is shown as the shaded region inside or on the curve. The point (x_0, y_0) is shown as a point inside the domain. An *open two-dimensional interval* is a rectangular region in the xy plane consisting of all points in the plane whose x coordinate is in an open interval (a, b) on the x axis and whose y coordinate is in an open interval (c, d) on the y axis.

Definition A *relative* or *local minimum* of a function $f(x, y)$ is a number m such that for some (x_0, y_0) in the domain D of $f(x, y)$, $f(x_0, y_0) = m$ and for some open two-dimensional interval containing (x_0, y_0), $f(x_0, y_0) \leq f(x_1, y_1)$ for any point

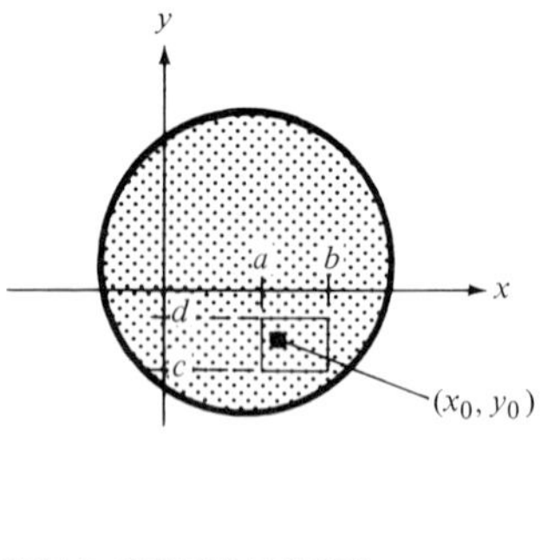

FIGURE 8-9

(x_1, y_1) that is both in D and in the open two-dimensional interval.

The point (x_0, y_0) *yields* the relative minimum. We may similarly define a *relative* or *local maximum* of a function. As before, we use the word "extremum" to refer to either a maximum or a minimum.

Clearly, any absolute extremum of $f(x, y)$ is also a relative extremum of the same type.

We can visualize a part of the surface of the earth as the graph of a function of two variables. The heights of hilltops are local maxima and the heights of the places where water first accumulates during a rainstorm are the local minima. For the continental United States, the height of Mount Whitney is the absolute maximum and the height of the lowest point in Death Valley is the absolute minimum.

A point (x, y) in the domain D of $f(x, y)$ is an *interior point* of D if there exists an open two-dimensional interval containing (x, y) and lying in D. In effect we are saying that a point is an interior point of D if all nearby points are also in D. The set of all interior points of D is called the *interior* of D. Points of D not in the interior of D are called *boundary* points of D.* In Figure 8-9, (x_0, y_0) is in the interior of D. The interior is the shaded portion inside the curve; the boundary of D is the curve itself.

If we are given a function $f(x, y)$ of two variables, we have techniques for finding all points that are in the interior of the domain of $f(x, y)$ and which yield potential relative extrema. These techniques are very much like those for functions of one variable. The basic observation is the following: If either $\partial f/\partial x$ or $\partial f/\partial y$ exists and is not 0 at a point (x_0, y_0) in the interior of the domain of $f(x, y)$, then $f(x, y)$ *cannot* have a relative extremum at (x_0, y_0). For example, suppose that $\partial f/\partial y|_{(x_0, y_0)}$ exists and is different from 0. When $\partial f/\partial y|_{(x_0, y_0)} > 0$, as in Figure 8-10, then for any small positive h, we have $f(x_0, y_0 + h) > f(x_0, y_0)$ and $f(x_0, y_0 - h) < f(x_0, y_0)$. If $\partial f/\partial y|_{(x_0, y_0)} < 0$, the inequalities would be reversed. In either event, $f(x_0, y_0)$ cannot be a relative extremum. The same type of analysis applies if $\partial f/\partial x$ exists and is different from zero. Thus we have the following property.

Property 8-1 If $f(x, y)$ has a relative extremum at a point (x_0, y_0) in the interior of the domain of $f(x, y)$, then *both* **(a)** and **(b)** must be true:

(a) Either $\left.\dfrac{\partial f}{\partial x}\right|_{(x_0, y_0)} = 0$ or $\dfrac{\partial f}{\partial x}$ does not exist at (x_0, y_0).

(b) Either $\left.\dfrac{\partial f}{\partial y}\right|_{(x_0, y_0)} = 0$ or $\dfrac{\partial f}{\partial y}$ does not exist at (x_0, y_0).

* There may be some points not in D that are also called boundary points of D, but we do not need to consider such points in this text.

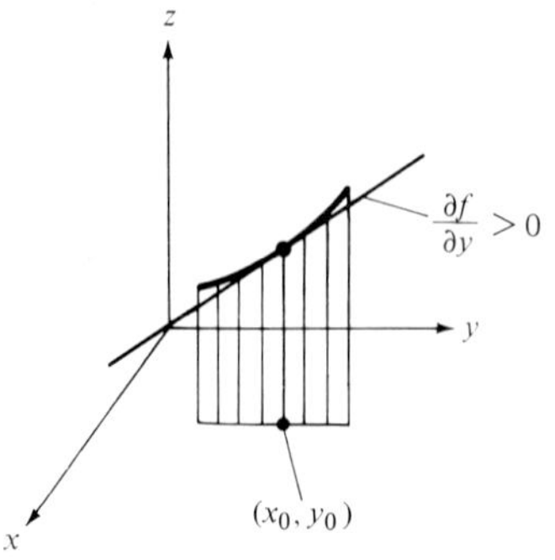

FIGURE 8-10

In many practical problems, we know that $\partial f/\partial x$ and $\partial f/\partial y$ both exist at all points in the domain of $f(x, y)$. In such problems we can find *all* possible points in the interior of the domain that yield relative extrema simply by finding all points (x_0, y_0) for which $\partial f/\partial x = 0$ and $\partial f/\partial y = 0$. In Section 8-5, we shall consider special techniques for finding some relative extrema that occur at points not in the interior of the domain.

Customarily, the set of all points (x_0, y_0) for which $\partial f/\partial x = 0$ is a curve or perhaps a line in the xy plane; similarly, for the set of all points for which $\partial f/\partial y = 0$. Thus, assuming $\partial f/\partial x$ and $\partial f/\partial y$ exist in the interior of the domain of $f(x, y)$, we can identify all points in the interior that *may* yield extrema by merely finding all points for which both $\partial f/\partial x$ and $\partial f/\partial y$ are zero. To do so usually requires us to solve two equations simultaneously. In simple cases, we frequently can find all points for which $\partial f/\partial x = 0$ and for which $\partial f/\partial y = 0$. And we can decide if they yield relative extrema and if so, which type they yield by analysis of the individual function.

Example. Find all relative maxima and minima of $f(x, y) = x^2 - xy + y^2 + 2y + 2$. The domain of $f(x, y)$ is the whole xy plane and thus every point is in the interior of the domain. We compute $\partial f/\partial x = 2x - y$ and $\partial f/\partial y = -x + 2y + 2$. Both $\partial f/\partial x$ and $\partial f/\partial y$ exist at all points of the xy plane. The set of all points for which $\partial f/\partial x = 0$ is the set of all points for which $2x - y = 0$; i.e., the set of all points on the line $y = 2x$. The set of all points for which $\partial f/\partial y = 0$ is the set of all points for which $-x + 2y + 2 = 0$; i.e., the set of all points on the line $y = \frac{1}{2}x - 1$. There is exactly one point for which both $\partial f/\partial x = 0$ and $\partial f/\partial y = 0$. It is the point of intersection of the lines $y = 2x$ and $y = \frac{1}{2}x - 1$.

The point of intersection of the two lines is found by solving the two equations simultaneously. By equating the two expressions for y, we get $2x = \frac{1}{2}x - 1$ or $x = -\frac{2}{3}$. When $x = -\frac{2}{3}$, then $y = -\frac{4}{3}$. The point of intersection is $(-\frac{2}{3}, -\frac{4}{3})$. This is the *only* point yielding a possible extremum.

We can decide that the point $(-\frac{2}{3}, -\frac{4}{3})$ does in fact yield both a relative and an absolute minimum by the following reasoning. For numerically large values of x or y (or both), the function $x^2 - xy + y^2 + 2y + 2$ is large positively, since the x^2 and/or y^2 terms dominate. (For $x = 1000$ and/or $y = 1000$ the function will be about 1,000,000, for example.) Hence we expect there to be a minimum value of the function. But the only possible minimum occurs at $(-\frac{2}{3}, -\frac{4}{3})$. We find it be substituting these coordinates for x and y and get $\frac{4}{9} - \frac{8}{9} + \frac{16}{9} - \frac{8}{3} + 2 = \frac{2}{3}$.

A WORD OF WARNING For functions of two variables, finding and verifying extrema are more complicated than for functions of one variable. To illustrate, consider $f(x, y) = x^3 + y^2$. To identify points at which relative maxima or minima might occur, we determine the first partial derivatives:

$$\frac{\partial f}{\partial x} = 3x^2$$

$$\frac{\partial f}{\partial y} = 2y$$

Both of them equal zero at the point (0, 0), i.e., at $x = 0$ and $y = 0$. Observe that $f(0, 0) = 0$. To investigate this point in more detail, we ask how $f(x, y)$ behaves when $x = 0$. Since $f(0, y) = 0^3 + y^2 = y^2$, we see that when y is very large positively or negatively, $f(0, y)$ is very large positively. Consequently, $f(0, 0)$ could not be a relative maximum but could possibly be a relative minimum.

Next we examine how $f(x, y)$ behaves when $y = 0$. Since $f(x, 0) = x^3 + 0^2 = x^3$, we see that when x is large positively, $f(x, 0)$ is large positively. But when x is large negatively, $f(x, 0)$ is also large negatively. Consequently, $f(0, 0)$ can be neither a relative maximum nor a relative minimum.

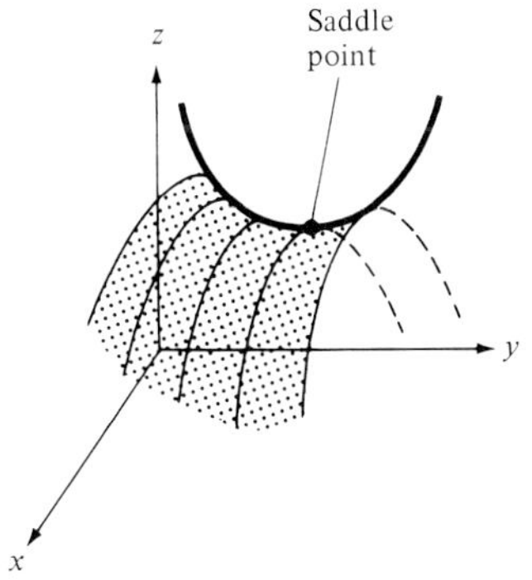

FIGURE 8-11

In fact, the graphs of functions of two variables can possess some interesting characteristics. For example, the graph in Figure 8-11 illustrates a "saddle point." Considering the graph in the x direction, the graph appears to possess a maximum at the saddle point, whereas when the graph is considered in the y direction, the graph appears to possess a minimum at the saddle point. For $f(x, y)$, neither a maximum nor a minimum occurs at the saddle point.

EXERCISES 8-3

In Exercises 1–12, **(a)** determine all points at which both first partial derivatives equal zero, and **(b)** using common sense and an analysis of the behavior of the function, determine if each such point yields a relative maximum, a relative minimum, or neither. Note that in each exercise the domain of $f(x, y)$ is the whole xy plane and each of $\partial f/\partial x$ and $\partial f/\partial y$ exists at every point of the xy plane.

1. $f(x, y) = x^2 + y^2 + 1$

2. $f(x, y) = 1 - x^2 - y^2$

3. $f(x, y) = 1 + x^2 - y^2$

4. $f(x, y) = y^4 - x^4$

5. $f(x, y) = (x + 1)^2 + (y - 1)^2 - 1$

6. $f(x, y) = (3x + 1)^2 + (1 - y)^2$

7. $f(x, y) = x^2 + xy + y^2 - x + 10$

8. $f(x, y) = x^2 - xy + y^2 + x - 30$

9. $f(x, y) = x^2 + y^3$

10. $f(x, y) = x^2 + y^3 + 3y^2 - 9y$

11. $f(x, y) = x(y^2 - 1)$ HINT: Observe how the sign of $f(x, y)$ varies near the two points for which $\partial f/\partial x = 0$ and $\partial f/\partial y = 0$.

12. $f(x, y) = (x^2 - 1)^2 + y^2$

8-4. FUNCTIONS OF MORE THAN TWO VARIABLES

In many practical problems in business or economics, we must deal with functions of more than two variables. For example, price will depend on production costs, marketing costs, demand estimates, competitors' prices, etc. We sometimes use a notation such as $f(x, y, z, w)$ to denote a function of four variables, for example.

When we must deal with more than a few variables, it is customary to denote the variables by use of subscripts, for example, $x_1, x_2, \ldots, x_n$. Thus we may write $f(x_1, x_2, \ldots, x_n)$ to indicate a function of the n variables $x_1, x_2, \ldots, x_n$. For example, $x_1^2 + x_2^2 + \cdots + x_n^2$ or $\sum_{i=1}^n x_i^2$ indicates a specific function of the n variables $x_1, x_2, \ldots, x_n$.

It is possible to define or explain what we mean by n-dimensional space; it is an analogue of three-dimensional space. Here $(x_1, x_2, \ldots, x_n)$ are the coordinates of a point relative to n fixed coordinate axes. Mathematicians, applied mathematicians, statisticians, and mathematical economists deal with n-dimensional space very frequently, very effectively, and very routinely. It sounds much more difficult than it really is. However, in this introductory treatment, we simply want to introduce some notation and a few of the ideas of the subject.

Let $f(x_1, \ldots, x_n)$ be a function of n variables. We define the partial derivative of f with respect to x_i, $\partial f/\partial x_i$, as the derivative of f with respect to x_i when all the other variables x_j are regarded as constants. We use the notation $\partial f/\partial x_i$ or $f_{x_i}(x_1, \ldots, x_n)$ or sometimes $f_i(x_1, \ldots, x_n)$ to denote such a partial derivative. It is defined formally as

$$\frac{\partial f}{\partial x_i} = \lim_{h \to 0} \frac{f(x_1, \ldots, x_{i-1}, x_i + h, x_{i+1}, \ldots, x_n) - f(x_1, \ldots, x_{i-1}, x_i, x_{i+1}, \ldots, x_n)}{h}$$

if this limit exists. There are n possible first partial derivatives of a function of n variables, namely, $\partial f/\partial x_1, \partial f/\partial x_2, \ldots, \partial f/\partial x_n$.

Since each first partial derivative is also a function of the n variables (although some may be formally missing in the formula used), we can define second partial derivatives as we did for a function of two variables in Section 8-2. Thus $\partial^2 f/\partial x_j\, \partial x_i$ means the partial derivative with respect to x_j of the first partial derivative with respect to x_i, $\partial[\partial f/\partial x_i]/\partial x_j$. As before, we expect $\partial^2 f/\partial x_j\, \partial x_i$ to be the same as $\partial^2 f/\partial x_i\, \partial x_j$ but this is not always the case. It *is* the case if these second partial derivatives are continuous functions in all the variables concerned.

Example 1. Let

$$f(x_1, \ldots, x_n) = \sum_{\substack{i,j=1 \\ i<j}}^{n} x_i^2 x_j^2$$

The notation

$$\sum_{\substack{i,j=1 \\ i<j}}^{n} x_i^2 x_j^2$$

means the sum of all possible terms such as $x_1^2 x_2^2$, $x_5^2 x_7^2$, etc., for which i and j are less than or equal to n and $i < j$. We do not have $x_1^2 \cdot x_1^2 = x_1^4$, for example. In other words, $f(x_1, \ldots, x_n)$ is the sum of all possible products $x_i^2 \cdot x_j^2$ for $1 \leq i < j$ and $1 \leq j \leq n$. For $n = 3$, we would have $f(x_1, x_2, x_3) = x_1^2 x_2^2 + x_1^2 x_3^2 + x_2^2 x_3^2$. For $n = 4$, we would have $x_1^2 x_2^2 + x_1^2 x_3^2 + x_1^2 x_4^2 + x_2^2 x_3^2 + x_2^2 x_4^2 + x_3^2 x_4^2$. For large n, we would have a great many terms.

Since the sigma notation occurs so frequently when we work with functions of several variables, we wish to consider how to obtain partial derivatives of functions expressed in the sigma notation. As an example, we shall consider

$$\frac{\partial}{\partial x_i}\left[\sum_{\substack{i, j=1 \\ i<j}}^{n} x_i^2 x_j^2\right]$$

Let us consider writing

$$\sum_{\substack{i, j=1 \\ i<j}}^{n} x_i^2 x_j^2$$

as the sum of two functions, the first with each term having x_i as a factor and the second with no term having x_i as a factor. As an example, we write

$$\begin{aligned} f(x_1, x_2, x_3, x_4) &= \sum_{\substack{i, j=1 \\ i<j}}^{4} x_i^2 x_j^2 \\ &= x_1^2 x_2^2 + x_1^2 x_3^2 + x_1^2 x_4^2 + x_2^2 x_3^2 + x_2^2 x_4^2 + x_3^2 x_4^2 \end{aligned}$$

Of the six terms, exactly three involve any particular x_i^2, for example, the first, fourth, and fifth involve x_2^2. Thus

$$\frac{\partial f(x_1, x_2, x_3, x_4)}{\partial x_2} = 2x_1^2 x_2 + 2x_2 x_3^2 + 2x_2 x_4^2 = \sum_{\substack{j=1 \\ j\neq 2}}^{4} 2x_2 x_j^2$$

all other terms being zero. The pattern we have observed is quite general. For

$$f(x_1, \ldots, x_n) = \sum_{\substack{i, j=1 \\ i<j}}^{n} x_i^2 x_j^2$$

we have

$$\frac{\partial f}{\partial x_i} = \sum_{\substack{j=1 \\ j\neq i}}^{n} 2x_i x_j^2$$

This sum for $\partial f/\partial x_i$ will involve only $(n-1)$ terms that are not zero, one for each j from 1 to n except for $j = i$.

The analysis given in Section 8-3 about maxima and minima for a function of two variables is essentially valid for a function of several variables. In other words, if a relative extremum exists at a given interior point $(a_1, a_2, \ldots, a_n)$, then for each i, $\partial f/\partial x_i$ evaluated at the point either must be zero or must fail to exist. For example, if $\partial f/\partial x_i$ exists and is positive at $(a_1, a_2, \ldots, a_n)$, then for every small $h > 0$, $f(a_1, a_2, \ldots, a_i + h, \ldots, a_n) > f(a_1, a_2, \ldots, a_i, \ldots, a_n)$ and $f(a_1, a_2, \ldots, a_i - h, \ldots, a_n) < f(a_1, a_2, \ldots, a_i, \ldots, a_n)$ and thus we can have no relative extremum at $(a_1, a_2, \ldots, a_n)$.

Thus if all first partial derivatives exist at all points in the domain of $f(x_1, x_2, \ldots, x_n)$, we can find all interior points that can yield relative extrema by solving the n equations $\partial f/\partial x_1 = 0$, $\partial f/\partial x_2 = 0$, ..., $\partial f/\partial x_n = 0$ simultaneously. The problems of solving such systems of equations and of checking to see which solutions actually yield extrema are beyond the scope of this book.

EXERCISES 8-4

Find $\partial f/\partial x$, $\partial f/\partial y$, $\partial f/\partial z$, and $\partial^2 f/\partial x\,\partial y$ in Exercises 1–6. Indicate the domain of $f(x, y, z)$ and of each of the partial derivatives if it is different from the domain of $f(x, y, z)$.

1. $f(x, y, z) = xy^2z^3$

2. $f(x, y, z) = xy + yz + xz$

3. $f(x, y, z) = \dfrac{y}{x^2 + z^2}$

4. $f(x, y, z) = x^2y + \dfrac{xy}{z}$

5. $f(x, y, z) = e^{xyz}$

6. $f(x, y, z) = \ln(1 + x^2 + y^2z^2)$

In Exercises 7–12, find an expression (using sigma notation if appropriate) for $\partial f/\partial x_i$ if $f(x_1, \ldots, x_n)$ is the indicated function. You should first try it for $n = 3$ or 4 and perhaps for $i = 1$ if you do not observe the pattern for i and n.

7. $\displaystyle\sum_{i=1}^{n} x_i^2$

8. $\displaystyle\sum_{\substack{i,j=1\\ i<j}}^{n} x_i x_j$

9. $\displaystyle\sum_{\substack{i,j=1\\ i<j}}^{n} x_i^2 x_j^3$

10. $\displaystyle\sum_{\substack{i,j=1\\ i<j}}^{n} (x_i + x_j)^2$

11. $\displaystyle\sum_{\substack{i,j=1\\ i<j}}^{n} e^{(x_i + x_j)}$

12. $\displaystyle\sum_{\substack{i,j=1\\ i<j}}^{n} e^{x_i x_j}$

In Exercises 13–16, it is possible to find all relative maxima and minima by observation, or alternatively to observe that there are not any. Find the maxima and minima and the points that yield them. Verify that the various first partial derivatives are zero at the points that yield the maxima or minima.

13. $x^2 + y^2 + z^2 + w^2$

14. $(x - 1)^4 + y^2 + (z - 3)^2 + 2$

15. $1 - (x^2y^2 + x^2z^2 + y^2z^2)$

16. $\dfrac{1}{x^2 + (y + 2)^2 + (z + 6)^2}$

8-5. CONSTRAINTS

When maximizing or minimizing functions of only one variable, we were often concerned with the function over a prescribed interval. For example, a typical problem would be to find the absolute maximum of the function $f(x) = 1 - x^2$ over the interval [3, 8]. For functions of two variables, the counterpart is to maximize the function $f(x, y)$ subject to one or more *constraints*. Specifying the interval for a function of one variable placed restrictions on the permissible values of x; similarly, specifying constraints for a function of two variables places restrictions on the permissible values of x and y.

In Section 8-3, we determined that the point $(-\frac{2}{3}, -\frac{4}{3})$ yielded an absolute minimum of the function $f(x, y) = x^2 - xy + y^2 + 2y + 2$. The function had no absolute maximum. The approach used was to determine the values of x and y at which both first partial derivatives $\partial f/\partial x$ and $\partial f/\partial y$ equal zero. Since $f(x, y)$ is defined for all x and y, and no restrictions were placed on the permissible values of x and y, this problem belongs to a class of problems known as *unconstrained* maximum (or minimum) problems.

We shall now illustrate a problem involving an *equality constraint*. As in the previous illustration, let $f(x, y) = x^2 - xy + y^2 + 2y + 2$. However, let the permissible values of x and y be restricted to those values satisfying the equation $x - y + 1 = 0$. This equation is called an *equality constraint*.

When the constraint is given by a "nice" equation, we can often rewrite the constraint giving y as a function of x, or alternatively, x as a function of y. In the above case, we can write the constraint as $y = x + 1$. We then substitute $x + 1$ for y in the function $f(x, y)$:

$$f(x, x + 1) = x^2 - x(x + 1) + (x + 1)^2 + 2(x + 1) + 2$$

The result is a function of x alone, which we shall denote as $f_1(x)$. Simplifying the above formula gives

$$f_1(x) = f(x, x + 1) = x^2 + 3x + 5$$

The graph of this function is illustrated in Figure 8-12.

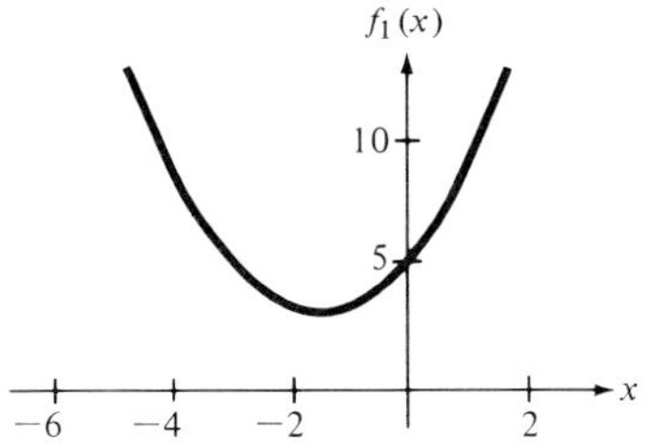

FIGURE 8-12

The maximum or minimum of this function can be obtained by using the first derivative to locate the critical points:

$$f'_1(x) = \frac{df_1(x)}{dx} = 2x + 3$$

Since $f'_1(x) = 0$ at $x = -\frac{3}{2}$ and only at $-\frac{3}{2}$, this is the unique critical point. By any one of the approaches suggested in Chapter 4, it can be ascertained that this point yields a relative and indeed an absolute minimum. This minimum is $(-\frac{3}{2})^2 + 3 \cdot (-\frac{3}{2}) + 5 = \frac{11}{4}$.

This value of y corresponding to $x = -\frac{3}{2}$ is $y = x + 1 = -\frac{3}{2} + 1 = -\frac{1}{2}$. Therefore, the absolute minimum of the function $f(x, y) = x^2 - xy + y^2 + 2y + 2$ subject to the constraint $x - y + 1 = 0$ occurs at the point $(-\frac{3}{2}, -\frac{1}{2})$ and is $\frac{11}{4}$.

When the constraint can be solved explicitly for either x or y, the process illustrated above can be applied. However, such a solution is not always possible. In those cases, we can try a technique known as Lagrange's method, which will be presented in Section 8-7.

In many important problems, the constraint is an inequality rather than an equality. For example, suppose that the constraint is

$$y \geq x + 1$$

In Figure 8-13 the line $y = x + 1$ is graphed. The unshaded region together with the line contains all values (x, y) that satisfy the constraint $y \geq x + 1$. For a fixed x, $y = x + 1$ is a point on the line; for $y > x + 1$ we have points above the line. This region is usually called the *feasible region*, because the values of x and y in this region do not violate the inequality constraint. In effect we regard the feasible region as a new restricted domain of our function. It is the set of points with which we are concerned. The shaded region in Figure 8-13 contains the points (x, y) that violate the constraint and thus are outside the feasible region.

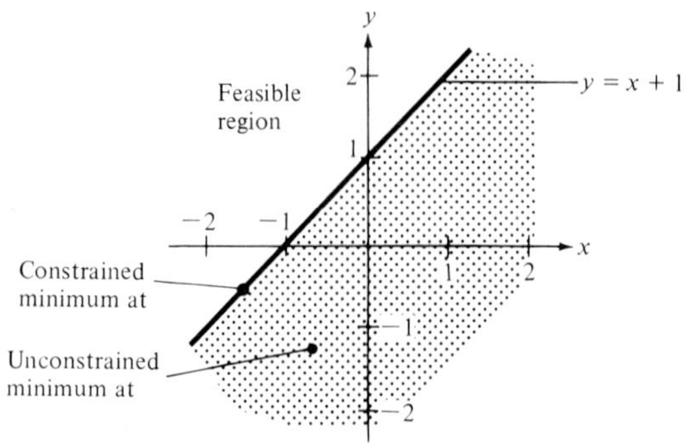

FIGURE 8-13

In such problems, the relative maxima or minima may occur at the following:

1. At points in the *interior* of the feasible region. In Figure 8-13, the interior of the region is the set of points for which $y > x + 1$. To find points in the interior yielding local extrema, we use the techniques of Section 8-3. We find all points of the interior for which $\partial f/\partial y$ is zero or is not defined *and* for which $\partial f/\partial y$ is zero or is not defined. Then we test the function at such points.
2. At points on the *boundary* of the feasible region. In Figure 8-13, the boundary is the line $y = x + 1$. To find points on the boundary yielding relative extrema, the equality constraint problem is solved, as illustrated previously in this section.

In other words, to solve a constrained extremum problem with one such inequality constraint, we must solve two other problems: (1) an unconstrained extremum problem for the interior of the feasible region, and (2) an extremum problem considering only the equality portion of the constraint.

REMARK: In using the techniques outlined above to find relative extrema in the interior or on the boundary of the feasible region, we customarily identify *all* possible points that can yield such extrema. Thus we know that there can be no relative extrema except at the few points we have identified.

To illustrate, as computed in Section 8-3, the *only* possible unconstrained extremum of $f(x, y) = x^2 - xy + y^2 + 2y + 2$ is at $(-\frac{2}{3}, -\frac{4}{3})$. This point is not in the feasible region of the constrained problem (see Figure 8-13). Therefore, in the interior of the feasible region, there can be *no* points that yield relative extrema, or hence absolute extrema. We have used our knowledge about the original unconstrained problem to guarantee that no extremum can occur in the interior of the region.

Next we solve the equality portion of the constrained minimization problem. In our example earlier in this section, we minimized the function $f(x, y) = x^2 - xy + y^2 + 2y + 2$ subject to the equality constraint $y = x + 1$. We concluded that this minimum is at $(-\frac{3}{2}, -\frac{1}{2})$ and is $\frac{11}{4}$. Since there are no relative minima within the feasible region, the only possible absolute minimum of the function $f(x, y) = x^2 - xy + y^2 + 2y + 2$ subject to the constraint $y \geq x + 1$ is at the point $(-\frac{3}{2}, -\frac{1}{2})$. Indeed $\frac{11}{4}$ is the absolute minimum of the function subject to the constraint. How do we conclude that? We first observe that for large x or large y, $f(x, y)$ is very large positively, since the x^2 or y^2 terms in $f(x, y)$ dominate. Therefore, we do not expect an absolute maximum. However, for no values of x or y is $f(x, y)$ very large negatively. Therefore, we do expect an absolute minimum. By the above process we have concluded that the only candidate for this absolute minimum is $\frac{11}{4}$, which occurs at $(-\frac{3}{2}, -\frac{1}{2})$.

Let us modify this problem by changing the constraint from $y \geq x + 1$ to $y \leq x + 1$. Now the shaded region in Figure 8-13 becomes the feasible region. The unconstrained minimum at $(-\frac{2}{3}, -\frac{4}{3})$ is in the feasible region, and it is the absolute minimum. The point $(-\frac{3}{2}, -\frac{1}{2})$ that yields the minimum along the constraint is *not* the absolute minimum, nor does it yield a relative minimum since points in the feasible region near $(-\frac{3}{2}, -\frac{1}{2})$ yield values smaller than the value of the function at $(-\frac{3}{2}, -\frac{1}{2})$.

In practice, a function may have any number of inequality constraints. For example, consider minimizing the function $f(x, y) = x^2 - xy + y^2 + 2y + 2$ subject to the following two constraints (the first inequality is that illustrated in Figure 8-13).

$$y \geq x + 1$$

$$x \geq 1$$

The feasible region is now reduced to the *unshaded* wedge illustrated in Figure 8-14 on page 340. The point of intersection of the two constraints is called a *vertex* of the feasible region.

As before, the unconstrained absolute minimum does not occur in the feasible region.

To locate the constrained minimum, we must consider the equality portion of *each* constraint. We first consider the equality constraint $y = x + 1$. After substituting into $f(x, y)$, we obtain, as before, $f_1(x) = f(x, x + 1) = x^2 + 3x + 5$. However, because of the other inequality constraint, $x \geq 1$, we are only interested in values of x for which $x \geq 1$. For these values, the minimum of $f_1(x)$ is at $x = 1$ and the corresponding value of y is $y = x + 1 = 1 + 1 = 2$. This point is located at the vertex of the feasible region in Figure 8-14. The minimum is $f_1(1) = 1^2 + 3 + 5 = 9$.

Next, we must consider the equality portion of the constraint $x \geq 1$. We substitute $x = 1$ into $f(x, y)$ to obtain

$$f_2(y) = f(1, y) = 1 - y + y^2 + 2y + 2 = y^2 + y + 3$$

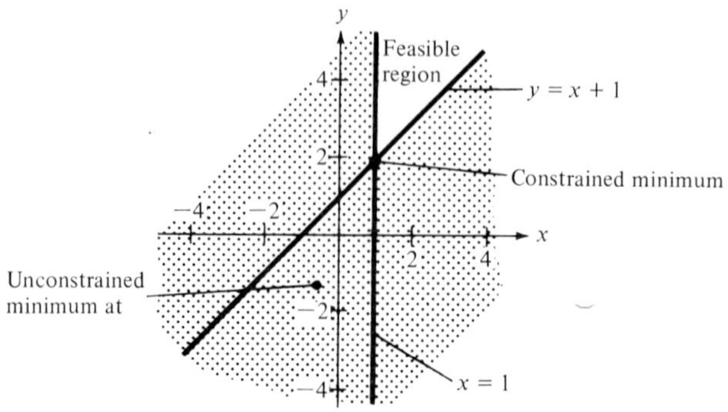

FIGURE 8-14

Because of the constraint $y \geq x + 1$, we are only concerned with values of y for which $y \geq 2$. For these values, the minimum of $f_2(y)$ is at $y = 2$ and has the value 9. Since for $y = 2$ we have $x = 1$, the resultant point (1, 2) is also seen to be the vertex of the feasible region in Figure 8-14. Since the minimum along both constraints is at this point, this is the only candidate for the minimum of the constrained minimization problem. Since $f(x, y)$ is large whenever x or y is large, it is plausible that there be a minimum over the feasible region. Indeed, by methods beyond the scope of this book, we can argue that $f(x, y)$ has a minimum. We know that $f(1, 2) = 9$ must be it since in this problem no point other than (1, 2) can yield an extremum.

In general, the constrained minimum need not necessarily occur at a vertex of the feasible region. For example, suppose the constraints were

$$y \geq x + 1$$

$$x \leq 1$$

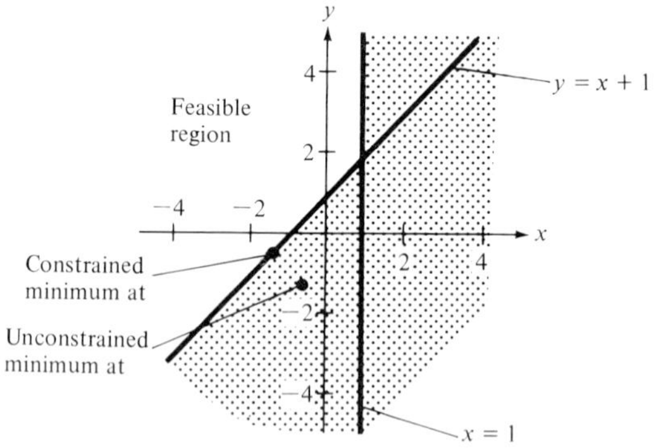

FIGURE 8-15

Now the feasible region is as illustrated in Figure 8-15. The constrained minimum is at $(-\frac{3}{2}, -\frac{1}{2})$, which is not a vertex of the feasible region. In Section 8-6, we shall study special types of functions for which the extrema, if any, always occur at vertices.

In this section, we have only considered *linear constraints*, that is constraints obtained from graphs that are straight lines. In general, constraints do not have to be linear. For example, the feasible region subject to the constraints $x \geq 0$, $y \geq 0$, and $xy \leq 1$ (or $y \leq 1/x$) is illustrated in Figure 8-16.

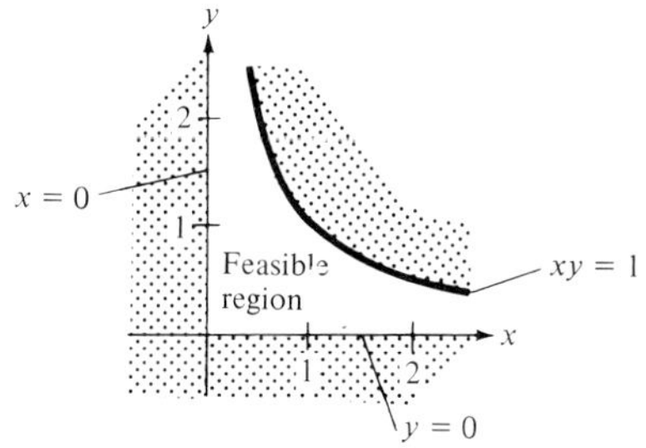

FIGURE 8-16

EXERCISES 8-5

1. In the early paragraphs of this section, the problem of minimizing $f(x, y) = x^2 \quad xy + y^2 + 2y + 2$ subject to the constraint $x - y + 1 = 0$ was solved by writing the constraint as $y = x + 1$ and substituting into $f(x, y)$. Solve the problem by writing the constraint as $x = y - 1$, substitute into $f(x, y)$, and minimize the resulting function of y. Your answer should be the same as that obtained earlier.

In Exercises 2–10, graph the feasible region, i.e., graphically identify the points satisfying the given inequalities. Note that the graph of $y \geq f(x)$ consists of the graph of $y = f(x)$ together with all points above it in the plane. Customarily, we shade the points *not* in the feasible region.

2. $x \geq 0$
$y \geq 0$
$x + 4y \leq 10$

3. $x \geq 1$
$y \geq 0$
$2x + 4y \leq 20$

4. $y \geq 0$
$y \leq x$
$x \leq 5$

5. $x \geq y$
$2y \geq x$

6. $x \geq 0$
$y \geq 0$
$x + y \leq 10$
$y \leq 3 + x$
$y + 5x \leq 40$

7. $x \geq 1$
$y \geq -2$
$y \leq 8 - x$
$y \leq 3 + x$
$x \leq 4$

8. $y \geq 0$
$y \leq 4 - x^2$

9. $y \geq x^3$
$y \leq x$
$x \geq 0$

10. $y \leq x^2$
$x \geq 0$
$y + x \leq 10$

In Exercises 11–16, determine all relative extrema of the function $f(x, y)$ subject to the equality constraint.

11. $f(x, y) = x^2 + xy \qquad y = x$

12. $f(x, y) = \dfrac{x}{y} \qquad x^2 = y - 1$

13. $f(x, y) = xy^2 \qquad x + y - 1 = 0$

14. $f(x, y) = x^2 + y^2 \qquad xy = 1$

15. $f(x, y) = (x + y)^2 \qquad x^2 - y = 2$

16. $f(x, y) = x^2 - 4y^3 \qquad y^2 = x + 4$

In Exercises 17–22, determine all relative extrema of the function $f(x, y)$ subject to the inequality constraint. Be sure to consider relative extrema occurring at interior points as well as on the boundary of the feasible region.

17. $f(x, y) = x^2 + y^2 \qquad x + y \leq 10$

18. $f(x, y) = (x + y)^2 + x \qquad x \leq y$

19. $f(x, y) = \dfrac{x}{y} \qquad x^2 \leq y - 1$

20. $f(x, y) = (x - 1)^2 + y^2 \qquad x \leq y;\ y \geq 0$

21. $f(x, y) = x^2 + (y + 1)^2 \qquad y \leq x;\ x \geq -2$

22. $f(x, y) = x^2 + y^2 \qquad y \leq 1/x;\ x \geq 0;\ y \geq 0$

8-6. LINEAR MAXIMUM-MINIMUM PROBLEMS WITH CONSTRAINTS

Many problems of practical interest involve finding the maximum or minimum of a function of the type

$$f(x, y) = ax + by$$

(where a and b are constants) subject to restrictions on the permissible values of x and y. "Blending" problems are of this type, and we shall use one of these as an example.

Suppose that you are a manufacturer of Pow-Pow enriched breakfast cereal. One serving of this wonderful product will supply an adult's minimum daily requirements for vitamin A and vitamin B_1. The basic ingredients of Pow-Pow contain negligible amounts of these vitamins, so they must be supplied by additives that we shall call X and Y. Suppose additive X costs 5 cents per gram and additive Y costs 2 cents per gram. If x and y are the grams of additives X and Y used per box of Pow-Pow, the total cost in cents of the additives per box of Pow-Pow is

$$f(x, y) = 5x + 2y$$

One way to represent such cost functions graphically is to graph the lines of equal cost. For example, suppose that we wish to graph the values of x and y for which the cost per box is 10 cents. Since $f(x, y) = 5x + 2y$, we have the equation

$$5x + 2y = 10$$

From the form of the equation, you should recognize that the graph would be a straight line. In Figure 8-17, lines corresponding to various fixed values

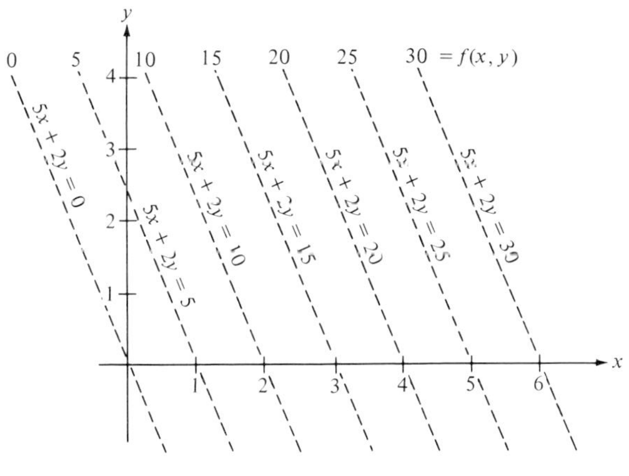

FIGURE 8-17

of $f(x, y)$ are graphed. Functions of the type $f(x, y) = ax + by$ are called *linear* functions.

The manufacturer would like, of course, to select x and y so as to minimize the cost, i.e., to minimize $f(x, y)$. You should recognize that $f(x, y) = 5x + 2y$ has no minimum, since smaller values of x and smaller values of y produce even smaller values of $f(x, y)$. This conclusion is easily confirmed by examining the first partial derivatives:

$$\frac{\partial f(x, y)}{\partial x} = \frac{\partial}{\partial x}(5x + 2y) = 5$$

$$\frac{\partial f(x, y)}{\partial y} = \frac{\partial}{\partial y}(5x + 2y) = 2$$

Since these partial derivatives are never equal to zero, there is no unconstrained maximum or minimum of $f(x, y) = 5x + 2y$.

However, not just any values of x and y are acceptable. By the very nature of the problem posed at the start of this section, negative values of x or y are meaningless. Therefore, x and y must satisfy the inequalities

$$x \geq 0$$

$$y \geq 0$$

These two equations constitute constraints on the acceptable or feasible solutions for x and y.

Our formulation of the problem is not yet complete. We are adding X and Y to enrich Pow-Pow. To determine how much of each to add, we need to know what percent (or fraction) of an adult's minimum daily vitamin requirements per serving of Pow-Pow is provided by a gram of each additive per box. Suppose that 1 gram of X supplies 20% of the minimum daily requirements for vitamin A and 30% of the minimum daily requirement for vitamin B_1. Similarly, suppose that we know that 1 gram of Y supplies 50% of the minimum daily requirement of vitamin A and 25% of the minimum daily requirement of vitamin B_1 (see Table 8-1).

TABLE 8-1

	Percent of minimum daily requirements per serving supplied by 1 gram of the respective additive per box	
Additive	Vitamin A	Vitamin B_1
X	20	30
Y	50	25

To meet exactly the daily vitamin requirements, the following two equations must be satisfied:

$$0.2x + 0.5y = 1$$

$$0.3x + 0.25y = 1$$

Of course, we are generally not concerned should we exceed the minimum daily requirements, so instead we use the inequality constraints

$$0.2x + 0.5y \ge 1$$
$$0.3x + 0.25y \ge 1$$

With the additional constraints $x \ge 0$ and $y \ge 0$, our problem now has four constraints, all of which are linear. The feasible region is the unshaded portion of the xy plane illustrated in Figure 8-18.

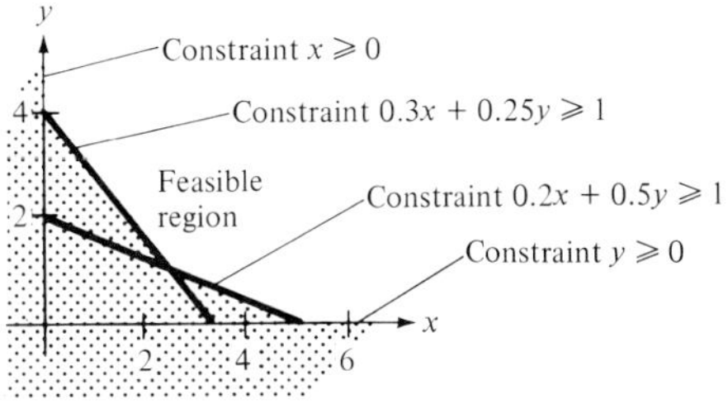

FIGURE 8-18

The minimum can be readily found by superimposing the graph of the cost function in Figure 8-17 and the graph of the feasible region in Figure 8-18. The result is shown in Figure 8-19. From this graph, we see that the minimum cost is at $y = 4$, $x = 0$, which gives a total cost of $f(0, 4) = 5 \cdot 0 + 2 \cdot 4 = 8$ cents. At this point, we supply $0.2x + 0.5y = (0.2) \cdot 0 + (0.5) \cdot 4 = 2$ or twice the minimum daily requirements of vitamin A. We supply $0.3x + 0.25y = (0.3) \cdot 0 + (0.25) \cdot 4 = 1$ or exactly the minimum daily requirements of vitamin B_1.

If we change the cost function so that additives X and Y both cost 5 cents per gram, the cost function becomes $f(x, y) = 5x + 5y$. Figure 8-20 illustrates the graph of the same constraints with this cost function. We now see that the minimum occurs at the intersection of the constraints

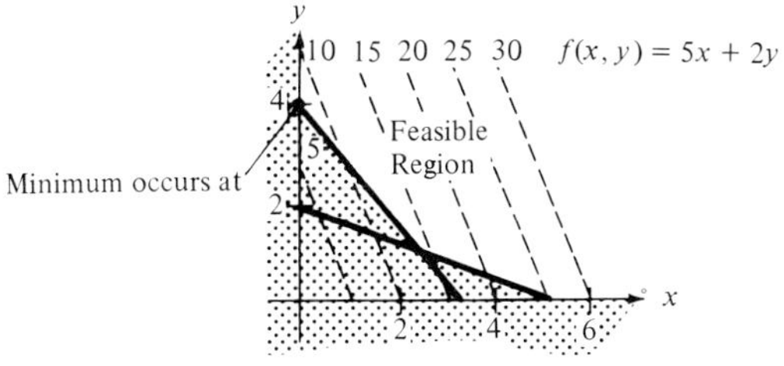

FIGURE 8-19

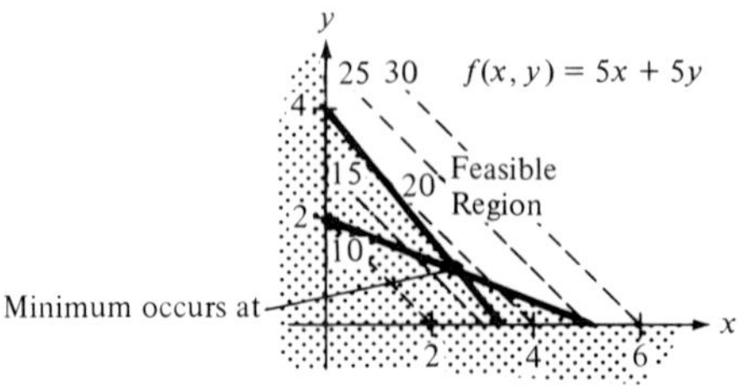

FIGURE 8-20

$0.3x + 0.5y \geq 1$ and $0.3x + 0.25y \geq 1$. This point is $x = 2\frac{1}{2}$, $y = 1$, and the cost is

$$f(2\tfrac{1}{2}, 1) = 5 \cdot 2\tfrac{1}{2} + 5 \cdot 1 = 17\tfrac{1}{2} \text{ cents}$$

Suppose next that X and Y cost 2 and 5 cents, respectively, making the cost function $f(x, y) = 2x + 5y$. Figure 8-21 illustrates the graph of the constraints with this cost function. In this case, the minimum occurs at any point along that portion of the boundary of the feasible region formed by the constraint $0.2x + 0.5y \leq 1$. That is, we can choose any value of x between $2\frac{1}{2}$ and 4 so long as we choose $y = 2 - 0.4x$ (obtained by solving $0.2x + 0.5y = 1$ for y). The cost is $f(x, 2 - 0.4x) = 2x + 5(2 - 0.4x) = 10$.

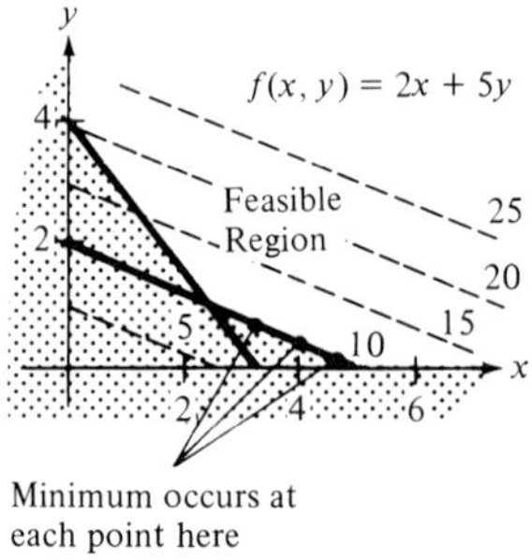

FIGURE 8-21

Whenever the cost function $f(x, y)$ is linear and all constraints are linear, the absolute extremum will *always* occur at a vertex of the feasible region, i.e., at one of the intersections of the constraints. In the unusual case, where a line of equal cost coincides with one of the boundaries of the feasible region, there may be no unique point yielding our extremum, but instead all points along this boundary may yield the same extremum.

Problems of this type are generally referred to as *linear programming* problems. When only two variables are involved, the extremum can be found by the graphical techniques we have used. When more than two variables are involved, such simple graphical techniques cannot usually be applied. However, the solution can be found by simply examining the points of intersection of the various constraints. A computational method known as the *simplex method* is available to solve such problems, and special programs to solve linear programming problems are available for all but very small computers.

EXERCISES 8-6

1. For the problem presented in the text, determine the minimum cost and the values of x and y that produce this cost if X costs 1 cent per gram and Y costs 10 cents per gram.

2. For the problem presented in the text, let $f(x, y) = 5x + 2y$. Assume for taste reasons that we must use 1 gram or more of X in each box of Pow-Pow. Determine the minimum values of x and y and the corresponding cost.

3. Minimize $f(x, y) = x + 10y$ subject to the following constraints:
$$y \geq 0$$
$$1 \leq x \leq 4$$
$$x + y \geq 5$$
Solve this problem using the graphical technique illustrated in Figures 8-19 to 8-21.

4. Minimize $f(x, y) = 3x + y$ subject to the following constraints:
$$5y - x \leq 50$$
$$y + 4x \geq 12$$
$$10y + x \geq 10$$
$$4x - y \geq 16$$
Solve this problem using the graphical technique illustrated in Figures 8-19 to 8-21.

5. Maximize $f(x, y) = y + 3x$ subject to the following constraints:
$$y \geq 0$$
$$x \geq 0$$
$$2y + x \leq 6$$
$$2x + y \leq 8$$
$$x \leq 3$$
Solve this problem graphically.

6. Maximize $f(x, y) = 4y + x$ subject to the following constraints:

$$y \geq 1$$
$$x \geq 0$$
$$2y + x \leq 8$$
$$2x + y \leq 12$$
$$x \geq y$$

Solve this problem graphically.

7. Suppose that our plant produces products X and Y. Product X sells for \$15 per unit and Y sells for \$5 per unit. These products are made from raw materials A and B. We have 60 units of A and 150 units of B. To make a unit of product X, one must use 2 units of A and 1 unit of B. To make product Y, we must use 1 unit of A and 4 units of B. How much of each product should we produce in order to maximize our profit, and what is our maximum profit? Determine your answers to the nearest integer values.

8. Our company uses two machines to manufacture products X and Y. Product X sells for \$15 per unit and Y sells for \$5 per unit. To produce product X requires 2 hours on the first machine and 4 hours on the second. To produce product Y requires 3 hours on each machine. A total of 120 hours is available on each machine. How much of each product should we produce in order to maximize profits, and what is the maximum profit?

9. Suppose that the constraints in both Exercise 7 and Exercise 8 apply to our production of X and Y. How much should we produce, and what is the maximum profit?

10. Our company has an advertising budget of \$200,000, which it spends on TV advertising and magazine advertising. Television advertising costs \$20,000 per unit and reaches approximately 10,000 people per unit. Magazine advertising costs \$12,000 per unit and reaches approximately 18,000 people per unit. For policy reasons, we require that at least 4 units of TV advertising be used and no more than 15 units of magazine advertising be used. Formulate this as a linear programming problem and solve it graphically.

*8-7. LAGRANGE'S METHOD

We shall now give an example from economics to illustrate the basic idea that an individual will spend his income in such a manner so as to derive the most utility or pleasure. Let's consider the simple case in which John derives pleasure from only two ways of spending money: for recreation and for clothing. Let x be the dollars spent on recreation, and let y be the dollars spent on clothes. Suppose that after John pays his basic weekly living expenses, he has \$50 to spend on these items. If he spends the entire \$50, then x and y must satisfy the following:

$$x + y = 50$$

This equation becomes a constraint on John's decisions as to how much money he will spend for recreation and how much for clothing.

To complete the formulation of this problem, we must relate John's pleasure or degree of happiness to x and y. Typical human behavior is that the first unit of a product brings a relatively large degree of pleasure, but the pleasure derived decreases as additional units are purchased. For recreation, suppose that we let the pleasure derived be related to the money spent on recreation by $\sqrt{x}$. The graph of this function in Figure 8-22 is consistent with our observation on human behavior.

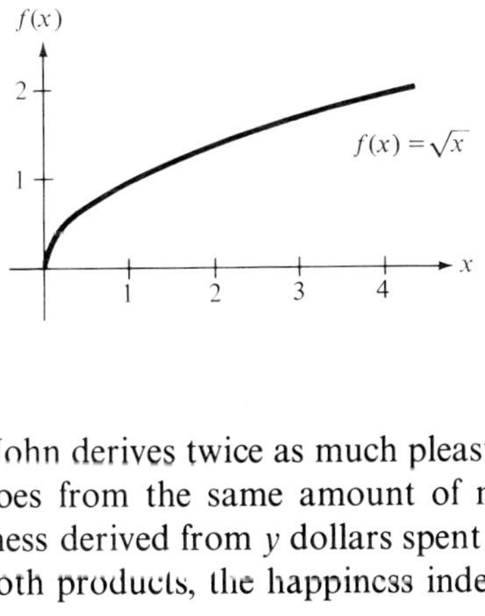

FIGURE 8-22

Suppose that John derives twice as much pleasure from money spent on clothing as he does from the same amount of money for recreation. Therefore, the happiness derived from y dollars spent on clothing would be given by $2\sqrt{y}$. For both products, the happiness index is

$$f(x, y) = \sqrt{x} + 2\sqrt{y}$$

John wishes to select the values of x and y that maximize this function subject to the constraint

$$x + y = 50$$

The function $f(x, y)$ is a measure of the total happiness that John derives from his money.

The solution to this particular problem is easily obtained by the approach illustrated in Section 8-4, namely, solving the constraint for x and substituting into $f(x, y)$ to obtain a function of y alone. Although we shall omit the steps, the result is that for maximal happiness, $x = 10$ and $y = 40$, yielding a maximal happiness index of $\sqrt{10} + 4\sqrt{10}$ or $5\sqrt{10}$.

The approach described in Section 8-4 applies only when the constraint can be readily solved for one of the variables (x or y). This is not always possible, and, in fact, is rarely possible for problems of practical interest. Therefore, we need another way to approach such problems.

One such method is that of Lagrange, a great French mathematician (1736–1813). Lagrange proved that for functions of the type we are considering, the maximum (or minimum) of the function $f(x, y)$ subject to the constraint $g(x, y) = 0$ could be obtained by finding the maximum (or minimum) of the following function (with no constraints):

$$F(x, y, \lambda) = f(x, y) - \lambda g(x, y)$$

This function is known as the *Lagrangian function*, and the variable λ is called a *Lagrange multiplier*, with λ being the Greek letter lambda.

We can use the techniques of Section 8-3 to locate the extrema of the Lagrangian function. We simply determine $\partial F(x, y, \lambda)/\partial x$, $\partial F(x, y, \lambda)/\partial y$, and $\partial F(x, y, \lambda)/\partial \lambda$ and set all three equal to zero:

$$\frac{\partial F(x, y, \lambda)}{\partial x} = \frac{\partial f(x, y)}{\partial x} - \lambda \frac{\partial g(x, y)}{\partial x} = 0$$

$$\frac{\partial F(x, y, \lambda)}{\partial y} = \frac{\partial f(x, y)}{\partial y} - \lambda \frac{\partial g(x, y)}{\partial y} = 0$$

$$\frac{\partial F(x, y, \lambda)}{\partial \lambda} = \qquad - g(x, y) \qquad = 0$$

In the last equation, we used the fact that $f(x, y)$ and $g(x, y)$ are constant functions of λ, and, therefore, $\partial f(x, y)/\partial \lambda = 0$ and $\partial[\lambda g(x, y)]/\partial \lambda = g(x, y)$. Observe that the last equation is simply the original constraint. These three equations involve three unknowns (x, y, and λ), and we can try to solve them.

When we maximize (or minimize) the Lagrangian function $F(x, y, \lambda) = f(x, y) - \lambda g(x, y)$, one of the three equations that must be satisfied is the constraint $g(x, y) = 0$. Furthermore, when the constraint is satisfied, we have $F(x, y, \lambda) = f(x, y) - \lambda \cdot 0 = f(x, y)$. Since we are only interested in those values of $f(x, y)$ when the constraint is satisfied, then the Lagrangian function $F(x, y, \lambda)$ equals $f(x, y)$ at all points of interest. Consequently, maximizing (or minimizing) $F(x, y, \lambda)$ under these conditions is equivalent to maximizing (or minimizing) $f(x, y)$ subject to the constraint $g(x, y) = 0$.

In the problem presented earlier involving John's happiness, our functions were

$$f(x, y) = \sqrt{x} + 2\sqrt{y}$$

$$g(x, y) = x + y - 50 = 0$$

Note that we have written the constraint $x + y = 50$ as $x + y - 50 = 0$ to obtain $g(x, y)$.

To use Lagrange's method to find the maximum of $f(x, y)$, we first form the Lagrangian:

$$\begin{aligned} F(x, y, \lambda) &= f(x, y) - \lambda g(x, y) \\ &= \sqrt{x} + 2\sqrt{y} - \lambda(x + y - 50) \end{aligned}$$

We now write the three partial derivatives and equate each to zero:

$$\frac{\partial F(x, y, \lambda)}{\partial x} = \frac{1}{2\sqrt{x}} - \lambda = 0$$

$$\frac{\partial F(x, y, \lambda)}{\partial y} = \frac{1}{\sqrt{y}} - \lambda = 0$$

$$\frac{\partial F(x, y, \lambda)}{\partial \lambda} = x + y - 50 = 0$$

To solve these equations, we may solve the first two for x and y in terms of λ:

$$x = \frac{1}{4\lambda^2}$$

$$y = \frac{1}{\lambda^2}$$

Substituting into the third equation (the constraint), we have

$$\frac{1}{4\lambda^2} + \frac{1}{\lambda^2} - 50 = 0$$

Solving for λ^2, we have

$$\lambda^2 = \tfrac{1}{40}$$

The values of x and y are given by

$$x = \frac{1}{4\lambda^2} = \frac{1}{4} \cdot 40 = 10$$

$$y = \frac{1}{\lambda^2} = 40$$

This solution agrees with the result cited earlier.

In our formulation we assumed that John would spend all of his money. This resulted in the equality constraint

$$x + y = 50$$

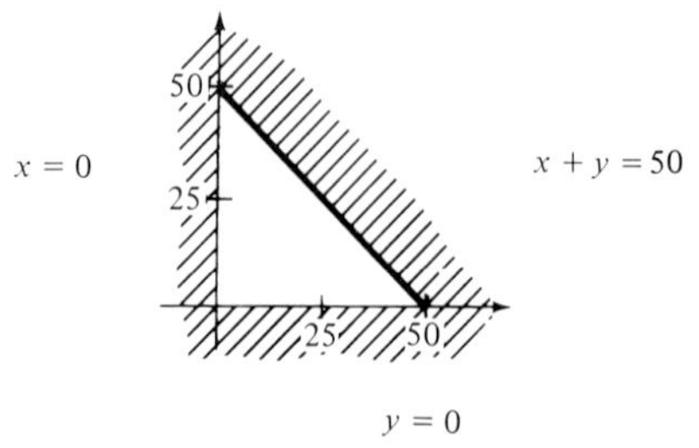

FIGURE 8-23

If John has the option of not spending all of his money, we would write the inequality constraint

$$x + y \leq 50$$

Since negative values of x and y are meaningless, the values of x and y that maximize $f(x, y)$ may be the coordinates of any point of the boundary on the lines $x = 0$, $y = 0$, or $x + y = 50$ illustrated in Figure 8-23, or any interior point of the region bounded by these lines. Lagrange's method can be extended to problems involving inequality constraints, but we shall not do so. Alternatively, one can first solve the unconstrained maximization problem. If the absolute maximum lies inside the permissible region, then we are finished. If not, then we must solve the constrained maximization problem.

EXERCISES 8-7

Use Lagrange's method to obtain all relative extrema of the function $f(x, y)$ subject to the equality constraint.

1. $f(x, y) = x^2 + xy$ $\qquad y = x$

2. $f(x, y) = \dfrac{x}{y}$ $\qquad x^2 = y - 1$

3. $f(x, y) = xy^2$ $\qquad x + y - 1 = 0$

4. $f(x, y) = x^2 + y^2$ $\qquad xy = 1$

5. $f(x, y) = (x - 1)^2 + y^2$ $\qquad x = y$

6. $f(x, y) = x^2 + y^2$ $\qquad y + x = 10$

7. $f(x, y) = (x + y)^2$ $\qquad x^2 - y = 2$

8. $f(x, y) = x^2 - 4y^3$ $\qquad y^2 = x + 4$

9. In Example 1 of Section 4-5, we considered a situation in which a farmer wished to fence a rectangular region using 600 meters of fence and enclose the maximum possible area. We let x be one dimension of the region and y be the other. The area $A = xy$ is a function of two variables, which we could have represented as $f(x, y)$. The constraint is given by $2x + 2y = 600$. The problem is to maximize $f(x, y) = xy$ subject to the constraint $2x + 2y = 600$. Use Lagrange's method to solve this problem, i.e., do not substitute the constraint into $f(x, y)$. Your answer should agree with that obtained in Section 4-5.

10. Our company has an advertising budget of \$200,000, which it spends on TV advertising and magazine advertising. Let t be the units of TV advertising, and let m be the units of magazine advertising. If TV advertising costs \$20,000 per unit and magazine advertising costs \$12,000 per unit, then t and m must satisfy the constraint

$$20t + 12m = 200$$

If t units of TV advertising reach $10{,}000 \cdot t^2$ people and m units of magazine advertising reach $18{,}000 \cdot m$ people, the number (in thousands) of people reached by our advertising is

$$f(t, m) = 10t^2 + 18m$$

We would like to maximize this function subject to the constraint.

(a) Use Lagrange's method to find the values of t and m that maximize $f(t, m)$.

(b) Solve the constraint for m and substitute the result for m in $f(t, m)$. The resulting function of t can be maximized by the techniques presented in Chapter 4. Your answers should agree with those obtained in part **(a)**.

REVIEW EXERCISES: CHAPTER 8

In Exercises 1–4, **(a)** identify the domain of the given function, **(b)** state where the function is positive, and **(c)** state where the function is zero.

1. $\dfrac{1}{xy}$

2. $\sqrt{x^2 + y^2 - 1}$

3. $\ln(x^2y^2)$

4. e^{xy}

In Exercises 5 and 6, identify the points where the given planes cross the coordinate axes.

5. $z = 3x - 4y - 12$

6. $z = x + 8y - 8$

In Problems 7 and 8, draw the x, y, and z axes and identify the points as indicated. Then draw the triangular region with these points as vertices to indicate the plane through the point.

7. $(2, 0, 0), (0, 3, 0), (0, 0, 1)$

8. $(1, 0, 0), (0, 2, 0), (0, 0, 3)$

In Exercises 9–14, find $\partial f/\partial x$, $\partial f/\partial y$, $\partial^2 f/\partial y\, \partial x$ and $\partial^2 f/\partial x^2$ for the given functions. Do not simplify your answers for the second partials.

9. $3x^2y + y$

10. $\dfrac{x}{y}$

11. $3xy + e^{xy}$

12. $\ln(1 + x^2 + y^2)$

13. $2xy - \dfrac{y}{x^2 + 1}$

14. $e^{x^2+y^2} - (x^2 + y^2)$

In Exercises 15 and 16, compute $\partial^2 f/\partial y\, \partial x$ and $\partial^2 f/\partial x\, \partial y$, identify their domains, and verify that they agree therein.

15. $f(x) = 2xy^2 - x^3y + e^{xy^2}$

16. $f(x) = x \ln(1 + y) - ye^{xy}$

In Exercises 17–20, find expressions in $\sum$ notation for $\partial f/\partial x_i$. You may wish to consider a case for an explicit sum for $n = 3$ or 4 to help you organize your thoughts.

17. $\sum\limits_{\substack{i, j=1 \\ i<j}}^{n} x_i^3 x_j^3$

18. $\sum\limits_{\substack{i, j=1 \\ i \neq j}}^{n} (2x_i - 3x_j^2)$

19. $\sum\limits_{\substack{i, j=1 \\ i \neq j}}^{n} x_i x_j^2$

20. $\sum\limits_{\substack{i, j=1 \\ i<j}}^{n} e^{-x_i y_i}$

In Exercises 21–24, find **(a)** all points at which $\partial f/\partial x = 0$ and $\partial f/\partial y = 0$ and **(b)** using common sense about numbers, determine whether such points yield absolute maxima, absolute minima or neither for $f(x, y)$.

21. $f(x, y) = x^2 - xy + 4y^2 - 2x - 2y + 6$

22. $f(x, y) = x^2 - y^2 - 3x - 4y - 2$

23. $f(x, y) = 4x^2 + xy - y^2 - 2y - 2$

24. $f(x, y) = 3x^2 + y^2 - 2x + y + 5$

In Exercises 25–28, graph the feasible regions determined by the given constraints by shading all points *not* in such regions.

25. $x + y \geq 2$
$x \geq 0$
$y \geq 1$

26. $x - y \leq 1$
$x \geq 1$
$y \geq 0$

27. $xy \geq 1$
$0 \leq x \leq 2$
$0 \leq y \leq 2$

28. $x + y \leq 3$
$2x - y \leq 4$
$y - 2x \leq 0$

In Exercises 29–32, find all absolute maxima (if any) and all absolute minima, if any, of the given functions $f(x, y)$ subject to the specified constraints. You may have to use some common sense about numbers.

29. $f(x, y) = (x - 1)^2 + y^2$
$x - y = 0$

30. $f(x, y) = xy + 2y^2 - x + 5$
$x = 2 + 2y$

31. $f(x, y) = x^2 + 4y^2 - 2x + 8y - 6$
$y \leq x + 1$

32. $f(x, y) = x^2 - 2xy + 4y - 4$
$y \geq x - 1$

33. By the method of the text for the graphical solution of a linear programming problem, find the minimum value of the function $2x + 3y$ subject to the constraints $x \geq 0$, $y \geq 0$ and $3x + y \geq 3$.

34. As in Problem 33, find the maximum value of $4x + y + 1$ subject to the constraints $x \geq 0$, $y \geq 1$, and $x + y \leq 5$.

In Exercises 35 and 36, use Lagrange's method to obtain all relative extrema of the given functions subject to the given equality constraint.

35. $f(x, y) = x^2 + y^3$ $\quad x^2 = 4 + y^2 - y^3$

36. $f(x, y) = xy^2 - 2xy - x + 2y$ $\quad x = y^2 + 1$

CHAPTER NINE

DIFFERENTIAL EQUATIONS

9-1. WHAT ARE DIFFERENTIAL EQUATIONS?

Equations that relate derivatives of a function to the function and/or the independent variable are called *differential equations*. As in previous sections, we may denote the independent variable as either x or t, the function by y, and its derivative by dy/dx, dy/dt, or y'. Examples are $dy/dx = x^2 + y^2$, $d^2y/dx^2 + x\,dy/dx - 2y = e^x$, $dy/dt = \ln(t + y)$, $y' = y$, $y' = x + 1$, and so on. We *solve* a differential equation by finding or exhibiting all functions that satisfy the differential equation (in the sense, as we later illustrate, that taking the derivative of the function and substituting into the equation leaves an equation true for all applicable values of x or t).

Consider first a differential equation of the form $y' = g(x)$ for some particular function $g(x)$. Such a differential equation can be solved directly by the process of integration; i.e., $y = \int g(x)\,dx$, assuming that the integral exists. Indeed, finding a function y whose derivative with respect to x is a given function $g(x)$ is simply finding the antiderivative of $g(x)$. Over any interval for which the integral exists, the usual answer can be expressed in the form $y = \int g(x)\,dx$ or as $G(x) + C$, where $G(x)$ is any function whose derivative is $g(x)$ over the interval.

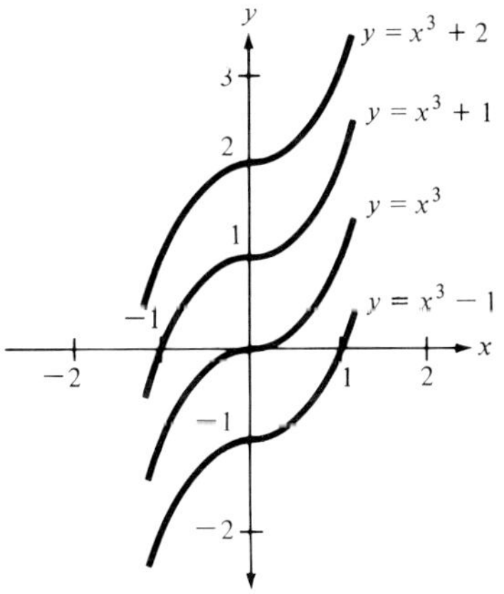

FIGURE 9-1

In Figure 9-1 we graph several particular solutions of the differential equation $y' = 3x^2$. Each solution is of the form $y = x^3 + C$. We have drawn the solutions for various values of C, specifically for $C = -1, 0, 1$, and 2. The function x^3 is defined for all x and by adding or subtracting the constant C to x^3 we can move the graph up or down C units in the vertical direction. For example, the graph of the function $x^3 + 1$ is 1 unit above the graph of x^3.

Finding a Solution Through a Specified Point We see that for any point (x_0, y_0) we can find a solution of $y' = 3x^2$ through the point (x_0, y_0). To find a solution of $y' = 3x^2$ through the point (2, 1), for example, we substitute 2 and 1 for x and y in $y = x^3 + C$ and thus find a value of C for which $1 = 2^3 + C$. Solving for C, we get $C = -7$. The desired solution is then $y = x^3 - 7$. It is a solution of the differential equation and it passes through the point (2, 1). Indeed, for the particular differential equation $y' = 3x^2$ and for any point of the plane, there is exactly one solution of the differential equation through the specified point and defined for all x. The solution can be formed by appropriately specifying C in $y = x^3 + C$.

In a sense, there are two points of view we can adopt: for a differential equation like $y' = 3x^2$, we can write a solution that involves an arbitrary constant C or we can arbitrarily specify a point and get a solution through that point by proper choice of C.

These dual points of view of (1) getting a solution involving an arbitrary constant or (2) specifying a point and getting an explicit solution through the point are valid in a much more general framework than for an equation like $y' = g(x)$ for which a general result is $G(x) + C$ for $G(x)$, an antiderivative of $g(x)$.

We call a differential equation of the form $dy/dx = f(x, y)$ involving only a first-order derivative, a *first-order differential equation.* As with the special case $y' = 3x^2$, we might expect a solution of a first-order differential equation to involve a single arbitrary constant. If we specify a point in the xy plane, we might expect there to be a unique solution through that point obtainable by a proper choice of C.

For many first-order equations such will be the case. For others there will be some exceptions. In Section 9-6, we shall give conditions guaranteeing unique solutions through specified points.

Let us now consider another example of a first-order differential equation and a solution of it. In a later section we shall see how to find the solution. For the moment we regard it as given.

Let $y' = x + y$. We cannot get a solution as before simply by integration since, not knowing y, we do not have a device for finding $\int (x + y)\,dx$. However, let us note that $y = e^x - x - 1$ is a solution of $y' = x + y$. Taking the derivative of the function $y = e^x - x - 1$, we get $y' = e^x - 1$. Substituting for y and for y' in

$$y' = x + y$$

we get

$$(e^x - 1) = x + (e^x - x - 1)$$

or, after simplifying,

$$e^x - 1 = e^x - 1$$

which is true for all x. Therefore, the function $e^x - x - 1$ is a solution.

To get a solution involving an arbitrary constant, we might be tempted to add a constant to this function, writing $y = e^x - x - 1 + C$. Unfortunately, this is not a solution unless $C = 0$. Since $y' = e^x - 1$, on substitution into $y' = x + y$, we get

$$e^x - 1 = x + (e^x - x - 1 + C)$$

or, after simplifying,

$$e^x - 1 = e^x - 1 + C$$

which is true *only* if $C = 0$. For this particular differential equation, an arbitrary constant may be introduced as $y = Ce^x - x - 1$, i.e., as a multiplier of e^x. Note that

$$y' = \frac{d(Ce^x - x - 1)}{dx} = Ce^x - 1$$

Substituting for y and for y' into

$$y' = x + y$$

we get

$$Ce^x - 1 = x + (Ce^x - x - 1)$$

or after simplifying

$$Ce^x - 1 = Ce^x - 1$$

For any value of C, $Ce^x - x - 1$ is a solution.

If we want to find the value of C for which the solution $Ce^x - x - 1$ will pass through the point $(1, -3)$, for example, we substitute 1 for x and -3 for y in the equation $y = Ce^x - x - 1$ getting $-3 = C \cdot e - 2$ or $C \cdot e = -1$ or $C = -1/e$. The particular solution through $(1, -3)$ is

$$y = -\frac{1}{e} \cdot e^x - x - 1 = -(e^{-1} \cdot e^x) - x - 1 = -e^{x-1} - x - 1$$

In Sections 9-2 to 9-4, we shall consider a few special types of differential equations for which we can actually find and exhibit solutions. In Section 9-6, we shall discuss a method of finding a numerical solution for a first-order differential equation. A more complete treatment of differential equations involves studying many more special procedures for particular types of differential equations as well as a study of more generally applicable numerical procedures. A more complete study also involves "existence theorems," i.e., establishing criteria that let us decide for a particular differential equation whether or not there exists a solution through a particular point. Such "existence theorems" are of particular importance in the computer age because they tell us whether or not it makes sense to look for a numerical solution.

REMARK: In this chapter we are concerned only with "ordinary differential equations" as distinct from "partial differential equations". An ordinary differential equation involves ordinary derivatives, whereas a partial differential equation involves partial derivatives as defined in Chapter 8. For example, with f understood to be a function of x and y, we could have

$$\frac{\partial^2 f}{\partial x^2} + \frac{\partial^2 f}{\partial y^2} = 0$$

We would seek all functions $f(x, y)$ satisfying that equation (which is a famous and classical one). The techniques for partial differential equations are quite different from those for ordinary differential equations.

EXERCISES 9-1

In Exercises 1–10, **(a)** verify by substitution that the given function is a solution of the differential equation, **(b)** find the value of C for which the graph of the function passes through the indicated point, and **(c)** write the solution through the specified point.

	Equation	Solution	Point
1.	$y' = 3x + 2y$	$y = Ce^{2x} - \frac{3}{2}x - \frac{3}{4}$	$(0, 4)$
2.	$y' = x^2 - x$	$y = \frac{x^3}{3} - \frac{x^2}{2} + C$	$(-1, 1)$
3.	$y' = y - 1$	$y = Ce^x + 1$	$(-1, 0)$
4.	$y' = 2xy$	$y = Ce^{x^2}$	$(2, -2)$
5.	$y' = \frac{y}{x \ln x}$	$y = C \ln x$	$(e, 2)$
6.	$y' = \frac{2y}{x \ln x}$	$y = C(\ln x)^2$	$(e, 2)$
7.	$y' = -\frac{y}{x}$	$y = \frac{C}{x}$	$(1, 3)$
8.	$y' = -2\frac{y}{x}$	$y = \frac{C}{x^2}$	$(1, 4)$
9.	$y' = \frac{1}{2x}(y - 1)$	$y = C\sqrt{x} + 1$	$(1, 2)$

10. $y' = \dfrac{y}{2\sqrt{x}}$ $\qquad y = Ce^{\sqrt{x}}$ $\qquad (1, e)$

9-2. SEPARABLE VARIABLES

A special type of first-order differential equation that we are able to reduce to a problem of integration is one in which the variables are separable, i.e., in which we can separate the variables as a product. Consider the first-order differential equation $y' = f(x, y)$ and suppose that $f(x, y)$ can be written as $g(x) \cdot h(y)$. Then we can write $y' = f(x, y)$ as $y' = g(x) \cdot h(y)$ and we can write $y'/h(y) = g(x)$. Regarding y as a function of x, we can integrate both sides of this equation with respect to x, getting

$$\int \frac{y'}{h(y)}\,dx = \int g(x)\,dx$$

Now we can employ the process of change of variables or substitution as discussed in Section 7-1. Since $dy = y'\,dx$, we write

$$\int \frac{y'\,dx}{h(y)} = \int \frac{dy}{h(y)} = \int g(x)\,dx$$

If we can integrate $\int [1/h(y)]\,dy$ as a function $H(y)$ and $\int g(x)$ as a function $G(x)$, then we can write the solution as

$$H(y) = G(x) + C$$

which gives us an equation involving one arbitrary constant with y implicitly given as a function of x.

Example 1. Solve $y' = x^2y^3 + y^3$ for a solution through $(1, 2)$. Since $x^2y^3 + y^3$ factors as $(x^2 + 1) \cdot y^3$, the variables are separable. We write the equation as $y'/y^3 = x^2 + 1$ and proceed to integrate with respect to x:

$$\int \frac{y'}{y^3}\,dx = \int (x^2 + 1)\,dx$$

We make the substitution $dy = y'\,dx$ to obtain

$$\int \frac{dy}{y^3} = \int (x^2 + 1)\,dx$$

We now have

$$-\frac{1}{2}\left(\frac{1}{y^2}\right) = \frac{x^3}{3} + x + C$$

By multiplying by $2y^2$ and adding 1 to both sides, we can write a general solution in the form

$$\frac{2x^3y^2}{3} + 2xy^2 + 2Cy^2 + 1 = 0$$

To find the solution through the point (1, 2), we substitute 1 and 2 for x and y, respectively, to determine C.

$$\frac{2 \cdot 1^3 \cdot 2^2}{3} + (2 \cdot 1 \cdot 2^2) + (2 \cdot 2^2 \cdot C) + 1 = 0$$

Therefore, $C = (-1 - 8 - \frac{8}{3})/8 = -\frac{35}{24}$, and the desired equation relating x and y is

$$\frac{2x^3y^2}{3} + 2xy^2 - \frac{35}{12}y^2 + 1 = 0$$

It may be verified, but is tedious to do, that for y given implicitly as a function of x by this equation, we have $y' = (x^2 + 1) \cdot y^3$.

Example 2. Find a solution of $y' = e^{-y}(x^3 - x)$ that passes through the point $(-1, 0)$. The equation $y' = e^{-y}(x^3 - x)$ can be written $y'/e^{-y} = x^3 - x$ or $e^y y' = x^3 - x$. Integrating both sides of the equation with respect to x, we have:

$$\int e^y y'\,dx = \int (x^3 - x)\,dx$$

We substitute $dy = y'\,dx$ to get

$$\int e^y\,dy = \int (x^3 - x)\,dx$$

We now integrate:

$$e^y = \frac{x^4}{4} - \frac{x^2}{2} + C$$

To determine the solution through the point $(-1, 0)$, we substitute and get

$$e^0 = \frac{(-1)^4}{4} - \frac{(-1)^2}{2} + C \quad \text{or} \quad 1 = \frac{1}{4} - \frac{1}{2} + C \quad \text{or} \quad C = \frac{5}{4}$$

Therefore, the desired solution is given by

$$e^y = \frac{x^4}{4} - \frac{x^2}{2} + \frac{5}{4}$$

By taking logs to the base e of both sides, we have

$$y = \ln\left(\frac{x^4}{4} - \frac{x^2}{2} - \frac{5}{4}\right) \qquad \text{or} \qquad y = \ln(x^4 - 2x^2 - 5) - \ln 4$$

as another form of the solution. This form yields y explicitly as a function of x. In Exercise 15, you are asked to verify this as a solution.

Example 3. Find the solution of $dy/dx = e^{x+y}$, which passes through (1, 1). We have $e^{x+y} = e^x \cdot e^y$ and therefore $y'/e^y = e^x$. On integrating with respect to x, we get $-e^{-y} = e^x + C$. On substituting $x = 1$ and $y = 1$, we have $-e^{-1} = e^1 + C$ or $C = -e - 1/e$. Therefore, the desired solution is

$$e^{-y} = e + 1/e - e^x$$

or, solving for y, we have $y = -\ln[e + (1/e) - e^x]$.

The symbolic flowchart in Figure 9-2 summarizes the procedure for solving differential equations by separating variables. The flowchart is for finding the so-called *general solution.* If a *particular solution* that passes through the point (x_0, y_0) is desired, then, using the notation in the flowchart, the arbitrary constant C would be determined from

$$C = H(y_0) - G(x_0)$$

There are many first-order differential equations in which the variables are *not* separable. Note the decision block in the flowchart in Figure 9-2. Here are a few of these equations.

$$y' = x + y$$

$$y' = x^2 + y^2$$

$$y' = xy - x + 1$$

$$y' = \frac{x - y}{x + y}$$

$$y' = \ln(x + y^3)$$

In later sections, we shall learn how to solve some of these equations. For others, there may be no solutions which can be expressed in simple form.

In working exercises in differential equations, there are frequently several different forms in which the answers can be expressed. For each of Example 2 and Example 3 above, we got both an exponential and a log form. It is customary to solve for y explicitly if it is fairly easy to do so. In checking your answers against those given in the text, you may have to

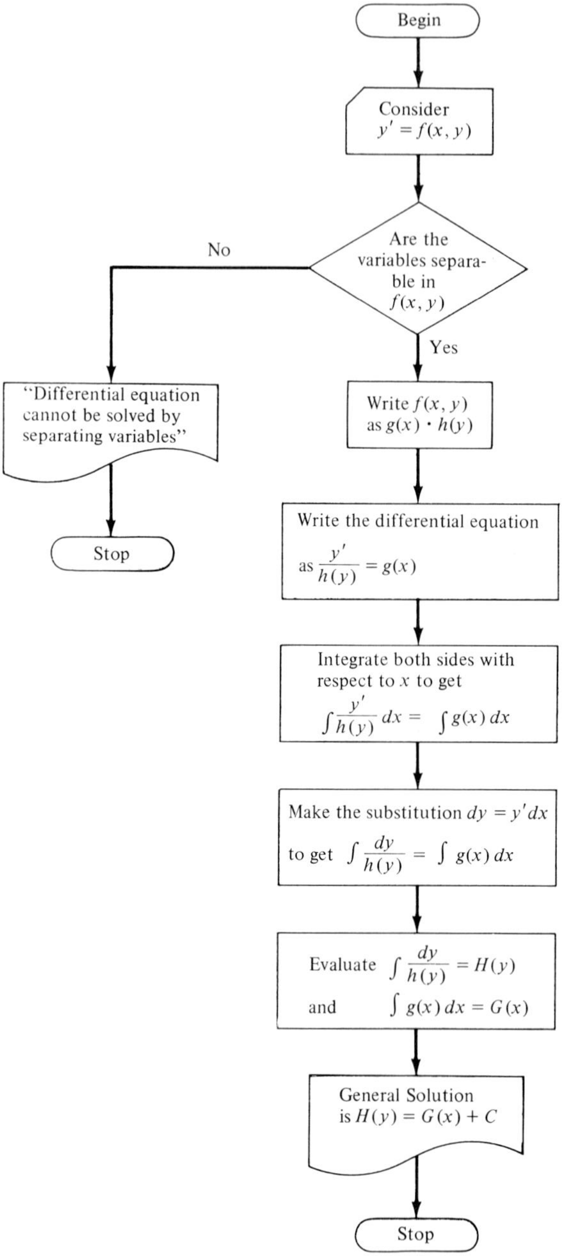
Begin
Consider
$y' = f(x, y)$
Are the variables separable in $f(x, y)$
No
Yes
"Differential equation cannot be solved by separating variables"
Stop
Write $f(x, y)$ as $g(x) \cdot h(y)$
Write the differential equation as $\frac{y'}{h(y)} = g(x)$
Integrate both sides with respect to x to get $\int \frac{y'}{h(y)} dx = \int g(x) dx$
Make the substitution $dy = y'dx$ to get $\int \frac{dy}{h(y)} = \int g(x) dx$
Evaluate $\int \frac{dy}{h(y)} = H(y)$ and $\int g(x) dx = G(x)$
General Solution is $H(y) = G(x) + C$
Stop

FIGURE 9-2

change the form you have. However, the ultimate test of any particular answer is that it works: It must pass through the point in question and when you compute the derivative, it must satisfy the differential equation.

EXERCISES 9-2

Write each of the functions given in Exercises 1–6 explicitly as a product of a function of x by a function of y. You may have to do some factoring. HINT: Recall that $e^{a+b} = e^a \cdot e^b$.

1. $f(x, y) = \dfrac{1 - x - e^x}{1 + y - e^y}$

2. $f(x, y) = (xye^{y+x^2})$

3. $f(x, y) = x(y + 1) - e^x(y + 1)$

4. $f(x, y) = x^2y - xy - y$

5. $f(x, y) = e^{x+3y} \cdot xy^3 \cdot \ln y$

6. $f(x, y) = \dfrac{\ln x - 2 + e^x}{y^2 - 2y + 5}$

For Exercises 7–14, find a solution of the given differential equation that passes through the indicated point. Verify your answer by differentiation.

7. $y' = xy \quad (0, 2)$

8. $y' = e^{-x-y} \quad (-1, 1)$

9. $y' = \dfrac{y+1}{x+1} \quad (0, 0)$

10. $y' = yx^2 + yx - y \quad (-1, e)$

11. $y' = xe^{x^2+y} \quad (1, 0)$

12. $y' - \dfrac{y}{x} \quad (e, c)$

13. $y' = \dfrac{1}{x^2y} \quad (1, 2)$

14. $y' = \dfrac{x^2 - 2x + 5}{y^3 - y^2 + 5y - 1} \quad (-1, -2)$

15. Verify that $y(x)$ defined by

$$y = \ln\left(\frac{x^4}{4} - \frac{x^2}{2} + \frac{5}{4}\right)$$

passes through $(-1, 0)$ and satisfies $y' = e^{-y}(x^3 - x)$. HINT: Use the form $e^y = (x^4/4) - (x^2/2) + (5/4)$.

16. Verify that $y(x)$ defined by $y = -\ln(e + (1/e) - e^x)$ satisfies $y' = e^{x+y}$ and passes through $(1, 1)$. HINT: Use the form $e^{-y} = e + (1/e) - e^x$.

9-3. FIRST-ORDER LINEAR DIFFERENTIAL EQUATIONS

A differential equation is called *linear* if it can be written in a form such as

$$y'' + P(x) \cdot y' + Q(x) \cdot y = R(x)$$

In such a form, the dependent variable y and its derivatives each appear to the first degree and in just one term, and no two of these appear in the same term. The equation above is called a *second-order* linear differential equation, since the highest derivative it involves is the *second* derivative. Here is a specific example:

$$y'' + (x^2 + 1)y' + 2y = 15x$$

There are very special devices for solving linear differential equations. In particular, we first observe how we can solve any *first-order* linear differential equation of the following form

$$y' + P(x) \cdot y = 0$$

If the appropriate integrals exist, we can separate the variables to find a function y that satisfies this differential equation and has an arbitrary constant in the solution.

There is one limitation to our procedure, however; we solve the equation only "up to integration"; i.e., we can exhibit a solution *if* we can find the integrals of the functions concerned. Specifically, we write a solution in terms of such integrals. In a sense, we regard the process of integration of a function as a more elementary process than that of solving a differential equation. Whenever we have reduced the latter problem to the former, we regard ourselves as finished with the solution of the differential equation, but we must proceed with the integration.

Returning to the first-order equation above, we have

$$y' + P(x) \cdot y = 0$$

In solving it, we regard the variables as separable and divide through by y and write

$$\frac{y'}{y} + P(x) = 0 \qquad \text{or} \qquad \frac{y'}{y} = -P(x)$$

As before, regarding each side of the latter equation as a function of x (which it is), we can write

$$\int \frac{y'}{y}\,dx = -\int P(x)\,dx$$

since if two functions of x are equal over an interval, their integrals with respect to x must also be equal or must differ by a constant. Integrating the left-hand side, we have $\int (y'/y)\,dx = \ln y + C_1$, since $d(\ln y)/dx = (1/y) \cdot y'$.

Therefore, we have

$$\ln y + C_1 = -\int P(x)\,dx$$

Noting that $C_1 = -\ln C$ for some number C, we have

$$\ln y - \ln C = -\int P(x)\,dx$$

or

$$\ln \frac{y}{C} = -\int P(x)\,d(x)$$

or

$$\frac{y}{C} = e^{-\int P(x)\,dx}$$

or

$$y = Ce^{-\int P(x)\,dx}$$

This formula is the usual formula for the general solution of the equation,

$$y' + P(x)y = 0$$

As long as $P(x)$ is a function whose integral exists, we have $y = Ce^{-\int P(x)\,dx}$ as a solution.

Here are some examples.

Example 1. $y' + 2y = 0$. Observe that $P(x) = 2$ for this example. By formula, $y = C \cdot e^{-\int 2\,dx}$ or $y = Ce^{-2x}$. This is a solution, since if $y = Ce^{-2x}$, $y' = -2Ce^{-2x}$ and, on substituting, we have $-2Ce^{-2x} + 2(Ce^{-2x}) = 0$. This differential equation can also be solved by the *method* of separating the variables rather than by the *formula*. We can write

$$\frac{y'}{y} = -2$$

$$\int \frac{y'}{y}\,dx = -\int 2\,dx$$

$$\ln y = -2x + C_1$$

$$y = e^{-2x+C_1} = e^{-2x} \cdot e^{C_1} = Ce^{-2x}$$

writing e^{C_1} as the constant C.

Example 2. $y' + x \cdot y = 0$. Observe that $P(x) = x$. Then, by formula,

$$y = Ce^{-\int x\,dx} = Ce^{-x^2/2}$$

To verify that $y = Ce^{-x^2/2}$ is a solution, we first determine $y' = d[Ce^{-x^2/2}]\,dx = -Cxe^{-x^2/2}$. We substitute for y and y' in

$$\begin{aligned} y' \quad &+ \quad xy \quad = 0 \\ -Cxe^{-x^2/2} &+ x \cdot [Ce^{-x^2/2}] = 0 \end{aligned}$$

This verifies that $Ce^{-x^2/2}$ is a solution of $y' + xy = 0$.

Alternatively, we can use the method of separation of variables:

$$\frac{y'}{y} = -x$$

$$\int \frac{y'}{y}\,dx = -\int x\,dx$$

$$\ln y = -\frac{x^2}{2} + C_1$$

$$y = e^{-(x2/2)+C_1} = e^{C_1} \cdot e^{-x2/2} = Ce^{-x2/2}$$

Finally, we note that if we specify a point (x_0, y_0) in the plane, then it is possible to find a solution of $y' + xy = 0$ through that point. For instance, if we want a solution through the point (1, 2), we have $2 = C \cdot e^{-1/2}$ or solving for C, we get $C = 2\sqrt{e}$. Therefore, a solution through (1, 2) is $y = 2\sqrt{e} \cdot e^{-x2/2}$. Since $e^{-x2/2}$ cannot be zero, we can always solve for C no matter what x is specified to be.

Example 3. Find a solution of $xy' + 3y = 0$, which passes through the point (2, 4). We could write $y' + (3/x)y = 0$ and solve by formula. Instead we give a solution by the method of separation of variables:

$$\frac{y'}{y} = \frac{-3}{x} \quad \text{and} \quad \int \frac{y'}{y}\,dx = -3\int \frac{dx}{x}$$

Therefore $\ln y = -3 \cdot \ln x + C_1$ or

$$\begin{aligned} y = e^{-3 \cdot \ln x + C_1} &= e^{-3 \cdot \ln x} \cdot e^{C_1} \\ &= (e^{\ln x})^{-3} \cdot e^{C_1} \\ &= (x)^{-3} \cdot C \end{aligned}$$

Therefore,

$$y = \frac{C}{x^3}$$

To find C, we let $x = 2$ and $y = 4$, getting $4 = C/8$ or $C = 32$. The desired solution is $y = 32/x^3$. To verify this function as a solution, we first compute y' and then substitute into $y' + (3/x)y = 0$

$$y' = \frac{d}{dx}\left[\frac{32}{x^3}\right] = -96 \cdot \frac{1}{x^4}$$

Therefore, on substituting we have

$$-96 \cdot \frac{1}{x^4} + \frac{3}{x} \cdot \frac{32}{x^3} = 0 \qquad \text{or} \qquad -\frac{96}{x^4} + \frac{96}{x^4} = 0$$

In the last example, we note that we can find the value of C whenever $x \neq 0$, i.e., for any point (x_0, y_0) with $x_0 \neq 0$, there will be a solution through the point. In general, we can find a solution through a specified point if on substituting the coordinates of the point into the general solution, we can solve for the arbitrary constant.

EXERCISES 9-3

In Exercises 1–10, find a solution involving an arbitrary constant (**a**) by formula and (**b**) by the method (of separation of variables) illustrated in the text. Your answers should agree. Finally, (**c**) find the solution through the indicated point, and (**d**) verify that solution. In Exercises 5 and 6, rewrite and solve as in Example 3.

1. $y' + y = 0 \quad (0, 2)$

2. $y' - 2y = 0 \quad (-1, 2)$

3. $y' + 2xy = 0 \quad (0, 1)$

4. $y' + 3x^2y = 0 \quad (1, 1)$

5. $xy' + y = 0 \quad (1, 1)$

6. $x^2y' + y = 0 \quad (2, -1)$

7. $y' + (x^2 + 1)y = 0 \quad (-1, -1)$

8. $y' - (x + 1)y = 0 \quad (0, -1)$

9. $y' - e^x \cdot y = 0 \quad (0, 2)$

10. $y' + e^x \cdot y = 0 \quad (0, -1)$

9-4. THE COMPLETE FIRST-ORDER LINEAR DIFFERENTIAL EQUATION

In the previous section, we showed how any first-order linear differential equation of the following form can be reduced to an integration problem. For

$$y' + P(x)y = 0$$

we have

$$y = Ce^{-\int P(x)\,dx}$$

as a general solution. We now address the question of solving the so-called *complete* first-order linear equation $y' + P(x) \cdot y = Q(x)$. In this context, we call $y' + P(x) \cdot y = 0$ the *reduced* equation.

Let $Y(x)$ be any function that satisfies the complete equation. Therefore, we must have

$$Y'(x) + P(x) \cdot Y(x) = Q(x)$$

In the previous section, we showed that the solution of the reduced equation

$$y' + P(x) \cdot y = 0$$

is $Ce^{-\int P(x)\,dx}$. Although we shall leave the proof to Exercise 13, it is not hard to show that $Ce^{-\int P(x)\,dx} + Y(x)$ must also satisfy the complete equation.

The observation that $Ce^{-\int P(x)\,dx} + Y(x)$ is a solution gives us a method for trying to solve any particular complete equation. We first solve the reduced equation to obtain $Ce^{-\int P(x)\,dx}$. We then try to find a function $Y(x)$ that satisfies the complete equation.

Once we have a function $Y(x)$ that satisfies the complete equation, then we can frequently use the arbitrary constant in $Ce^{-\int P(x)\,dx} + Y(x)$ to find a solution through a particular point.

To find some such function $Y(x)$ satisfying the complete equation, we study what is known as the "method of undetermined coefficients." We shall, however, restrict our attention to cases in which $P(x)$ is a constant and $Q(x)$ is a polynomial or involves functions of the exponential type. We begin with two examples and will conclude with some suggestions as to how $Y(x)$ can be chosen by examining $Q(x)$.

Example 1. Find a solution of $y' + y = x + 3$ through the point (0, 4). We first seek any solution of the complete equation and "guess" at the form of an answer. Suppose that there is a solution of the form $Y(x) = Ax + B$. Then, since $Y(x) = Ax + B$, we have, on differentiating, $Y'(x) = A$ and thus on substituting for y and y' in the complete equation

$$y' + y = x + 3$$

we get

$$(A) + (Ax + B) = x + 3$$

This is a statement of equality between two polynomials. The polynomial on the left is equal to the one on the right *if and only if* (a) the constant terms are the same, (b) the first-degree terms (if any) are the same, etc. Therefore,

$$A + B = 3$$

since the constant terms must be the same. And

$$A = 1$$

since the first degree terms must be the same. There are no higher-degree terms. Hence $A = 1$ and $B = 2$, giving $x + 2$ as a solution.

Since, by the techniques of Section 9.3, we know that $y = Ce^{-x}$ is a solution of the reduced equation, we have $y = Ce^{-x} + x + 2$ as a general solution of $y' + y = x + 2$ involving a constant, C. Finally, in finding a solution through (0, 4), we determine the constant C by substituting (0, 4) into the general solution, getting $4 = C \cdot e^0 + 0 + 2$ or $4 = C \cdot 1 + 2$ or $C = 2$. Hence $y = 2e^{-x} + x + 2$ is the desired solution.

Example 2. Find a solution of $y' - 3y = 5e^x - e^{2x}$ through the point $(0, -1)$. We first seek any particular solution of the complete equation. We *guess* or *try* a solution of the form $Y(x) = Ae^x + Be^{2x}$, since e^x and e^{2x} occur on the right-hand side of the equation above.

On substituting $Y(x) = Ae^x + Be^{2x}$ and $Y'(x) = Ae^x + 2Be^{2x}$ for y and y' in the complete equation

$$y' - 3y = 5e^x - e^{2x}$$

we get

$$(Ae^x + 2Be^{2x}) - 3(Ae^x + Be^{2x}) = 5e^x - e^{2x}$$

On collecting terms and writing them on the left, we have

$$(A - 3A - 5)e^x + (2B - 3B + 1)e^{2x} = 0$$

or

$$(-2A - 5)e^x + (-B + 1)e^{2x} = 0$$

The coefficients of each of e^x and e^{2x} must be zero. Therefore, we get

$$-2A - 5 = 0 \quad \text{and} \quad -B + 1 = 0$$

or

$$A = -\tfrac{5}{2} \quad \text{and} \quad B = 1$$

Our desired solution is $Y(x) = -\frac{5}{2}e^x + e^{2x}$.

Since, by the techniques of the previous section, $y = Ce^{3x}$ is known to be a solution of the reduced equation, we have $y = Ce^{3x} - \frac{5}{2}e^x + e^{2x}$ as a solution of the complete equation, and this solution involves an arbitrary constant.

To find a solution through $(0, -1)$, we substitute $(0, -1)$ for (x, y) and get

$$-1 = C - \tfrac{5}{2} + 1 \quad \text{or} \quad C = \tfrac{1}{2}$$

The desired solution through (0, −1) is

$$y = \tfrac{1}{2}e^{3x} - \tfrac{5}{2}e^{x} + e^{2x}.$$

For the equation $y' + P(x)y = Q(x)$, with $P(x)$ being constant, we shall now consider how to guess the form of $Y(x)$ from an inspection of $Q(x)$. In Example 1, observe that $Q(x) = x + 3$ and we were successful with $Y(x) = Ax + B$. In Example 2, observe that $Q(x) = 5e^x - e^{2x}$ and we were successful with $Y(x) = Ae^x + Be^{2x}$. What if $Q(x) = 3e^x - 8$? We might try $Ae^x + B$. What this implies is that we should include a term in $Y(x)$ corresponding to each type of term in $Q(x)$. However, we find, for example, that if $Q(x) = x$, we should try $Y(x) = Ax + B$, and that if $Q(x) = xe^x$, we should try $Y(x) = Axe^x + Be^x$. Therefore, in certain cases, we need to include additional terms.

Table 9-1 will suffice for all examples in this book, but we may need to add the various forms together as suggested above.

TABLE 9-1

$Q(x)$	$Y(x)$
a constant	A
x or $x +$ a constant	$Ax + B$
x^n or any polynomial whose highest power is x^n	$A_n x^n + A_{n-1}x^{n-1} + \cdots + A_1 x + A_0$
a constant e^{ax}	Ae^{ax}
xe^{ax} or $xe^{ax} + e^{ax}$	$Axe^{ax} + Be^{ax}$

From an inspection of the last two entries, it should be apparent how to handle cases like $Q(x) = x^2e^{ax}$. Also if e^{ax} is a solution of the reduced equation, and e^{ax} is a term of $Q(x)$, then we need to try a solution of the form xe^{ax}, but we do not give exercises of this type.

The method of undetermined coefficients is applicable in various cases in which $P(x)$ is not a constant or $Q(x)$ has other forms, but we do not consider such cases in this text.

EXERCISES 9-4

For the following linear, first-order differential equations find **(a)** the general solution of the reduced equation, **(b)** a solution of the complete equation, **(c)** a solution of the complete equation involving an arbitrary constant, and **(d)** a solution through the specified point.

1. $y' - 2y = x + 4 \quad (0, 1)$ **2.** $y' + y = 2 - x \quad (1, 3)$

3. $y' + 3y = 2x + 1 \quad (0, -1)$ **4.** $y' - y = 3x + 2 \quad (-1, 0)$

5. $y' - 3y = e^x - 2e^{-x} \quad (0, 2)$ **6.** $y' + 2y = 4e^{2x} \quad (1, -1)$

7. $y' + y = e^{-2x} + x - 1 \quad (0, 3)$ HINT: Try a solution of the form $Ae^{-2x} + Bx + D$.

8. $y' + 4y = x^2 \quad (-1, 1)$ HINT: Try a solution of the form $Ax^2 + Bx + D$.

9. $y' - 4y = 4x^2 - 1 \quad (2, 2)$ HINT: Try a solution of the form $Ax^2 + Bx + D$.

10. $y' - y = e^{2x} - 3e^{3x} + 6 \quad (0, -3)$ HINT: Try a solution of the form $Ae^{2x} + Be^{3x} + D$.

11. $y' + 2y = 3xe^x - 2e^x \quad (0, 4)$ HINT: Try a solution of the form $Axe^x + Be^x$.

12. $y' + y = 2xe^x - e^{3x} \quad (0, -2)$ HINT: Try a solution of the form $Axe^x + Be^x + De^{3x}$.

13. By substituting $y = Ce^{-\int P(x)\,dx} + Y(x)$ into the differential equation $y' + P(x)y = Q(x)$ and collecting terms, verify that $Ce^{-\int p(x)\,dx} + Y(x)$ is a solution if $Y(x)$ is a solution.

9-5. WHERE DO DIFFERENTIAL EQUATIONS COME FROM?

In the world of business and in many aspects of today's society, we are concerned with what will happen in the future. For example, we would like to know what this country's balance of payments will be for the next 5 or 10 years. Similarly, a corporation's stockholders would like to know how the company's assets will change over the next few years.

One approach is simply to extrapolate from past data. For example, if our company's assets for the past 5 years are given by the indicated points in Figure 9-3, we would be tempted to assume past trends will continue and predict that the company's assets will follow the dashed curve. Although beyond the scope of this text, techniques exist for finding a function that "best" represents past data. It is then assumed that this function will be accurate in the future.

An alternative to this approach is to perform a more fundamental analysis of the problem. Using the company's assets as the example, we would attempt to determine what factors cause the company's assets to change, and then quantitatively relate the rate of change of the company's assets to these factors. The result of such an analysis is usually a differential equation. The resulting equation is based upon a more fundamental under-

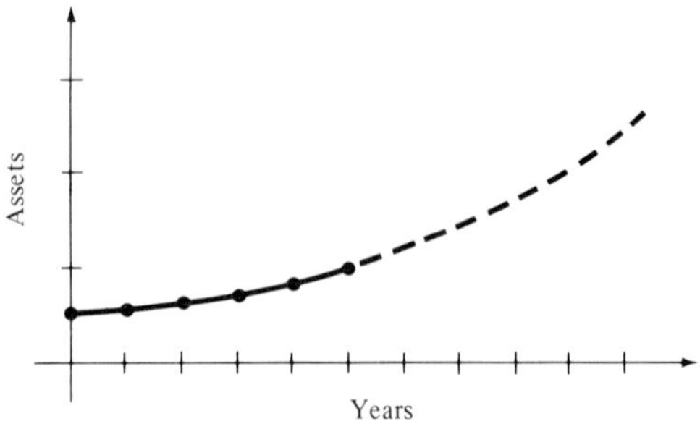

FIGURE 9-3

standing of the problem; consequently, one would usually have more confidence in its predictions for the future as compared to simple extrapolation from past experience.

As our first example, we shall consider the future world population. Our objective is to obtain a function, say $W(t)$, that gives the world population as a function of time. Customarily, we cannot derive an explicit relationship for the function $W(t)$. Instead, we write an expression relating the rate of change of $W(t)$ to those factors that affect $W(t)$, namely, the birth rate and death rate. With a known current population, the result is a differential equation whose solution is the desired function $W(t)$.

For this case, we ask what is $dW(t)/dt$? It is the rate of change of the number of people in the world with respect to time. For world population, we can say "During any period of time, the number of births less the number of deaths equals the change in the number of people in the world over this period of time." For example, if in one year, there were 150 million births and 80 million deaths, then the world population increased by 70 million that year.

The mathematical relationship is most conveniently expressed in terms of rates: "The rate of births minus the rate of deaths equals the rate of change in the number of people in the world." Since the latter is $dW(t)/dt$, we write

$$\frac{dW(t)}{dt} = \begin{Bmatrix}\text{rate of}\\ \text{births}\end{Bmatrix} - \begin{Bmatrix}\text{rate of}\\ \text{deaths}\end{Bmatrix}$$

We analyze the "rate of births" and "rate of deaths" concepts more carefully. In the sense of the equation above, the two rates are comparable in concept and we discuss only the former. The "rate of births" is the product of the "rate per person" by the number of people. The "rate per person" is usually the "rate per person per year," since such a rate must be identified with a time interval. The "rate per person per year" is usually compiled from census figures or estimates from governmental agencies concerned with such matters. With a differential equation we are really properly concerned with instantaneous rates rather than average yearly rates. In a sense, the annual rate is the slope of a secant line, while the instantaneous rate is the slope of a tangent. If functions are not changing rapidly, then the slope of the secant line should be close to the slope of the tangent line, justifying our use of an annual rate as an instantaneous rate.

In what follows, we let r_b and r_d denote the "birth and death rates per person per year." Then returning to our differential equation, we have

$$\text{rate of births } = r_b \cdot W(t)$$

$$\text{rate of deaths} = r_d \cdot W(t)$$

Substituting into our equation for $dW(t)/dt$, we get

$$\frac{dW(t)}{dt} = r_b W(t) - r_d W(t)$$

For convenience, we often let $R = r_b - r_d$ and write

$$\frac{dW(t)}{dt} = R \cdot W(t)$$

For the formulation to be complete, we need a boundary or initial condition. If the world population in 1975 were 4 billion, the boundary condition could be

$$W(1975) = 4$$

where W is in billions of people and t is in years A.D. Alternatively and more commonly, we would let t be the time in years *past* 1975, in which case our boundary condition would be

$$W(0) - 4$$

To solve the differential equation, we could observe that it is a first-order, linear differential equation with the variables being separable. We can use the techniques in either Section 9-2 or 9-3 to solve it. Thus $W(t) = Ce^{Rt}$. Using time measured since 1975, we get $4 = C \cdot 1$ and thus $W(t) = 4e^{Rt}$.

As our second example, we shall consider an expanded version of the miser's retirement fund considered in Section 6-6. In the original problem we stated that the miser retired with \$200,000 and that he expected to withdraw \$5000 per year for living expenses. However, we further stated that his withdrawal would be adjusted to reflect an inflation rate of 5% compounded continuously, and thus the withdrawal rate would be $(5000)e^{0.05t}$.

As an extension to this problem, suppose that the miser deposits his money very conservatively into an account that draws only 3% interest compounded continuously. If we let the principal at time t be $p(t)$, then the interest is drawn at a continuous rate of $0.03p(t)$. We want to formulate an equation that can be solved for $p(t)$.

To formulate the equation, we shall examine the principal $p(t)$ in the miser's savings account. The rate of change of principal, $dp(t)/dt$, is given by

$$\frac{dp(t)}{dt} = \left\{\begin{matrix}\text{rate of}\\ \text{input of}\\ \text{money into}\\ \text{the account}\end{matrix}\right\} - \left\{\begin{matrix}\text{rate of}\\ \text{output of}\\ \text{money from}\\ \text{the account}\end{matrix}\right\}$$

That is, the difference between the rate in and the rate out equals the rate of change of the amount of money in the account.

The input of money into the miser's account is the interest drawn:

$$\text{rate of input} = \text{interest drawn} = 0.03p(t)$$

The output of money from the miser's account is due to the withdrawals:

$$\text{rate of output} = \text{withdrawals} = (5000) \cdot e^{0.05t}$$

The rate of change of principal $p(t)$ is given by the following differential equation:

$$\frac{dp(t)}{dt} = 0.03p(t) - (5000)e^{0.05t}$$

This equation can be rearranged to

$$\frac{dp(t)}{dt} - 0.03p(t) = -(5000)e^{0.05t}$$

This equation can be solved for $p(t)$, using the techniques in Section 9-4. With \$200,000 at time $t = 0$, the miser's money would last about 145 years.

The ability to formulate differential equations only comes with practice. The key to success is to identify a specific quantity whose rate of change can be examined. This quantity will generally be confined in some manner. For example, in the world population problem, we were concerned only with the rate of change of the number of people *on earth*. We do not consider those on Mars. If we are concerned with the population of a country, then we examine the rate of change of the number of people *in that country*. In such a case, immigration and emigration rates, as well as birth and death rates, would have to be considered. In the example of the miser's retirement, we examine the rate of change of the money *in his bank account*.

In essence, we are concerned with the rate of change of some quantity *within some specific confines*. We often call the specific confines the *system*. To formulate a differential equation, we examine the rates of entry and exit of the quantity under consideration across the boundaries of the system. We generally have some input and output terms. These are inserted appropriately into the general equation

$$\frac{d}{dt}\left\{\begin{matrix}\text{quantity}\\ \text{within the}\\ \text{system}\end{matrix}\right\} = \left\{\begin{matrix}\text{rate at which}\\ \text{the quantity}\\ \text{enters the system}\end{matrix}\right\} - \left\{\begin{matrix}\text{rate at which}\\ \text{the quantity}\\ \text{leaves the system}\end{matrix}\right\}$$

In other words, we simply take input terms as positive and output terms as negative.

EXERCISES 9-5

1. A shipper of produce finds that the rate of decay of his produce is given by $0.1Q(t)$ where $Q(t)$ is the amount of produce on hand at time t. The shipper started with 1000 units of the produce. Develop a differential equation and boundary condition that can be solved for $Q(t)$.

HINT: $dQ(t)/dt = -\{\text{rate of decay of the produce}\}$.

2. Let us modify the above problem slightly. A grocer purchases 100 units of produce. From past experience, he sells the produce at a rate of 10 units per day. The rate of spoilage in units per day is given by $0.1tQ(t)$, where $Q(t)$ is the quantity of produce on hand at time t. Develop a differential equation and boundary condition that can be solved for $Q(t)$.

HINT: $dQ(t)/dt = -\{\text{rate of output of produce}\}$ where both decay and sales are "outputs."

3. Suppose that an item of equipment initially costs \$1,000,000 and depreciates at the rate of 10% per year. Write a differential equation and boundary condition that can be solved for $V(t)$, the value of the equipment as a function of time.

HINT: The rate of loss of value is $0.1V(t)$, and is an output term.

4. In the first example in this section, we considered the world population. Let us now consider the population of a single country. To describe the population of a country, we must also consider immigration (people entering the country) and emigration (people leaving the country). Suppose the immigration is given by $0.000001W(t)$, where $W(t) = 4{,}000{,}000{,}000e^{0.02t}$ is the world population and t is the time past 1975. Suppose that the emigration is given by $0.00003P(t)$, where $P(t)$ is the population of the country. Let r_b and r_d be as given in the text, and let the population of the country be 10,000,000 people in 1975. Obtain a differential equation and boundary condition that can be solved for $P(t)$.

HINT: $dP(t)/dt$ = {rate of input of people into the country} – {rate of output of people from the country} where "input" includes births and immigration and "output" includes deaths and emigration.

5. Let us make the following change to the miser's retirement problem as considered in this section. Suppose that independent of his savings, the miser receives a retirement income of \$2000 per year for the rest of his life. This \$2000 is the actual dollar income and is not adjusted to compensate for inflation. Derive the differential equation and boundary condition that can be solved for the money in his account.

6. Suppose that our country's known energy reserves total 150 units at this time. We are now consuming energy at a rate of 8 units per year. However, we expect this to increase according to the formula $8e^{0.02t}$, where t is time in years. Suppose that we are discovering new energy sources at a constant rate of 5 units per year. Develop a differential equation and boundary conditions that can be solved to give the units of our known energy reserves as a function of time.

   HINT: Write an equation for the rate of change of our energy reserves.

7. Consider an airplane that leaves New York with a gross weight of 200,000 pounds, of which 50,000 pounds is fuel. In flight, the airplane's engines consume fuel at the rate of 0.001 pounds per minute for each pound of gross weight (including fuel). We wish to write an equation that can be solved for the fuel on board the plane as a function of time.

   HINT: Write an equation for the rate of change of the fuel in the airplane's tanks.

8. Suppose that Soap-Suds Cola Company decides to change the carton in which shoppers take home a six-pack. At the time the new carton is introduced, Soap-Suds estimates that it has 465,000 cartons on the grocery shelves. Past data indicate that purchases are fairly constant at 125,000 cartons per week. For economic reasons, Soap-Suds chooses not to exchange all of the cartons on the shelf, but instead to supply only new cartons in all further deliveries to the grocery stores. Therefore, if we define the system as the grocery shelves, the input of new cartons is 125,000 per week.

   Let us assume that the new and old cartons are visibly identical, and thus the shoppers should not preferentially take one over the other. Under these conditions, the new cartons would be removed at the rate of $125{,}000[N(t)/465{,}000]$ cartons per week, where $N(t)$ is the number of new cartons on the grocery shelves.

Derive a differential equation and a boundary condition that can be solved for $N(t)$.

HINT: Consider the rate of change of the number of new cartons on the grocery shelves.

9-6. NUMERICAL SOLUTIONS OF DIFFERENTIAL EQUATIONS

We have presented a few of many known techniques for solving certain types of differential equations. However, there are many differential equations for which there is no known method for obtaining a solution expressible in terms of integrals of elementary functions of the type considered in this text. Using only those techniques presented earlier in this chapter, we are unable to solve an equation like

$$y' + y^2 = e^{-x}$$

for a solution passing through (0, 2), i.e., for a solution for which $y(0) = 2$.

The equation is first order but is not linear. The variables are not separable.

However, we can solve such an equation numerically. Indeed, one of the most important uses of computers in applications of mathematics is to obtain *approximate* (numerical) solutions of such equations.

We may write the equation as

$$y' = e^{-x} - y^2$$

specifying that $y(0) = 2$. Thus we may regard the equation as a first-order differential equation. The more general form is

$$y' = f(x, y), \qquad y(0) = y_0$$

where y_0 is a constant. We call the condition $y(0) = y_0$ an *initial* or *boundary* condition. In effect, it specifies the beginning point $(0, y_0)$ of our desired solution. In the example above, $f(x, y) = e^{-x} - y^2$ and $y_0 = 2$.

Before trying to compute a numerical solution, we first should ask the question, "Does the equation have a solution through the specified point?" If it does, it makes sense to compute the solution; if it does not, we may waste a lot of expensive computer time.

Fortunately, it is easy to state usable conditions under which we know there is a solution.

Existence Theorem Let $f(x, y)$ be a continuous function and suppose that $\partial f/\partial y$ is also continuous in some open two-dimensional interval of the xy plane that contains (x_0, y_0). Then there is a unique solution to $y' = f(x, y)$ through (x_0, y_0) and defined for all x near x_0.

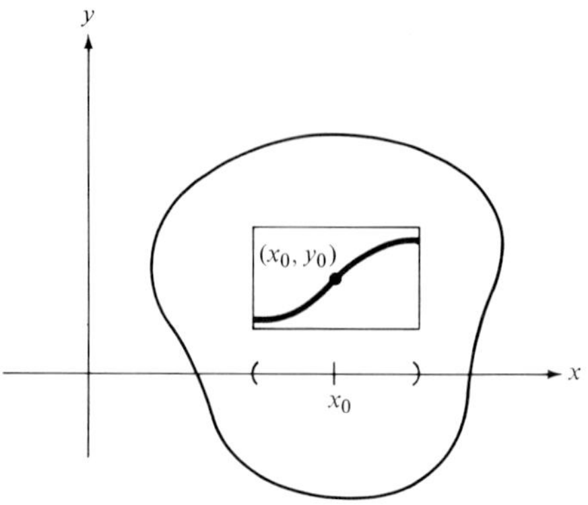

FIGURE 9-4

A graphical interpretation of this theorem is illustrated in Figure 9-4.

The function $f(x, y) = e^{-x} - y^2$ is continuous for all points of the xy plane and $\partial f/\partial y = -2y$ is also. For any point in the plane, specifically the point (0, 2), there will be a unique solution through that point and whose graph is defined for all x near $x = 0$.

A solution of the differential equation would give y as a function of x, and the function might be defined for all x, or merely for all x near $x = 0$. In most problems of practical interest, we are concerned with a solution only over a finite interval, say, the interval $[0, x_F]$. In Figure 9-5, we illustrate the solution of a differential equation over such an interval.

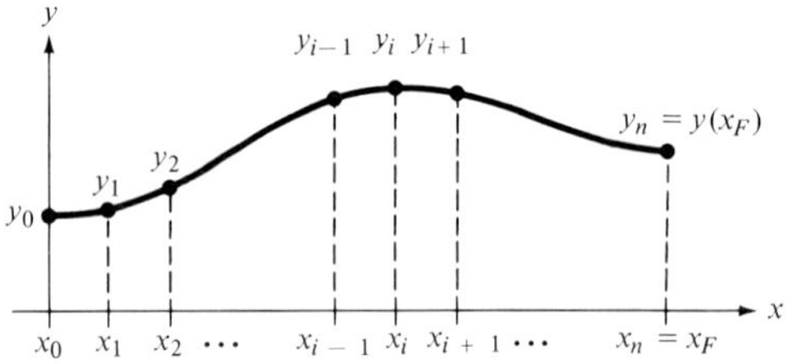

FIGURE 9-5

When using a computer, we do not expect to obtain a solution in the form of y as a function of x defined in terms of known functions. Instead we only approximate the solution at a finite number of points. These points may be obtained by dividing the interval $[0, x_F]$ into some number, say n, of subintervals of equal size. With n subintervals of equal length, the length of each subinterval is x_F/n. We denote this length by h. For convenience, let $x_0 = 0$ be the left-hand end point of the first subinterval, and x_1 be the right-hand end point of the first subinterval. Using this notation, the right-hand end point of subinterval i is x_i. These are illustrated in Figure 9-5.

We designate $f(x_i)$ as y_i. Our objective is to develop a process by which the computer can be used to calculate $y_1, y_2, \ldots, y_i, \ldots, y_n$. As we shall see, we cannot in general obtain points exactly on the graph of an explicit solution for y as a function of x, but we can obtain points as close to the graph of the explicit solution as we like.

We note that the value of y_0 is given by the initial condition:

$$y_0 = y(0)$$

We also note that we know $x_0, x_1, \ldots, x_i, \ldots, x_n$; in our example, $x_i = i \cdot h$ for each i from 0 to n. Let us now consider how we might calculate y_1.

Since we know x_0 and y_0, we can evaluate $f(x, y)$ at (x_0, y_0), i.e., we can calculate $f(x_0, y_0)$. Since the differential equation is of the form $y' = f(x, y)$, we then know the value of the derivative y' at x_0. It is $f(x_0, y_0)$. We shall designate it as y'_0:

$$y'_0 = f(x_0, y_0)$$

In other words, we know the slope of the line tangent to the graph of the solution at the point (x_0, y_0).

We shall note that the derivative y'_0 may be approximated by the equation

$$y'_0 \approx \frac{y_1 - y_0}{h}$$

From the definition,

$$y'_0 = y'\Big|_{x=x_0} = \lim_{h \to 0} \frac{y(x_0 + h) - y(x_0)}{h} = \lim_{h \to 0} \frac{y_1 - y_0}{h}$$

For any small value of h, we get the approximate equation above. Solving for y_1, we can write the equation as

$$y_1 \approx y_0 + hy'_0$$

or since $y'_0 = f(x_0, y_0)$, we have, on substituting $f(x_0, y_0)$ for y'_0,

$$y_1 \approx y_0 + h \cdot f(x_0, y_0)$$

This equation gives us an approximate value of y_1 based on the known quantities x_0 and y_0 and chosen value h.

Using the differential equation $y' = e^{-x} - y^2$ with initial condition $y(0) = 2$, we would write

$$y_1 \approx y_0 + h \cdot (e^{-x_0} - y_0^2)$$

where $y_0 = 2$ and $x_0 = 0$. For $h = 0.1$, we have

$$y_1 \approx 2 + 0.1 \cdot (e^{-0} - 2^2) = 2 - 0.3 = 1.7$$

Therefore, we have $y_1 = y(0.1) \approx 1.7$.

Once we have a value for y_1, we can calculate an approximate value for y_2 by the same approach. Identical reasoning gives the equation

$$y_2 \approx y_1 + h \cdot f(x_1, y_1)$$

For $h = 0.1$, $x_1 = 0.1$, and $y_1 = 1.7$, as calculated above, we have $y_2 \approx y_1 + h \cdot f(x_1, y_1) = 1.7 + (0.1) \cdot (e^{-0.1} - (1.7)^2) \approx 1.50$.

We can generalize to obtain an equation relating y_i to y_{i-1} and x_{i-1}:

$$y_i \approx y_{i-1} + h \cdot f(x_{i-1}, y_{i-1})$$

This equation is actually a recursion relationship in which we interpret "$\approx$" as "$=$."

The flowchart in Figure 9-6 illustrates the computational sequence to be used in giving a numerical solution to a differential equation of the type

$$y' = f(x, y)$$

$$y(0) = y_0$$

The input block accepts a numerical value for n (the number of subintervals), x_F (the right end point of the interval over which the solution is to be obtained), and y_0 (the initial condition). We then calculate $h = x_F/n$, assign 0 to x_0, and assign 1 to i. We then enter a loop to calculate successive values for x_i and y_i. In this loop we first calculate y'_{i-1} as $f(x_{i-1}, y_{i-1})$, and then calculate $y_i \leftarrow y_{i-1} + h \cdot y'_{i-1}$ and $x_i \leftarrow x_{i-1} + h$. We finally add 1 to i, and compare i to n to see if another iteration of the loop is necessary.

For a given differential equation, the function $f(x, y)$ in the recursion relationship $y_i = y_{i-1} + h \cdot f(x_{i-1}, y_{i-1})$ can be replaced by a formula involving x_{i-1} and y_{i-1}. For example, consider the differential equation

$$y' + y^2 = e^{-x}$$

for which $f(x, y) = e^{-x} - y^2$. Therefore, $f(x_{i-1}, y_{i-1})$ is given by

$$f(x_{i-1}, y_{i-1}) = e^{-x_{i-1}} - (y_{i-1})^2$$

The equation for y_i is

$$y_i = y_{i-1} + h \cdot e^{-x_{i-1}} - h \cdot (y_{i-1})^2$$

This could be easily incorporated into the flowchart in Figure 9-6.

Of course, the smaller the value of h, the better the approximation $y'_{i-1} = (y_i - y_{i-1})/h$. However, to make h smaller, we must increase the number of subintervals (i.e., n is larger). Since we have more subintervals, more calculations are required to obtain the approximation to the solution over the interval $[0, x_F]$. As in all numerical procedures, the trade-off is effort for accuracy. You may obtain any degree of accuracy, provided that you are willing to expend sufficient effort.

How does one obtain an estimate of the accuracy of the solution? In practice, the simplest way is to compute approximate solutions for two different values of h. If the difference between the two approximate solutions

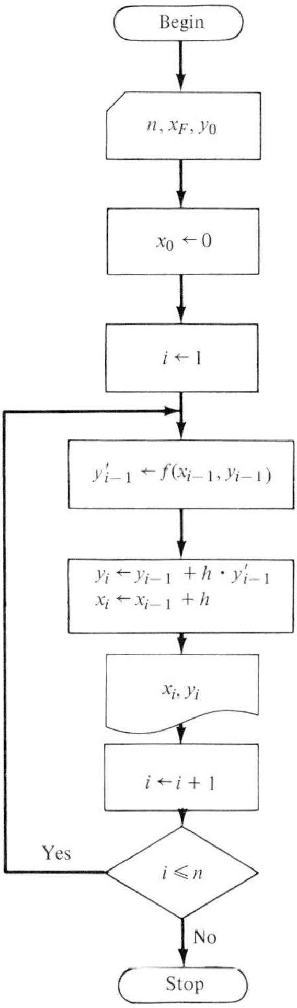

FIGURE 9-6

is significant in comparison to the accuracy with which the solution is desired, another approximation is computed, using an even smaller value of h.

The method we have presented is probably the simplest of all numerical techniques for solving differential equations. It goes by various names, including the Euler (pronounced "oiler") method and finite differences. The attractive feature of the method is its simplicity. There exist a large number of other techniques for solving differential equations, and most of these techniques are computationally more efficient than the Euler method. Especially in problems whose solution requires substantial computer time, consideration should be given to using one of these other methods.

EXERCISES 9-6

For the differential equations in Exercises 1–5, (**a**) write as $y' = f(x, y)$ and determine $\partial f/\partial y$, and (**b**) use the existence theorem to determine if a solution exists near the given point.

1. $y' + y = x^2 - 1 \quad (1, 5)$

2. $y' + y^2 = \dfrac{1}{x} \quad (3, 0)$

3. $y' + \dfrac{x}{y} = 0 \quad (-1, 1)$

4. $y' + \sqrt{y} = x^2 \quad (2, 3)$

5. $y' + y^2 = \sqrt{x} \quad (1, 9)$

For Exercises 6–10, (**a**) write as $y' = f(x, y)$; (**b**) write an expression for y'_0; (**c**) for $h = 0.1$, calculate an approximation for y_1; and (**d**) write an equation relating y_i to y_{i-1} and x_{i-1}.

6. $y' - \sqrt{y} = x \quad y(0) = 2$

7. $y' - \dfrac{x}{y} = 2 \quad y(0) = 1$

8. $y' - y^2 = \sqrt{x} \quad y(0) = 1$

9. $y' - x^2 y = e^{-x} + 1 \quad y(0) = 3$

10. $y' - y^4 = x^2 \quad y(0) = 1$

For Exercises 11–12, (**a**) divide the interval [0, 1] into 10 subintervals and calculate an approximation to the solution by the technique of the text, (**b**) solve the equation by one of the techniques presented earlier in this chapter, and (**c**) compare your answer for $y(1)$.

11. $y' + 2y = 0 \quad y(0) = 1$

12. $y' + y = 1 \quad y(0) = 0$

REVIEW EXERCISES: CHAPTER 9

For Exercises 1–10, (**a**) verify that the given function is a solution of the differential equation, and (**b**) find the value of C for which the graph of the function passes through the indicated point.

	Differential equation	Solution	Point
1.	$y' = \dfrac{1}{x}$	$y = \ln(Cx)$	$(e, 4)$
2.	$y' = e^x + 1$	$y = e^x + x + C$	$(1, e)$
3.	$y' = y - x + 1$	$y = x + Ce^x$	$(1, 1 + e)$

4. $y' = x^2 - 2y$ $\quad y = Ce^{-2x} + \frac{1}{2}x^2 - \frac{1}{2}x + \frac{1}{4}$ $\quad (0, 1)$

5. $y' = e^x - 3y$ $\quad y = Ce^{-3x} + \frac{1}{4}e^x$ $\quad (0, 4)$

6. $y' = -\dfrac{y}{2x}$ $\quad y = \dfrac{C}{\sqrt{x}}$ $\quad (1, 4)$

7. $y' = \dfrac{2y}{x \ln x^2}$ $\quad y = C \ln (x^2)$ $\quad (e, 4)$

8. $y' = -\dfrac{y + 4}{x}$ $\quad y = \dfrac{C}{x} - 4$ $\quad (1, 0)$

9. $y' = \dfrac{y}{2\sqrt{x}}$ $\quad y = Ce^{\sqrt{x} - 2}$ $\quad (4, 3)$

10. $y' = 2x(y + 2)$ $\quad y = Ce^{x^2} - 2$ $\quad (0, 8)$

For Exercises 11–28, find a solution of the differential equation that passes through the indicated point.

11. $y' = (x + 1)y$ $\quad (-1, 2)$

12. $y' = \dfrac{y + 1}{x}$ $\quad (1, 0)$

13. $y' = \dfrac{y}{x - 1}$ $\quad (1 + e, e^2)$

14. $y' = ye^x$ $\quad (1, e)$

15. $y' = \dfrac{e^x}{y}$ $\quad (0, \sqrt{2})$

16. $y' = \dfrac{x}{y^2 + 1}$ $\quad (4, 3)$

17. $y' - 2y = 0$ $\quad (0, 4)$

18. $y' - 2xy = 0$ $\quad (1, 4)$

19. $y' - e^{-x}y = 0$ $\quad (0, e^2)$

20. $xy' - y = 0$ $\quad \left(1, \dfrac{1}{\sqrt{e}}\right)$

21. $y' + 2y = 3x + 1$ $\quad (0, 0)$

22. $y' - y = x^2 + 3$ $\quad (1, 1)$

23. $y' - 3y = 2e^x + 3$ $\quad (1, -1)$

24. $y' + 3y = 3e^x$ $\quad (0, 2)$

25. $y' - 2y = xe^x - 2e^x$ $\quad (-1, 0)$

26. $y' - y = xe^{2x} + 3e^{2x}$ $\quad (0, 0)$

27. $y' - 4y = 4x - e^x$ $\quad (0, 1)$

28. $y' + 4y = e^x + 3x$ $\quad (0, 0)$

APPENDIX: TABLES I-VI

TABLE I. VALUES OF $(1+i)^n$

Year	1%	2%	3%	4%	5%	6%	7%	8%	9%	10%
1	1.0100	1.0200	1.0300	1.0400	1.0500	1.0600	1.0700	1.0800	1.0900	1.1000
2	1.0201	1.0404	1.0609	1.0816	1.1025	1.1236	1.1449	1.1664	1.1881	1.2100
3	1.0303	1.0612	1.0927	1.1249	1.1576	1.1910	1.2250	1.2597	1.2950	1.3310
4	1.0406	1.0824	1.1255	1.1699	1.2155	1.2625	1.3108	1.3605	1.4116	1.4641
5	1.0510	1.1041	1.1593	1.2167	1.2763	1.3382	1.4026	1.4693	1.5386	1.6105
6	1.0615	1.1262	1.1941	1.2653	1.3401	1.4185	1.5007	1.5869	1.6771	1.7716
7	1.0721	1.1487	1.2299	1.3159	1.4071	1.5036	1.6058	1.7138	1.8280	1.9487
8	1.0829	1.1717	1.2668	1.3686	1.4774	1.5938	1.7182	1.8509	1.9926	2.1436
9	1.0937	1.1951	1.3048	1.4233	1.5513	1.6895	1.8385	1.9990	2.1719	2.3579
10	1.1046	1.2190	1.3439	1.4802	1.6289	1.7908	1.9672	2.1589	2.3674	2.5937
11	1.1157	1.2434	1.3842	1.5395	1.7103	1.8983	2.1049	2.3316	2.5804	2.8531
12	1.1268	1.2682	1.4258	1.6010	1.7959	2.0122	2.2522	2.5182	2.8127	3.1384
13	1.1381	1.2936	1.4685	1.6651	1.8856	2.1329	2.4098	2.7196	3.0658	3.4523
14	1.1495	1.3195	1.5126	1.7317	1.9799	2.2609	2.5785	2.9372	3.3417	3.7975
15	1.1610	1.3459	1.5580	1.8009	2.0789	2.3966	2.7590	3.1722	3.6425	4.1772
16	1.1726	1.3728	1.6047	1.8730	2.1829	2.5404	2.9522	3.4259	3.9703	4.5950
17	1.1843	1.4002	1.6528	1.9479	2.2920	2.6928	3.1588	3.7000	4.3276	5.0545
18	1.1961	1.4282	1.7024	2.0258	2.4066	2.8543	3.3799	3.9960	4.7171	5.5599
19	1.2081	1.4568	1.7535	2.1068	2.5270	3.0256	3.6165	4.3157	5.1417	6.1159
20	1.2202	1.4859	1.8061	2.1911	2.6533	3.2071	3.8697	4.6609	5.6044	6.7275

The above table permits values of $(1+i)^n$ to be read directly for $n \leq 20$. For $n > 20$, we can proceed as illustrated by the following example:

$$\begin{aligned}(1+0.04)^{25} &= (1+0.04)^{20} \cdot (1+0.04)^5 \\ &= 2.1911 \cdot 1.2167 \\ &= 2.6659\end{aligned}$$

where $(1+0.04)^{20}$ and $(1+0.04)^5$ can be read directly from the table.

Unfortunately, there is no convenient way to use the table for values of i greater than 10%.

TABLE II. VALUES OF $(1 - i)^n$

Year	1%	2%	3%	4%	5%	6%	7%	8%	9%	10%
1	0.9900	0.9800	0.9700	0.9600	0.9500	0.9400	0.9300	0.9200	0.9100	0.9000
2	0.9801	0.9604	0.9409	0.9216	0.9025	0.8836	0.8649	0.8464	0.8281	0.8100
3	0.9703	0.9412	0.9127	0.8847	0.8574	0.8306	0.8044	0.7787	0.7536	0.7290
4	0.9606	0.9224	0.8853	0.8493	0.8145	0.7807	0.7481	0.7164	0.6857	0.6561
5	0.9510	0.9039	0.8587	0.8154	0.7738	0.7339	0.6957	0.6591	0.6240	0.5905
6	0.9415	0.8858	0.8330	0.7828	0.7351	0.6899	0.6470	0.6064	0.5679	0.5314
7	0.9321	0.8681	0.8080	0.7514	0.6983	0.6485	0.6017	0.5578	0.5168	0.4783
8	0.9227	0.8508	0.7837	0.7214	0.6634	0.6096	0.5596	0.5132	0.4703	0.4305
9	0.9135	0.8337	0.7602	0.6925	0.6302	0.5730	0.5204	0.4722	0.4279	0.3874
10	0.9044	0.8171	0.7374	0.6648	0.5987	0.5386	0.4840	0.4344	0.3894	0.3487
11	0.8953	0.8007	0.7153	0.6382	0.5688	0.5063	0.4501	0.3996	0.3544	0.3138
12	0.8864	0.7847	0.6938	0.6127	0.5404	0.4759	0.4186	0.3677	0.3225	0.2824
13	0.8775	0.7690	0.6730	0.5882	0.5133	0.4474	0.3893	0.3383	0.2935	0.2542
14	0.8687	0.7536	0.6528	0.5647	0.4877	0.4205	0.3620	0.3112	0.2670	0.2288
15	0.8601	0.7386	0.6333	0.5421	0.4633	0.3953	0.3367	0.2863	0.2430	0.2059
16	0.8515	0.7238	0.6143	0.5204	0.4401	0.3716	0.3131	0.2634	0.2211	0.1853
17	0.8429	0.7093	0.5958	0.4996	0.4181	0.3493	0.2912	0.2423	0.2012	0.1668
18	0.8345	0.6951	0.5780	0.4796	0.3972	0.3283	0.2708	0.2229	0.1831	0.1501
19	0.8262	0.6812	0.5606	0.4604	0.3774	0.3086	0.2519	0.2051	0.1666	0.1351
20	0.8179	0.6676	0.5438	0.4420	0.3585	0.2901	0.2342	0.1887	0.1516	0.1216

The above table permits values of $(1 - i)^n$ to be read directly for $n \leq 20$. For $n > 20$, we can proceed as illustrated by the following example:

$$\begin{aligned}(1 - 0.04)^{25} &= (1 - 0.04)^{20} \cdot (1 - 0.04)^{5} \\ &= 0.4420 \cdot 0.8154 \\ &= 0.3604\end{aligned}$$

where $(1 - 0.04)^{20}$ and $(1 - 0.04)^{5}$ can be read directly from the above table.

Unfortunately, there is no easy way to use the table for values of i greater than 10%.

TABLE III. VALUES OF e^x FOR SMALL VALUES OF x

x	0.00	0.01	0.02	0.03	0.04	0.05	0.06	0.07	0.08	0.09
0.0	1.0000	1.0101	1.0202	1.0305	1.0408	1.0513	1.0618	1.0725	1.0833	1.0942
0.1	1.1052	1.1163	1.1275	1.1388	1.1503	1.1618	1.1735	1.1853	1.1972	1.2092

TABLE IV. VALUES OF e^x FOR LARGER VALUES OF x

x	.0	.1	.2	.3	.4	.5	.6	.7	.8	.9
.0	1.0000	1.1052	1.2214	1.3499	1.4918	1.6487	1.8221	2.0138	2.2255	2.4596
1.0	2.7183	3.0042	3.3201	3.6693	4.0552	4.4817	4.9530	5.4739	6.0496	6.6859
2.0	7.3891	8.1662	9.0250	9.9742	11.0232	12.1825	13.4637	14.8797	16.4446	18.1741
3.0	20.0855	22.1980	24.5325	27.1126	29.9641	33.1155	36.5982	40.4473	44.7012	49.4024
4.0	54.5982	60.3403	66.6862	73.6998	81.4509	90.0171	99.4843	109.9471	121.5104	134.2898
5.0	148.4132	164.0219	181.2722	200.3368	221.4064	244.6919	270.4264	298.8674	330.2996	365.0375

The above table permits e^x to be read directly for $0 \leq x \leq 5.9$. For values of x outside this range, we proceed as follows:

For $x < 0$, we use the property that $e^{-x} = 1/e^x$. For example,

$$e^{-1.6} = \frac{1}{e^{1.6}} = \frac{1}{4.9530} = 0.2019$$

where $e^{1.6}$ can be read directly from the table.

For $x > 5.9$, we use the property that $e^{a+b} = e^a \cdot e^b$. To do this, we must express x as a sum of two numbers whose values are each less than 5.9. For example,

$$e^{7.2} = e^5 \cdot e^{2.2} = 148.4132 \cdot 9.0250 = 1339.4$$

where e^5 and $e^{2.2}$ can be read directly from the table. For large values of x, more than two terms may be required. For example, $e^{17.8} = e^{(5+5+5+2.8)} = e^5 \cdot e^5 \cdot e^5 \cdot e^{2.8}$ where e^5 and $e^{2.8}$ can be read from the table.

Actually, there are several ways to use our short table for larger values of x, and the above approach is only one possibility.

TABLE V. VALUES OF $\ln x$ FOR VALUES OF x NEAR ONE

x	0.00	0.01	0.02	0.03	0.04	0.05	0.06	0.07	0.08	0.09
1.0	0.0000	0.0100	0.0198	0.0296	0.0392	0.0488	0.0583	0.0677	0.0770	0.0862
1.1	0.0953	0.1044	0.1133	0.1222	0.1310	0.1398	0.1484	0.1570	0.1655	0.1740

TABLE VI. VALUES OF ln x

x	.0	.1	.2	.3	.4	.5	.6	.7	.8	.9
1.0	.0000	.0953	.1823	.2624	.3365	.4055	.4700	.5306	.5878	.6419
2.0	.6931	.7419	.7885	.8329	.8755	.9163	.9555	.9932	1.0296	1.0647
3.0	1.0986	1.1314	1.1632	1.1939	1.2238	1.2528	1.2809	1.3083	1.3350	1.3610
4.0	1.3863	1.4110	1.4351	1.4586	1.4816	1.5041	1.5261	1.5476	1.5686	1.5892
5.0	1.6094	1.6292	1.6487	1.6677	1.6864	1.7047	1.7228	1.7405	1.7579	1.7750
6.0	1.7918	1.8083	1.8245	1.8405	1.8563	1.8718	1.8871	1.9021	1.9169	1.9315
7.0	1.9459	1.9601	1.9741	1.9879	2.0015	2.0149	2.0281	2.0412	2.0541	2.0669
8.0	2.0794	2.0919	2.1041	2.1163	2.1282	2.1401	2.1518	2.1633	2.1748	2.1861
9.0	2.1972	2.2083	2.2192	2.2300	2.2407	2.2513	2.2618	2.2721	2.2824	2.2925
10.0	2.3026	2.3125	2.3224	2.3321	2.3418	2.3514	2.3609	2.3702	2.3795	2.3888

The above table permits ln x to be read directly for $1 \leq x \leq 10.9$. For values of x outside this range, we proceed as follows:

For $0 < x < 1$, we use the property that $\ln x = -\ln(1/x)$. For example,

$$\ln 0.4 = -\ln \frac{1}{0.4} = -\ln 2.5 = -0.9163,$$

where ln 2.5 can be read directly from the table.

For $x > 10.9$, we first note that we can write x as $y \cdot 10^n$, where $1 \leq y \leq 10$. We then note that

$$\begin{aligned} \ln x = \ln(y \cdot 10^n) &= \ln y + \ln(10^n) \\ &= \ln y + n \ln(10) \\ &= \ln y + 2.3026n \end{aligned}$$

For example,

$$\begin{aligned} \ln(58) = \ln(5.8 \cdot 10) &= \ln(5.8) + \ln(10) \\ &= 1.7579 + 2.3026 \\ &= 4.0605 \end{aligned}$$

ANSWERS TO SELECTED EXERCISES

EXERCISES 1-1

1. 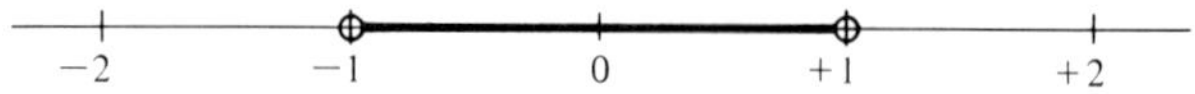

3. 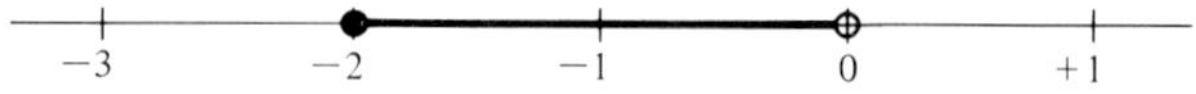

5. 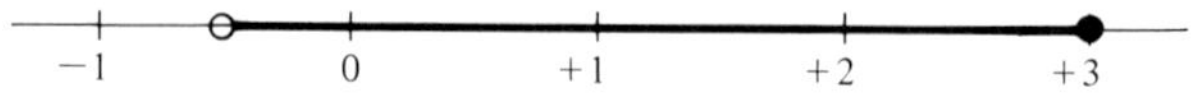

7. $-1 < x < 1$ **9.** $-2 \le x < 0$ **11.** $-\frac{1}{2} < x \le 3$

13. (**a**) 1
(**b**) $y - 2 = 1 \cdot (x - 1)$
(**c**) $y - 4 = 1 \cdot (x - 3)$
(**d**) $y = x + 1$
(**e**) 5
(**f**) -5
(**g**)

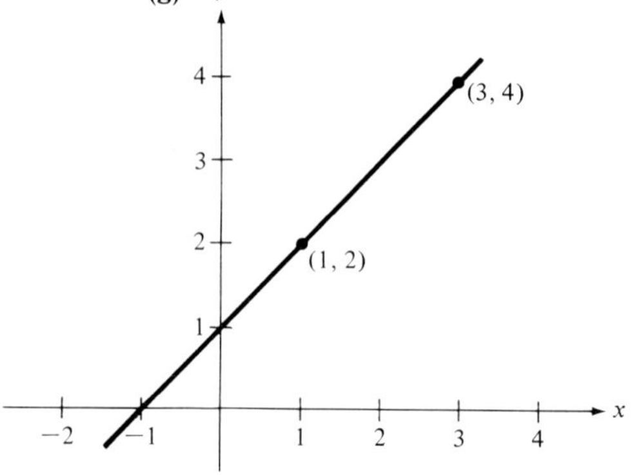

15. (**a**) 1
(**b**) $y - 5 = 1 \cdot (x - 2)$
(**c**) $y - 4 = 1 \cdot (x - 1)$
(**d**) $y = x + 3$
(**e**) 7
(**f**) -7
(**g**)

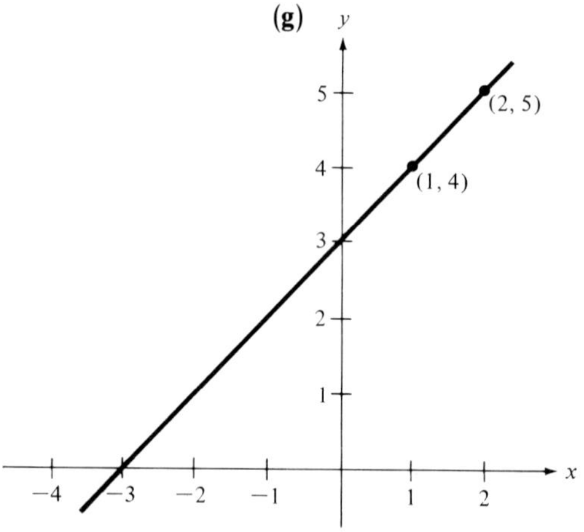

17. (**a**) $-\frac{3}{2}$
(**b**) $y - 2 = -\frac{3}{2} \cdot (x + 4)$
(**c**) $y + 1 = -\frac{3}{2} \cdot (x + 2)$
(**d**) $y = -\frac{3}{2}x - 4$
(**e**) -10
(**f**) 0

(g)

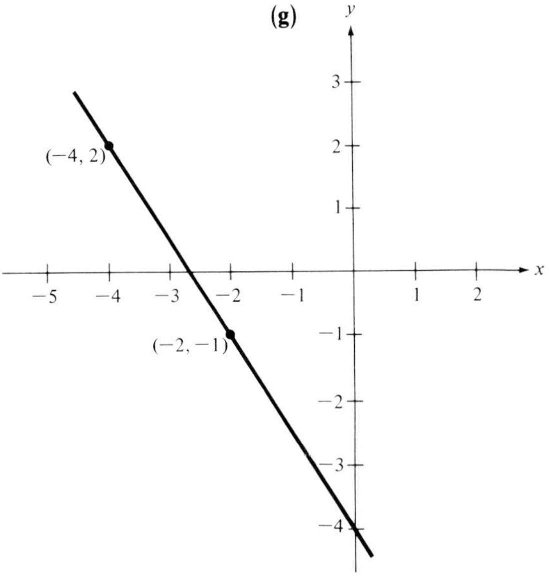

19. **(a)** Undefined
(b) $x = 1$
(c) $x = 1$
(d) Not possible
(e) No such point
(f) 1

(g)

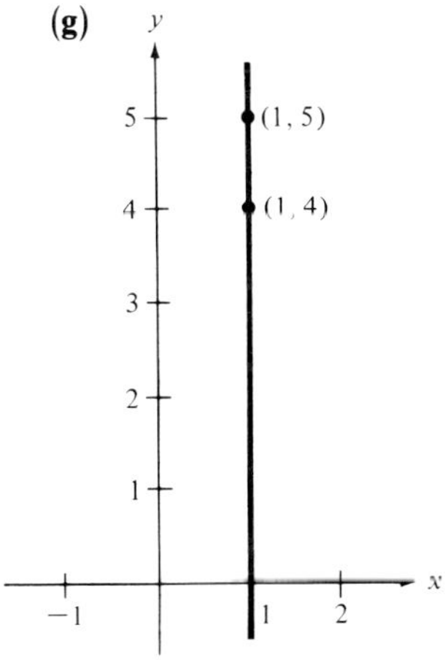

21. **(a)** 0
(b) $y + 1 = 0$
(c) $y + 1 = 0$
(d) $y = -1$
(e) -1
(f) No such point

(g)

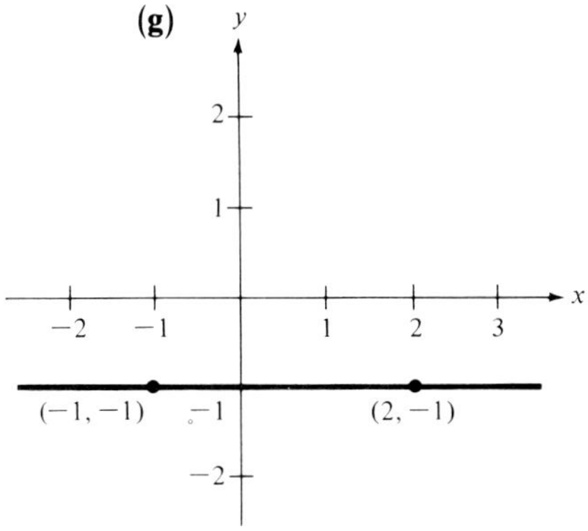

23. **(a)** $y - 2 = 2 \cdot (x + 1)$
(b) $y = 2x + 4$
(c) 20
(d) -6

(e)

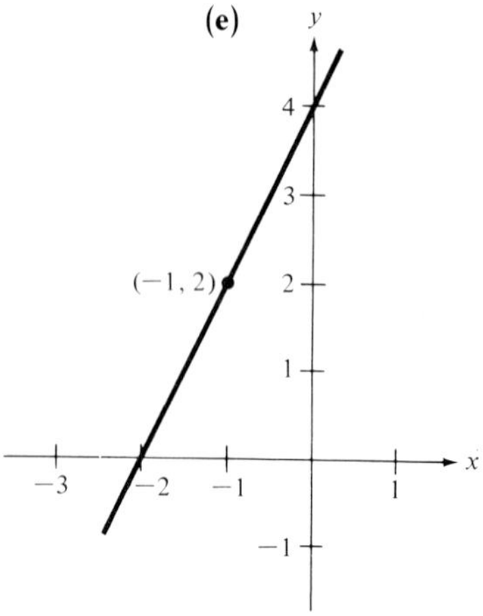

25. **(a)** $y + 1 = -\frac{1}{2} \cdot (x + 2)$
(b) $y = -\frac{1}{2}x - 2$
(c) -6
(d) 12

(e)

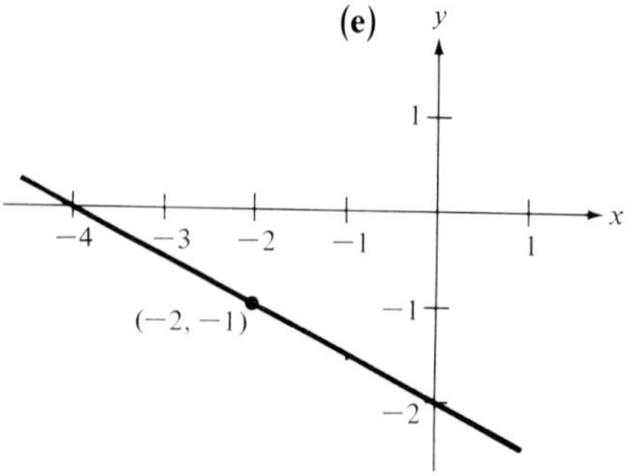

27. **(a)** $y - 2 = 0$
(b) $y = 2$
(c) 2
(d) No such point

(e)

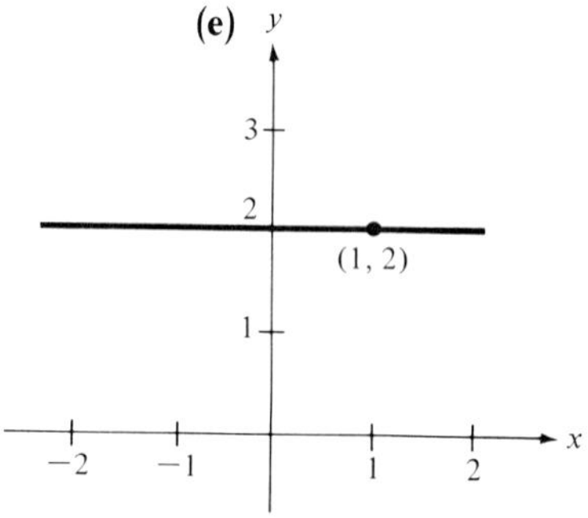

29. **(a)** $\frac{5}{4}$
(b) $y - 3 = \frac{5}{4}(x - 2)$
(c) $y = \frac{5}{4}x + \frac{1}{2}$

31. **(a)** $-\frac{4}{7}$
(b) $y - 0 = -\frac{4}{7} \cdot (x - 4)$
(c) $y = -\frac{4}{7}x + \frac{16}{7}$

33. **(a)** $-\frac{1}{3}$
(b) $y + 2 = -\frac{1}{3}(x - 3)$
(c) $y = -\frac{1}{3}x - 1$

EXERCISES 1-2

1. $\frac{1}{3}$ **3.** 3 **5.** 1 **7.** All x **9.** $x \neq 0$ **11.** $x \geq 3$

13. $x \geq 1$ and $x \leq -1$ **15.** 1, -1 **17.** -2

19.

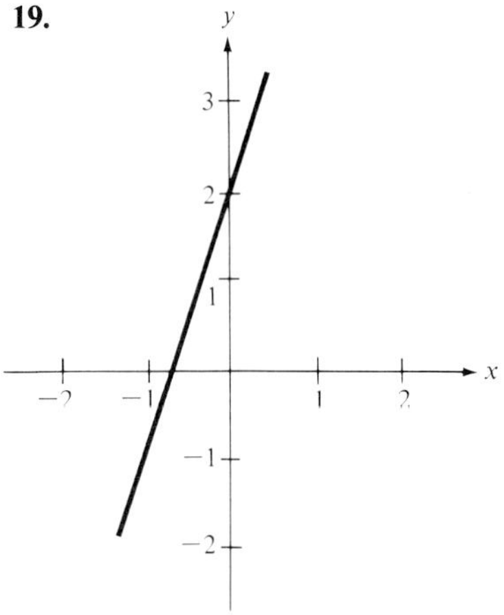

21.

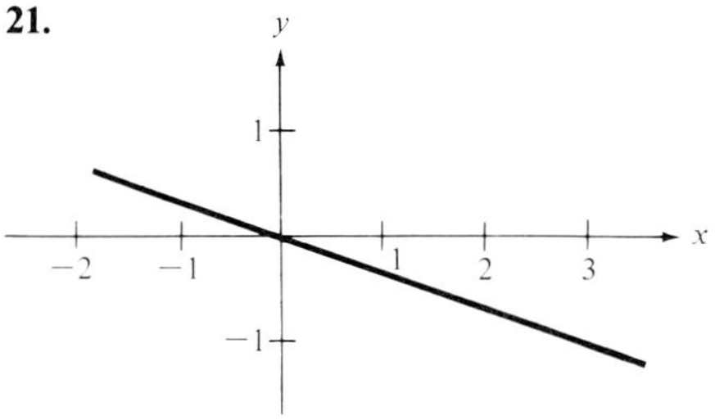

23.

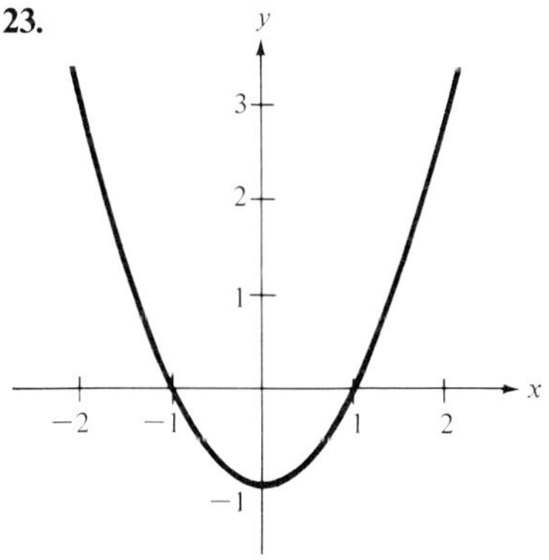

25.

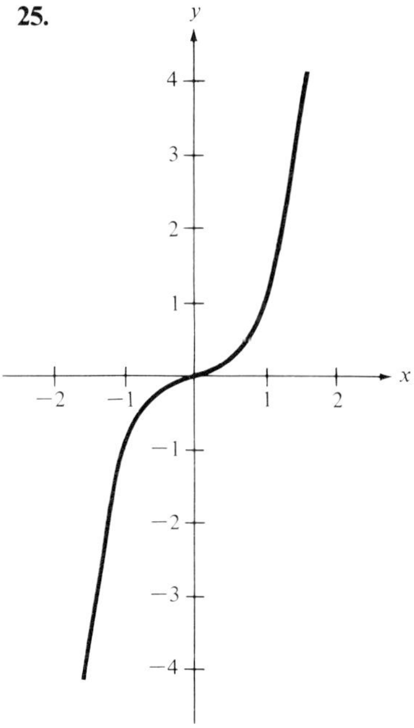

27.

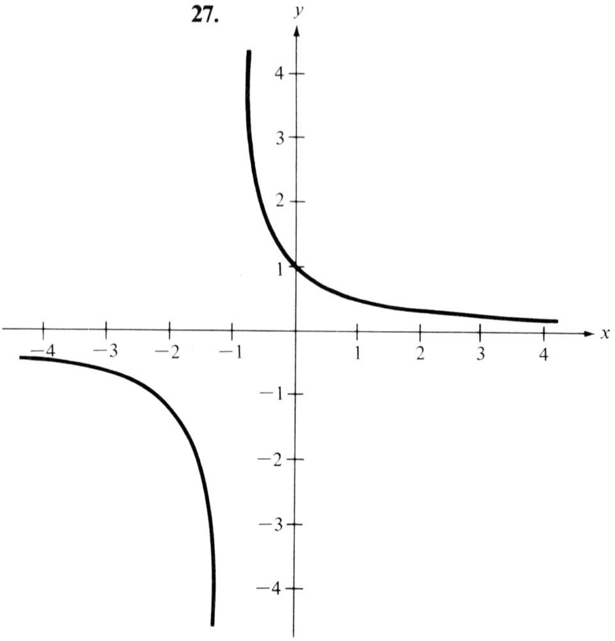

29.

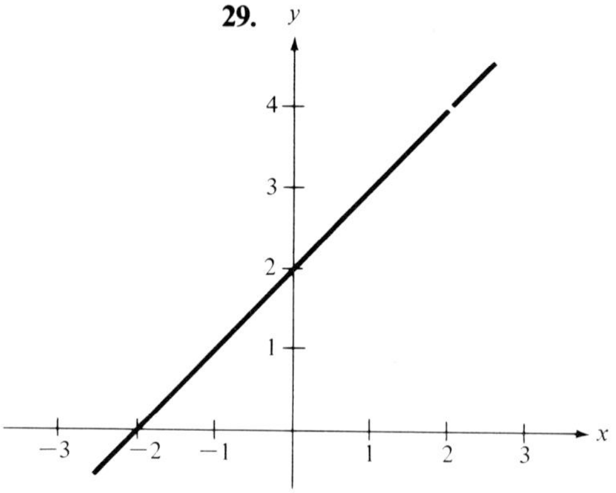

EXERCISES 1-3

1. **(a)** Point x_1 is to the left of point x_0
(b) Point x_1 is to the right of point x_0
(c) Point x_1 is identical to point x_0
(d) Yes

3. **(a)** c is not in interval (a, b)
(b) c is in interval (a, b)
(c) c is not in interval (a, b)
(d) c is not in interval (a, b)
(e) c is not in interval (a, b)
(f) Yes

5.

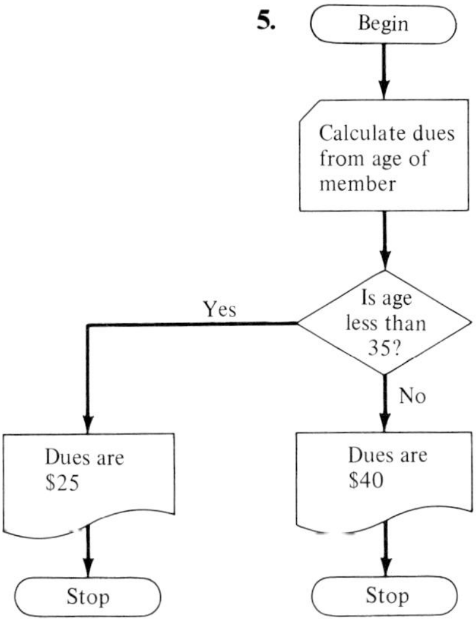

Begin
Calculate dues from age of member
Is age less than 35?
Yes
No
Dues are $25
Dues are $40
Stop
Stop

7.

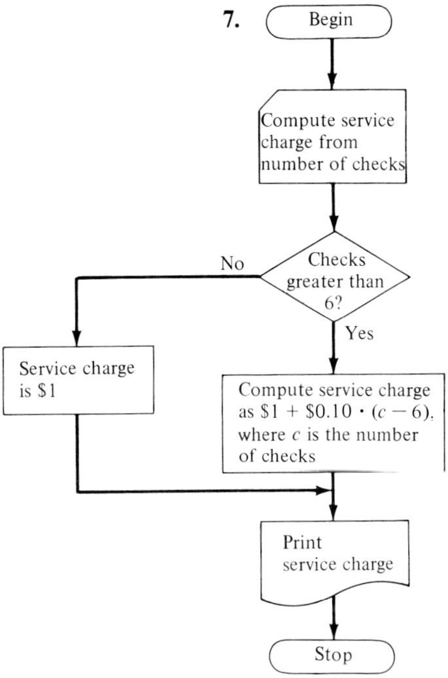

Begin
Compute service charge from number of checks
Checks greater than 6?
No
Yes
Service charge is $1
Compute service charge as $1 + $0.10 · (c − 6), where c is the number of checks
Print service charge
Stop

9.

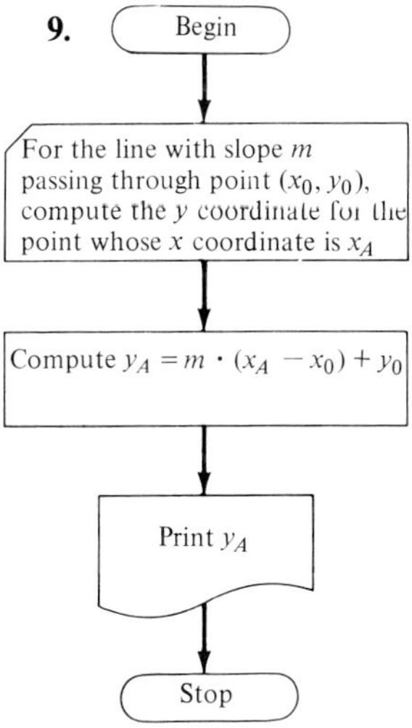

Begin
For the line with slope m passing through point (x0, y0), compute the y coordinate for the point whose x coordinate is xA
Compute yA = m · (xA − x0) + y0
Print yA
Stop

11.

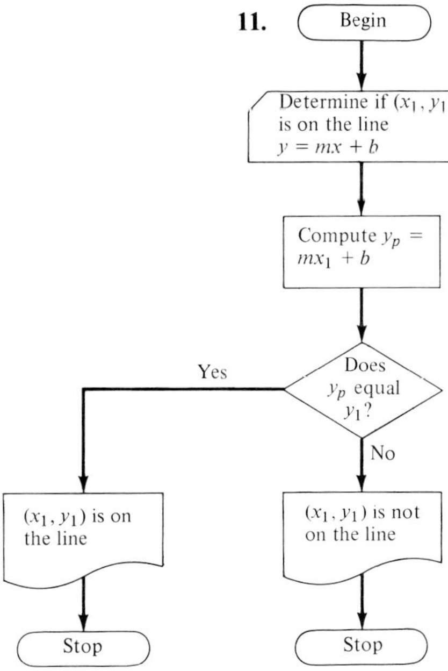

Begin
Determine if (x1, y1) is on the line y = mx + b
Compute yp = mx1 + b
Does yp equal y1?
Yes
No
(x1, y1) is on the line
(x1, y1) is not on the line
Stop
Stop

EXERCISES 1-4

1. 13 **3.** 2.5 **5.** 0.7 **7.** No **9.** Yes

11. (**a**) 1
(**b**) 6
(**c**) 12

13. (**a**) 2.8 is positive
(**b**) 0 is zero
(**c**) −18.1 is negative
(**d**) Yes

15.

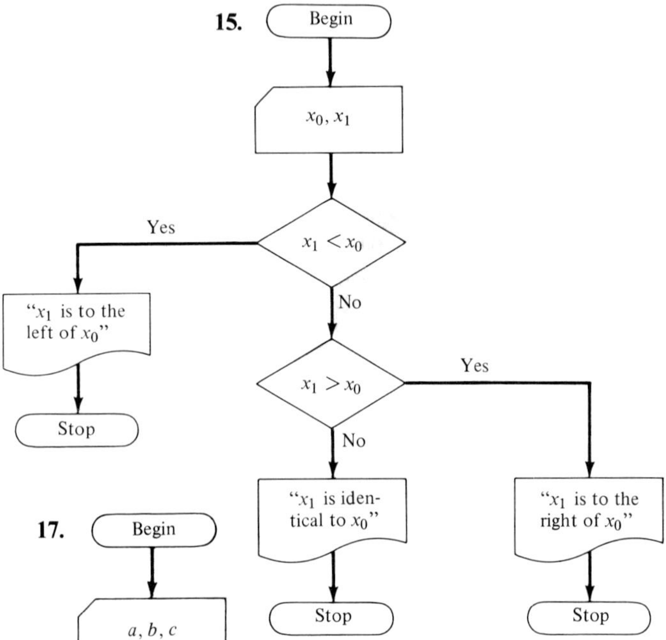

17.

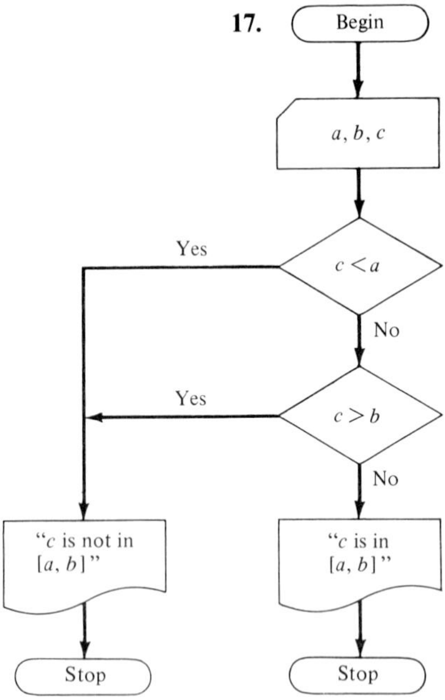

19.

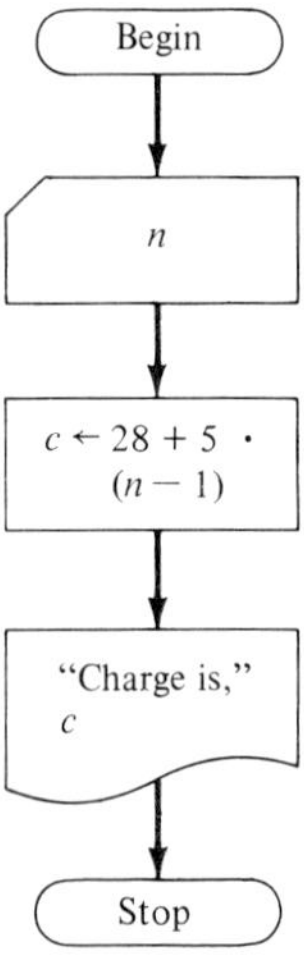

21.

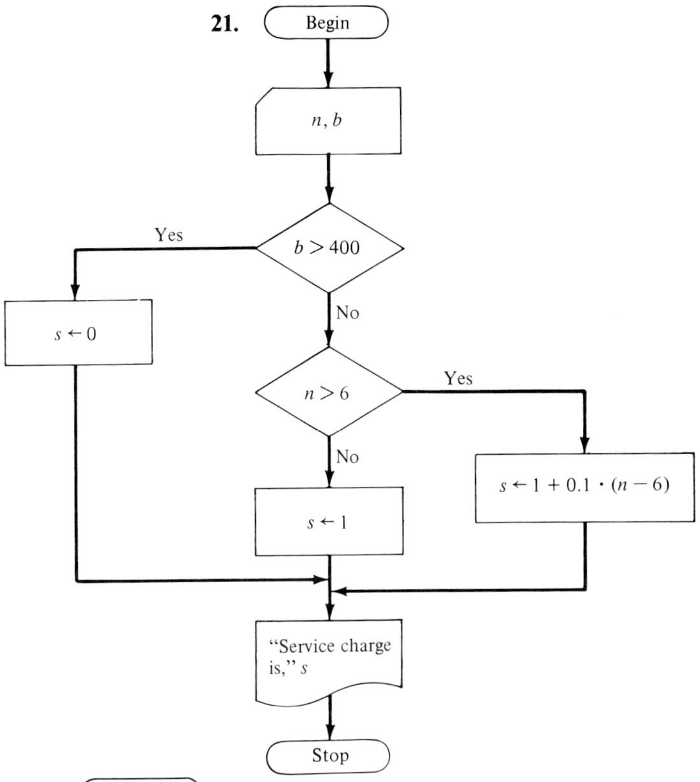
Begin
n, b
b > 400
Yes
No
s ← 0
n > 6
Yes
No
s ← 1 + 0.1 · (n − 6)
s ← 1
"Service charge is," s
Stop

23.

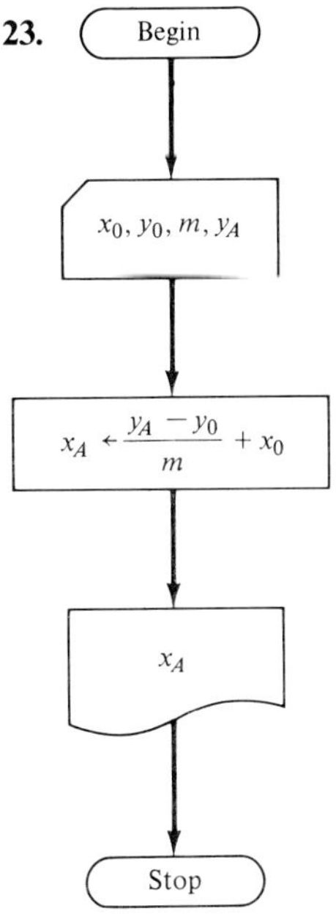
Begin
x0, y0, m, yA
xA ← (yA − y0)/m + x0
xA
Stop

25.

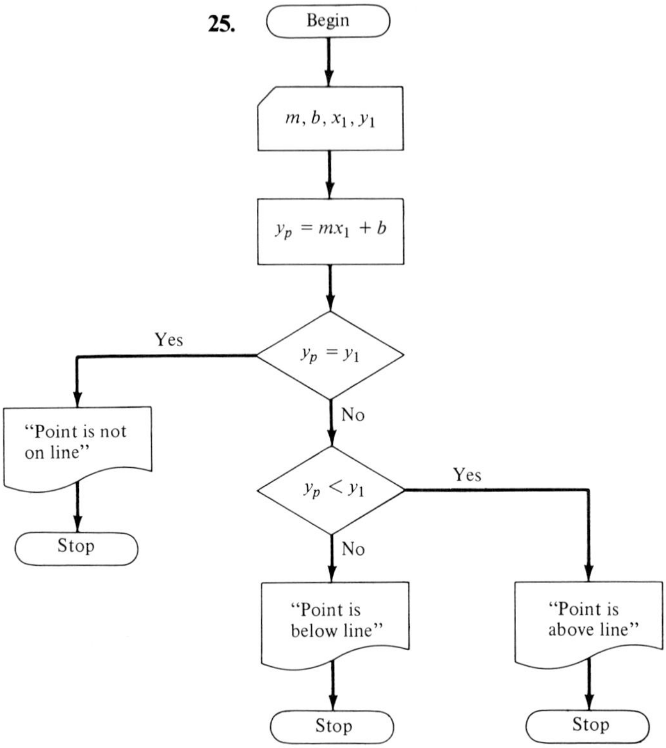

CHAPTER 1. REVIEW EXERCISES

1. $y - 5 = 2(x - 1)$ **3.** $y = 5$ **5.** $x = 1$ **7.** $y - 4 = -5(x + 1)$

9. $y + 2 = -(x + 3)$ **11.** All x **13.** All x **15.** All x

17. $\frac{1}{2}$, 1 **19.** 1

21.

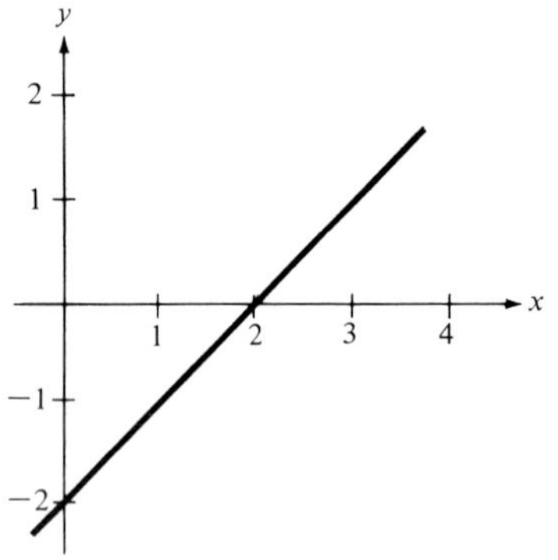

23.

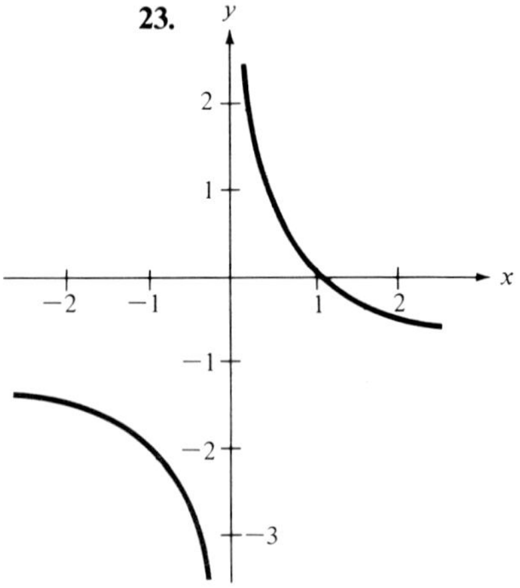

25.

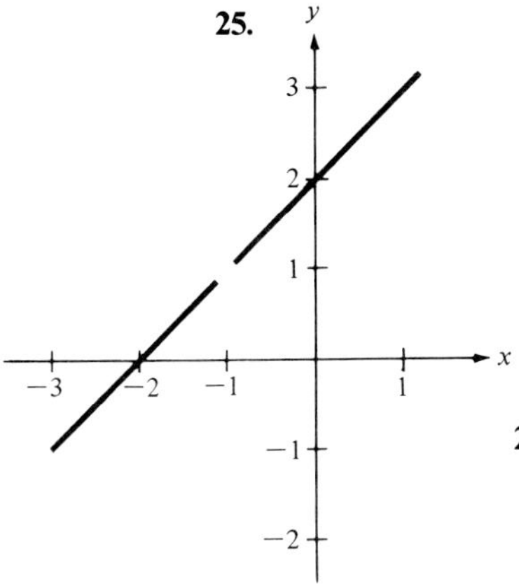

y
3
2
1
−1
−2
x
−3
−2
−1
1

−2

27.

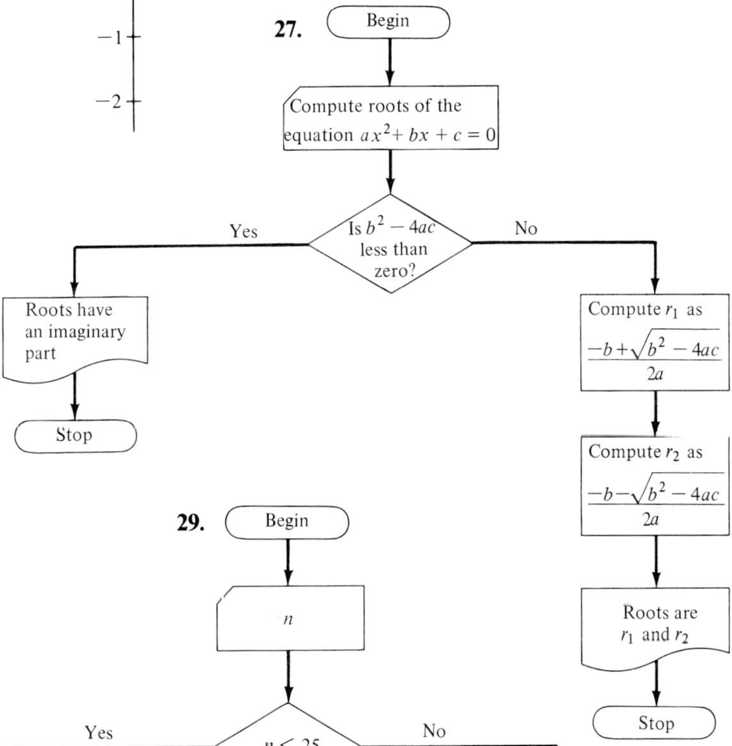

Begin
Compute roots of the equation $ax^2 + bx + c = 0$
Is $b^2 - 4ac$ less than zero?
Yes
No
Roots have an imaginary part
Stop
Compute r_1 as $\frac{-b + \sqrt{b^2 - 4ac}}{2a}$
Compute r_2 as $\frac{-b - \sqrt{b^2 - 4ac}}{2a}$
Roots are r_1 and r_2
Stop

29.

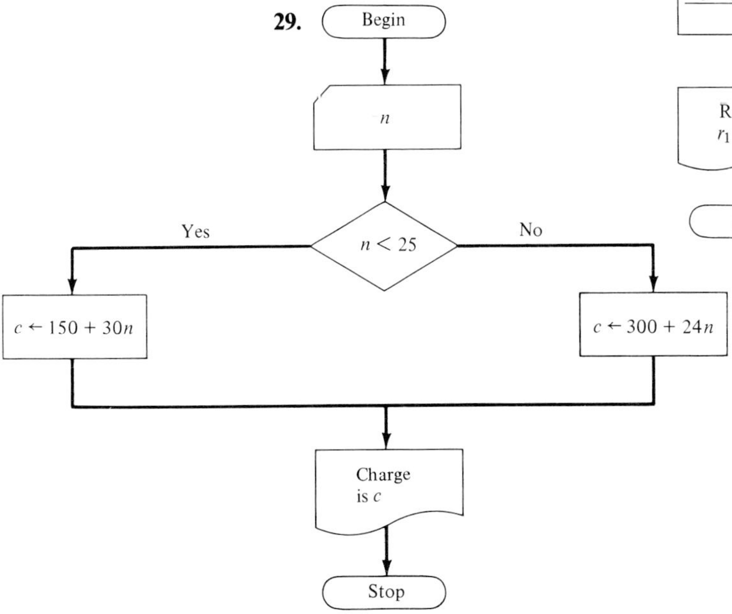

Begin
n
$n < 25$
Yes
No
$c \leftarrow 150 + 30n$
$c \leftarrow 300 + 24n$
Charge is c
Stop

31.

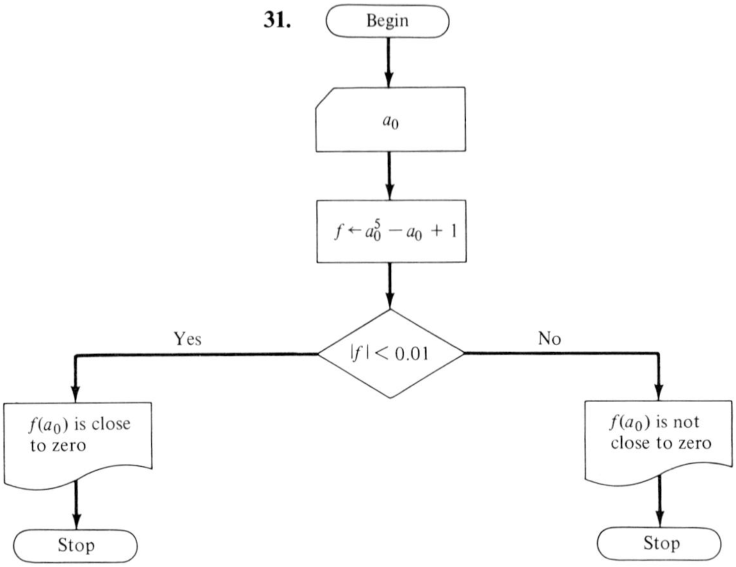

EXERCISES 2-1

1. **(a)** $\frac{1}{2}, \frac{1}{4}, \frac{1}{6}, \frac{1}{8}, \frac{1}{10}, \frac{1}{12}$

(b)

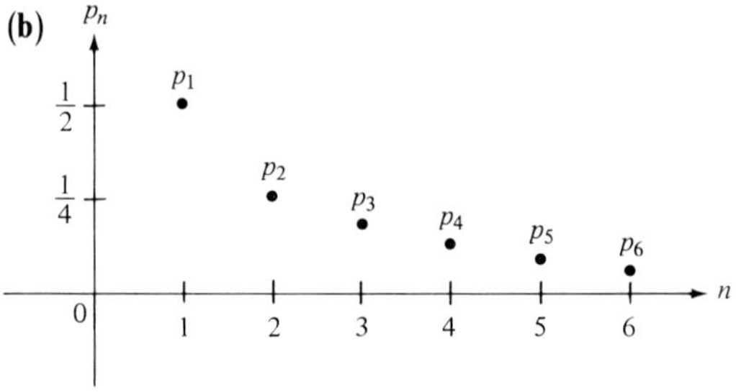

(c)

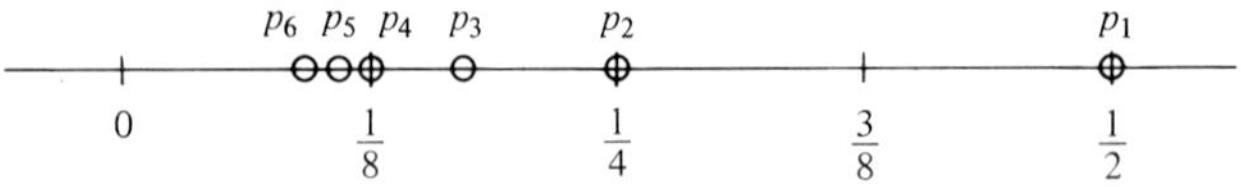

3. **(a)** 1, 4, 9, 16, 25, 36

(b)

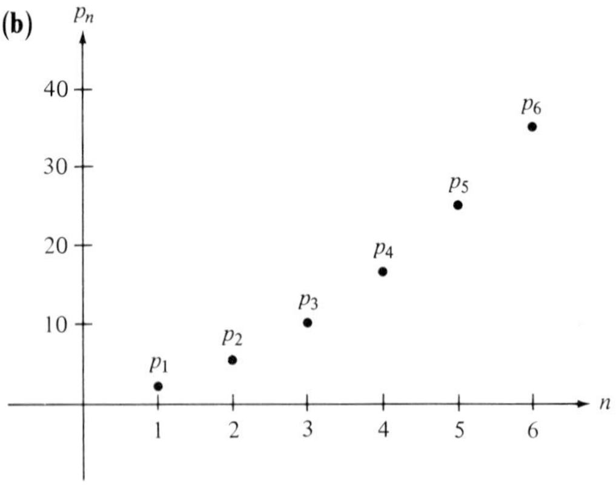

(c)

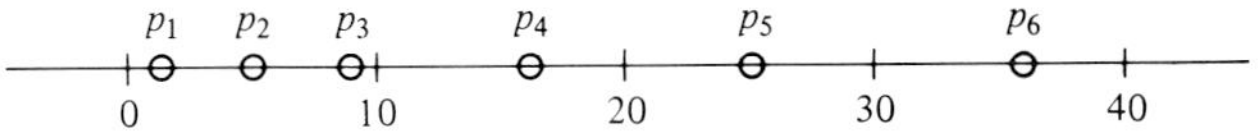

5. **(a)** $\frac{1}{2}, \frac{1}{4}, \frac{1}{8}, \frac{1}{16}, \frac{1}{32}, \frac{1}{64}$ **(b)**

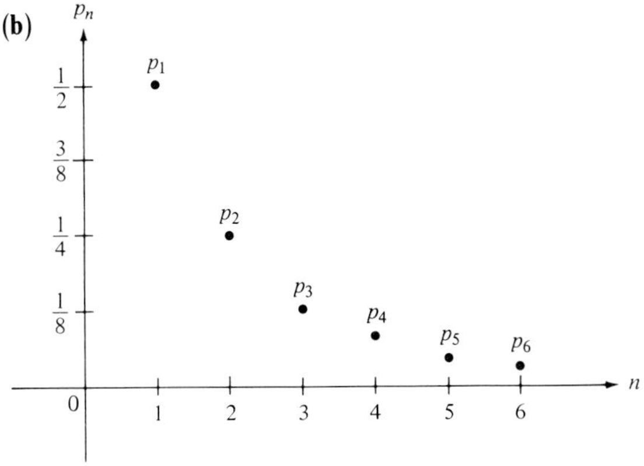

(c)

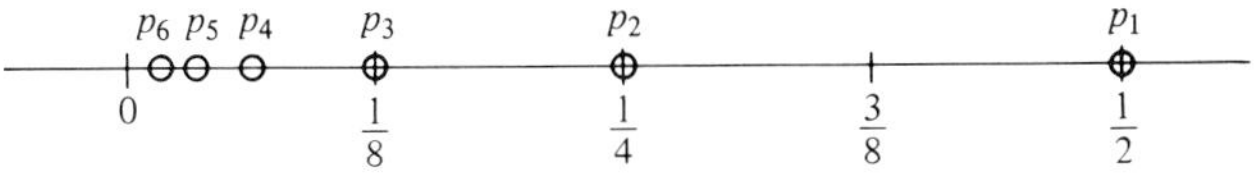

7. **(a)** $-\frac{1}{2}, \frac{1}{3}, -\frac{1}{4}, \frac{1}{5}, -\frac{1}{6}, \frac{1}{7}$ **(b)**

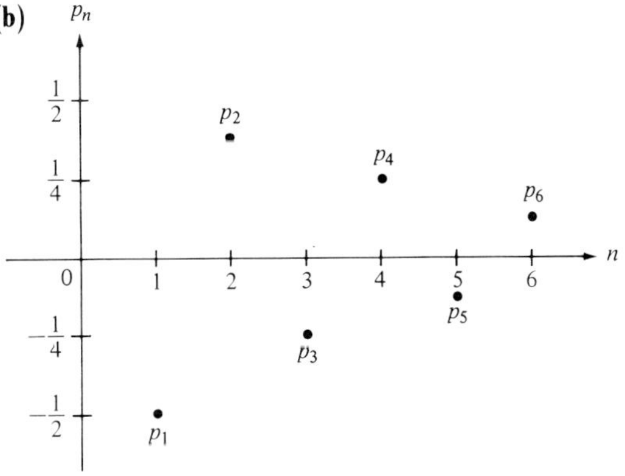

(c)

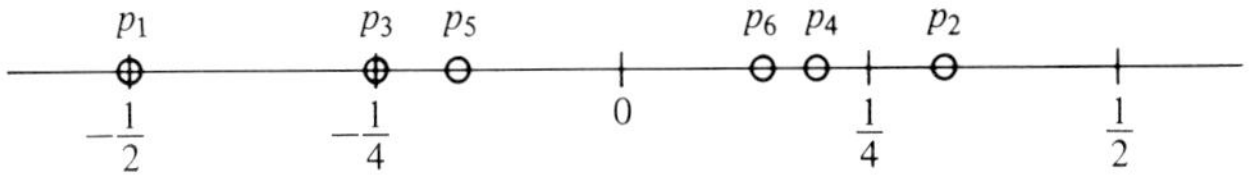

9. **(a)** $0, \frac{1}{4}, \frac{2}{9}, \frac{3}{16}, \frac{4}{25}, \frac{5}{36}$ **(b)**

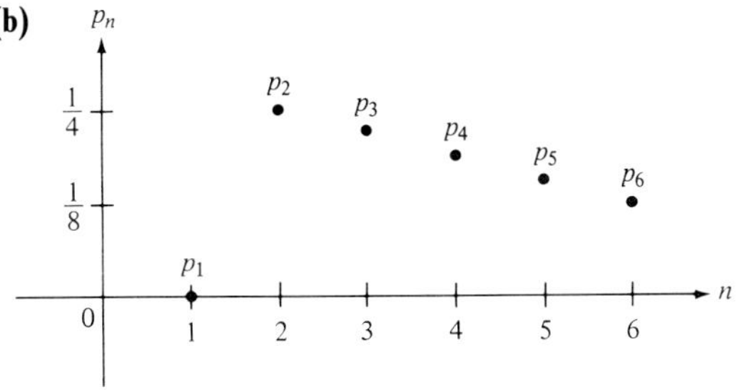

(c)

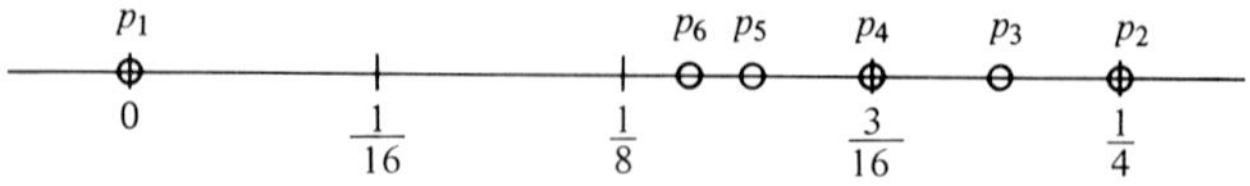

11. **(a)** $1, \frac{1}{3}, \frac{1}{5}, \frac{1}{7}, \frac{1}{9}, \frac{1}{11}$

(b)

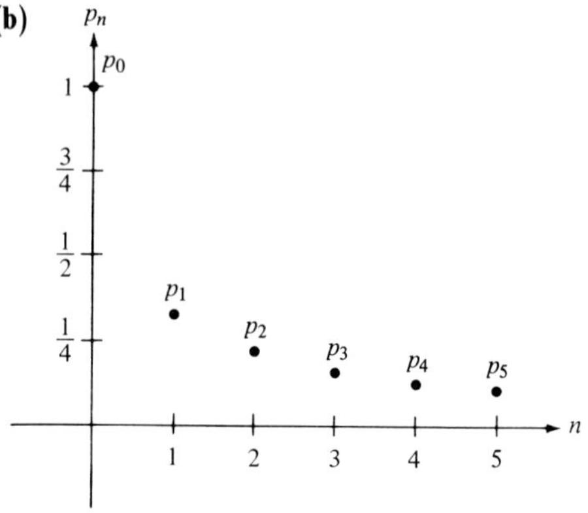

(c)

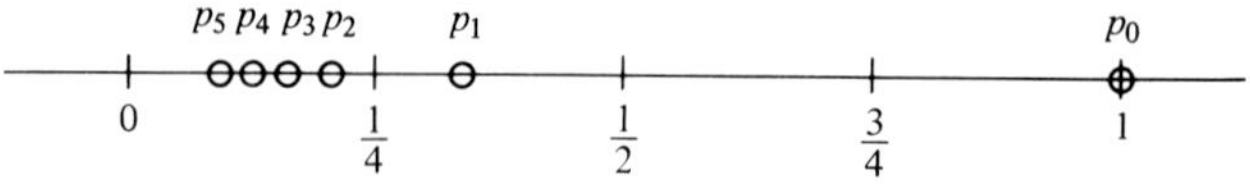

13. **(a)** 0, 1, 8, 27, 64, 125

(b)

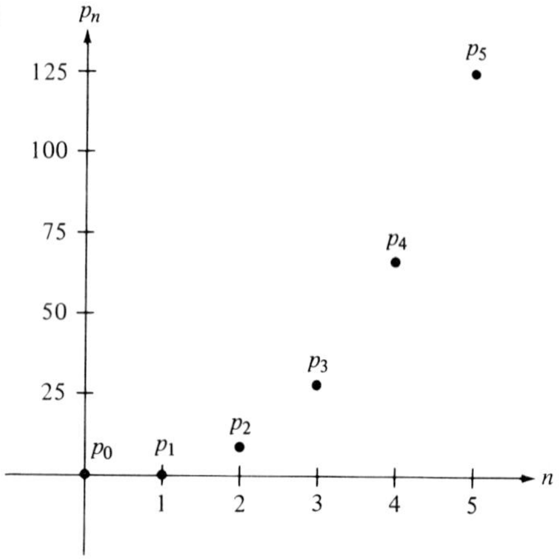

(c)

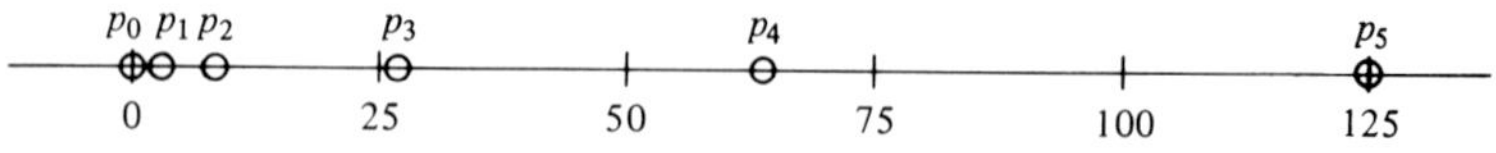

15. **(a)** $1, -2, 4, -8, 16, -32$

(b)

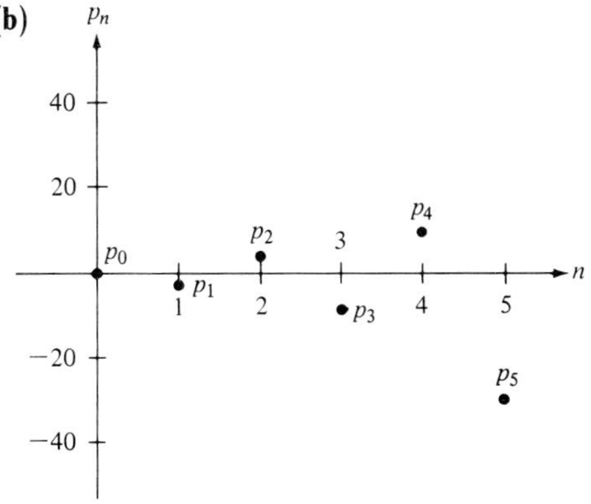

(c)

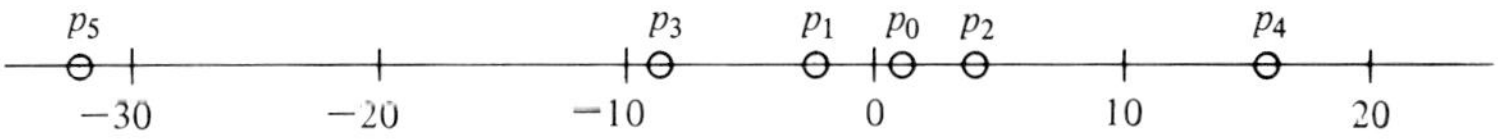

17. **(a)** $1, -\frac{1}{2}, \frac{1}{3}, -\frac{1}{4}, \frac{1}{5}, -\frac{1}{6}$

(b)

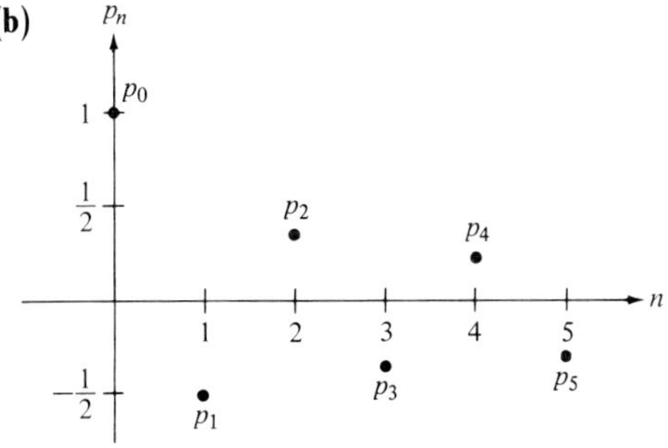

(c)

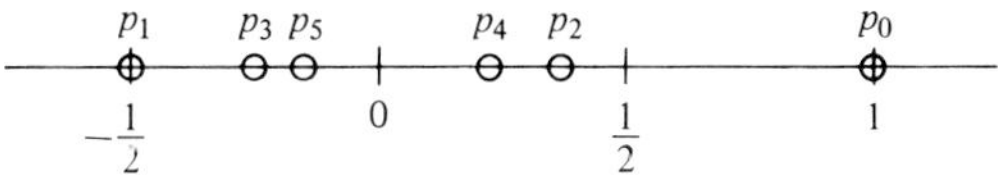

19. **(a)** $-1, 0, \frac{1}{5}, \frac{1}{5}, \frac{3}{17}, \frac{2}{13}$

(b)

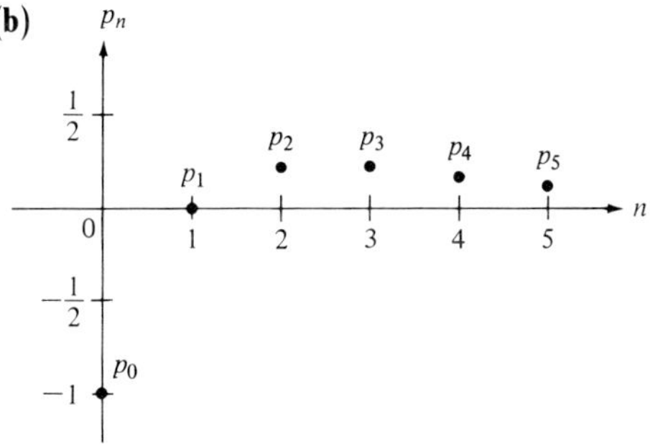

(c)

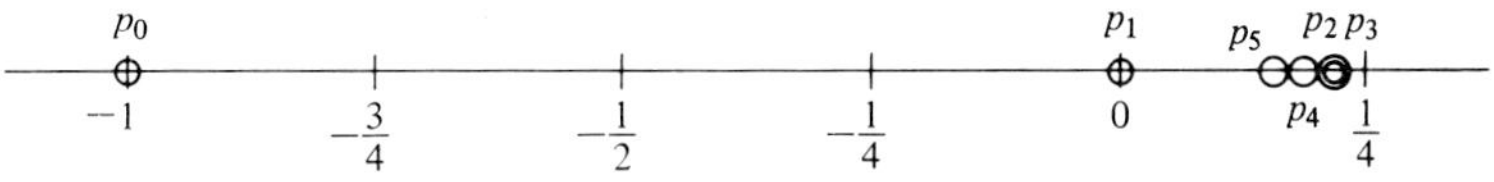

21. $p_n = 2n$ **23.** $p_n = 2n + 1$ **25.** $p_n = 1 - 2n$ **27.** $p_n = (-1)^n \cdot n$

29. $p_n = -(-2)^n$

31. $p_n = \dfrac{n+5}{n+3}$

33. $p_n = -(-n)^3$

35. (a) 1.4, 1.41, 1.414, 1.4142, 1.41421, 1.414213, 1.4142135
(b) 4, 1, 4, 2, 1, 3, 5

37. (23) $p_n = 2n + 3$
(27) $p_n = (-1)^{n+1} \cdot (n+1)$
(29) $p_n = (-1)^n \cdot 2^{n+1}$

EXERCISES 2-2

1. 1, 2, 3, 4, 5, 6

3. 2, 6, 18, 54, 162, 486

5. 0, 1, 3, 6, 10, 15

7. $p_{n-1} = n(n-2)$
$p_{n+1} = n(n+2)$

9. $p_{n-1} = (-1)^{n-1} \cdot (3n-3)$
$p_{n+1} = (-1)^{n+1} \cdot (3n+3)$

11. $q_n = \dfrac{n-1}{n} \cdot q_{n-1}$

$q_1 = \dfrac{1}{2}$

13. $q_n = \dfrac{n+2}{n} \cdot q_{n-1}$

$q_0 = 2$

15. $q_n = -\dfrac{n}{n+1} q_{n-1}$

$q_0 = 1$

17. $p_n = n,\ n = 1, 2, 3 \cdots$

19. $p_n = (-3)^n,\ n = 0, 1, 2 \cdots$

21. $p_n = p_{n-1} + 2$
$p_1 = 3$

23. $p_n = p_{n-1} - 1$
$p_1 = 2$

EXERCISES 2-3

1. (a) $n = 1, 2, 3$
(b) $n = 1,\ p_1 = 2$
$n = 2,\ p_2 = 4$
$n = 3,\ p_3 = 8$
(c) Block 1. Read a value (2) for x
Block 2. Assign 1 to n
Block 3. Value 1 currently assigned to n is less than 3; proceed to Block 4
Block 4. Compute $2^1 = 2$; assign to p_1
Block 5. Print $n = 1$ and $p_1 = 2$
Block 6. Add 1 to n, giving a value of 2 for n
Block 3. Value 2 currently assigned to n is less than 3; proceed to Block 4
Block 4. Compute $2^2 = 4$; assign to p_2
Block 5. Print $n = 2$ and $p_2 = 4$
Block 6. Add 1 to n, giving a value of 3 for n

Block 3. Value 3 currently assigned to n is equal to 3; proceed to Block 4
Block 4. Compute $2^3 = 8$; assign to p_3
Block 5. Print $n = 3$ and $p_3 = 8$
Block 6. Add 1 to n, giving a value of 4 for n
Block 4. Value 4 currently assigned to n is greater than 3; proceed to "Stop" block

(d) Change the statement in Block 3 to $n \leq 5$
(e) Change the statement in Block 2 to $n \leftarrow 3$ and the statement in Block 3 to $n \leq 6$

3. **(a)** $n = 1, 2, 3, 4$
(b) $n = 1, p_1 = 3$
$n = 2, p_2 = 5$
$n = 3, p_3 = 7$
$n = 4, p_4 = 9$
(c) Change the statement in Block 3 to $n \leq 6$
(d) Change the statement in Block 2 to $n \leftarrow 20$ and the statement in Block 3 to $n \leq 30$

5. **(a)** $n = 1, 2$
(b) $n = 1, p_1 = 3$
$n = 2, p_2 = 5$
(c) Yes

7. **(a)** $n = 1, 2, 3$
(b) $n = 1, p_1 = 2$
$n = 2, p_2 = 4$
$n = 3, p_3 = 8$
(c) Change the statement in Block 6 to $n \leq 5$
(d) Change the statement in Block 2 to $n \leftarrow 3$ and the statement in Block 6 to $n \leq 6$
(e) Yes

9.

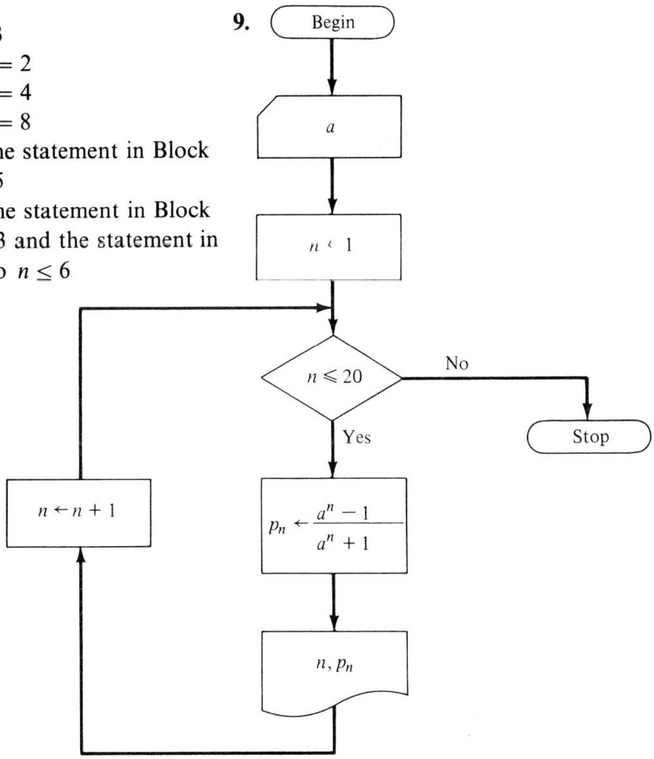

11.

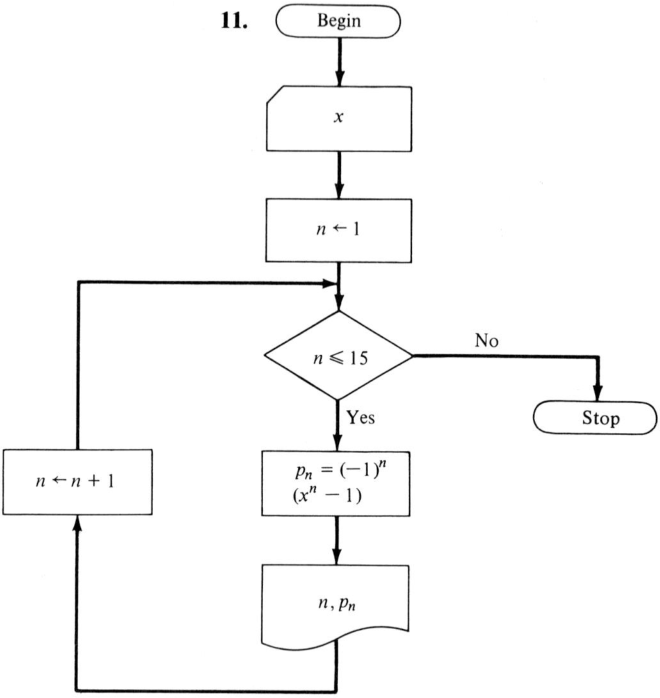
Begin
x
n ← 1
n ≤ 15
No
Yes
Stop
n ← n + 1
p_n = (−1)^n (x^n − 1)
n, p_n

13.

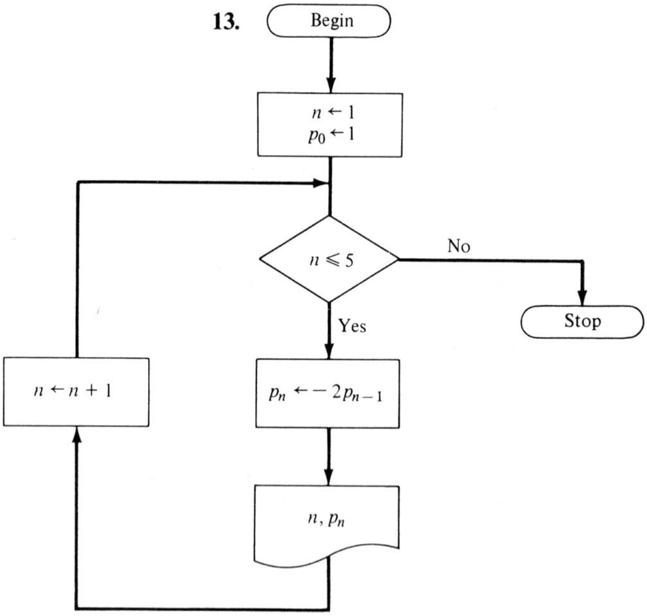
Begin
n ← 1
p_0 ← 1
n ≤ 5
No
Yes
Stop
n ← n + 1
p_n ← −2p_{n−1}
n, p_n

15.

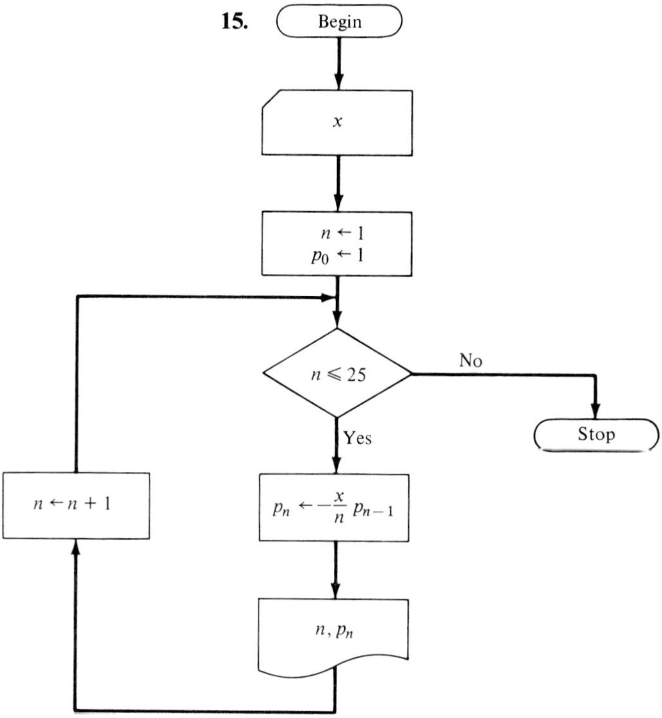

EXERCISES 2-4

1. Converges; 4 **3.** Converges; -1 **5.** Converges; -1

7. Converges; 0 **9.** Converges; 0 **11.** Diverges; neither $+\infty$ nor $-\infty$

13. Diverges; ∞ **15.** Converges; 0 **17.** Diverges; ∞

19. Diverges; ∞ **21.** Converges; 0 **23.** Converges; 0

EXERCISES 2-5

1. 2 **3.** 6 **5.** 1 **7.** 2 **9.** 2 **11.** 2 **13.** $\frac{1}{3}$ **15.** 4

17. 1 **19.** 0 **21.** 2 **23.** $-\infty$ **25.** 0 **27.** 2

29. -1 **31.** Diverges **33.** Diverges **35.** Converges

EXERCISES 2-6

1. 2 **3.** Does not exist **5.** $-\frac{1}{2}$ **7.** ∞ **9.** ∞ **11.** 45

13. 2 **15.** -3 **17.** -6 **19.** 4 **21.** -2 **23.** 4 **25.** 2

27. $\frac{1}{3}$ **29.** ∞ **31.** Does not exist

33. Limit exists and is $9\frac{1}{3} \approx 9.33$
$f(2.98) = 9.22$
$f(2.99) = 9.27$
$f(3.01) = 9.39$
$f(3.02) = 9.45$

35. Limit exists and is 4
$f(-1.02) = 4.08$
$f(-1.01) = 4.04$
$f(-0.99) = 3.96$
$f(-0.98) = 3.92$

37. Limit does not exist because terms approach $+\infty$
$f(0.98) = 2500$
$f(0.99) = 10{,}000$
$f(1.01) = 10{,}000$
$f(1.02) = 2500$

EXERCISES 2-7

1. None **3.** $+3, -3$ **5.** None **7.** Continuous for all $x \geq -1$

9. None **11.** 1 **13.** Yes **15.** No **17.** No

19. $p(x) + q(x) = 2x^4$
$p(x) \cdot q(x) = x^8 - 1$

21. $[x_2, x_3]$; $[x_1, x_3]$ **23.** -1

EXERCISES 2-8

1. **(a)** No

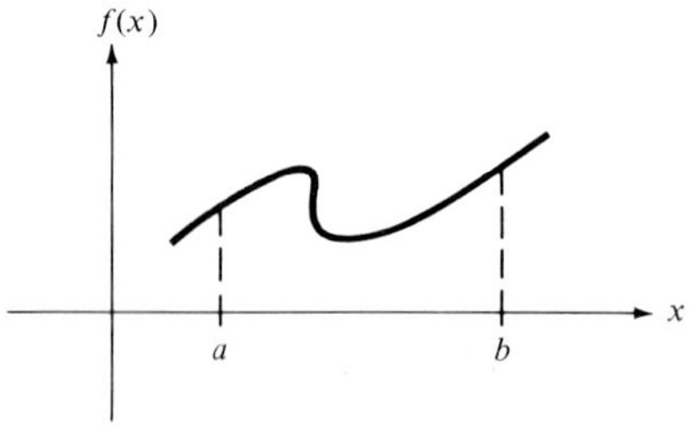

(b) Yes

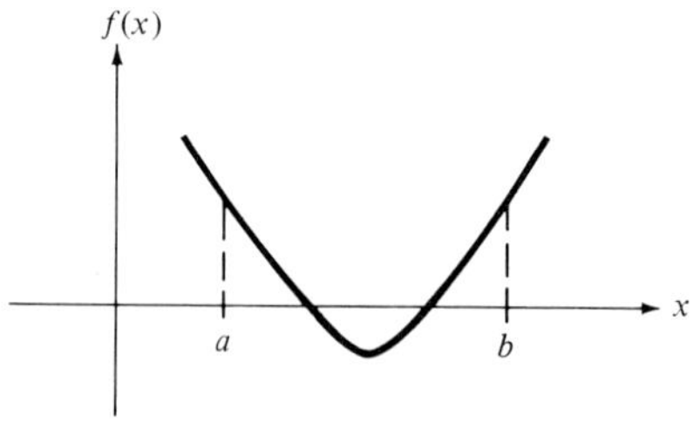

3. $[-2, -1]$ and $[1, 2]$

5. $f(a_{n-1}) \cdot f(z_{n-1}) < 0$

7.

n	a_n	b_n	z_n	$f(z_n)$
0	0	2	1	−4
1	1	2	1.5	2.59
2	1	1.5	1.25	−1.95
3	1.25	1.5	1.37	−0.17
4	1.37	1.5	1.43	0.98
5	1.37	1.43	1.40	0.38
6	1.37	1.40	1.385	0.10

9.

n	a_n	b_n	z_n	$f(z_n)$
0	0	1	0.5	−0.75
1	0.5	1	0.75	0.06
2	0.5	0.75	0.62	−0.38
3	0.62	0.75	0.68	−0.18
4	0.68	0.75	0.71	−0.08
5	0.71	0.75	0.73	−0.01

11.

n	a_n	b_n	z_n	$f(z_n)$
0	0.25	1	0.62	−0.77
1	0.25	0.62	0.43	−0.24
2	0.25	0.43	0.34	0.28
3	0.34	0.43	0.38	0.01
4	0.38	0.43	0.40	−0.1
5	0.38	0.40	0.39	−0.05

CHAPTER 2. REVIEW EXERCISES

1. $1, \frac{1}{4}, \frac{1}{9}, \frac{1}{16}, \frac{1}{25}, \frac{1}{36}$

3. $-\frac{1}{2}, \frac{1}{4}, -\frac{1}{6}, \frac{1}{8}, -\frac{1}{10}, \frac{1}{12}$

5. 0, 2, 0, 2, 0, 2

7. 0.2, 0.04, 0.008, 0.0016, 0.00032, 0.000064

9. $1, 1, 2, \frac{3}{2}, \frac{8}{3}, \frac{15}{8}, \frac{48}{15}$

11. $q_n = q_{n-1} + 3;\ q_0 = 5$

13. $z_n = -5z_{n-1};\ z_0 = 1$

15. $b_n = \left(\frac{n}{n-1}\right)^2 b_{n-1};\ b_1 = 1$

17. $c_n = \frac{1}{n(n+1)(n+2)}$, n positive

19. **(a)** 1, 3, 5, 7, 9
(b) Change the statement in Block 1 to $n \leftarrow 2$ and the statement in Block 2 to $n \leq 10$
(c) Change the statement in Block 1 to $n \leftarrow 0$, the statement in Block 4 to $n \leftarrow n + 5$, and the statement in Block 2 to $n \leq 25$.
(d) Change the statement in Block 1 to $n \leftarrow 9$, the statement in Block 4 to $n \leftarrow n - 2$, and the statement in Block 2 to $n \geq 1$.

21. -2 **23.** ∞ **25.** ∞ **27.** 0 **29.** 0

31. **(a)** 20
(c) -2
(e) $\frac{2}{3}$
(g) 2

33. 0 **35.** 0 **37.** 0 **39.** $\frac{1}{8}$ **41.** $x = 0$

43. $x = -3$; $x = 1$ **45.** $x = 0$; $x = -1$ **47.** No **49.** No

51. **(a)** $V_2 = V_1\left(1 + \frac{i}{2}\right)^2 = V_0\left(1 + \frac{i}{2}\right)^4$

$V_3 = V_2\left(1 + \frac{i}{2}\right)^2 = V_0\left(1 + \frac{i}{2}\right)^6$

$V_4 = V_3\left(1 + \frac{i}{2}\right)^2 = V_0\left(1 + \frac{i}{2}\right)^8$

(b) $V_n = V_0\left(1 + \frac{i}{2}\right)^{2n}$

(c) $V_n = V_{n-1}\left(1 + \frac{i}{2}\right)^2$

53. **(a)** $V_2 = V_1(1 + i) - D$
$V_3 = V_2(1 + i) - D$
$V_4 = V_3(1 + i) - D$

(b) $V_n = V_{n-1}(1 + i) - D$

EXERCISES 3-1

1. **(a)** 9
(b) 6
(c) $y - 2 = 6(x - 1)$

3. **(a)** $3 - h$
(b) 3
(c) $y + 1 = 3(x + 2)$

5. **(a)** $-\frac{1}{8}$
(b) $-\frac{1}{4}$
(c) $y - \frac{1}{2} = -\frac{1}{4}(x - 2)$

7. **(a)** -6
(b) -3
(c) $y + 2 = -3x$

9. **(a)** 21
(b) 9
(c) $y - 3 = 9(x - 1)$

11. **(a)** -0.2488
(b) -0.25
(c) 0.0012

13. (a) -2.0304
(b) -2
(c) 0.0304

15. (a) 9.009
(b) 9
(c) 0.009

EXERCISES 3-2

1. (a) 1
(b) All x

3. (a) $3x^2$
(b) All x

5. (a) $-\dfrac{2}{x^3}$
(b) All $x \neq 0$

7. (a) $-\dfrac{1}{2x\sqrt{x}}$
(b) All $x > 0$

9. (a) $-\dfrac{2}{(x-1)^3}$
(b) All $x \neq 1$

11. (a) $-\dfrac{2}{x^2}$
(b) All $x \neq 0$

13. (a) $2x + 1$
(b) 5
(c) $y - 6 = 5(x - 2)$

15. (a) $-\dfrac{1}{(x+3)^2}$
(b) -1
(c) $y - 1 = -(x + 2)$

17. (a) $1 - \dfrac{3}{x^2}$
(b) $\frac{1}{4}$
(c) $y - \frac{7}{2} = \frac{1}{4}(x - 2)$

19. (a) $-\dfrac{1}{(x-1)^2}$
(b) -1
(c) $h = 0.1$: approximation $= -1.111$; error $= 0.111$
$h = 0.01$: approximation $= -1.01$; error $= 0.01$

21. (a) $-\dfrac{4}{(x-2)^2}$
(b) $-\frac{4}{9} \approx -0.4444$
(c) $h = 0.1$: approximation $= -0.460$; error $= -0.016$
$h = 0.01$: approximation $= -0.45$; error $= 0.006$

EXERCISES 3-3

1. $9t^2$ 3. $-80t^{-11}$ 5. $5t^4 + 16t^3 - 4t$ 7. $2x + 5$

9. $3x^2 + 20x + 18$

11. $-270x^4 + 144x^3 + 3$

13. $\dfrac{8}{(4-x)^2}$

15. $\dfrac{-x^2 + 5}{(x^2 + 2x + 5)^2}$

17. $\dfrac{-32x}{(8 + x^2)^2}$

19. $3x^2 + 36x + 104$

21. $\dfrac{x^2 + 2x + 9}{-(x-1)^2(x+5)^2}$

23. $h'(x) = k_1 f'(x) + k_2 g'(x)$

25. $u'(x) = f'(x)g(x)h(x) + f(x)g'(x)h(x) + f(x)g(x)h'(x)$

EXERCISES 3-4

1. $e'(t) = -20t + 280$

3. $a'(t) = 0.1 + \dfrac{2}{(t+1)^2}$

5. $\text{M.C.} = C'(q) = 1 + \dfrac{1}{(q+1)^2}$

   $\text{M.R.} = R'(q) = 2$

   $\text{M.P.} = \text{M.R.} - \text{M.C.} = 1 - \dfrac{1}{(q+1)^2}$

7. $\dfrac{1}{(q+1)^2}$

9. (a) $e(\frac{1}{2}) = 2$

   (b) $e(p) = \dfrac{1}{1-p}$

11. $\dfrac{-4}{(p+1)^2}$

13. $\dfrac{ds}{di} = \dfrac{d}{di}\left(\sqrt{\dfrac{i}{1000}}\right) = \dfrac{1}{\sqrt{1000}}\dfrac{d}{di}(\sqrt{i}) = \dfrac{1}{2\sqrt{1000i}}$

EXERCISES 3-5

1. $3(x^4 - 2x)^2(4x^3 - 2)$

3. $\dfrac{1}{x^2}$

5. $1 + \dfrac{1}{2\sqrt{x}}$

7. $1 + \dfrac{1}{x^2}$

9. $6t(t^2+1)^2 + \dfrac{2t}{(t^2+1)^2}$

11. $16(t^3 + t^2 - 5)^{15}(3t^2 + 2t)$

13. $-\dfrac{(t-t^2)^2(5t^2+4t-3)}{(t+1)^2}$

15. $\frac{4}{3}(2t^3 - t^2 + 5t)^{1/3}(6t^2 - 2t + 5)$

17. (a) $s(t) = \sqrt{\dfrac{6}{1000} + \dfrac{t}{4000} - \dfrac{0.01}{t+10}}$

   (b) $\dfrac{ds}{dt} = \dfrac{\frac{1}{4000} + [0.01/(t+10^2)]}{2\sqrt{\frac{6}{1000} + (t/400) - [0.01/(t+10)]}}$

   (c) $\dfrac{\frac{1}{4} + [10/(t+10)^2]}{2000\sqrt{\frac{6}{1000} + (t/400) - [0.01/(t+10)]}}$

EXERCISES 3-6

1. **(a)** $f'(x) = 3x^2 - 4x + 5$
$f''(x) = 6x - 4$
$f'''(x) = 6$
(b) $f'(2) = 9$
$f''(2) = 8$
$f'''(x) = 6$

3. **(a)** $f'(x) = 6x^2 + \frac{1}{x^2}$

$f''(x) = 12x - \frac{2}{x^3}$

$f'''(x) = 12 + \frac{6}{x^4}$

(b) $f'(-1) = 7$
$f''(-1) = -10$
$f'''(-1) = 18$

5. **(a)** $f'(x) = \frac{3}{2}x^{1/2} - 1$
$f''(x) = \frac{3}{4}x^{-1/2}$
$f'''(x) = -\frac{3}{8}x^{-3/2}$
(b) $f'(4) = 2$
$f''(4) = \frac{3}{8}$
$f'''(4) = -\frac{3}{64}$

7. **(a)** $f'(x) = -\frac{1}{x^2} + \frac{4}{x^3} - \frac{9}{x^4}$

$f''(x) = \frac{2}{x^3} - \frac{12}{x^4} + \frac{36}{x^5}$

$f'''(x) = -\frac{6}{x^4} + \frac{48}{x^5} - \frac{180}{x^6}$

(b) $f'(1) = -6$
$f''(1) = 26$
$f'''(1) = -138$

9. $f'(x) = 4(x^2 - x + 1)^3 \cdot (2x - 1)$
$f''(x) = 12(x^2 - x + 1)^2 \cdot (2x - 1)^2 + 8(x^2 - x + 1)^3$

11. $f'(x) = \dfrac{3x^2 - 2}{2\sqrt{x^3 - 2x + 4}}$

$f''(x) = \dfrac{3x^4 - 24x^2 + 48x - 4}{4(x^3 - 2x + 4)^{3/2}}$

13. $f'(2) = \frac{2}{9}$
$f''(2) = -\frac{4}{27}$

15. $f'(0) = 1$
$f''(0) = 0$

17. At $t = 1$, growth rate $= \frac{5}{36}$
At $t = 4$, growth rate $= \frac{2}{9}$
At $t = 1$, rate of change in growth rate $= \frac{2}{27}$
At $t = 4$, rate of change in growth rate $= \frac{1}{108}$

EXERCISES 3-7

3. **(a)** $y' = \dfrac{x}{y}$

(b) $y = \dfrac{x}{\sqrt{x^2 + 1}}$

5. **(a)** $y' = -\sqrt{\dfrac{y}{x}}$

(b) $y' = -\dfrac{1 - \sqrt{x}}{\sqrt{x}}$

7. **(a)** $y' = -\dfrac{x}{y}$

(c) $y'\Big|_{(-3,4)} = \frac{3}{4}$

(d) $y - 4 = \frac{3}{4}(x + 3)$

9. **(a)** $y' = -1$

(c) $y'\Big|_{(1,0)} = -1$

(d) $y = -(x - 1)$

11. **(a)** $y' = \dfrac{1 - 2xy^3}{3x^2y^2}$

(c) $y'\Big|_{(1,1)} = -\frac{1}{3}$

(d) $y - 1 = -\frac{1}{3}(x - 1)$

13. **(a)** $y' = -\dfrac{y^3 + 4xy}{3xy^2 + 2x^2 - 12y}$

(c) $y'\Big|_{(2,1)} = -\frac{9}{2}$

(d) $y - 1 = -\frac{9}{2}(x - 2)$

15. **(a)** $y' = \dfrac{y}{x}$

(c) $y'\Big|_{(8,6)} = \frac{3}{4}$

(d) $y - 6 = \frac{3}{4}(x - 8)$

17. **(a)** $y' = \dfrac{y^2 - x^2}{2xy}$

(c) $y'\Big|_{(1,0)} = \infty$ (tangent line is vertical)

(d) $x = 1$

EXERCISES 3-8

1. $\dfrac{dy}{dt} = 2x\dfrac{dx}{dt} - 2\dfrac{dx}{dt}$; $\dfrac{dy}{dt}\Big|_{t_0} = 10$

3. $\dfrac{dy}{dt} = \dfrac{2 + \sqrt{x}}{2(1 + \sqrt{x})^2}\dfrac{dx}{dt}$; $\dfrac{dy}{dt}\Big|_{t_0} = \dfrac{0.3}{8}$

5. $\dfrac{dy}{dt} = \dfrac{x\dfrac{dx}{dt} + u\dfrac{du}{dt}}{\sqrt{x^2 + u^2}}$; $\dfrac{dy}{dt}\Big|_{t_0} = -\dfrac{1}{5}$

7. $\dfrac{dw}{dt} = \dfrac{0.1375}{(3.75)^2} \approx 0.00978$

CHAPTER 3. REVIEW EXERCISES

1. **(a)** $4x^3$
(b) $y - 1 = 4(x - 1)$

3. **(a)** $\dfrac{1}{2\sqrt{x}}$

(b) $y - 1 = \frac{1}{2}(x - 1)$

5. **(a)** $4(x - 1)^3$
(b) $y - 1 = -4x$

7. **(a)** $\dfrac{-x^2 - 4x + 1}{(x^2 - x - 1)^2}$

(b) $y - 4 = -11(x - 2)$

9. **(a)** $\dfrac{3x + 2}{2\sqrt{x + 1}}$

(b) $y - 6 = \frac{11}{4}(x - 3)$

11. $9t^8$

13. $-120x^{-21} - 2x$

15. $-\dfrac{\sqrt{3}}{(z + 1)^2}$

17. $\dfrac{3t^2 - 2t - 1}{2\sqrt{t^3 - t^2 - t - 1}}$

19. $-25z^4(z^5 - 1)^{-6}$

21. $2t + 1 - \dfrac{2}{(t-1)^2}$

23. $\dfrac{2z^2 + z - 1}{\sqrt{z^2 - 1}}$

25. $\dfrac{2(x + \sqrt{x^2 - 1}) - 2x[(1 + x/(\sqrt{x^2-1})]}{(x + \sqrt{x^2 - 1})^2}$

27. $\text{M.C.} = \dfrac{1}{2\sqrt{q+1}}$

$\text{M.R.} = \dfrac{10(1-q)}{(q+1)^3}$

$\text{M.P.} = \text{M.R.} - \text{M.C.} =$

$\dfrac{10(1-q)}{(q+1)^3} - \dfrac{1}{2\sqrt{q+1}}$

29. $f'(x) = -7(x+3)^{-8}$
$f''(x) = 56(x+3)^{-9}$
$f'''(x) = -504(x+3)^{-10}$

31. $f'(x) = 8x(x^2 + 2)^3$
$f''(x) = (56x^2 + 16)(x^2 + 2)^2$
$f'''(x) = x(x^2 + 2)(336x^2 + 288)$

33. $f'(x) = 4(x+2)^{-2}$
$f''(x) = -8(x+2)^{-3}$
$f'''(x) = 24(x+2)^{-4}$

35. **(a)** $y' = \dfrac{2xy + y^4}{2x^2 - xy^3}$
(b) $-\frac{1}{3}$
(c) $y + 1 = -\frac{1}{3}(x - 1)$

37. **(a)** $y' = \dfrac{y}{x}$
(b) $\frac{1}{2}$
(c) $y - 1 = \frac{1}{2}(x - 2)$

39. **(a)** $y' = -\dfrac{10(x+y)^4 + y}{10(x+y)^4 + x}$
(b) $-\frac{3}{4}$
(c) $y + 1 = -\frac{3}{4}(x - 2)$

EXERCISES 4-1

1. Let $f(x)$ be a function with specified domain D. If there is a number c in D such that $f(c) \le f(x)$ for all x in D, then $f(c)$ is called the *absolute minimum* of $f(x)$ in D

3. Absolute maximum is $\frac{1}{4}$; occurs at $x = 0$
Absolute minimum is $\frac{1}{8}$; occurs at $x = 2$ and $x = -2$

5. No absolute maximum
Absolute minimum is 1; occurs at $x = 0$

7. Absolute maximum is 1; occurs at $x = 0$
No absolute minimum

9. Absolute maximum is $\sqrt[3]{2}$; occurs at $x = -2$
Absolute minimum is 0; occurs at $x = 0$

11. $[-\frac{1}{3}, \frac{8}{3}]$

13. $[\sqrt[3]{-2}, \sqrt[3]{7}]$

15. **(a)** Relative maxima: $f(x_1)$ occurs at x_1; $f(x_3)$ occurs at x_3; $f(x_5)$ occurs at x_5
Relative minima: $f(x_2)$ occurs at x_2; $f(x_4)$ occurs at x_4
(b) Absolute maximum: $f(x_5)$ occurs at x_5
Absolute minimum: $f(x_4)$ occurs at x_4

17. (a) Relative maxima: $f(x_3)$ occurs at x_3; $f(x_5)$ occurs at x_5
Relative minima: $f(x_2)$ occurs at x_2; $f(x_4)$ occurs at x_4
(b) Absolute maximum: none
Absolute minimum: $f(x_4)$ occurs at x_4

19. (a) Relative maxima: $f(x_3)$ occurs at x_3
Relative minima: $f(x_4)$ occurs at x_4
(b) Absolute maximum: none
Absolute minimum: $f(x_4)$ occurs at x_4

21.

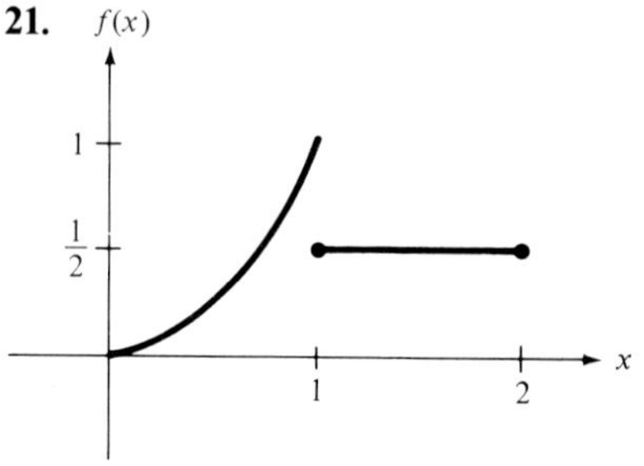

Absolute maximum: none
Absolute minimum: 0

EXERCISES 4-2

1. (a) All x
(b) $x = 0$
(c) No relative extrema

3. (a) All $x \neq \frac{1}{2}$
(b) No critical points
(c) No relative extrema

5. (a) All x
(b) $x = 0$
(c) No relative extrema

7. (a) $x \geq 1$ and $x \leq -1$
(b) $x = 1$ and $x = -1$
(c) Relative minima at $x = 1$ and $x = -1$; $f(1) = f(-1) = 0$; no relative maxima

9. (a) All $x \neq 1$
(b) No critical points
(c) No relative extrema

11. (a) $-\frac{1}{2}$, -2, 8
(b) -2 is a relative maximum; $f(-2) = 2$
$-\frac{1}{2}$ is a relative minimum; $f(-\frac{1}{2}) = -\frac{1}{4}$
8 is a relative maximum; $f(8) = 72$
(c) Absolute maximum $= 72$
Absolute minimum $= -\frac{1}{4}$

13. **(a)** 0, 3, 5
(b) 0 is a relative maximum; $f(0) = 2$
3 is a relative minimum; $f(3) = -7$
5 is a relative maximum; $f(5) = -3$
(c) Absolute maximum = 2
Absolute minimum = -7

15. **(a)** -5, -3, 1, 5
(b) -5 is a relative minimum; $f(-5) = -12$
-3 is a relative maximum; $f(-3) = 20$
1 is a relative minimum; $f(1) = -12$
5 is a relative maximum; $f(5) = 148$
(c) Absolute maximum = 148
Absolute minimum = -12

17. **(a)** -1, 3
(b) -1 is a relative minimum; $f(-1) = 1$
3 is a relative minimum; $f(3) = \frac{1}{9}$
(c) No absolute maximum
Absolute minimum = $\frac{1}{9}$

19. **(a)** -2, 0, 3
(b) -2 is a relative minimum; $f(-2) = \frac{1}{3}$
0 is a relative maximum; $f(0) = -1$
3 is a relative minimum; $f(3) = \frac{1}{8}$
(c) No absolute maximum
No absolute minimum

EXERCISES 4-3

1. $x = 3$ is a relative minimum

3. $x = 1$ is a relative minimum
$x =$ 2 is a relative maximum

5. $x = 0$ is a relative maximum
$x = 1$ is a relative minimum
$x = -1$ is a relative minimum

7. $x = 0$ is a relative maximum

9. $x = -4$ is a relative maximum

11. **(a)** All x
(b) Increasing for $x > -2$
Decreasing for $x < -2$
(c) -2
(d) Concave upward everywhere
(e) None
(f) $x = -2$ is a relative minimum

13. (a) All x
(b) Decreasing for $-3 < x < 1$; increasing elsewhere
(c) $x = -3$, $x = 1$
(d) Concave upward for $x > -1$; concave downward for $x < -1$
(e) $x = -1$
(f) $f(-3)$ is a relative maximum; $f(1)$ is a relative minimum

15. (a) All x
(b) Decreasing for $x < -1$ and $0 < x < 1$; increasing for $-1 < x < 0$ and $x > 1$
(c) $x = 0$, $x = 1$, $x = -1$
(d) Concave upward for $x < -1/\sqrt{3}$ and $x > 1/\sqrt{3}$; concave downward for $-1/\sqrt{3} < x < 1/\sqrt{3}$
(e) $x = 1/\sqrt{3}$ and $x = -1/\sqrt{3}$
(f) $f(1)$ is a relative minimum; $f(-1)$ is a relative minimum; $f(0)$ is a relative maximum

EXERCISES 4-4

1. Has a minimum but no maximum
3. Has a maximum but no minimum
5. Has a maximum and minimum
7. No maximum or minimum
9. Has a minimum but no maximum
11. Absolute maximum $= 16$; absolute minimum $= 0$
13. Absolute minimum $= 9$; no absolute maximum
15. No absolute maximum; no absolute minimum
17. No absolute maximum; no absolute minimum
19. Absolute maximum $= f(10) = 970$
Absolute minimum $= f(-10) = -970$
21. Absolute maximum $= f(0) = 1$
Absolute minimum $= f(2) = \frac{1}{3}$
23. Absolute maximum $= f(1) = 1$
Absolute minimum $= f(\frac{1}{6}) = -\frac{13}{12}$
25. Absolute minimum $= f(0) = 1$; absolute maximum $= f(-2) = f(2) = 5$

EXERCISES 4-5

1. Maximum product $= 15 \cdot 15 = 225$
3. Maximum occurs at $x = 15$, $y = 15$
9. $x = 30$; minimum cost $= 80$

11. Cost $= \dfrac{600(20 + 0.2v)}{100} + \dfrac{8 \cdot 600}{v}$; v for minimum cost = 90 kilometers per hour

13. $x = 200$; $b = 200$

15. $x = 250$; $R = 1385$

EXERCISES 4-6

1. (**a**) 1, -1
(**b**) $f(1) = -1$ is relative minimum
$f(-1) = 2$ is relative maximum
(**c**) Concave upward for $x > 0$
Concave downward for $x < 0$
(**d**)

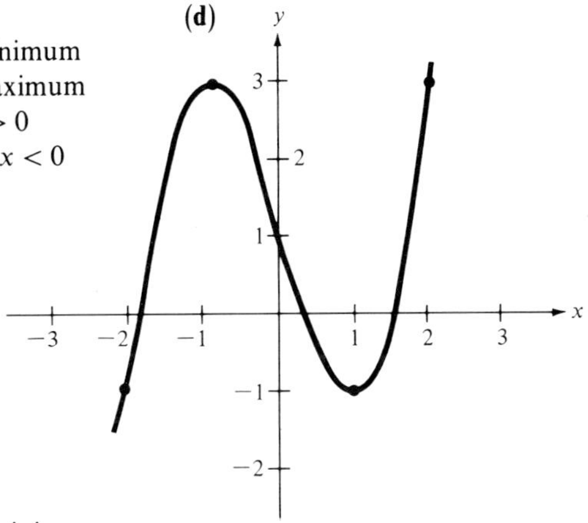

3. (**a**) No critical points
(**b**) No local maximum or minimum
(**c**) Concave upward for $x > 0$; concave downward for $x < 0$
(**d**)

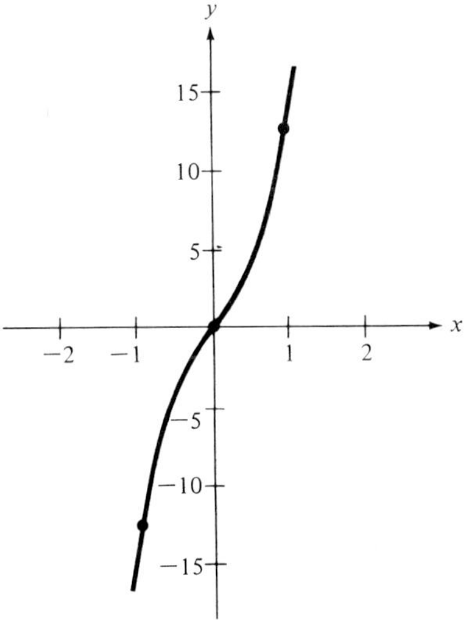

5. (**a**) $x = 0$
(**b**) $f(0) = -1$ is relative minimum
(**c**) Concave upward everywhere

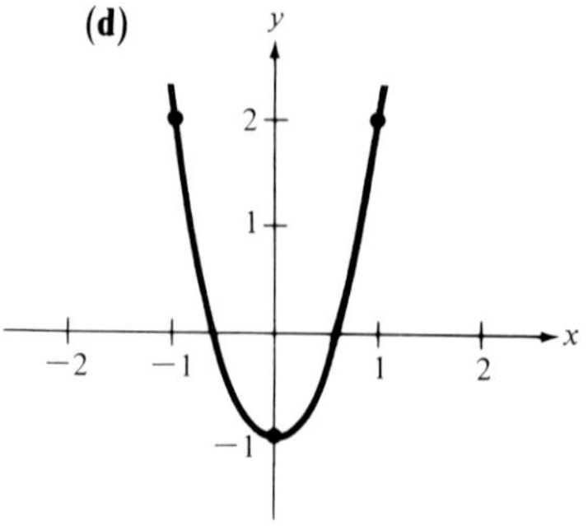

7. **(a)** $x = 0$
(b) $f(0) = 0$ is relative maximum
(c) Concave downward everywhere

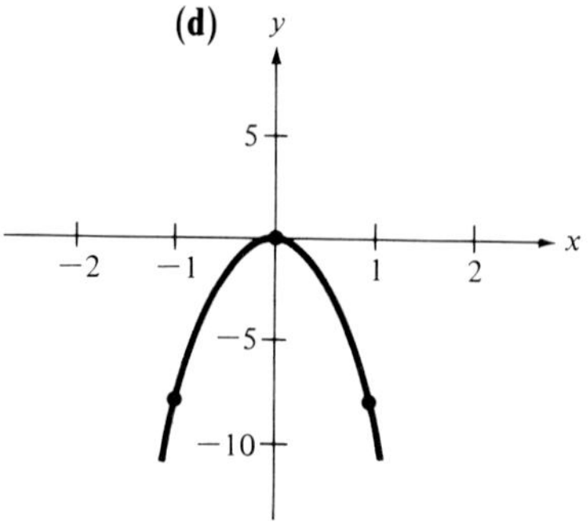

9. **(a)** $x = 1$
(b) $f(1) = 1$ is neither relative maximum nor relative minimum
(c) Concave upward for $x > 1$; concave downward for $x < 1$

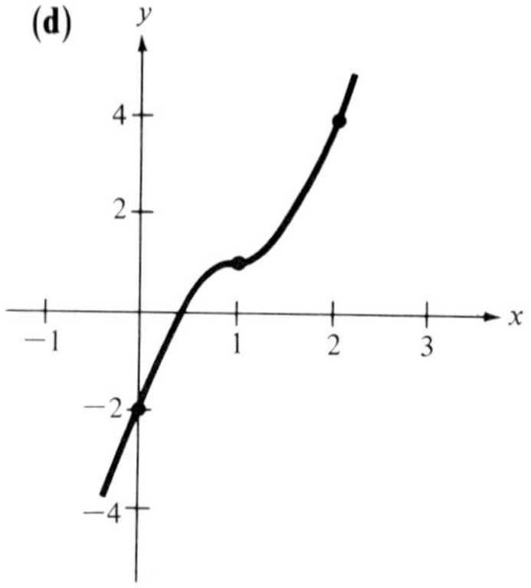

11. **(a)** $x = 1$; $x = -\frac{1}{3}$
(b) $f(-\frac{1}{3}) = -\frac{5}{27}$ is relative minimum
$f(1) = 1$ is relative maximum
(c) Concave upward for $x < \frac{1}{3}$; concave downward for $x > \frac{1}{3}$

(d)

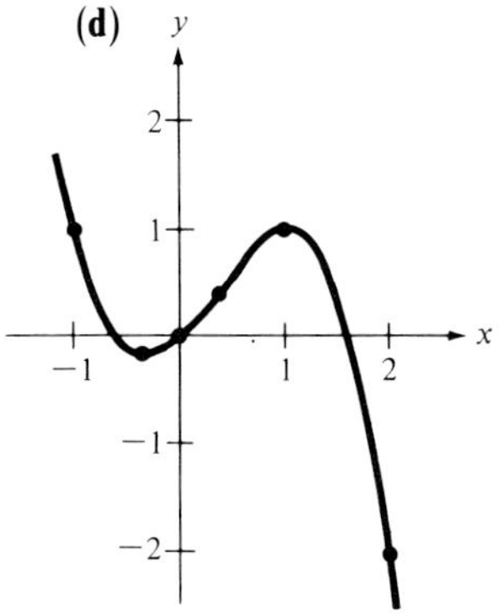

EXERCISES 4-7

1. $f'(x) = 2x - 2 = 2(x - 1)$

n	a_n	$f(a_n)$	$f'(a_n)$	$a_n - \dfrac{f(a_n)}{f'(a_n)}$
1	4	4	6	$\frac{10}{3}$
2	$\frac{10}{3}$	$\frac{4}{9}$	$\frac{14}{3}$	3.24
3	3.24	0.0176		

3. $f'(x) = 3x^2 - 4x - 3$

n	a_n	$f(a_n)$	$f'(a_n)$	$a^n - \dfrac{f(a_n)}{f'(a_n)}$
1	0	1	−3	$\frac{1}{3}$
2	$\frac{1}{3}$	−0.185	−4	0.287
3	0.287	0.002		

5. $f'(x) = 3x^2 - 4x - 3$

n	a_n	$f(a_n)$	$f'(a_n)$	$a_n - \dfrac{f(a_n)}{f'(a_n)}$
1	−2	−9	17	−1.47
2	−1.47	−2.09	7.89	−1.21
3	−1.21	−0.07		

7. $f(x) = x^2 - 5; f'(x) = 2x$

n	a_n	$f(a_n)$	$f'(a_n)$	$a_n - \dfrac{f(a_n)}{f'(a_n)}$
1	2	−1	4	2.25
2	2.25	0.0625	4.5	2.236
3	2.236	−0.0003	4.472	2.236

9. $f(x) = x^5 - 2; f'(x) = 5x^4$

n	a_n	$f(a_n)$	$f'(a_n)$	$a_n - \dfrac{f(a_n)}{f'(a_n)}$
1	1	−1	5	1.2
2	1.2	0.488	10.4	1.153
3	1.153	0.0377	8.837	1.149
4	1.149	0.0026	8.71	1.149

EXERCISES 4-8

1.

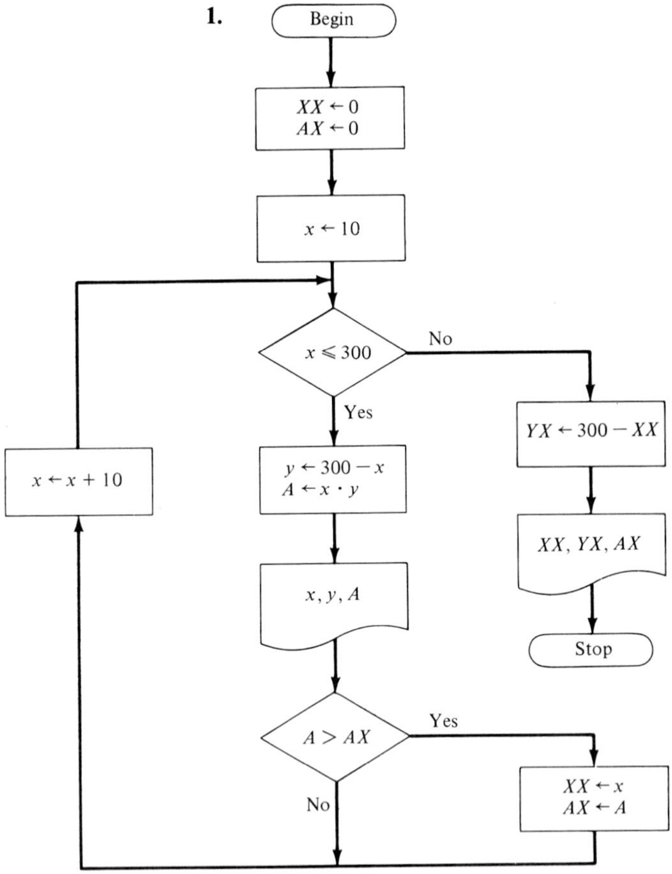

3.

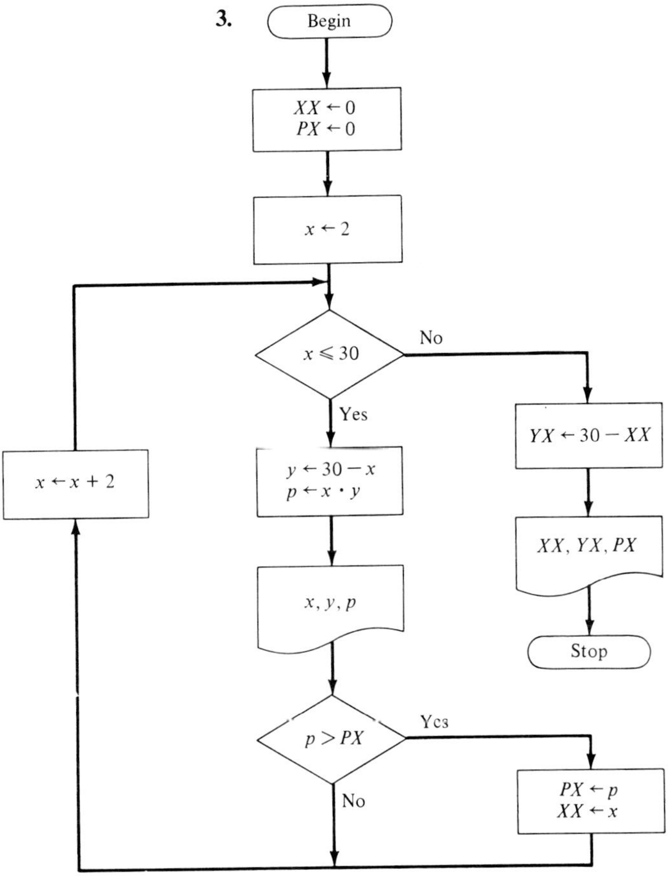

Begin
XX ← 0
PX ← 0
x ← 2
x ≤ 30
No
Yes
YX ← 30 − XX
y ← 30 − x
p ← x · y
x ← x + 2
XX, YX, PX
x, y, p
Stop
p > PX
Yes
No
PX ← p
XX ← x

5.

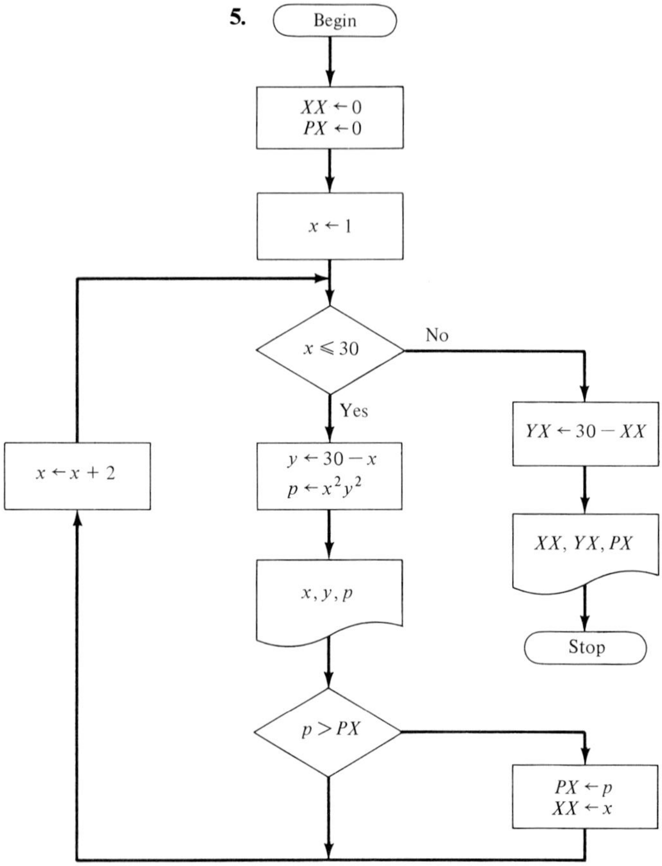
Begin
XX ← 0
PX ← 0
x ← 1
x ⩽ 30
No
Yes
YX ← 30 − XX
y ← 30 − x
p ← x²y²
x ← x + 2
XX, YX, PX
x, y, p
Stop
p > PX
PX ← p
XX ← x

7.

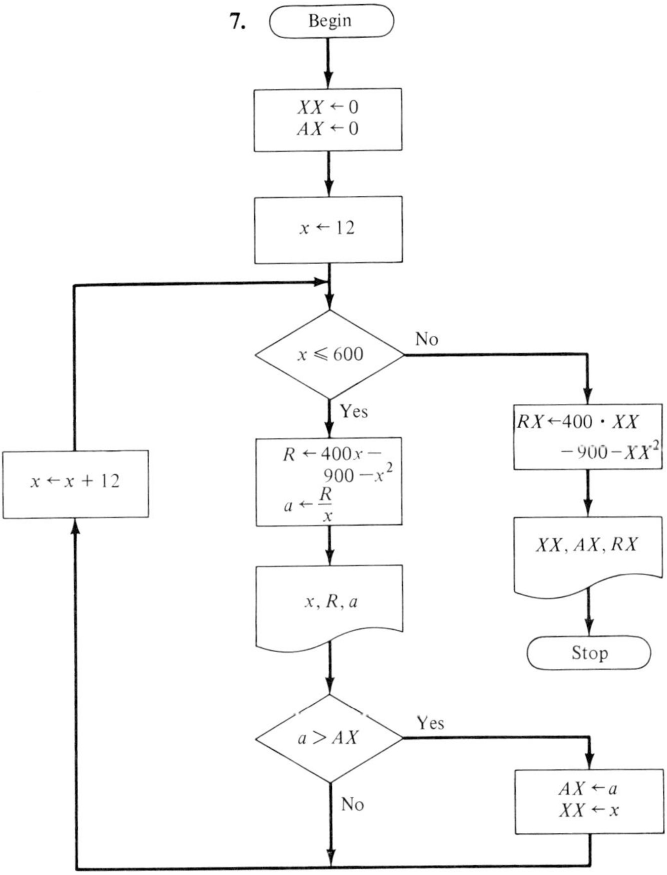

CHAPTER 4. REVIEW EXERCISES

1. **(a)** Increasing
   (b) Concave upward
   (c) None

3. **(a)** Increasing
   (b) Concave upward
   (c) $x = 0$

5. **(a)** Decreasing
   (b) Concave downward
   (c) $x = \dfrac{1}{\sqrt{6}}$; $x = -\dfrac{1}{\sqrt{6}}$

7. **(a)** Increasing
   (b) Concave upward
   (c) $x = \frac{4}{3}$

9. **(a)** Increasing
   (b) Concave upward
   (c) $x = 1$

11. (**a**) $-5, -3, 1, 5$
(**b**) $x = -5$ is relative minimum
$x = -3$ is relative maximum
$x = 1$ is relative minimum
$x = 5$ is relative maximum
(**c**) Absolute maximum = 160
Absolute minimum = 0

13. (**a**) 1
(**b**) $x = 1$ is neither
(**c**) No absolute maximum; no absolute minimum

15. (**a**) $-2, -1, 0$
(**b**) $x = -2$ is relative minimum
$x = -1$ is relative maximum
$x = 0$ is relative minimum
(**c**) Absolute minimum = -9
No absolute maximum

17. (**a**) $-1, 0, 1$
(**b**) $x = 1$ is relative maximum
$x = 0$ is relative minimum
$x = 1$ is relative maximum
(**c**) Absolute maximum = 2
Absolute minimum = $\sqrt{3}$

19. (**a**) None
(**b**) No relative maxima or minima
$x = 4$ is relative minimum
(**c**) No absolute maximum
No absolute minimum

21. (**a**) Relative minima: $f(x_2)$, occurs at x_2; $f(x_4)$, occurs at x_4
Relative maxima: $f(x_1)$, occurs at x_1; $f(x_3)$, occurs at x_3; $f(x_5)$, occurs at x_5
(**b**) Absolute minimum: $f(x_4)$, occurs at x_4
Absolute maximum: $f(x_1)$, occurs at x_1

23. (**a**) Relative minima: None
Relative maxima: $f(x_3)$, occurs at x_3
(**b**) Absolute minimum: None
Absolute maximum: $f(x_3)$, occurs at x_3

25. (**a**) Relative minima: None
Relative maxima: None
(**b**) Absolute minimum: None
Absolute maximum: None

27.

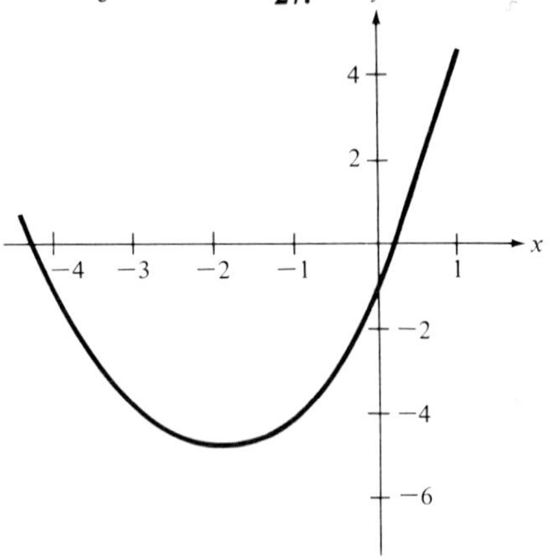

29.

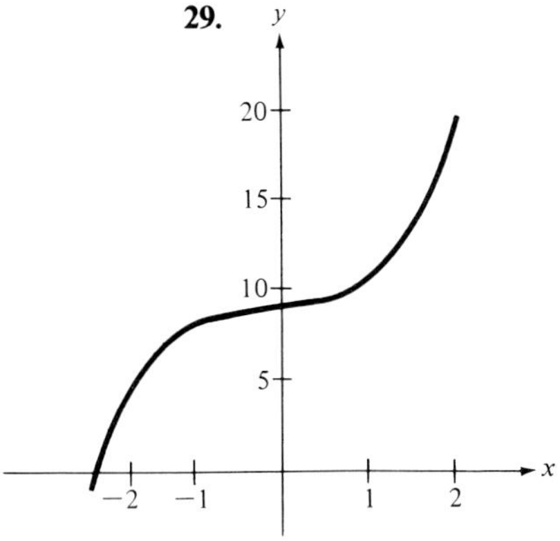
y
20
15
10
5
x
−2
−1
1
2

31.

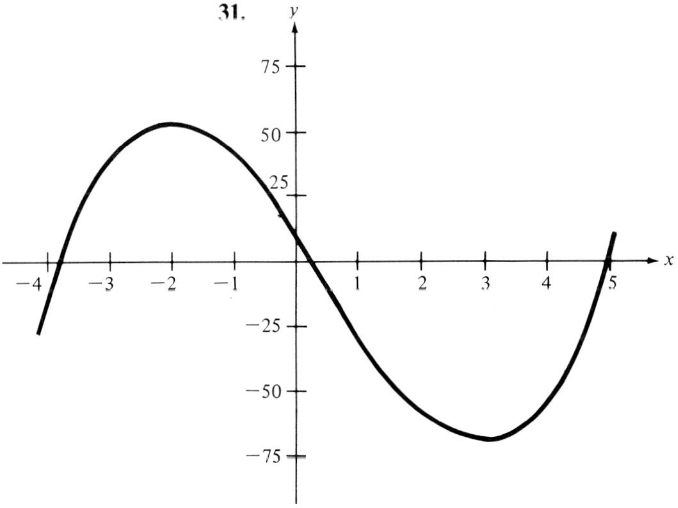
y
75
50
25
x
−4
−3
−2
−1
1
2
3
4
5
−25
−50
−75

33.

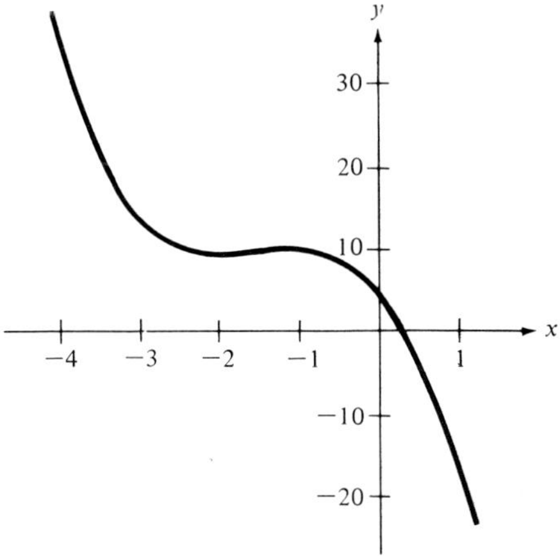
y
30
20
10
x
−4
−3
−2
−1
1
−10
−20

EXERCISES 5-1

1. 1.480 **3.** 4.661 **5.** 10.835 **7.** 0.6648 **9.** 0.5132

11. 17 to 18 **13.** 9 to 10 **15.** 16 to 17 **17.** 8 to 9

19. (**a**) Decay of 4%
(**b**) Decay of 4%

21. (**a**) Decay of 36%
(**b**) Decay of 36%

23. He lost. In fact, if he played the game $2n$ times, his final amount is $(1.5)^n \cdot (0.5)^n = (0.75)^n$ times his initial amount.

25. Doubling life = 10 years; 32,000,000 in 1990; 256,000,000 in 2020; 512,000,000 in 2030; $(200{,}000{,}000) \cdot (1.02)^{25} \approx 328{,}000{,}000$ U.S. population in 2020; $(200{,}000{,}000)(1.02)^{30} \approx 362{,}000{,}000$ U.S. population in 2030; college students exceed U.S. population sometime between 2020 and 2030.

27. Doubling life at 8% is 9 years; $\frac{347}{9} = 38.5$ doublings; say 38;

$$24 \times 2^{38} = 24 \times 2^{10} \times 2^{10} \times 2^{10} \times 2^{8} \approx 24 \times 1000 \times 1000 \times 1000 \times 250 = \$6{,}000{,}000{,}000{,}000$$

EXERCISES 5-2

1. 1024 **3.** $1024 \times 1.414 = 1448$ **5.** 0.0000000001

7.

x	e^x	e^{-x}
−2	0.135	7.39
−1	0.368	2.72
0	1	1
1	2.72	0.368
2	7.39	0.135

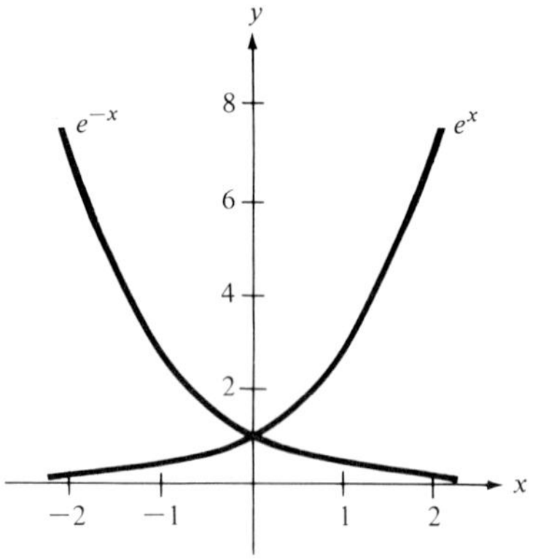

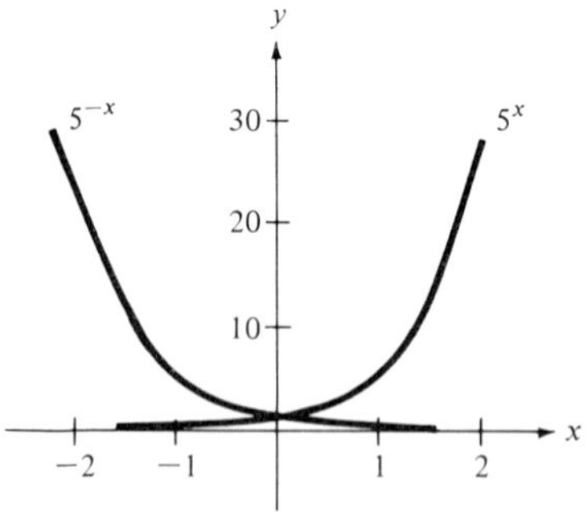

9.

x	5^x	5^{-x}
−2	0.04	25
−1	0.2	5
0	1	1
1	5	0.2
2	25	0.04

11. Increasing for $a > 1$
Decreasing for $0 < a < 1$
Constant for $a = 1$

13. 2.718283075; 2.71828184

15. Increase by 4; $4^1 = 4$; $4^2 = 16$

17. Increase by 10; $10^1 = 10$; $10^2 = 100$

19. Decrease by 4; $4^{-1} = \frac{1}{4}$; $4^{-2} = \frac{1}{16}$

21. Decrease by 10; $10^{-1} = \frac{1}{10}$; $10^{-2} = \frac{1}{100}$

23. $\frac{1}{3}$

25. 4

EXERCISES 5-3

1. e^2 3. $x^2 + 1$ 5. 17 7. $x^2 + 1$ 9. $2x - 5$ 11. $e^{3 \ln 10}$

13. $e^{e \ln 3}$ 15. $e^{(x+1) \ln 4}$ 17. $e^{2 \ln x}$ 19. 3 21. 2 23. 0

25.

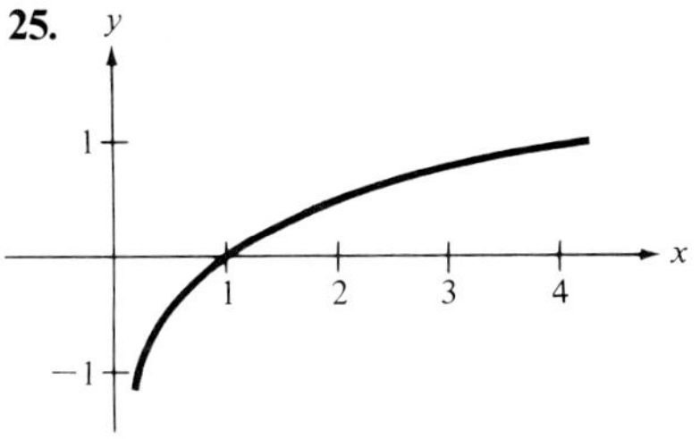

27.

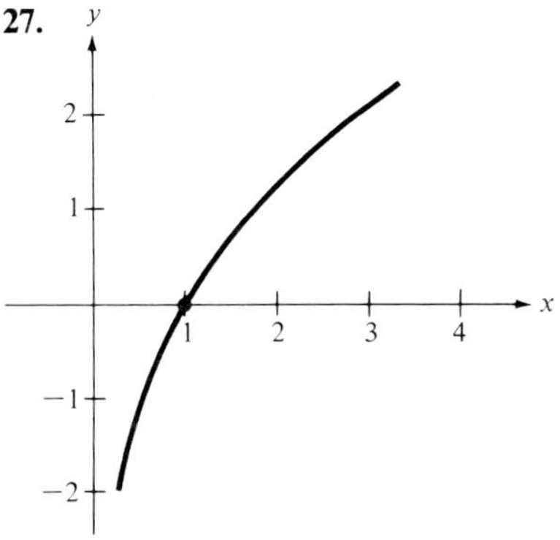

29. 3.17

31. 0.318

33. (a) $y = x^2$
(b) $x \geq 0$; $y \geq 0$

35. (a) $y = \sqrt{4 - x^2}$
(b) $0 \leq x \leq 2$; $0 \leq y \leq 2$

37. (a) $y = \sqrt[5]{x + 3}$
(b) All x; all y

39. $M = a^{\log_a M}$
$N - a^{\log_a N}$

$a^{\log_a (N \cdot M)} = M \cdot N = (a^{\log_a N}) \cdot (a^{\log_a M}) = a^{\log_a N + \log_a M}$

Therefore, $\log_a (N \cdot M) = \log_a N + \log_a M$

EXERCISES 5-4

1. $\dfrac{1}{x \ln 3}$

3. $1 + \dfrac{1}{x}$

5. $3x^2 - \dfrac{1}{x \ln 2}$ **7.** $\ln x + 1$ **9.** $\ln x + \dfrac{1+x}{x}$ **11.** $\dfrac{2}{2x+3}$

13. $-\dfrac{1}{x}$ **15.** $\dfrac{3(\ln x)^2}{x}$ **17.** $\dfrac{2}{x \ln 10}$ **19.** $\dfrac{2}{x+1} + \dfrac{\frac{1}{2}}{x+4}$

21. (**a**) $x = \dfrac{1}{e}$ is a local minimum

$f\left(\dfrac{1}{e}\right) = -\dfrac{1}{e}$

(**b**)

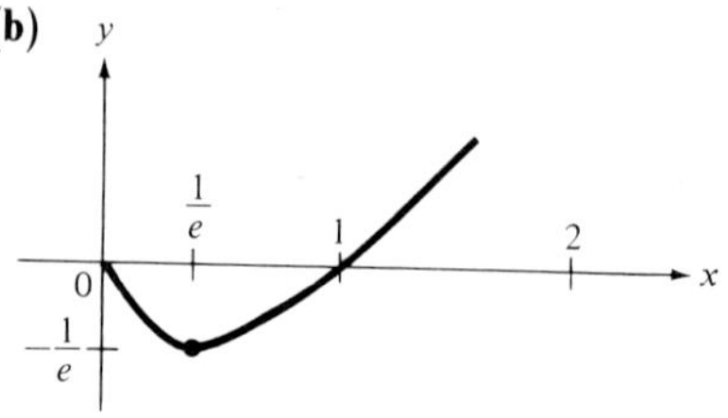

23. (**a**) $x = 0$ is a relative minimum; $f(0) = 0$
(**b**)

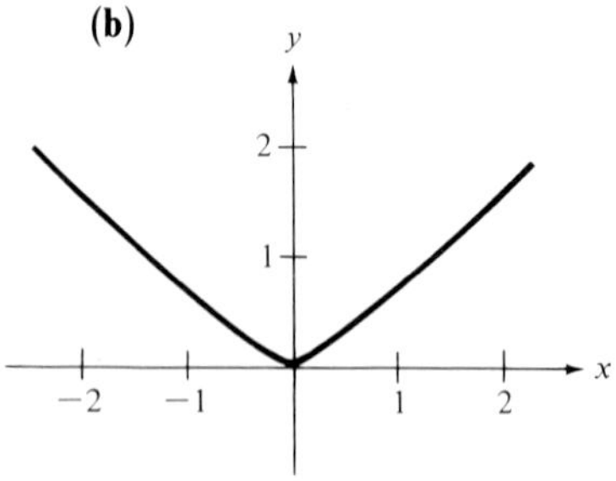

25. $y = 0$ **27.** $x = 3$

EXERCISES 5-5

1. $f'(x) = -e^{-x}$ **3.** $f'(x) = 2^x \cdot (\ln 2)$ **5.** $f'(x) = -\dfrac{3}{x^2} \exp\left(\dfrac{1}{x}\right)$

7. $e^x + \dfrac{1}{x}$ **9.** $f'(x) = e^x \cdot (x+1)$, $f'(-1) = 0$ **11.** $f'(x) = 1$, $f'(e-1) = e$

13. $f'(x) = 2x(\ln 10)10^{x^2}$, $f'(0) = 0$ **15.** $f'(x) = (2x-1)(\ln 3) \cdot 3^{x^2-x}$, $f'(0) = -\ln 3$

17. $f'(x) = \frac{1}{4}e^x$

19. $f'(x) = \left[\ln (x+1) + \dfrac{x}{x+1}\right](x+1)^x$

21. (**a**) Local maximum at $x = 1$; $f(1) = \dfrac{1}{e}$

(b)

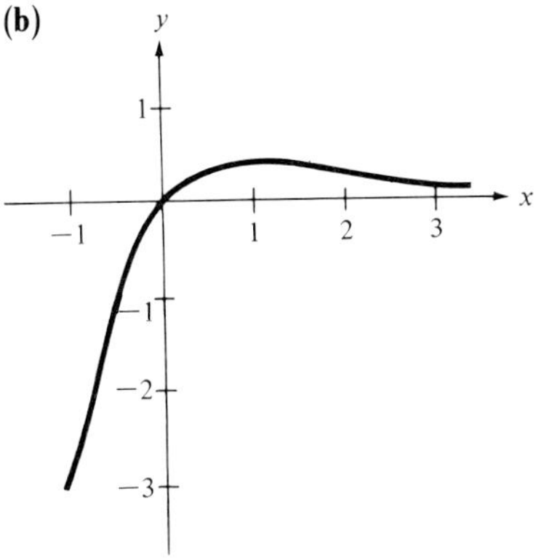

23. **(a)** Local maximum at $x = 0$; $f(0) = 1$

(b)

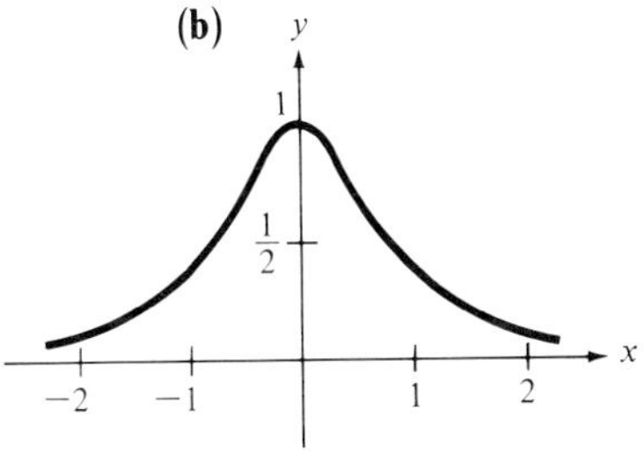

25. $y = x$ **27.** $\frac{1}{4}e^{-q/4}$

EXERCISES 5-6

1. **(a)** 10.25%
(b) 7 years

3. **(a)** 8.16%
(b) 8.24%

5. 2.02% **7.** 9.42% **9.** 1.98% **11.** 8.62%

13. For continuous growth processes, the amount of growth increases with time; for continuous decay processes, the amount of decay decreases with time.

15. 7.69% **17.** 19.72% **19.** 21.04% per year

21. 6.93% for 10-year half-life
13.86% for 5-year half-life

CHAPTER 5. REVIEW EXERCISES

1. 23 **3.** 10 **5.** 23 **7.** 10 **9.** 35 **11.** 10 **13.** 69

15. 12 **17.** 3.92 **19.** 7.70 **21.** 2.02% **23.** 6.18%

25. $\frac{1}{x} + \frac{2x}{x^2+1}$ **27.** $\frac{3}{(3x+5)(\ln 10)}$ **29.** $2e^{2x} + 2xe^{x^2}$

31. $2(\ln 10) \cdot 10^{2x}$ **33.** $y - e^{-6} = -3e^{-6}(x-2)$ **35.** $x < 2$

37.

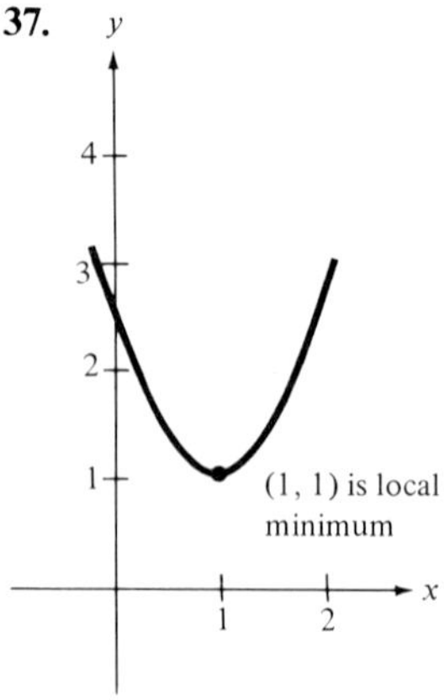

39. $2(1 - x^2)$ **41.** 0

43. $x + 1$ **45.** $+2$ or -2

47. $-\frac{3}{2}$ **49.** 60

51. 11 days

EXERCISES 6-1

1. $\sum_{j=1}^{n} (x_j)^2$ **3.** $\sum_{j=1}^{n} j^2$ **5.** $\sum_{j=1}^{n} j(j+1)$ **7.** $\sum_{j=1}^{n} \frac{1}{j}$

9. $1 \cdot 3 + 2 \cdot 5 + 3 \cdot 7 + 4 \cdot 9 + 5 \cdot 11 + 6 \cdot 13$

11. $\frac{2}{2} + \frac{3}{4} + \frac{4}{6} + \frac{5}{8} + \frac{6}{10}$

13. $4 + 4 + 4 + 4 + 4 + 4 + 4 + 4 + 4 + 4$

15. $\frac{1}{2} \sum_{j=1}^{5} j^2$ **17.** $\frac{1}{2} \sum_{j=1}^{20} (x_j + 3) = \frac{1}{2} \left[\sum_{j=1}^{20} (x_j) \right] + 30$

21.

j	x_j	$f(x_j)$
1	0	1
2	$\frac{1}{2}$	$1\frac{1}{4}$
3	1	2
4	$1\frac{1}{2}$	$3\frac{1}{4}$

$\sum_{j=1}^{n} f(x_j)\, \Delta x = \frac{15}{4}$

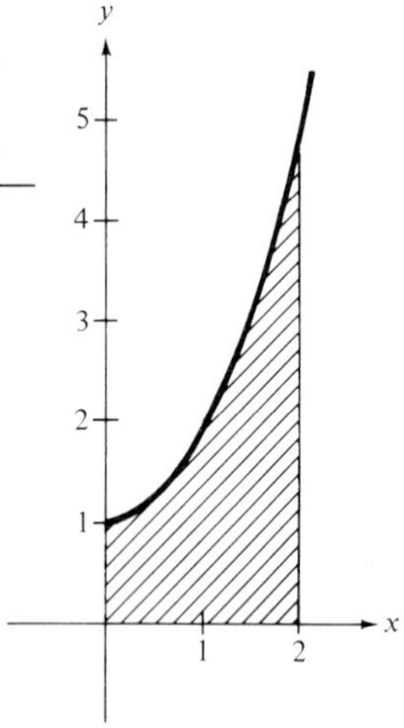

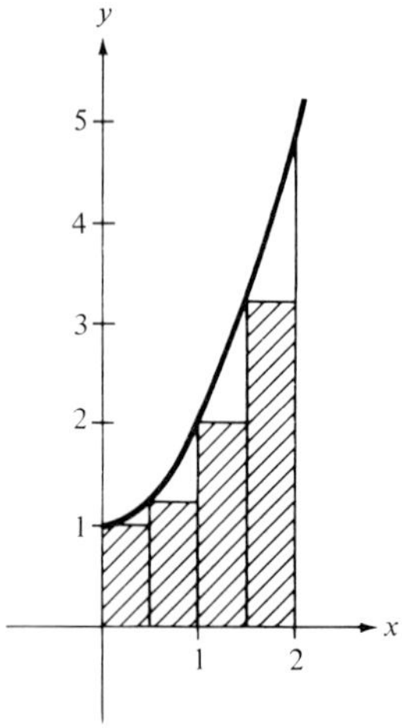

23.

j	x_j	$f(x_j)$
1	-1	4
2	$-\frac{1}{2}$	$4\frac{1}{2}$
3	0	5
4	$\frac{1}{2}$	$5\frac{1}{2}$

$$\sum_{j=1}^{n} f(x_j)\,\Delta x = 9\tfrac{1}{2}$$

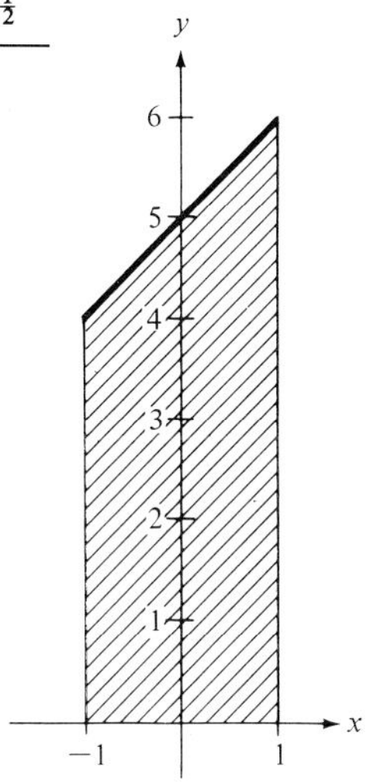

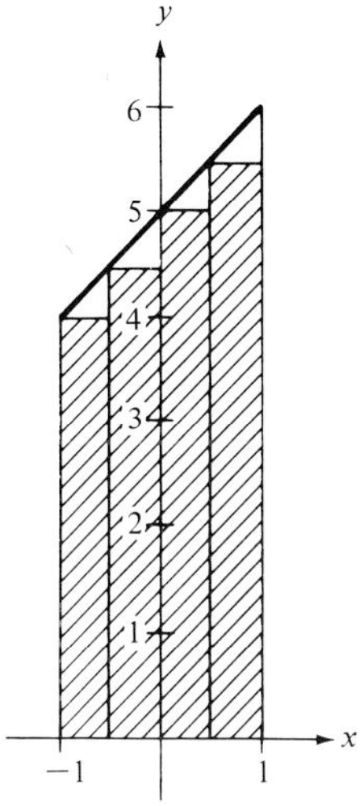

25.

j	x_j	$f(x_j)$
1	0	1
2	$\frac{1}{2}$	1.65
3	1	2.72
4	$1\frac{1}{2}$	4.48

$$\sum_{i=1}^{n} f(x_j)\,\Delta x = 4.925$$

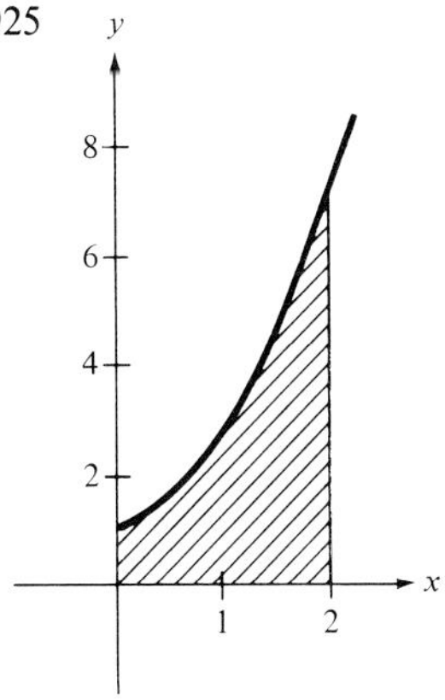

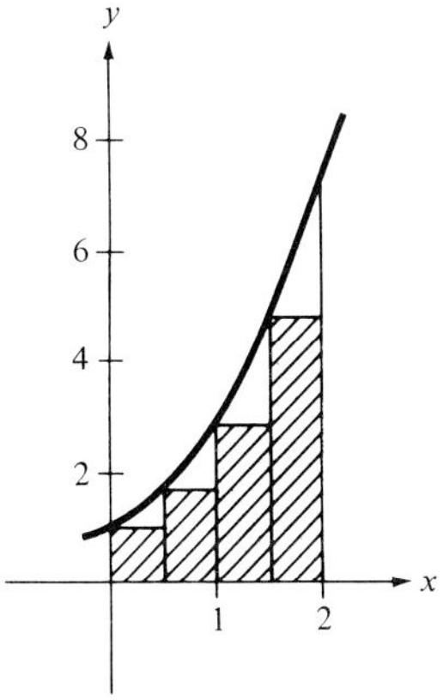

27.

j	x_j	$f(x_j)$
1	0	0
2	1	0.69
3	2	1.61
4	3	2.30

$$\sum_{j=1}^{n} f(x_j)\,\Delta x = 4.60$$

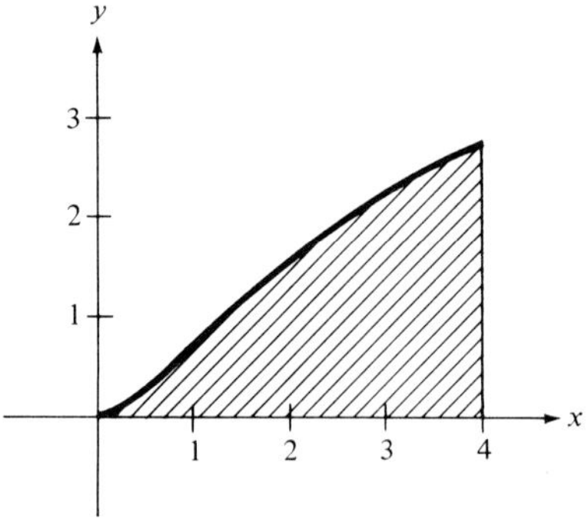

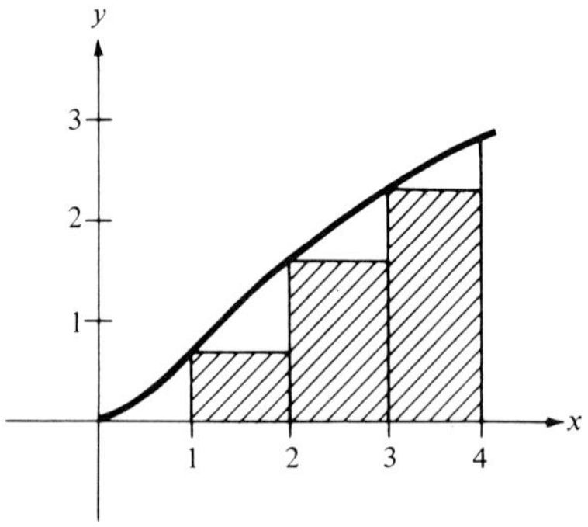

29. (a)

j	x_j	$f(x_j)$
1	$\frac{1}{2}$	$1\frac{1}{4}$
2	1	2
3	$1\frac{1}{2}$	$3\frac{1}{4}$
4	2	5

$$\sum_{j=1}^{n} f(x)\,\Delta x = 5\tfrac{3}{4}$$

(b)

j	x_j	$f(x_j)$
1	$\frac{1}{4}$	$1\frac{1}{16}$
2	$\frac{3}{4}$	$1\frac{9}{16}$
3	$1\frac{1}{4}$	$2\frac{9}{16}$
4	$1\frac{3}{4}$	$3\frac{1}{16}$

$$\sum_{j=1}^{n} f(x)\,\Delta x = 4\tfrac{5}{8}$$

EXERCISES 6-2

1. (a) 4; 0
   (b) 2

3. (a) 8; −8
   (b) 0

5. (a) 12; −12
   (b) 2

7. (a) 0; −4
   (b) −2

9. (a)

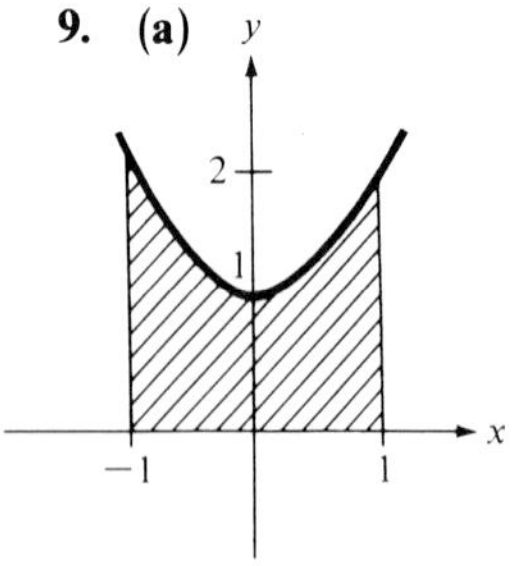

(b) 4; 2
(c) 3

11. (**a**)

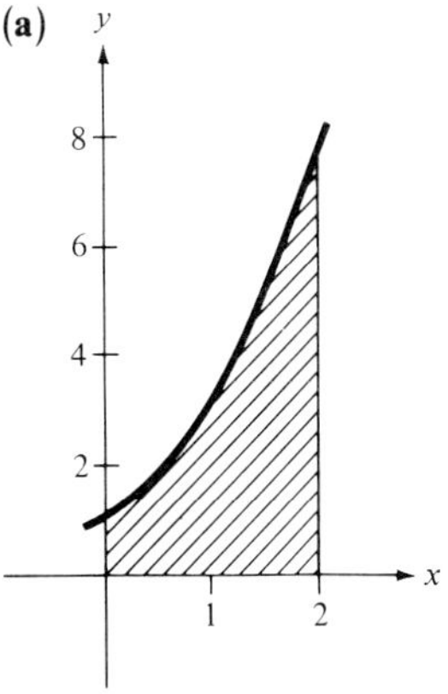

(**b**) 14.8; 2

(**c**) $\dfrac{1 + 7.4}{2} \cdot 2 = 8.4$

13. (**a**)

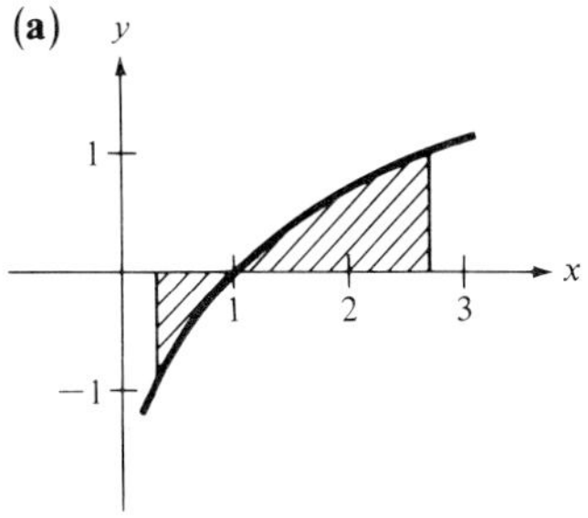

(**b**) 2.35; -2.35

(**c**) $-\dfrac{0.6}{2} + \dfrac{1.7}{2} = \dfrac{1.1}{2} = 0.55$

15. (**a**)

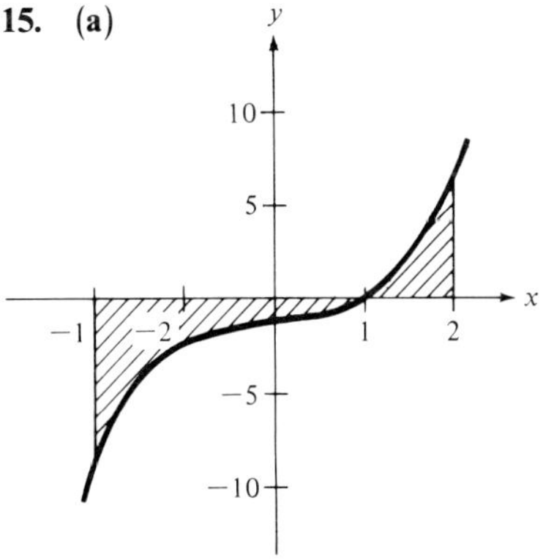

(**b**) 28; -36
(**c**) -4

17. (**a**)

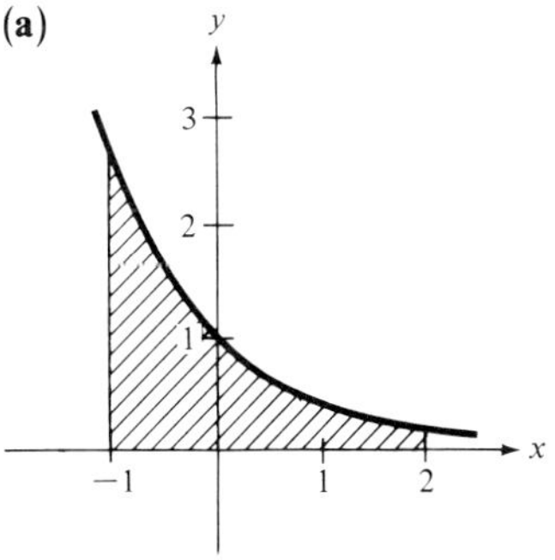

(**b**) 5.1; 0.42

(**c**) $1 + \dfrac{2.7}{2} + \dfrac{1.2}{2} = 3.35$

19. $3\int_0^1 x^4\,dx - 2\int_0^1 x^3\,dx + 3\int_0^1 x\,dx + 5\int_0^1 dx$

21. $4\int_{-1}^2 x^5\,dx - 3\int_{-1}^2 x^3\,dx - 2\int_{-1}^2 x\,dx + 6\int_{-1}^2 dx$

23. Yes **25.** -5 **27.** $5 + 2 = 7$

EXERCISES 6-3

1. (**a**) $f(x)$ is a polynomial, and thus is continuous for all x
   (**c**) 2

3. **(a)** $f(x)$ is a polynomial, and thus is continuous for all x
(c) 6

5. **(a)** $\sqrt{x}$ is continuous for $x > 0$; $1/x^2$ is continuous for $x \neq 0$; consequently, $f(x)$ is continuous for $x > 0$
(c) -0.72

7. **(a)** $f(x)$ is continuous for all x, since x and e^x are continuous for all x
(c) $3e^4 + 1$

9. **(1)** $x^3 + 8$
(3) $x^2 + x + 8$
(5) $-\frac{1}{x} - \frac{2}{3}x^{3/2} + 8$
(7) $xe^x - e^x + 8$

11. 12 **13.** $\sqrt{\frac{2}{3}}$ **15.** 300 **17.** $e - 1$

EXERCISES 6-4

1. $\frac{x^{11}}{11} + C$
3. $-\frac{1}{3x^3} + C$
5. $-2x^{-1/2} + C$
7. $\frac{e^{3x}}{3} + C$
9. $\frac{x^4}{4} - x^3 + \frac{2}{3}x^{3/2} + C$
11. $3 \ln x + \frac{2}{3x^3} - \frac{5}{4x^4} + C$
13. $\frac{2}{5}x^{5/2} + \frac{2}{3}x^{3/2} + 2x^{1/2} + C$
15. $\frac{2^x}{\ln 2} - \frac{2^{3x}}{3 \ln 2} + C$
17. $-\frac{1}{10}k + C$
19. $\frac{2}{11}x^{44} + C$
21. $-\frac{10}{18}x^{6/5} + C$
23. $-8 \ln x + C$
25. $-\frac{5}{22}x^{11/5} + C$
27. $\frac{1}{4}(1 + x)^4 + C$
29. $\frac{1}{3}(x + 3)^3 + C$

EXERCISES 6-5

1. 20 **3.** $e + \frac{1}{e} - 2$ **5.** $\frac{143}{15}$ **7.** $\frac{5}{6} - \frac{e^{-2}}{2} - \frac{e^{-3}}{3}$ **9.** 23

11. ln 2 **13.** $\frac{1}{x^{1/3}}$ is not continuous at $x = 0$

15. $\frac{1}{x^{1/5}}$ is not continuous at $x = 0$

17. **(a)** $F(x) = x^3 - \frac{x^2}{2} + x + 4$
(c) 12

19. **(a)** $F(x) = \frac{x^2}{2} + e^x - 5\frac{1}{2} - e$

(c) $e^2 - 3\frac{1}{2} - e$

21. $C = 20{,}000 + \int_0^x (2 + 0.1t)\, dt = 20{,}000 + 2x + 0.05x^2$

23. **(a)** $\int_0^2 3000e^{-t/2}\, dt \approx 3793$

(b) 1207

(c) $\int_0^t 3000e^{-x/2}\, dx = 6000(1 - e^{-t/2})$

(d) 3.6

EXERCISES 6-6

1. $P_1(T) = 43.33 - 8T + 0.3T^2 + 0.00667T^3$

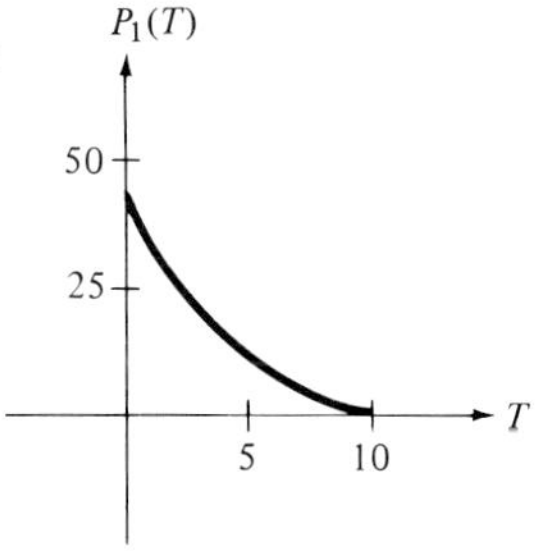

3. Second: $\int_1^2 5000e^{0.05t}\, dt = \$5390$

Tenth: $\int_9^{10} 5000e^{0.05t}\, dt = \$8041$

Twentieth: $\int_{19}^{20} 5000e^{0.05t}\, dt = \$13{,}257$

5. 96

7. $\int_0^{10} 100{,}000(1 - e^{-t/2})\, dt = \$801{,}350$

9. $\int_0^T \frac{e^{0.02t}}{40}\, dt = \frac{e^{0.02T} - 1}{0.8}$; consumed in 29.4 years

EXERCISES 6-7

1.

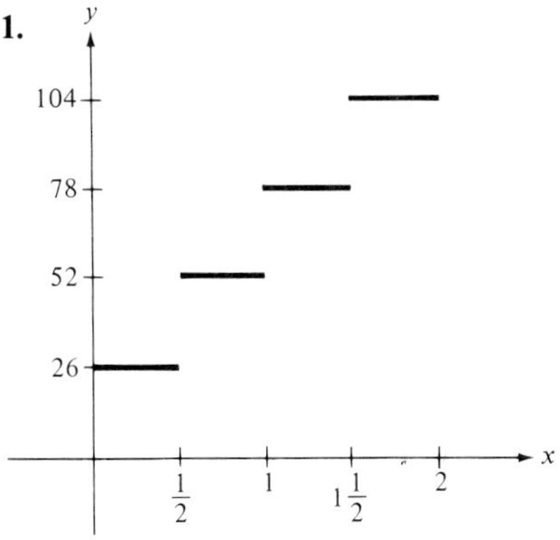

3. $\frac{17}{6}$ 5. $\frac{2}{3}\sqrt{2} - \frac{4}{3}$ 7. $\frac{2027}{12} \approx 168.9$ 9. $e + 1 - \ln 2 \approx 3.025$ 11. 0

EXERCISES 6-8

1. $S_1 = 1 \cdot 3$
$S_2 = 1 \cdot 3 + 2 \cdot 5$
$S_3 = 1 \cdot 3 + 2 \cdot 5 + 3 \cdot 7$
$S_4 = 1 \cdot 3 + 2 \cdot 5 + 3 \cdot 7 + 4 \cdot 9$
$S_5 = 1 \cdot 3 + 2 \cdot 5 + 3 \cdot 7 + 4 \cdot 9 + 5 \cdot 11$
$S_6 = 1 \cdot 3 + 2 \cdot 5 + 3 \cdot 7 + 4 \cdot 9 + 5 \cdot 11 + 6 \cdot 13$

3. $S_1 = \frac{2}{2}$
$S_2 = \frac{2}{2} + \frac{3}{4}$
$S_3 = \frac{2}{2} + \frac{3}{4} + \frac{4}{6}$
$S_4 = \frac{2}{2} + \frac{3}{4} + \frac{4}{6} + \frac{5}{8}$
$S_5 = \frac{2}{2} + \frac{3}{4} + \frac{4}{6} + \frac{5}{8} + \frac{6}{10}$

5.
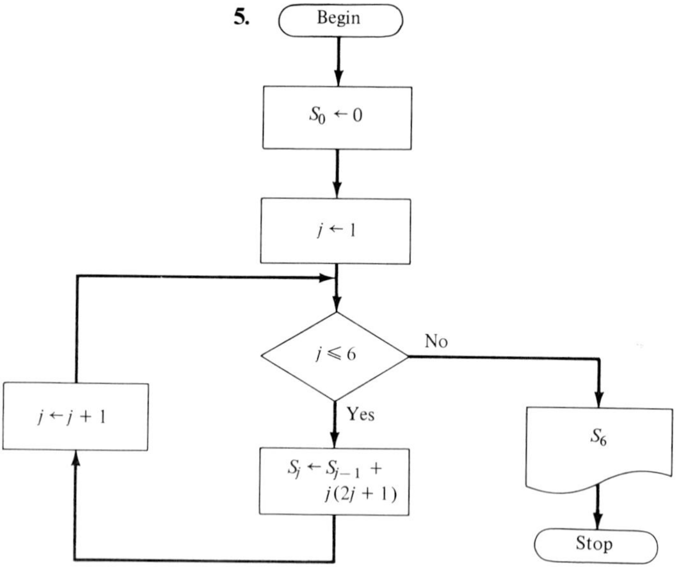

7.

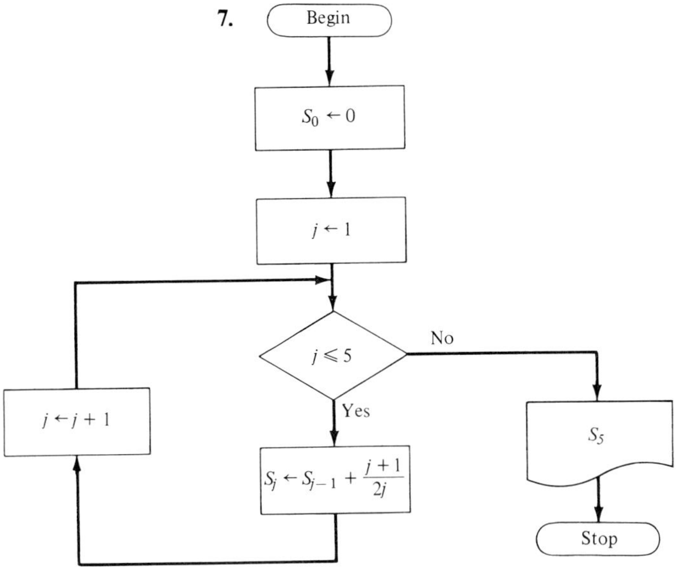
Begin
$S_0 \leftarrow 0$
$j \leftarrow 1$
$j \leqslant 5$
No
Yes
$j \leftarrow j + 1$
$S_j \leftarrow S_{j-1} + \frac{j+1}{2j}$
S_5
Stop

9.

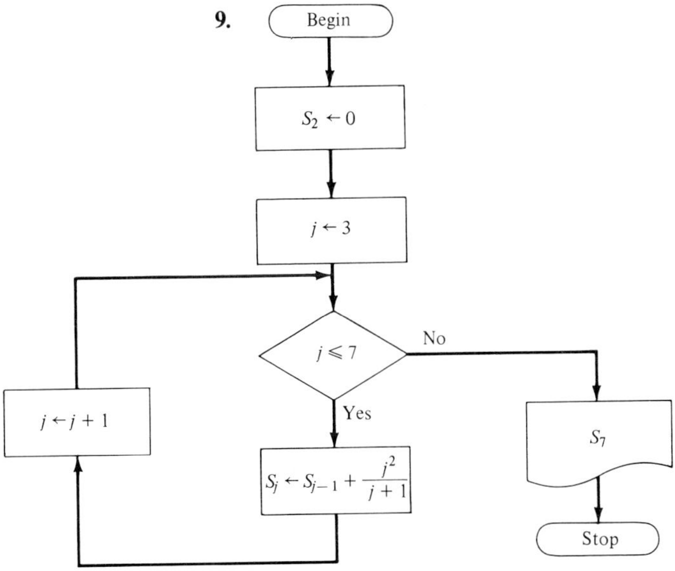
Begin
$S_2 \leftarrow 0$
$j \leftarrow 3$
$j \leqslant 7$
No
Yes
$j \leftarrow j + 1$
$S_j \leftarrow S_{j-1} + \frac{j^2}{j+1}$
S_7
Stop

11. Change the statement in Block 6 to $x_j \leftarrow a + (j - 1) \cdot \Delta x$

13.

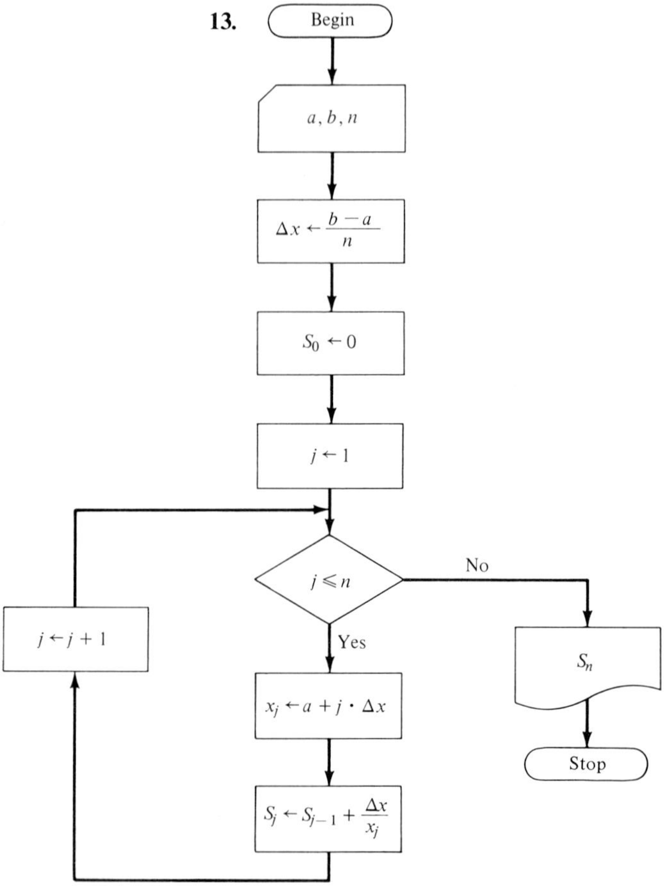

15.

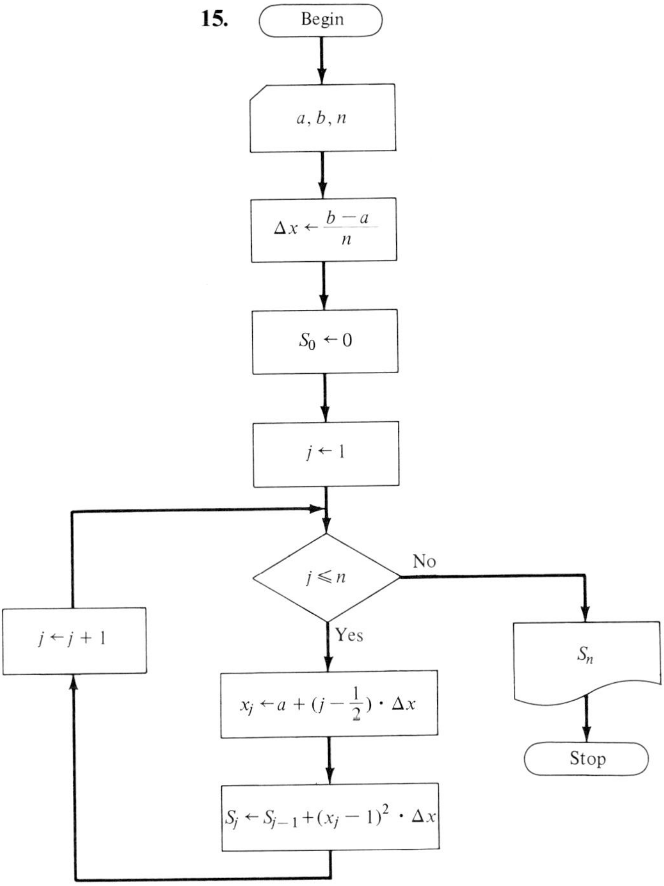
Begin
a, b, n
$\Delta x \leftarrow \frac{b-a}{n}$
$S_0 \leftarrow 0$
$j \leftarrow 1$
$j \leqslant n$
No
Yes
$j \leftarrow j + 1$
$x_j \leftarrow a + (j - \frac{1}{2}) \cdot \Delta x$
$S_j \leftarrow S_{j-1} + (x_j - 1)^2 \cdot \Delta x$
S_n
Stop

17.

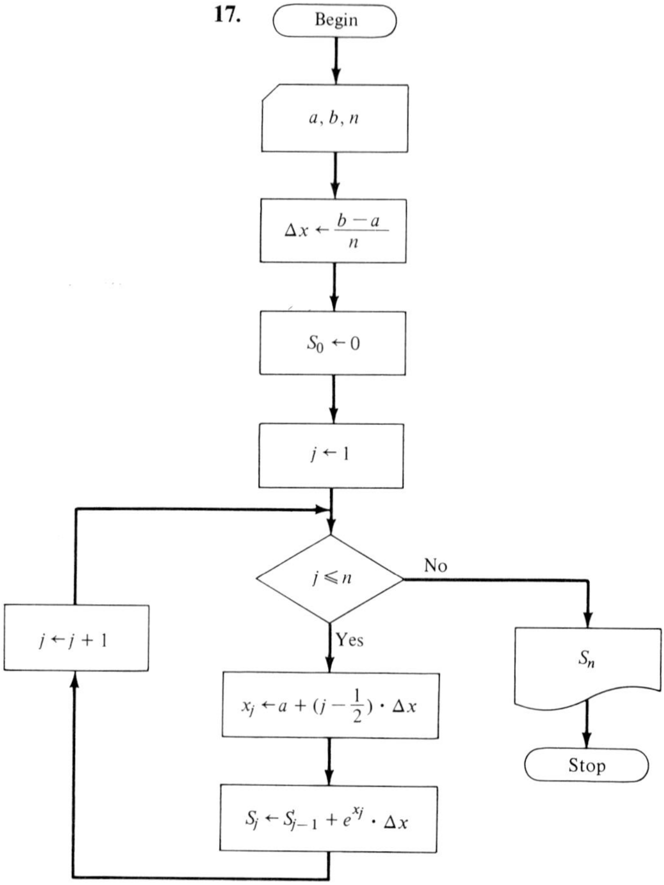

CHAPTER 6. REVIEW EXERCISES

1. $1 - e^{-2}$ **3.** 0 **5.** $e - \frac{e^3}{3} - \frac{2}{3}$ **7.** $-\frac{2}{3}$ **9.** $\frac{55}{6}$ **11.** $\frac{1}{\ln 2}$

13. $\ln 2 - \frac{1}{2}$ **15.** $\ln 3 + 3\frac{1}{9}$ **17.** -1 **19.** 1 **21.** 0 **23.** 2

25. Upper bound $= 12$
Lower bound $= 6$

27. Upper bound $= 48$
Lower bound $= 24$

29. 2 **31.** -1 **33.** 0 **35.** 6

37. $\int (f(x) \cdot g(x))\,dx$ is not easily related to $\int f(x)\,dx$ and $\int g(x)\,dx$

39.

j	x_j	$e^{x_j/2}$
1	1.2	1.82
2	1.4	2.01
3	1.6	2.23
4	1.8	2.46
5	2.0	2.72
6	2.2	3.00
7	2.4	3.32
8	2.6	3.67
9	2.8	4.06
10	3.0	4.48
		29.77

$$\sum_{j=1}^{10} e^{x_j/2}\,\Delta x = (29.77)\cdot(0.2) = 5.96$$

41. 86 **43.** 70 **45.** $\sum_{j=1}^{n} j\cdot(j+1)$ **47.** $\sum_{j=1}^{n}\frac{1}{j}$ **49.** $\sum_{j=1}^{n}\frac{2^j}{2j+1}$

51. $(1{,}000{,}000)(e^{0.52}-1)\approx 682{,}000$ **53.** $\frac{\ln 3}{2}$

EXERCISES 7-1

1. $\frac{e^{x^2-1}}{2}+C$ **3.** $\frac{(x^2-x+2)^5}{5}+C$ **5.** $\frac{(\ln x)^2}{2}+C$ **7.** $\frac{2}{3}\sqrt{x^3-2}+C$

9. $\frac{3}{2}e^{x^2}+C$ **11.** $\frac{2}{5}(x^3+x+2)^{5/2}+C$ **13.** $\frac{(x-x^3)^6}{6}+C$

15. $e^{x^3+2x}+C$ **17.** $\frac{1}{3}(8-2\sqrt{2})$ **19.** $\frac{1}{10}$ **21.** $\frac{\ln 4}{6}$ **23.** 324

EXERCISES 7-2

1. $\frac{2x-1}{4}e^{2x}+C$ **3.** $\frac{1}{2}(x^2-1)e^{x^2}+C$ **5.** $-(x+4)e^{-x}+C$

7. $\frac{2\ln(x+1)-1}{4}(x+1)^2+C$ **9.** 1 **11.** 2 **13.** $2-5e^{-1}$

15. **(a)** $5\ln 5-4\ln 4-1$
(b) $6\ln 6-5$

EXERCISES 7-3

1. 1 **3.** Does not exist **5.** $\frac{1}{2}$ **7.** -1

9. $\lim_{h\to\infty} \int_{-h}^{h} x^3\,dx = 0$

$\lim_{h\to\infty} \int_{0}^{h} x^3\,dx = \lim_{h\to\infty} \left(\frac{h^4}{4}\right)$, which does not exist

11. $\int_1^{\infty} \frac{dx}{x^m} = \lim_{h\to\infty} \int_1^h \frac{dx}{x^m} = \frac{1}{m-1} \lim_{h\to\infty} (1 - h^{1-m})$, which exists only when $m > 1$

13. $\int_1^{\infty} g(x)\,dx$ does not exist

EXERCISES 7-4

1. ∞ **3.** 0 **5.** $-\infty$ **7.** $-\infty$

9. Not continuous at $x = 0$; value of integral is 2

11. Not continuous at $x = 1$; integral does not exist

13. Not continuous at $x = -2$; value of integral is -2

15. Not continuous at $x = 0$; integral does not exist for $m > 1$

17. **(a)** Exists for $0 < m < 1$
(b) Does not exist for $m \geq 1$

19. Not continuous at $x = 0$; integral does not exist

21. Not continuous at $x = 0$; value of integral is -1

EXERCISES 7-5

1. $\int_0^8 (62.4)\left(\frac{\pi}{4}\right)\frac{(12-x)^3}{4}\,dx = \frac{62.4\pi}{64}(12^4 - 4^4) \approx 62{,}731$

3. $\int_0^{15} (62.4)(250)x(15-x)\,dx = (62.4)(250)(15)^3(\frac{1}{2} - \frac{1}{3}) \approx 8{,}775{,}000$

5. $\int_0^{150} \left(\frac{\pi}{4}\right)\frac{(500)^2}{100}(80 - 0.002x^2)\,dx = \frac{\pi}{4}\frac{(500)^2}{100}(150)(65) \approx 19{,}144{,}125$

7. $\int_0^{80} 40{,}000e^{0.01x}\,dx = 4{,}000{,}000(e^{0.8} - 1) \approx 4{,}902{,}163$

CHAPTER 7. REVIEW EXERCISES

1. $-\frac{1}{2(x^2+2)} + C$

3. $xe^{x-1} - e^{x-1} + C$

5. $-3x^2e^{-x} - 6xe^{-x} - 6e^{-x} + C$

7. $\frac{4}{3}(x^2+4)^{3/2} + C$

9. $-\frac{2 \ln x}{x} - \frac{2}{x} + C$

11. $x^2e^{x+1} - 2xe^{x+1} + 2e^{x+1} + C$

13. $\frac{(\ln 2)^2}{2}$

15. $\frac{2e^3}{9} + \frac{e^2}{4} + \frac{13}{36}$

17. 0 **19.** $e - e^{1/3}$ **21.** $1 - 7e^{-6}$ **23.** $\frac{1}{4}$ **25.** Does not exist

27. $\frac{1}{3}$ **29.** Does not exist **31.** Does not exist **33.** Does not exist

35. Does not exist

EXERCISES 8-1

1. (**a**) All x; all y
(**b**) Only at the point (0, 0)

3. (**a**) All x; all y
(**c**) Is never zero

5. (**a**) All (x, y) satisfying $x^2 + y^2 \leq 4$; this is the set of points on or inside a circle of radius 2 and centered at the origin
(**b**) All (x, y) satisfying $x^2 + y^2 = 4$; this is the set of points on a circle of radius 2 centered at the origin

7. (**a**) All x; all y
(**b**) Only at the point (0, 0)

9. (0, 0, −6), (0, −3, 0), (2, 0, 0)

11. (0, 0, −10), (0, −2, 0), (5, 0, 0)

13. (0, 0, 6), (0, 2, 0), (3, 0, 0)

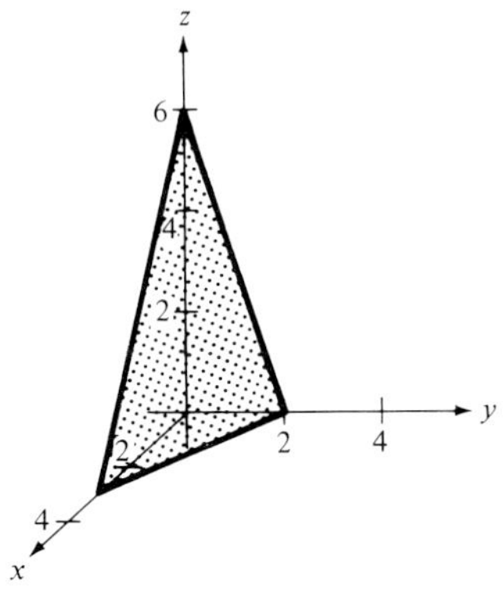

15. (0, 0, 4), (0, 2, 0), (4, 0, 0)

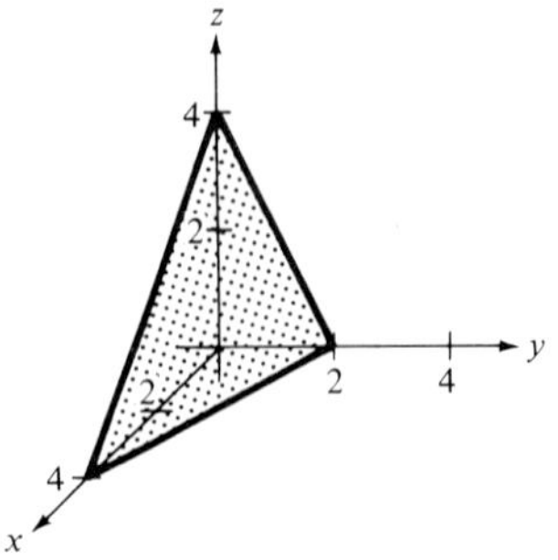

17.

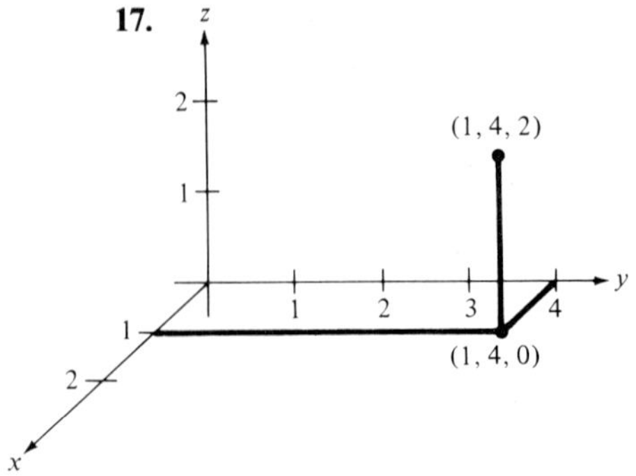

19.

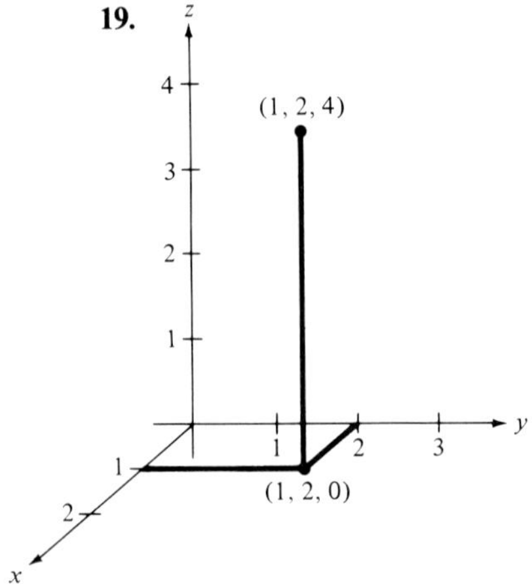

EXERCISES 8-2

1. $\dfrac{\partial f}{\partial x} = y^2;\ \left.\dfrac{\partial f}{\partial x}\right|_{(1,\,-2)} = 4$

$\dfrac{\partial f}{\partial y} = 2xy;\ \left.\dfrac{\partial f}{\partial y}\right|_{(1,\,-2)} = -4$

3. $\dfrac{\partial f}{\partial x} = \dfrac{2}{2x + y^2};\ \left.\dfrac{\partial f}{\partial x}\right|_{(1,\,1)} = \dfrac{2}{3}$

$\dfrac{\partial f}{\partial y} = \dfrac{2y}{2x + y^2};\ \left.\dfrac{\partial f}{\partial y}\right|_{(1,\,1)} = \dfrac{2}{3}$

5. $\dfrac{\partial f}{\partial x} = y - y(\ln 3)3^{xy};\ \left.\dfrac{\partial f}{\partial x}\right|_{(-1,\,1)} = 1 - \dfrac{\ln 3}{3}$

$\dfrac{\partial f}{\partial y} = x - x(\ln 3)3^{xy};\ \left.\dfrac{\partial f}{\partial y}\right|_{(-1,\,1)} = -1 + \dfrac{\ln 3}{3}$

7. $\dfrac{\partial f}{\partial x} = 2xy - 4(1 + y^2);\ \left.\dfrac{\partial f}{\partial x}\right|_{(-2,\,2)} = -28$

$\dfrac{\partial f}{\partial y} = x^2 - 8xy;\ \left.\dfrac{\partial f}{\partial y}\right|_{(-2,\,2)} = 36$

9. $\dfrac{\partial f}{\partial x} = (1 + 2x + 2y)e^{2x+3y}$; $\left.\dfrac{\partial f}{\partial x}\right|_{(-1,\,1)} = e$

$\dfrac{\partial f}{\partial y} = (1 + 3x + 3y)e^{2x+3y}$; $\left.\dfrac{\partial f}{\partial y}\right|_{(-1,\,1)} = e$

11. $\dfrac{\partial f}{\partial x} = 2y$

$\dfrac{\partial f}{\partial y} = 2x - 2y$

$\dfrac{\partial^2 f}{\partial x^2} = 0$

$\dfrac{\partial^2 f}{\partial y^2} = -2$

$\dfrac{\partial^2 f}{\partial x\,\partial y} = 2$

$\dfrac{\partial^2 f}{\partial y\,\partial x} = 2$

13. $\dfrac{\partial f}{\partial x} = ye^x$

$\dfrac{\partial f}{\partial y} = e^x$

$\dfrac{\partial^2 f}{\partial x^2} = ye^x$

$\dfrac{\partial^2 f}{\partial y^2} = 0$

$\dfrac{\partial^2 f}{\partial x\,\partial y} = e^x$

$\dfrac{\partial^2 f}{\partial y\,\partial x} = e^x$

15. $\dfrac{\partial f}{\partial x} = \dfrac{1}{x + y}$

$\dfrac{\partial^2 f}{\partial x^2} = -\dfrac{1}{(x + y)^2}$

$\dfrac{\partial^2 f}{\partial x\,\partial y} = -\dfrac{1}{(x + y)^2}$

$\dfrac{\partial f}{\partial y} = \dfrac{1}{x + y}$

$\dfrac{\partial^2 f}{\partial y^2} = -\dfrac{1}{(x + y)^2}$

$\dfrac{\partial^2 f}{\partial y\,\partial x} = -\dfrac{1}{(x + y)^2}$

17. $\dfrac{\partial f}{\partial x} = -\dfrac{4y}{(x - 2y)^2}$: defined everywhere except along the line $y = \dfrac{x}{2}$

$\dfrac{\partial f}{\partial y} = \dfrac{4x}{(x - 2y)^2}$: defined everywhere except along the line $y = \dfrac{x}{2}$

19. $\dfrac{\partial f}{\partial x} = \dfrac{-x}{\sqrt{4 - x^2 - y^2}}$; defined for all (x, y) satisfying $x^2 + y^2 < 4$

$\dfrac{\partial f}{\partial y} = \dfrac{-y}{\sqrt{4 - x^2 - y^2}}$; defined for all (x, y) satisfying $x^2 + y^2 < 4$

21. $\dfrac{\partial f}{\partial x} = 2x$; defined everywhere

$\dfrac{\partial f}{\partial y} = 2y$; defined everywhere

$\dfrac{\partial f}{\partial z} = 2z$; defined everywhere

23. $\dfrac{\partial f}{\partial x} = \dfrac{-x}{\sqrt{4 - x^2 - y^2 - z^2}}$; defined for (x, y, z) satisfying $x^2 + y^2 + z^2 < 4$

$\dfrac{\partial f}{\partial y} = \dfrac{-y}{\sqrt{4 - x^2 - y^2 - z^2}}$; defined for (x, y, z) satisfying $x^2 + y^2 + z^2 < 4$

$\dfrac{\partial f}{\partial z} = \dfrac{-z}{\sqrt{4 - x^2 - y^2 - z^2}}$; defined for (x, y, z) satisfying $x^2 + y^2 + z^2 < 4$

EXERCISES 8-3

1. **(a)** (0, 0)
(b) Relative minimum

3. **(a)** (0, 0)
(b) Neither

5. **(a)** $(-1, 1)$
(b) Relative minimum

7. **(a)** $(\frac{2}{3}, -\frac{1}{3})$
(b) Relative minimum

9. **(a)** (0, 0)
(b) Neither

11. **(a)** $(0, 1)$; $(0, -1)$
(b) $(0, 1)$ is neither
$(0, -1)$ is neither

EXERCISES 8-4

1. $\dfrac{\partial f}{\partial x} = y^2z^3$ $\qquad \dfrac{\partial f}{\partial y} = 2xyz^3$

$\dfrac{\partial f}{\partial z} = 3xy^2z^2$ $\qquad \dfrac{\partial^2 f}{\partial x\, \partial y} = 2yz^3$

Domain of $f(x, y, z)$ and all of the above partial derivatives is all points (x, y, z)

3. $\dfrac{\partial f}{\partial x} = \dfrac{-2xy}{(x^2 + z^2)^2}$ $\qquad \dfrac{\partial f}{\partial y} = \dfrac{1}{x^2 + z^2}$

$\dfrac{\partial f}{\partial z} = \dfrac{-2zy}{(x^2 + z^2)^2}$ $\qquad \dfrac{\partial^2 f}{\partial x\, \partial y} = \dfrac{-2x}{(x^2 + z^2)^2}$

Domain of $f(x, y, z)$ and all of the above partial derivatives is all points (x, y, z) for which either $x \neq 0$ or $z \neq 0$

5. $\dfrac{\partial f}{\partial x} = yze^{xyz}$ $\qquad \dfrac{\partial f}{\partial y} = xze^{xyz}$

$\dfrac{\partial f}{\partial z} = xye^{xyz}$ $\qquad \dfrac{\partial^2 f}{\partial x\, \partial y} = (z + xyz^2)e^{xyz}$

Domain of $f(x, y, z)$ and the above partial derivatives is all points (x, y, z)

7. $2x_i$

9. $\sum_{j=1}^{i-1} 3x_i^2 x_j^2 + \sum_{j=i+1}^{n} 2x_i x_j^3$

11. $\sum_{\substack{j=1 \\ j \neq i}}^{n} e^{(x_i + x_j)}$

13. Relative minimum occurs at (0, 0, 0, 0); $f(0, 0, 0, 0) = 0$

15. Relative maximum occurs at (0, 0, 0); $f(0, 0, 0) = 1$

EXERCISES 8-5

3.

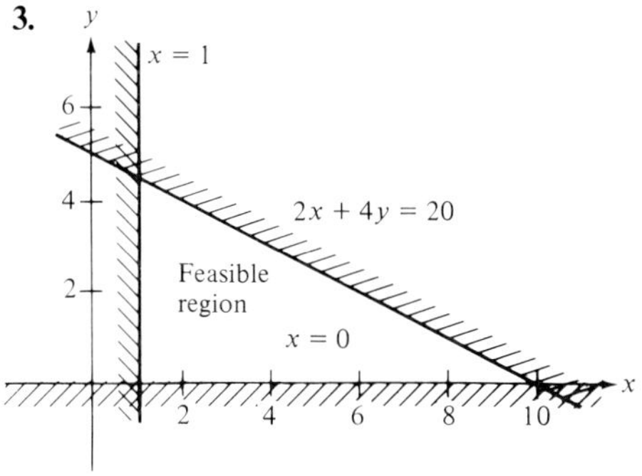

5.

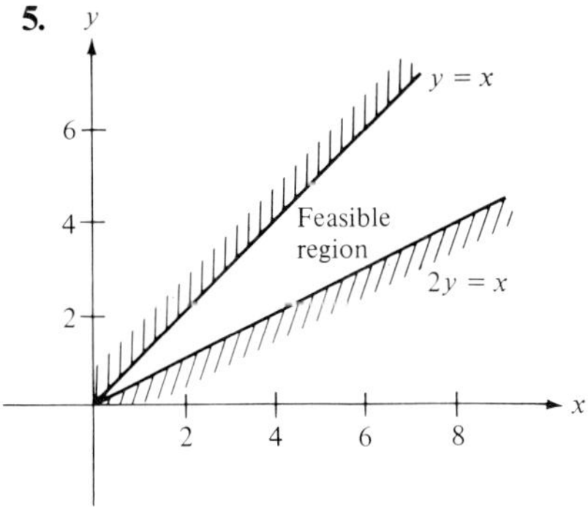

7.

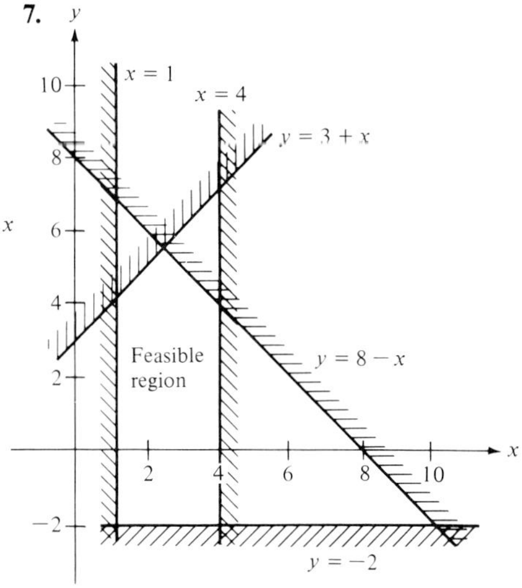

9.

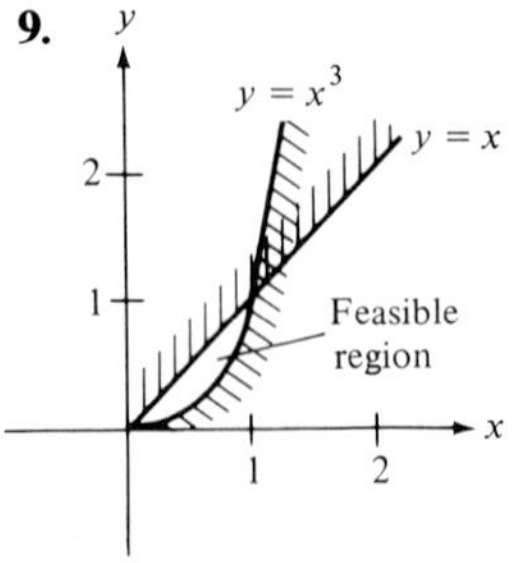

11. (0, 0) **13.** $(1, 0); (\frac{1}{3}, \frac{2}{3})$

15. $(-\frac{1}{2}, -\frac{7}{4}), (-2, 2), (1, -1)$

17. (0, 0), (0, 0) **19.** $(1, 2), (-1, 2)$

21. $(0, -1)$

EXERCISES 8-6

1. Minimum is at (5, 0); $f(5, 0) = 5$ **3.** Minimum is at (4, 1); $f(4, 1) = 14$

5. Maximum occurs at $(3, \frac{3}{2}); f(3, \frac{3}{2}) = 10\frac{1}{2}$

7. Produce 30 units of x, no units of y; $f(30, 0) = 450$

9. Minimum is at $x = 30$, $y = 0$; $f(30, 0) = 450$

EXERCISES 8-7

1. (0, 0) **3.** $(1, 0), (\frac{1}{3}, \frac{2}{3})$ **5.** $(\frac{1}{4}, \frac{1}{4})$ **7.** $(-\frac{1}{2}, -\frac{7}{4}), (-2, 2), (1, -1)$

9. $x = 150$; $y = 150$

CHAPTER 8. REVIEW EXERCISES

1. (**a**) All (x, y) for which $x \neq 0$ and $y \neq 0$
(**b**) Positive when $x > 0$ and $y > 0$ or when $x < 0$ and $y < 0$
(**c**) Never zero

3. (**a**) All (x, y) for which $x \neq 0$ and $y \neq 0$
(**b**) Positive when $x^2y^2 > 1$
(**c**) Zero when $x^2y^2 = 1$

5. $(4, 0, 0), (0, -3, 0), (0, 0, -12)$

7.

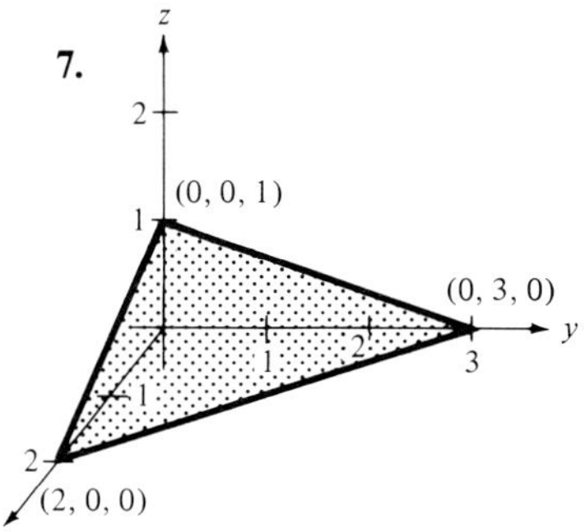

9. $\dfrac{\partial f}{\partial x} = 6xy$

$\dfrac{\partial f}{\partial y} = 3x^2 + 1$

$\dfrac{\partial^2 f}{\partial y \partial x} = 6x$

$\dfrac{\partial^2 f}{\partial x^2} = 6y$

11. $\dfrac{\partial f}{\partial x} = 3y + ye^{xy}$

$\dfrac{\partial f}{\partial y} = 3x + xe^{xy}$

$\dfrac{\partial^2 f}{\partial y\, \partial x} = 3 + e^{xy} + xye^{xy}$

$\dfrac{\partial^2 f}{\partial x^2} = y^2 e^{xy}$

13. $\dfrac{\partial f}{\partial x} = 2y + \dfrac{2xy}{(x^2+1)^2}$

$\dfrac{\partial f}{\partial y} = 2x - \dfrac{1}{x^2+1}$

$\dfrac{\partial f}{\partial y\, \partial x} = 2 + \dfrac{2x}{(x^2+1)^2}$

$\dfrac{\partial^2 f}{\partial x^2} = \dfrac{2y(x^2+1) - 8x^2 y}{(x^2+1)^3}$

15. $\dfrac{\partial f}{\partial x} = 2y^2 - 3x^2 y + y^2 e^{xy^2}$

$\dfrac{\partial^2 f}{\partial y\, \partial x} = 4y - 3x^2 + 2ye^{xy^2} + 2xy^3 e^{xy^2}$; domain is all x, y

$\dfrac{\partial f}{\partial y} = 4xy - x^3 + 2xye^{xy^2}$

$\dfrac{\partial^2 f}{\partial x\, \partial y} = 4y - 3x^2 + 2ye^{xy^2} + 2xy^3 e^{xy^2}$; domain is all x, y

17. $\displaystyle\sum_{\substack{j=1 \\ j \neq i}}^{n} 3x_i^2 x_j^3$

19. $\displaystyle\sum_{\substack{j=1 \\ j \neq i}}^{n} x_j^2 + 2x_i x_j$

21. $x = \frac{6}{5}$; $y = \frac{2}{5}$; absolute minimum

23. $x = \frac{2}{17}$; $y = -\frac{18}{17}$; neither

25.

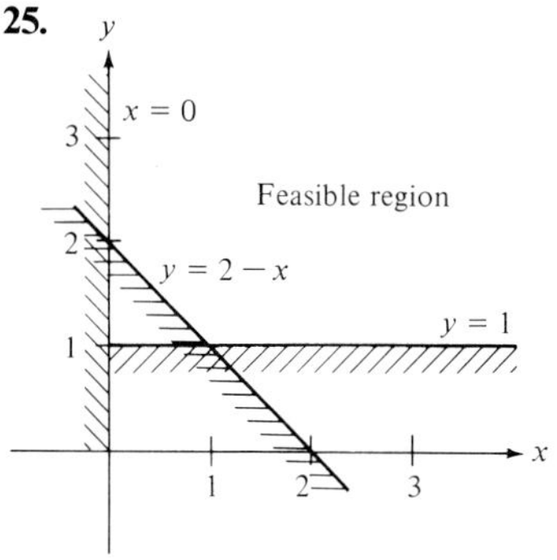

27.

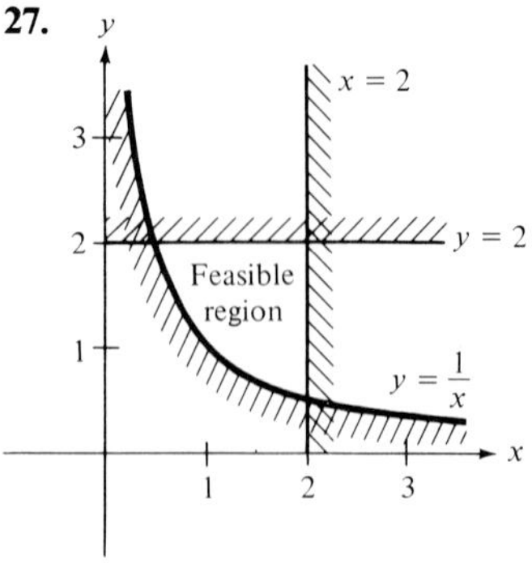

29. $(\frac{1}{2}, \frac{1}{2})$ is an absolute minimum

31. Absolute minimum is at $(1, -1)$

33. Minimum is at $x = 1$, $y = 0$
Minimum value is 2

35. $(2, 0)$; $(-2, 0)$

EXERCISES 9-1

1. **(b)** $C = 4\frac{3}{4}$
(c) $y = \frac{19}{4}e^{2x} - \frac{3}{2}x - \frac{3}{4}$

3. **(b)** $C = -e$
(c) $y = 1 - e^{x+1}$

5. **(b)** $C = 2$
(c) $y = 2 \ln x$

7. **(b)** $C = 3$
(c) $y = \dfrac{3}{x}$

9. **(b)** $C = 1$
(c) $y = \sqrt{x} + 1$

EXERCISES 9-2

1. $(1 - x - e^x) \cdot \left(\dfrac{1}{1 + y - e^y}\right)$

3. $(y + 1) \cdot (x - e^x)$

5. $(xe^x) \cdot (y^3 e^{3y} \ln y)$

7. $y = 2e^{x^2/2}$

9. $y = x$

11. $y = -\ln\left(1 + \dfrac{e}{2} - \dfrac{e^{x^2}}{2}\right)$

13. $y = \sqrt{6 - \dfrac{2}{x}}$

EXERCISES 9-3

1. **(a, b)** $y = Ce^{-x}$
(c) $y = 2e^{-x}$

3. **(a, b)** $y = Ce^{-x^2}$
(c) $y = e^{-x^2}$

5. **(a, b)** $y = \dfrac{C}{x}$
(c) $y = \dfrac{1}{x}$

7. **(a, b)** $y = Ce^{-(x^3/3)-x}$
(c) $y = -e^{-x^3/3)-x-4/3}$

9. **(a, b)** $y = C \exp(e^x)$
(c) $y = 2 \exp(e^x - 1)$

EXERCISES 9-4

1. **(a)** $y = Ce^{2x}$

   (b) $Y(x) = -\frac{x}{2} - \frac{9}{4}$

   (c) $y = Ce^{2x} - \frac{x}{2} - \frac{9}{4}$

   (d) $y = \frac{13}{4}e^{2x} - \frac{x}{2} - \frac{9}{4}$

3. **(a)** $y = Ce^{-3x}$
   (b) $Y(x) = \frac{2}{3}x + \frac{1}{9}$
   (c) $y = Ce^{-3x} + \frac{2}{3}x + \frac{1}{9}$
   (d) $y = -\frac{10}{9}e^{-3x} + \frac{2}{3}x + \frac{1}{9}$

5. **(a)** $y = Ce^{3x}$
   (b) $Y(x) = -\frac{1}{2}e^{x} + \frac{1}{2}e^{-x}$
   (c) $y = Ce^{3x} - \frac{1}{2}e^{x} + \frac{1}{2}e^{-x}$
   (d) $y = 2e^{3x} - \frac{1}{2}e^{x} + \frac{1}{2}e^{-x}$

7. **(a)** $y = Ce^{-x}$
   (b) $Y(x) = -e^{-2x} + x - 2$
   (c) $y = Ce^{-x} - e^{-2x} + x - 2$
   (d) $y = 6e^{-x} - e^{-2x} + x - 2$

9. **(a)** $y = Ce^{4x}$

   (b) $Y(x) = -x^2 - \frac{x}{2} + \frac{1}{8}$

   (c) $y = Ce^{4x} - x^2 - \frac{x}{2} + \frac{1}{8}$

   (d) $y = \frac{55}{8}e^{4x-8} - x^2 - \frac{x}{2} + \frac{1}{8}$

11. **(a)** $y = Ce^{-2x}$
   (b) $Y(x) = xe^{x} - e^{x}$
   (c) $y = Ce^{-2x} + xe^{x} - e^{x}$
   (d) $y = 5e^{-2x} + xe^{x} - e^{x}$

EXERCISES 9-5

1. $\frac{dQ(t)}{dt} = -0.1Q(t)$; $Q(0) = 1000$

3. $\frac{dV(t)}{dt} = -0.1V(t)$
   $V(0) = 1{,}000{,}000$

5. $\frac{dp(t)}{dt} = 0.03p(t) + 2000 - (5000)e^{0.05t}$
   $p(0) = 200{,}000$

7. $\frac{dF(t)}{dt} = -0.001(F(t) + 150{,}000)$
   $F(0) = 50{,}000$

EXERCISES 9-6

1. **(a)** $y' = x^2 - 1 - y = f(x, y)$

   $\frac{\partial f}{\partial y} = -1$

   (b) Solution exists near (1, 5)

3. **(a)** $y' = -\frac{x}{y} = f(x, y)$

   $\frac{\partial f}{\partial y} = \frac{x}{y^2}$

   (b) Solution exists near (−1, 1)

5. (a) $y' = \sqrt{x} - y^2 = f(x, y)$

$$\frac{\partial f}{\partial y} = -2y$$

(b) Solution exists near (1, 9)

7. (a) $y' = f(x, y) = 2 + \frac{x}{y}$
(b) $y'(0) = 2$
(c) $y_1 = 1.2$
(d) $y_{i+1} = y_i + h\left(2 + \frac{x_i}{y_i}\right)$

9. (a) $y' = f(x, y) = x^2y + e^{-x} + 1$
(b) $y'(0) = 2$
(c) $y_1 = 3.2$
(d) $y_{i+1} = y_i + h(x_i^2 y_i + e^{-xi} + 1)$

11. (a)

i	x_i	y_i	y_i'	y_{i+1}
0	0	1	−2	0.8
1	0.1	0.8	−1.6	0.64
2	0.2	0.64	−1.28	0.512
3	0.3	0.512	−1.02	0.41
4	0.4	0.41	−0.82	0.328
5	0.5	0.328	−0.656	0.262
6	0.6	0.262	−0.524	0.210
7	0.7	0.210	−0.420	0.168
8	0.8	0.168	−0.336	0.134
9	0.9	0.134	−0.268	0.107
10	1.0	0.107	−0.214	0.086

(b) $y = e^{-2x}$
(c) $y(1) = e^{-2} = 0.135$ as compared to 0.107 by the numerical method

CHAPTER 9. REVIEW EXERCISES

1. (b) e^3 3. (b) 1 5. (b) $3\frac{3}{4}$ 7. (b) 2 9. (b) 3

11. $y = 2e^{(x^2/2)+x+1/2]}$ 13. $y = e(x-1)$ 15. $y = \sqrt{2}\,e^{x/2}$

17. $y = 4e^{2x}$ 19. $\ln y = 3 - e^{-x}$ 21. $\frac{1}{4}e^{-2x} + \frac{3}{2}x - \frac{1}{4}$

23. $\frac{1}{e^2}e^{3x} - e^x - 1$ 25. $-e \cdot e^{2x} - xe^x + e^x$ 27. $-\frac{1}{22}e^{4x} - x - \frac{1}{4} + \frac{1}{3}e^x$

INDEX